AF803532

# Algorithms for Intelligent Systems

**Series Editors**

Jagdish Chand Bansal, Department of Mathematics, South Asian University, New Delhi, Delhi, India

Kusum Deep, Department of Mathematics, Indian Institute of Technology Roorkee, Roorkee, Uttarakhand, India

Atulya K. Nagar, School of Mathematics, Computer Science and Engineering, Liverpool Hope University, Liverpool, UK

This book series publishes research on the analysis and development of algorithms for intelligent systems with their applications to various real world problems. It covers research related to autonomous agents, multi-agent systems, behavioral modeling, reinforcement learning, game theory, mechanism design, machine learning, meta-heuristic search, optimization, planning and scheduling, artificial neural networks, evolutionary computation, swarm intelligence and other algorithms for intelligent systems.

The book series includes recent advancements, modification and applications of the artificial neural networks, evolutionary computation, swarm intelligence, artificial immune systems, fuzzy system, autonomous and multi agent systems, machine learning and other intelligent systems related areas. The material will be beneficial for the graduate students, post-graduate students as well as the researchers who want a broader view of advances in algorithms for intelligent systems. The contents will also be useful to the researchers from other fields who have no knowledge of the power of intelligent systems, e.g. the researchers in the field of bioinformatics, biochemists, mechanical and chemical engineers, economists, musicians and medical practitioners.

The series publishes monographs, edited volumes, advanced textbooks and selected proceedings.

More information about this series at http://www.springer.com/series/16171

Hari Mohan Dubey · Manjaree Pandit ·
Laxmi Srivastava · Bijaya Ketan Panigrahi
Editors

# Artificial Intelligence and Sustainable Computing

## Proceedings of ICSISCET 2020

Springer

*Editors*
Hari Mohan Dubey
Department of Electrical Engineering
Madhav Institute of Technology
and Science
Gwalior, India

Manjaree Pandit
Department of Electrical Engineering
Madhav Institute of Technology
and Science
Gwalior, India

Laxmi Srivastava
Department of Electrical Engineering
Madhav Institute of Technology
and Science
Gwalior, India

Bijaya Ketan Panigrahi
Department of Electrical Engineering
Indian Institute of Technology Delhi
Delhi, India

ISSN 2524-7565 ISSN 2524-7573 (electronic)
Algorithms for Intelligent Systems
ISBN 978-981-16-1219-0 ISBN 978-981-16-1220-6 (eBook)
https://doi.org/10.1007/978-981-16-1220-6

© The Editor(s) (if applicable) and The Author(s), under exclusive license to Springer Nature Singapore Pte Ltd. 2022
This work is subject to copyright. All rights are solely and exclusively licensed by the Publisher, whether the whole or part of the material is concerned, specifically the rights of translation, reprinting, reuse of illustrations, recitation, broadcasting, reproduction on microfilms or in any other physical way, and transmission or information storage and retrieval, electronic adaptation, computer software, or by similar or dissimilar methodology now known or hereafter developed.
The use of general descriptive names, registered names, trademarks, service marks, etc. in this publication does not imply, even in the absence of a specific statement, that such names are exempt from the relevant protective laws and regulations and therefore free for general use.
The publisher, the authors and the editors are safe to assume that the advice and information in this book are believed to be true and accurate at the date of publication. Neither the publisher nor the authors or the editors give a warranty, expressed or implied, with respect to the material contained herein or for any errors or omissions that may have been made. The publisher remains neutral with regard to jurisdictional claims in published maps and institutional affiliations.

This Springer imprint is published by the registered company Springer Nature Singapore Pte Ltd.
The registered company address is: 152 Beach Road, #21-01/04 Gateway East, Singapore 189721, Singapore

# Organizing Committee

## Coordinators

Dr. Manjaree Pandit, MITS, Gwalior, India
Dr. Laxmi Srivastava, MITS, Gwalior, India

## Organising Secretary

Dr. Pratesh Jayaswal, Professor, MITS, Gwalior, India
Dr. Vijay Bhuria, Assistant Professor, MITS, Gwalior, India

## Technical Committee

Dr. Hari Mohan Dubey, Associate Professor, MITS, Gwalior
Dr. Akhilesh Tiwari, Professor, MITS, Gwalior, Gwalior

## Session Chairs

(Session 2 & 3 in Parallel; Session 5 & 6 in Parallel)

## *Day-I (18th December 2020)*

**Session-I:** Dr. Harish Sharma, Dr. R S Jadon & Dr. Sulochana Wadhwani, MITS, Gwalior, India.
**Session-II:** Dr. Pradyumn Chaturvedi, NIT, Nagpur, India ; Dr. C. S. Mavi & Dr. Amit Aherwar, MITS, Gwalior, India.
**Session-III**: Dr. Nitin Mallik, The NorthCap University, Gurgaon, India, Dr. Pratesh Jayaswal & Dr Akhilesh Tiwari, MITS, Gwalior, India.

## *Day-II (19th December 2020)*

**Session-IV**: Dr. Arvind Jain, NIT, Agartala, India; Dr. M. K. Trivedi & Dr. A. K. Wadhwani, MITS, Gwalior, India.
**Session-V:** Dr. Ravi Shankar, NIT Patna, India; Dr. Manish Dixit & Dr. Manoj Gaur, MITS, Gwalior, India.
**Session-VI**: Dr. Jawar Singh, IIT, Patna, India; Dr. Anmol Ratan Saxena, NIT, Delhi, India; Dr. P. K. Singhal & Dr. Vijay Bhuria, MITS, Gwalior, India.

## Session Support

Paper presentation Session-I: Ms. Rishika Shah, Mr. Pushpendra Singh
Paper presentation Session-II: Ms. Rajni Mourya, Mr. Vikas Thakur
Paper presentation Session-III: Mr. Arvind Singh Tomar, Mr. Shubham Sharma
Paper presentation Session-IV: Mr. Kamal Sharma, Mr. Shubham Chitransh
Paper presentation Session-V: Ms. Priyanka Gupta, Ms. Sharadhha Dubey, Mr. Vinay Kumar Tatkaliya
Paper presentation Session-VI: Ms. Nupur Verma, Ms. Aditi Tiwari, Mr. Vimal Tiwari

## Session Conduction

Prof. Prabhakar Sharma
Dr. R. R. Singh Makwana
Mr. Atul Chauhan
Ms. Hemlata Arya
Mr. Arun Rana

Ms. Aishwarya
Mr. Feyaz Ahmad Warsi
Mrs. Priyanka Bhadoria

## Core and Technical Support

Mr. Nikhil Paliwal
Ms. Poonam Lodhi
Mr. Rajesh Sharma
Mr. Vinod Sharma

## Advisory Committee Members

Dr. Daizhong Su, Nottingham Trent University Burton Street, Nottingham, UK
Dr. Joze Balic, University of Maribor Smetanova, Slovenia
Dr. J. Paulo Davim, University of Aveiro, Portugal
Dr. Felix T. S. Chan, The Hong Kong Polytechnic University, Hong Kong
Dr. Ajith Abraham, Machine Intelligence Research Labs, Washington, USA
Dr. P. N. Suganthan, Nanyang Technological University, Singapore
Dr. Duc T. Pham, University of Birmingham, Birmingham, UK
Dr. I .K. Bhat, Former Vice-chancellor, Manav Rachna University, Delhi, India
Dr. S. G. Deshmukh, IIT, Delhi & Former Director, IIITM, Gwalior, India
Dr. Andrew Kusiak, University of Iowa, Iowa City, USA
Dr. D. P. Kothari, Former Director I/c, IIT, Delhi, India
Dr. Ali Cemal Benim, Duesseldorf University of Applied Sciences, Germany
Dr. S. N. Singh, Vice-chancellor, MMM University, Gorakhpur, India
Dr. R. A. Gupta, Vice-chancellor, Rajasthan Technical University, Kota, India
Dr. A. P. Mittal, Former, Member Secretary, AICTE, New Delhi, India
Dr. B. K. Panigrahi, IIT Delhi, India
Dr. G. A. Vijayalakshmi Pai, PSG College of Technology, Coimbatore, Tamil Nadu, India
Dr. Aparajita Ojha, PDPM IIITDM, Jabalpur, India
Dr. Shekhar Verma, IIIT, Allahabad, India
Dr. Vikas Rastogi, Delhi Technological University, Delhi, India
Dr. C. Patvardhan, Dayalbagh Educational Institute, Agra, India
Dr. Kapil Sharma, Delhi Technological University, Delhi, India
Dr. D. K. Chaturvedi, Dayalbagh Educational Institute, Agra, India
Dr. Lillie Dewan, NIT, Kurukshetra, India
Dr. Sumam Mary Idicula, Cochin University of Science and Technology, Cochin, India

Dr. Surender Reddy Salkuti, Woosong University, Republic of Korea, South Korea
Dr. Amit Shrivastava, Delhi Technological University, Delhi, India
Dr. A. K. Saxena, Dayalbagh Educational Institute, Agra, India
Dr. S. C. Choube, UIT, RGPV, Bhopal, India
Prof. Shekhar Verma, IIIT, Allahabad, India
Dr. Sanjay Agrawal, NITTTR, Bhopal, India
Er. Sandeep Gandi, General Manager (QS) Tata Projects Ltd, India
Dr. R. K. Mishra, IIT, BHU, Varanasi, India
Er. Ashutosh Chincholikar, Smart Controls India Ltd (Member Zeppelin Group), India.
Dr. K. V. Arya, ABV-IIITM, Gwalior, India
Dr. Rajesh Kumar, Malaviya National Institute of Technology Jaipur, India
Dr. Aditya Trivedi, ABV-IIITM, Gwalior, India
Dr. H. M. Suryawanshi, VNIT, Nagpur, India
Dr. P. K. Singh, ABV-IIITM, Gwalior, India
Dr. Manohar Singh, Central Power Research Institute, Bengaluru, India
Dr. M. M. Tripathi, Delhi Technological University, Delhi, India
Dr. Majid Jameel, Jamia Millia Islamia, New Delhi, India
Dr. Surekha Bhanot, BITS, Pilani, India
Dr. R. N. Sharma, NIT, Hamirpur, India

## Programme Committee Members

Dr. Trapti Jain, *IIT Indore, India*
Dr. Jawar Singh, *IIIT, Patna, India*
Dr. Pushpendra Singh, *Delhi Technological University (DTU), Delhi*
Dr. Tripta Thakur, *MANIT, Bhopal, India*
Dr. Abhishek Asthana, *Sheffield Hallam University, U.K*
Dr. Rolf Crook, *University of Leeds, UK*
Dr. Sudhir Singh Bhadoria, *RGPV, Bhopal, India*
Dr. A. K. Wadhwani , *MITS, Gwalior, India*
Dr. Rajesh Kumar, *Delhi Technological University (DTU), Delhi, India*
Dr. Ravichandra Babu, Director, *NPTI, Shivpuri, India*
Er. Vinod Katare, General Manager, *MPMKVVCL, Gwalior, India*
Dr. Kirti Pal Singh, *Gautam Buddha University, Noida, India*
Dr. C. S. Malvi, *MITS, Gwalior, India*
Dr. Y. Kumar, *MANIT, Bhopal, India*
Dr. Sonali Agarwal, *IIIT, Allahabad, India*
Dr. Manisha Sharma, *NIT, Hamirpur, India*
Dr. Perminderjit Singh, *Punjab Engineering College, Chandigarh, India*
Dr. Manish Dixit, *MITS, Gwalior, India*
Dr. Kamal Raj Pardasani, *MANIT, Bhopal, India*
Dr. Arvind Jain, *NIT, Agartala, India*

Dr. Debashis Chatterjee, *Jadavpur University, Kolkata, India*
Dr. Sarita Singh Bhadoria, *RGPV, Bhopal, India*
Dr. Shailaja Kumari M., *NIT, Warangal, India*
Dr. R. S. Thakur, *MANIT Bhopal*
Dr. P. K. Singhal, *MITS, Gwalior, India*
Dr. Sanjay Tiwari, *MITS, Gwalior, India*
Dr. V. P. Vishwakarma, *GGSIPU, Delhi, India*
Dr. Jayshri Vajpai, *M B M Engineering College, J N V University Jodhpur, Rajasthan, India*
Dr. Amit Singhal, *JK Lakshmipat University, Jaipur India*
Dr. Anmol Ratan Saxena, *NIT Delhi, New-Delhi, India*
Dr. Himmat Singh Ahirwar, *MITS, Gwalior, India*
Dr. Adikanda Parida, *Regional Institute of Science and Technology, Nirjuli, Arunachal Pradesh, India*
Dr. Sulochana Wadhwani, *MITS, Gwalior, India*
Dr. D. K. Saini, *University of Petroleum & Energy Studies, Dehradun, India*
Dr. R. Kansal, *MITS, Gwalior, India*
Dr. Arvind Jain, *NIT, Agartala, India*
Dr. R. K. Gupta, *MITS, Gwalior, India*
Dr. S. K. Jain, *MITS, Gwalior, India*
Dr. Akhilesh Tiwari, *MITS, Gwalior, India*
Dr. Urmila Kar, *NITTTR, Kolkata, India*
Dr. M. K. Trivedi, *MITS, Gwalior, India*
Dr. Manoj Gaur, *MITS, Gwalior, India*
Dr. Nitin Mallik, *The NorthCap University, Gurgaon, India*
Dr. Taruna Jain, *Barkatullah University, Bhopal, India*
Dr. Laxmi Shrivastav, *MITS, Gwalior, India*
Dr. Pradhuman Chaturvedi, *VNIT, Nagpur, India*
Dr. Shishir Dixit, *MITS, Gwalior, India*
Dr. Vandana Vikas Thakre, *MITS, Gwalior, India*
Dr. Amit Aherwar, *MITS, Gwalior, India*

## Reviewers

Dr. Amit Aherwar, *MITS, Gwalior, India*
Dr. Jyoti Vimal, *MITS, Gwalior, India*
Dr. Nitin Singh, *MNNIT, Allahabad, India*
Dr. Vineet Shekhar, *BIT, Sindri, India*
Dr. Vikram Saini, *MITS, Gwalior, India*
Dr. Shishir Dixit, *MITS, Gwalior, India*
Er. Bhavna Rathore, *MITS, Gwalior, India*
Dr. S. C. Dutta, *BIT, Sindri, India*
Er. Deep Kishore Parsediya, *MITS, Gwalior, India*

Er. Saurabh Kumar Rajput, *MITS, Gwalior, India*
Dr. Ramakoteswararao Alla, *NITK Surathkal, Mangalore, India*
Dr. Shailendra Dwivedi, *MANIT, Bhopal, India*
Dr. Preeti, *SVSU, Meerut, India*
Dr. Anshu Chaturvedi, *MITS, Gwalior, India*
Dr. Sulochana Wadhwani, *MITS, Gwalior, India*
Dr. Punjan Dohare, *MITS, Gwalior, India*
Er. Punit Kumar Johari, *MITS, Gwalior, India*
Er. Shubham Sharma, *MITS, Gwalior, India*
Er. Vimal Tiwari, *MITS, Gwalior, India*
Dr. Jitu Kujur, *BIT, Sindri, India*
Er. Salil Madhav Dubey, *MITS, Gwalior, India*
Mr. Praveen Bansal, *MITS, Gwalior, India*
Dr. Nekram Rawal, *MNNIT, Allahabad, UP, India*
Dr. Anmol Ratan Saxena, *NIT, Delhi, India*
Dr. Himmat Singh, *MITS, Gwalior, India*
Dr. Ramjeeprasad Gupta, *BIT, Sindri, India*
Dr. Rakesh Singh Jadon, *MITS, Gwalior, India*
Dr. Amit Baghel, *IIT, Guwahati, India*
Er. Nikhil Paliwal, *MITS, Gwalior, India*
Dr. Nitin Mallik, NorthCap University, *Gurgaon, India*
Dr. Gajanan Pandey, *BBAU, Lucknow, India*
Dr. Nishikant Kisku, *BIT, Sindri, Jharkhand, India*
Dr. Preeti Jain, JEC, Jabalpur, India
Dr. Ravishankar, *NIT, Patna, India*
Dr. Rajeev Gupta, *MNNIT, Allahabad, India*
Dr. Manjaree Pandit, *MITS, Gwalior, India*
Dr. Amritanshu Pandey, *IIT (BHU), Varanasi, India*
Dr. Hari Mohan Dubey, *MITS, Gwalior, India*
Er. Arun Kumar Singh, *IIT, Roorkee, India*

# Preface

During the COVID times, organizing an International conference has been quite challenging. Conventionally, conferences are meant to serve as platforms for meeting people, for exchanging ideas, for collaborating and getting inspiration through face to face communication!!

Organizing this virtual conference was a completely new experience. But with the support from our charismatic patrons, eminent experts, enthusiastic participants, diligent faculty, staff & research scholars of the institute, finally experts around the world, authors and faculty and scholars of the host institute were connected electronically in the virtual domain, for the Inaugural and 14 other Sessions of the two-day AICTE, India & TEQIP-III sponsored *Second International Conference on "Sustainable and Innovative Solutions for Current Challenges in Engineering & Technology"*.

The concept of sustainability is central to all development; the term becomes more relevant by each passing day as the very existence of life on this planet is endangered due to the continuous and mindless exploitation of nature and natural resources due to population explosion, economic growth, infrastructure development, lifestyle changes and many other such factors.

There is an urgent need to develop technologies that are sustainable. There is a need for experts from different domains to join hands, come together and create models and systems which are able to produce sustainable solutions to current and emerging problems for the benefit of the society. Innovations and advancements which have taken place in intelligent computing paradigm over the last few decades have provided smart solutions for almost all kinds of societal problems. Therefore, the focus of this conference is towards interdisciplinary applications of Intelligent and sustainable computing.

Most practical real world problems are usually ill defined, noisy, uncertain and complex which make them difficult to solve using traditional computational techniques and algorithms. Computational intelligence based techniques are found to deal with real problems and situations very effectively due to their model free structure, learning ability and a flexible approach. The interdisciplinary applicability of this paradigm makes it very attractive to researchers.

Right from infrastructure management to data mining, ICT, pattern recognition, image & video processing, healthcare informatics, bioinformatics, renewable energy pricing, scheduling and dispatch, internet of things, big data analysis, real-time operating systems, smart homes and devices, electric vehicles, computer integrated manufacturing, biomedical engineering etc. are domains where machine learning techniques and computational intelligence are extensively being used for designing, developing, manufacturing, controlling, analysing and optimizing diverse engineering systems and processes.

Technology is progressing at a very fast pace, which is putting a great burden on natural resources, creating heavy damage to the environment. The price of development is being paid in the form of climate change, hazardous pollution levels and degradation of the environment, which is endangering the very existence of life of the planet earth. There is a need for technocrats, academicians, researchers and all other stakeholders to come together and deliberate on this issue and try to find intelligent solutions to the problem. Scientific discoveries and technological innovations taking place for fulfilling the requirements of the power and energy sector, manufacturing sector, transportation, industrial automation, waste disposal etc. now must pointedly focus on the ecological footprint. *Madhav Institute of technology & Science, Gwalior is committed to this theme and therefore this is the second international conference being organized in this series.*

The first AICTE sponsored ICSISCET-2019 was conducted last year on 02–03 November 2019. Over 148 papers were received, 88 papers were accepted after rigorous review for presentation under 10 tracks; 16 sessions were conducted; 10 expert sessions were conducted; 24 experts contributed as sessions chairs and over 100 faculty and scholars participated. *As an environment friendly initiative and the sustainability theme, the best practice of giving e-certificates was started in the institute.*

The conference proceeding with 51 papers was published by Springer {Proceedings in Adaptation Learning and Optimization (PALO) series} entitled *Intelligent Computing Applications for Sustainable Real-World Applications.*

One edited volume was also published with 09 papers selected in the domain *Nature Inspired Optimization for Electrical Power System* by Springer series Algorithms for Intelligent Systems (AIS). Both these books have been very well received by the researchers and show more than 5600 and 1500 downloads respectively (within six months of publication) till this report is being compiled.

In this Second ICSISCET-2020:

- Over 102 papers were received; 69 papers were accepted after rigorous review for presentation under 4 tracks; 38 were selected for publication; 08 expert sessions were conducted; Six paper presentation sessions were held; 18 experts contributed as sessions chairs.
- Single blind review was carried out with minimum 2 and maximum 3 reviewers. The authors were asked to submit the marked copy of revised paper to highlight the changes incorporated on account of reviewer's suggestions.

- The selected papers from the conference will be published in the Springer Book Series Algorithms for Intelligent systems (AIS), entitled *Artificial Intelligence and Sustainable Computing—Proceedings of ICSISCET 2020.* ( https://www.springer.com/series/16171).

Gwalior, India Hari Mohan Dubey
Gwalior, India Manjaree Pandit
Gwalior, India Laxmi Srivastava
Delhi, India Bijaya Ketan Panigrahi

# Acknowledgement

The organizing chairs of the second International Conference on "Sustainable and Innovative Solutions for Current Challenges in Engineering & Technology" gratefully acknowledge financial support from the *All India Council for Technical Education (AICTE), New Delhi*, India *(Sanction order F. No. 67-4/IDC/GOC/Policy-6/ 2019-20 dated 25.06.2020) & Technical Education Quality Improvement Programme (TEQIP), Phase III (A Government of India Project assisted by the World Bank).*

The organizers also acknowledge cooperation of the *Delhi Technological University (DTU), Delhi, India,* under the twinning arrangement in TEQIP-III.

The *conference patron Prof. K. K. Aggarwal, Chairman NBA*, was the guiding force behind the conference. He also kindly accepted to deliver the keynote address, for which the organizers are eternally indebted.

*Dr. P. N. Suganthan of Nanyang Technological University, Singapore, Dr. Anuradha Ranasinghe of Liverpool Hope University, United Kingdom, Dr. Swagatam Das* of Indian Statistical Institute, Kolkata, West Bengal, India, *Dr. Carlos A. Coello Coello of CINVESTAV-IPN, Mexico, Dr. R. V. Rao* of NIT, Surat, Gujrat, India, *Dr. Sumantra Dutta Roy* of IIT, Delhi, India, *Dr. J. C. Bansal* of South Asian University, New Delhi, India, and *Dr. B. K. Panigrahi* of IIT, Delhi, India deserve a special mention for sparing their valuable time for delivering the *Technical Sessions.*

The organizers are grateful for the constant inputs and support from our Conference Chairs *Prof. R. V. Rao of NIT, Surat, India Dr. J. C. Bansal of South Asian University, New Delhi*, and *Dr. Mukhtiar Singh of DTU, Delhi, India.* Without their support the conference could not have been conducted at this scale.

*Mr. Aninda Bose* and *Mr. Radhakrishnan*, Editors from Springer Nature, deserves special thanks for helping and guiding the organizers throughout this journey which will culminate in publishing of the proceedings.

The organizing team is grateful to *Mr. Prashant Mehta, IAS & Former Director General of the Academy of Administration, Bhopal, India* for sparing time to inaugurate the conference.

Sincere thanks to Alumnus and Secretary of *The Scindia Engineering College Society, Gwalior, Er. Ramesh Agrawal*, for his constant encouragement and support. The conference could not have been possible without the support, motivation and guidance of the *Director, MITS, Gwalior, Dr. R. K. Pandit.*

Thanks are also due to all the *esteemed reviewers* for their time and significant contribution in maintaining the quality of the papers.

The contribution of *internal and external experts* as *Session Chairs* and the *Session Support Teams* during the two days was crucial for the effective conduction of the conference. They played a key role in conference conduction by giving valuable comments for improving the quality of the paper and by ensuring that all reviewer comments were incorporated into the revised papers, before publication. The organizers are grateful for their support and efforts in conference conduction.

During the COVID times, organizing an International conference in virtual mode was quite challenging. The hard work and efforts of the *Conference Core Team, Dr. Pratesh Jayaswal, Dr. Vijay Bhuria, Dr. R. R. Singh, Mr. Atul Chauhan, Mr. Nikhil Paliwal, Ms. Poonam Singh, Ms. Hemlata Arya, Mr. Arun Rana, Ms. Aishwarya, Mr. Rajesh Sharma and Mr. Vinod Sharma* is sincerely acknowledged. The members have worked relentlessly and have left no stone unturned to make the e-conference a reality.

Thanks are also due to media persons, guests, authors and all those who have directly or indirectly contributed in organizing and conducting this conference.

# Contents

# About the Editors

**Dr. Hari Mohan Dubey** is Associate Professor in the Department of Electrical Engineering, MITS, Gwalior (India). He received his M.E. degree in Electrical Engineering from MITS, Gwalior, in 2002, and Ph.D. degree in power systems from Rajiv Gandhi Proudyogiki Vishwavidyalaya, Bhopal, India, in 2017. His areas of research include the study of computational intelligence algorithms and their applications to electrical engineering, particularly, renewable integration, distributed generation, power system planning, operation, and control. He has more than 18 years of teaching and research experience. He has published more than 70 research papers published in various reputed international journals/conferences. He is a member of professional bodies like IEEE (USA), IE (India), IETE (India), and ISTE (India). He is also actively associated with various prestigious international journals as a reviewer.

**Dr. Manjaree Pandit** is currently Dean (Academics) & Professor in the Department of Electrical Engineering, MITS, Gwalior, India. She has completed her graduation and postgraduation from Maulana Azad National Institute of Technology, Bhopal, India, and doctorate from Jiwaji University, Gwalior, India. She has a teaching and research experience of more than 34 years. Her research areas are AI, machine intelligence, evolutionary computation, and electrical power system operation and optimization with renewable energy resources. She is an author and a recognized reviewer for many international publications. She has received national awards twice from ISTE, India, and a research award from UGC, India. Dr. Pandit has received the 'Top Peer Reviewer' award from Publons, Clarivate Analytics for being in the top 1% of reviewers in Engineering for the year 2017–2018 & 2018–2019, and also in cross-field for the year 2018–2019. She has completed 09 centrally funded research projects. She is a senior member of IEEE, USA, and Fellow of Institution of Electronics & Telecommunication Engineers, and Institution of Engineers (India). Dr. Pandit has organized many sponsored technical events like national/international conferences, seminars, workshops, and short-term training programmes. She has published more than 150 research papers in peer-reviewed international journals & conference proceedings and has authored 22 book chapters.

**Dr. Laxmi Srivastava** received her B.E. and Ph.D. in Electrical Engineering from the University of Roorkee (presently IIT Roorkee) India and M.Tech. in Electrical Engineering from IIT, Kanpur. She joined the Department of Electrical Engineering, Madhav Institute of Technology & Science, Gwalior, in 1984, and is currently working as Professor and Head of this department. She has a teaching and research experience of more than 36 years and has completed more than 12 sponsored research & development projects. She is the recipient of the prestigious Khosla Research award from the University of Roorkee, India (Now IIT, Roorkee). She has been on the review panel of prestigious international journals for the last 20 years. Her research is motivated by ANN, fuzzy logic and evolutionary computing-based techniques for power system security assessment, optimization, control, distributed generation FACTS controllers, and power quality. She is a co-author of two book chapters and has published around 160 papers in peer-reviewed international journals and conference proceedings, which are quite well cited.

**Dr. Bijaya Ketan Panigrahi** is Professor in the Department of Electrical Engineering at the Indian Institute of Technology Delhi. He received his M.Tech. degree in Electrical Engineering from University College of Engineering, Burla, Sambalpur, India, in 1995, and Ph.D. degree from Sambalpur University in 2004. His primary research areas include power quality, flexible alternating current transmission system (FACTS) devices, power system protection, artificial intelligence (AI) applications to power systems, and nature-inspired computing. He is actively working on numerous projects for industry and government. He is a member of various professional societies, including the IEEE (USA). He is serving as a chief editor to the International Journal of Power and Energy Conversion and also associated with many international journals of IEEE, IET, and Elsevier, in different capacities. He has more than 28 years of teaching and research experience. He is a renowned researcher in the field of evolutionary computing and has authored more than 600 research papers published in various reputed international journals and conference proceedings. His publications are highly cited among the research community in overall disciplines of engineering and science and also have 3 patents in his credits. He is the recipient of the 2002 Young Scientist Award, Department of Science and Technology, the Government of Orissa, and the 2015 Fellowship Award from the Indian National Academy of Engineering (INAE).

# Chapter 1
# Control of On-Board Electric Vehicle Charger

**Aakash Kumar Seth and Mukhtiar Singh**

## 1 Introduction

The fuel-based vehicles are very much responsible for increasing pollution level and produce substantial amount of carbon monoxide, nitrogen oxide and other pollutions [1]. Therefore, a clean and pollution-less alternative is required to save the environment. The electric-based vehicles are best suitable to minimize these issues. They are clean and eco-friendly and due to that they gained lot of attention from industries, government and consumers. Electric vehicles (EVs) are efficient, low carbon footprints, cheaper to maintain and less noise pollution; however, they suffer from battery life, cost and charging time [2–4].

In the last few years, lot of advancement has been done in individual components of EV, such as electric machine, batteries, drive train and on-board charger. The on-board charging can be of two types, i.e. conductive or inductive (wireless) [5]. With the emerging trend of smart grid, on-board charger may exchange the energy with grid. They are not limited up to the charging of battery but may support grid by supplying electrical energy back to utility, and this mode of operation is known as vehicle to grid (V2G) [6]. This mode is very useful to support the grid in case of emergency or blackout. The on-board charging is Level-1 (slow) charging and requires the power outlets of 230 V available in our houses. Generally, they are rated around 6–7 kW and takes 8–12 h to fully charge the battery pack [7, 8]. Furthermore, the on-board charger has some power quality features, such as compensation of reactive power (inductive or capacitive), harmonic filtering and correction of power factor [9].

The reactive power compensation devices such as static synchronous compensator, capacitor banks etc. are installed at load end to minimize the line losses [10]. Therefore, the on-board charger is very much suitable for aforementioned application

A. K. Seth (✉) · M. Singh
Delhi Technological University, Delhi, India

© The Author(s), under exclusive license to Springer Nature Singapore Pte Ltd. 2022 
H. M. Dubey et al. (eds.), *Artificial Intelligence and Sustainable Computing*,
Algorithms for Intelligent Systems, https://doi.org/10.1007/978-981-16-1220-6_1

and curtail down the installation cost of these devices. Furthermore, this compensation do not require the battery energy; therefore, there is no effect on state of charge (SOC) of battery and its life span. However, it will definitely affect the DC-link capacitor as number of charge–discharge cycles is increased [11].

Lot of topologies of on-board EV charger has been presented in the literature by many researchers [12, 13]. However, in general two-stage topology is used widely for an on-board EV charger. The topology has two power converters, i.e. AC to DC and DC to DC are connected in cascade [14]. Since the charger has two power converters, two separate controllers are required. Both the controllers have separate reference inputs, however, it is more feasible if reference inputs are less.

Among the various control techniques and topologies [15–17], a two-stage on-board bidirectional EV charger based on proportional integral (PI) controller is presented in this paper. The charger contains a fully controlled single-phase bidirectional AC–DC converter followed by a bidirectional buck-boost DC–DC converter which is able to operate in all four quadrants of active–reactive power plane. The charger can transfer the active power in both directions as well as compensate reactive power of nearby load. The charger has two separate controllers, i.e. each for AC to DC and DC to DC converter. The controller of first stage (i.e. AC to DC) is employed to control active–reactive power, whereas DC-link voltage and battery current are controlled by the second-stage controller. The first-stage controller has two *PI* controllers as they are simple in design and easy to implement. The two *PI* regulators are utilized to regulate active–reactive power, and the analog hysteresis controller is for grid current control. The grid current is controlled in actual periodic form and output pulses of hysteresis control are given to AC to DC converter. For the control design of second stage, two PI regulators are employed, i.e. each for controlling the DC-link voltage and battery current. Here, pulse width modulation (PWM) technique is used to generate the pulses for second-stage converter. Therefore, in presented control strategy five quantities are controlled, namely, real power ($P$), reactive power ($Q$), grid current ($I_g$), DC-link voltage ($V_{dc}$) and battery current ($I_{\text{bat}}$). However, the controller has three reference inputs and rests are derived from those signals.

The above-mentioned on-board EV charger and its controller have been simulated in MATLAB/Simulink environment. The performance has been analyzed by sudden change of active–reactive power command in all possible ways. The simulation study shows that controller works successfully under step change in reference command and providing good transient and steady-state response.

## 2 System Description

The on-board EV charger topology used in this paper to analyze the performance of controller is shown in Fig. 1. The first stage consists of single-phase full-bridge rectification and bidirectional buck-boost DC to DC conversion in second stage. The purpose of second stage is to provide galvanic isolation between grid and battery.

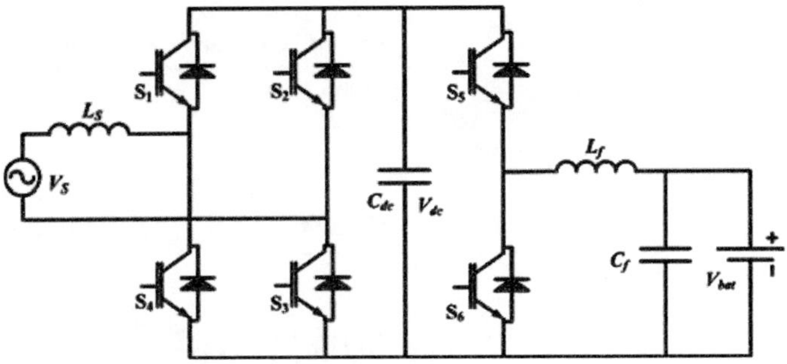


**Fig. 1** On-board EV charger

The main concentration of this paper is to investigate the performance of controller over the different variations in active–reactive power commands. Therefore, all the efforts have been on design and implementation of charger controller and analysis of results.

The grid voltage is 230 V (RMS) for single-phase supply and having the peak voltage of 325 V. Generally, the battery pack voltage of EV is around 350 V, therefore it is suggested that the DC-link voltage must be greater than battery voltage and peak of grid voltage [18]. Here, DC-link voltage is maintained at 400 V and SOC of battery is taken 50% so that the charge–discharge behavior can be analyzed. Furthermore, the total harmonic distortion (THD) in grid current must be within the permissible limit of 5% in all working modes.

## 3 Controller Design

The on-board EV charger operates in all over active–reactive power plane. The active power send back to grid is denoted by negative sign and vice versa. Whereas positive reactive power means compensation of inductive reactive power and negative in case of capacitive. Both active and reactive powers are measured at AC side of the charger.

Figure 2 shows the controller design for both stages. To regulate the active–reactive power in outer loop of AC to DC converter control, the first active–reactive power is calculated at AC side in αβ frame. The α component is actual signal ($I_\alpha = I_g$, $V_\alpha = V_s$), where orthogonal (β) axis components ($I_\beta$, $V_\beta$) are found by applying quarter delay function in α component [19]. After finding αβ components of voltage and current, instantaneous active–reactive theory is applied to calculate active–reactive power for single phase [20].

$$P = \frac{1}{2}\left(V_\alpha \cdot I_\alpha + V_\beta \cdot I_\beta\right) \quad (1)$$

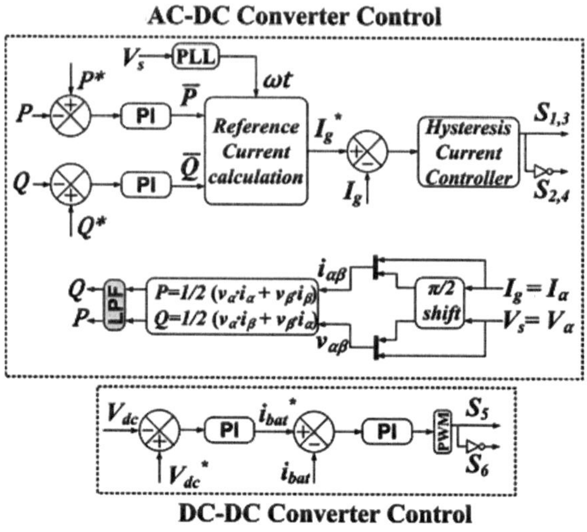


**Fig. 2** Control design of on-board EV charger

$$Q = \frac{1}{2}\left(V_\alpha \cdot I_\beta + V_\beta \cdot I_\alpha\right) \tag{2}$$

These calculated active ($P$) and reactive ($Q$) powers are passed through a low-pass filter (LPF) to minimize the ripple content.

The EV charger is commanded by controller to charge the battery pack according to customer requirements. Moreover, the utility can also set reference charging command with smart coordination between grid and charger during peak hours. In this case, the power grid must provide some benefits to customers. The reactive power command of controller is directly requested by grid. However, the charger can compensate reactive power within the limit of its rating.

To track the reference active–reactive power commands, first these are compared with computed active–reactive powers. The errors between them are minimized by two separate *PI* regulators in outer loop of controller. The active–reactive power regulators generate $\bar{P}$ and $\bar{Q}$, respectively, as follows:

$$\bar{P} = \left(K_{p1} + \frac{K_{i1}}{s}\right)\left(P^* - P\right) \tag{3}$$

$$\bar{Q} = \left(K_{p2} + \frac{K_{i2}}{s}\right)\left(Q^* - Q\right) \tag{4}$$

Now, these outputs are utilized to generate the reference grid current corresponding to active–reactive power command. It is calculated as follows:

$$\varphi = \tan^{-1}\left(\frac{\bar{Q}}{\bar{P}}\right) \tag{5}$$

$$I = \frac{\bar{P}}{V_s \cdot \cos\varphi} \tag{6}$$

$$I_g^* = \sqrt{2}I \cdot \sin(\omega t - \varphi) \tag{7}$$

The $\omega t$ is grid phase angle and calculated by phase locked loop (PLL).

Now, this reference signal is compared with actual grid current and error is minimized by analog hysteresis controller in inner loop. Further, the pulses generated by inner loop are given to full bridge AC to DC converter.

The second-stage controller has also two control loops. In the outer loop, the sensed DC-link voltage is compared with reference value and error is minimized by *PI* controller. The inner control loop generates reference of battery current and this reference is compared with actual sensed battery current. Further, inner loop regulator produces the duty ratio for second stage and pulses are generated by PWM technique.

$$i_{\text{bat}}^* = \left(K_{p3} + \frac{K_{i3}}{s}\right)\left(V_{dc}^* - V_{dc}\right) \tag{8}$$

$$d = \left(K_{p4} + \frac{K_{i4}}{s}\right)\left(i_{\text{bat}}^* - i_{\text{bat}}\right) \tag{9}$$

## 4 Simulation Results

The parameters of on-board EV charger depicted in Fig. 1 are listed in Table 1. Here, the charger is rated as 6.6 KVA as most of the on-board EV chargers rating are 6.6 KVA. The nominal voltage of battery pack is taken as 350 V and SOC is

**Table 1** System parameters

| Parameters | Symbol | Value |
|---|---|---|
| Charger apparent power | $S$ | 6.6 KVA |
| Grid voltage | $V_g$ | 230 V(RMS) |
| Grid frequency | $F$ | 50 Hz |
| Interfacing inductance | $L_s$ | 2 mH |
| DC-link capacitance | $C_{dc}$ | 330 µF |
| DC-link voltage | $V_{dc}$ | 400 V |
| DC–DC filter inductance | $L_f$ | 500 µH |
| DC–DC filter capacitance | $C_f$ | 200 µF |
| Converter switching frequency | $f_s$ | 20 kHz |
| Battery voltage | $V_{\text{bat}}$ | 350 V |

50%. However, the actual battery voltage may be higher than the nominal voltage and it depends on SOC. A line inductor of 2 mH is connected at the AC side to minimize harmonics in grid current. To authenticate the performance of controller, a test scenario has been created in MATLAB/Simulink with different values of active–reactive power. A total of eight modes has been developed in which the charger performs only one operation, i.e. either charging/discharging or reactive power compensation in first four modes. In the remaining four modes, the charger performs both operations simultaneously (Table 2).

Figure 3 shows the simulation results of SOC, battery voltage, battery current, DC-link voltage and active–reactive power during all modes. Here, the DC-link voltage is maintained at 400 V and the battery voltage is around 375 V; however, a slight variation occurs in battery voltage during the changeover. The battery current varies according to the active power command and it is negative during charging and vice versa. The SOC increases during charging of battery and decreases during

**Table 2** Simulation scenario

| Mode | Active power (KW) | Reactive power (KVAR) | Apparent power (KVA) | Time (s) |
|---|---|---|---|---|
| 1 | 6.6 | 0 | 6.6 | 0–1 |
| 2 | −6.6 | 0 | 6.6 | 1–2 |
| 3 | 0 | 6.6 | 6.6 | 2–3 |
| 4 | 0 | −6.6 | 6.6 | 3–4 |
| 5 | 4 | 5.25 | 6.6 | 4–5 |
| 6 | 5 | −4.31 | 6.6 | 5–6 |
| 7 | −5.5 | 3.65 | 6.6 | 6–7 |
| 8 | −2.5 | −6.11 | 6.6 | 7–8 |

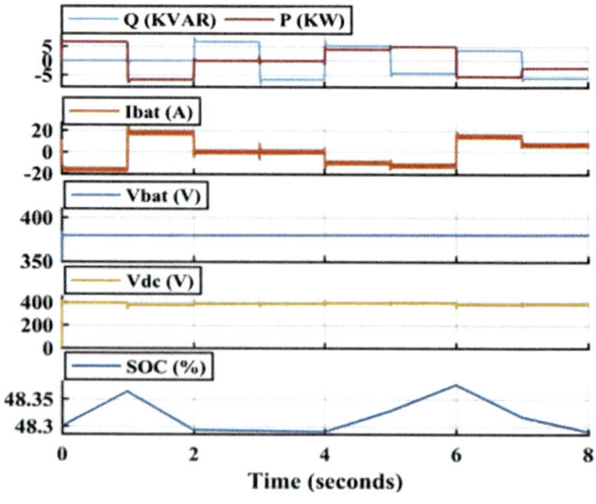


**Fig. 3** Simulation result of active–reactive power, battery current ($I_{bat}$), battery voltage ($V_{bat}$), DC-link voltage ($V_{dc}$) and SOC

discharging. There is no effect on battery SOC during the compensation of reactive power and it can be seen in modes 3 and 4.

The zoomed version of transition from one mode to another is shown in Fig. 4. Here, the voltage and current are shown with active–reactive power. As shown in Fig. 4a, the active power varies from +6.6 KW to −6.6 KW. There is no reactive power operation, therefore the grid current is in phase with voltage in mode 1 and out of phase in mode 2. In the transition of Fig. 4b, active power varies from −6.6 KW to zero and reactive power is +6.6 KVAR from zero. The current is lagging by 90° from out of phase during this changeover. The transition shown in Fig. 4c is opposite of Fig. 4a. In this, the reactive power command changes from +6.6 KVAR to −6.6 KVAR without any active power operation. Therefore, the current changes from lagging to leading by 90°. Figure 4d shows both active−reactive operation; in this, battery is charging at the rate of 4 KW and the leftover capacity, i.e. 5.25 KVAR, is utilized for reactive power compensation. Since the charger is compensating inductive reactive power while charging, the current is lagging by 53°. In Fig. 4e, the battery is charging at the rate of 5 KW and the compensating capacitive reactive power is of 4.31 KVAR. Therefore, the grid current is leading this time by 40°. Similarly, Fig. 4f and g shows the inductive and capacitive reactive power operation, respectively, while discharging of battery. Therefore, the current is lagging and leading by 146° and 112° during modes 7 and 8, respectively.

The simulation results show that the above-mentioned controller works under all possible conditions of active–reactive power plane. The controller has performed successfully with sudden change of references and settle quickly with zero steady-state error.

## 5 Conclusion

The paper presents control architecture of an on-board EV charger. The controller commands to charger and it can perform any active–reactive operation within the limit of its rating. The controller is based on *PI* regulators as they are simple in design and easy to implement. The charger has two power converters and they both have separate controller. Three quantities are controlled by first-stage controller in which the reference of one quantity is derived by other two input references. Similarly, two quantities are controlled by second-stage controller in which the reference of one quantity is derived by other input reference. Therefore, the charger has three input references and two references are derived by others inputs. A 6.6 KVA on-board EV charger is simulated in MATLAB/Simulink toolbox. For optimal operation, complete charger rating is utilized. The charger can charge the battery pack as per the customer desire rate and rest of the capacity is utilized in compensating of reactive power of local load in order to reduce line losses.

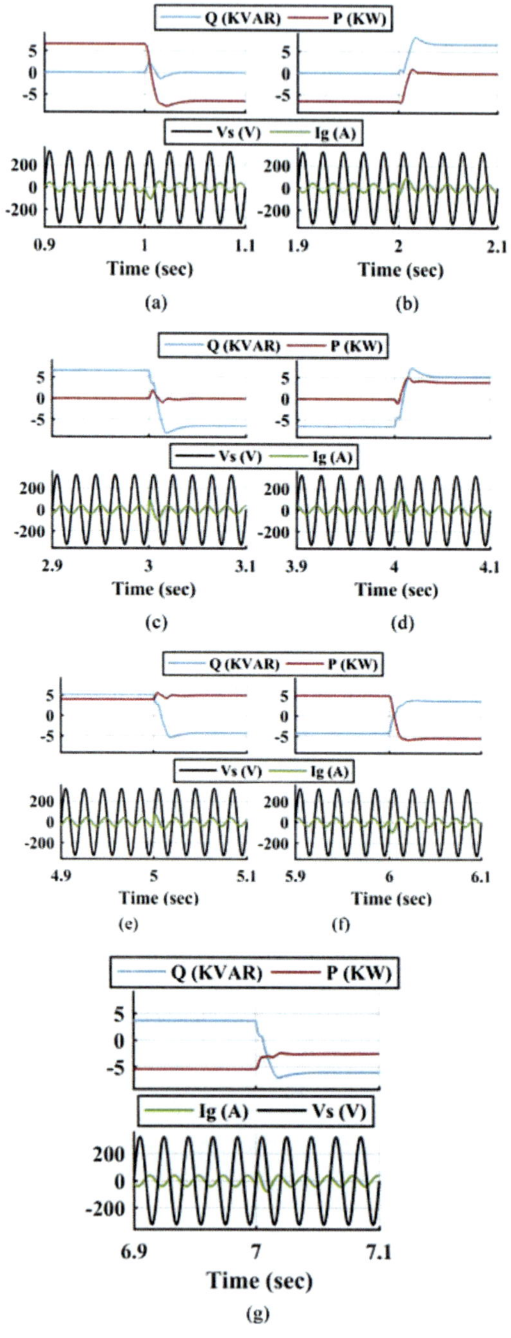


**Fig. 4** Zoomed transition from **a** mode 1–2, **b** mode 2–3, **c** mode 3–4, **d** mode 4–5, **e** mode 5–6, **f** mode 6–7 and **g** mode 7–8

## References


1. Williamson SS, Rathore AK, Musavi F (2015) Industrial electronics for electric transportation: current state-of-the-art and future challenges. IEEE Trans Indus Electron 62(5):3021–3032
2. Gautam D, Musavi F, Edington M, Eberle W, Dunford WG (2012) A zero voltage switching full-bridge dc-dc converter with capacitive output filter for a plug-in-hybrid electric vehicle battery charger. In: Proceedings of IEEE applied power electronics conference and exposition, pp 1381–1386
3. Ramadass P, Haran B, White R, Popov B (2002) Performance study of commercial $LiCoO_2$ and spinel-based Li-ion cells. J Power Sourc 111(2):210–220
4. Cassani PA, Williamson SS (2009) Feasibility analysis of a novel cell equalizer topology for plug-in hybrid electric vehicle energy-storage systems. IEEE Trans Veh Technol 58(8):3938–3946
5. Rathore AK, Prasanna UR (2012) Comparison of soft-switching voltage-fed and current-fed bi-directional isolated Dc/Dc converters for fuel cell vehicles. In: Proceedings of IEEE international symposium on industrial electronics, pp 252–257
6. Salmasi FR (2007) Control strategies for hybrid electric vehicles: Evolution, classification, comparison, and future trends. IEEE Trans Veh Technol 56(5):2393–2404
7. Chan CC (2007) The state of the art of electric, hybrid, and fuel cell vehicles. In: Proceedings of IEEE 95(4):704–718
8. Khaligh A, Li Z (2010) Battery, ultracapacitor, fuel cell, and hybrid energy storage systems for electric, hybrid electric, fuel cell, and plug-in hybrid electric vehicles: state of the art. IEEE Trans Veh Technol 59(6):2806–2814
9. Yilmaz M, Krein PT (2013) Review of battery charger topologies, charging power levels and infrastructure for plug-in electric and hybrid vehicles. IEEE Trans Power Electron 28(5):2151–2169
10. Chan CC, Bouscayrol A, Chen K (2010) Electric, hybrid, and fuel-cell vehicles: architectures and modeling. IEEE Trans Veh Technol 59(52):589–598
11. Clement-Nyns K, Haesen E, Driesen J (2010) The impact of charging plug-in hybrid electric vehicles on a residential distribution grid. IEEE Trans Power Syst 25(1):371–380
12. Asthana P, Verma V (2020) PV fed current controlled low stress high gain converter for battery charging applications. In: 2020 IEEE International conference on power electronics, smart grid and renewable energy (PESGRE 2020), Cochin, India, pp 1–4
13. Khaligh A, Dusmez S (2012) Comprehensive topological analysis of conductive and inductive charging solutions for plug-in electric vehicles. IEEE Trans Veh Technol 61(8):3475–3489
14. Wickramasinghe Abeywardana DB, Acuna P, Hredzak B, Aguilera RP, Agelidis VG (2018) Single-phase boost inverter-based electric vehicle charger with integrated vehicle to grid reactive power compensation. IEEE Trans Power Electron 33(4):3462–3471
15. Seth AK, Singh M (2020) Resonant controller of single-stage off-board EV charger in G2V and V2G modes. IET Power Electron 13(5):1086–1092. Accessed 8 Apr 2020
16. Cortés P, Rodríguez J, Antoniewicz P, Kazmierkowski M (2008) Direct power control of an AFE using predictive control. IEEE Trans Power Electron 23(5):2516–2523
17. He T, Zhu J, Lu DD, Zheng L (2019) Modified model predictive control for bidirectional four-quadrant EV chargers with extended set of voltage vectors. IEEE J Emerg Sel Top Power Electron 7(1):274–281
18. Zhou X, Lukic S, Bhattacharya S, Huang A (2009) Design and control of grid-connected converter in bi-directional battery charger for Plug-in hybrid electric vehicle application. In: 2009 IEEE vehicle power and propulsion conference, Dearborn, MI, pp 1716–1721
19. Kisacikoglu MC, Ozpineci B, Tolbert LM (2010) Examination of a PHEV bidirectional charger for V2G reactive power compensation. In: Proceedings of IEEE APEC Exposition, Palm Springs, CA, USA, 21–25 February 2010, pp 458–465
20. Hartmann M, Friedli T, Kolar JW (2011) Three-phase unity power factor mains interfaces of high power EV battery charging systems. In: Proceedings of the power electronics for charging electric vehicles ECPE workshop, Valencia, Spain, 21–22 March 2011, pp 1–66

# Chapter 2
# NSGA-III-Based Time–Cost–Environmental Impact Trade-Off Optimization Model for Construction Projects

**Aditi Tiwari, Kamal Sharma, and Manoj Kumar Trivedi**

## 1 Introduction

Construction projects produce large quantities of dust and greenhouse gases during their execution phase. Although time and cost are two crucial success factors of construction project planning [1], varying amount of resources utilization in project affects the time, cost and also the environmental impact of project [2]. Hence, the time, cost and environmental impact of project are not only important but also conflicting objectives of construction projects.

Over the past decades, a significant amount of attention has been given to develop the time–cost trade-off optimization models [3–5]. Despite of their undeniable benefits, these studies largely undermine the environmental impact of project. Therefore, for the construction managers, optimizing the trade-off between time, cost and environmental impact is increasingly important.

A construction project consists of many interconnected activities [6]. Each activity must be completed through one of its available alternatives. However, each alternative of an activity differs in completion time, completion cost and environmental impact due to variation in resource utilization amount and there are also several delivering methods of a project based on possible combinations of available alternatives of activities. Therefore, in order to minimize the time, cost and environmental impact of project, it is required to allocate an optimum alternative to each activity of project.

Deb and Jain [7] reported several difficulties in multi-objective optimization such as a large number of non-dominated solutions in population members. Then, Deb and Jain [7] introduced the non-dominated genetic algorithm III (NSGA-III)

A. Tiwari · K. Sharma (✉) · M. K. Trivedi
Department of Civil Engineering, Madhav Institute of Technology & Science, Gwalior, India

M. K. Trivedi
e-mail: manojtrivedi@mitsgwalior.in

© The Author(s), under exclusive license to Springer Nature Singapore Pte Ltd. 2022 
H. M. Dubey et al. (eds.), *Artificial Intelligence and Sustainable Computing*,
Algorithms for Intelligent Systems, https://doi.org/10.1007/978-981-16-1220-6_2

after replacing the crowding distance concept of NSGA-II by the reference point-based selection approach. This paper therefore applies NSGA-III to solve the TCEIT optimization problem.

Based on the above, this paper presents a NSGA-III-based TCEIT optimization model that focuses on the simultaneous optimization of three objectives, i.e. time, cost and environmental impact of construction project. In the remaining part of this paper, literature review is presented followed by problem formulation and NSGA-III-based TCEIT optimization model. Then, the applicability of proposed model is demonstrated using a case study project along validation and comparison of model. In the final section of paper, conclusion and scope for future work is presented.

## 2 Literature Review

Literature study shows deterministic, heuristic and meta-heuristic methods of project scheduling [8], of which meta-heuristic methods were found as more effective multi-objective optimization methods. In this order, Feng et al. [3] developed a time–cost trade-off model (TCT) based on genetic algorithm and Pareto-front approach. Xiong and Kuong [9] proposed multi-objective ant colony optimization-based TCT model. Xiong and Kuang [9] proposed multi-objective particle swarm optimization-based TCT model. Nevertheless, the original PSO still need some modification for guaranteed convergence to a local optimum [10]. Fallah-Mehdipour and his co-authors [11] demonstrated the efficiency of NSGA-II over MOPSO in solving the TCT problems. Then, Shahriari [12] developed a NSGA-II-based discrete TCT model. These studies were limited to bi-objective optimization because the likelihood of discovering the non-dominated solutions decreases exponentially as the number of conflicting objectives increases [13].

On the other hand, in today's world, many construction projects are being executed simultaneously. Construction projects are generally carried out under the open environment and release large amount of dust and greenhouse gases to the environment. Therefore, air pollution caused by construction projects can adversely affect the human life. Hence, in order to minimize the environmental impact with the time and cost of project, few researchers developed the time–cost–environmental impact trade-off models. Marzouk and his co-authors [14] considered the environment pollution alongside time and cost and proposed a NSGA-based multi-objective optimization framework; however, this framework was limited to planning phase of project. To fill this gap, Ozcan-Deniz and his co-authors [2] developed NSGA-II-based TCEIT model in which environmental impact of entire projects' phases is evaluated in terms of greenhouse gas emissions. Furthermore, Cheng and his co-authors [15] also solved TCEIT problem using the opposition-based multiple objective differential evolution.

The process of multi-objective optimization faces several problems such as large number of non-dominated solutions in population members, expensive computation of diversity and performance metrics, difficult visualization of higher-dimensional trade-off front and inefficient recombination operator. As mentioned earlier, in order to address the difficulties of multi-objective optimization. Deb [7] introduced NSGA-III after replacing the NSGA-II's crowding distance concept by reference point-based selection approach. Deb [7] tested the NSGA-III in optimizing 3–15 objectives and two practical problems. Among all such problems, NSGA-III was found to be able to find a well-converged and well-diversified set of Pareto-optimal solutions. Therefore, an attempt is made to use the NSGA-III in TCEIT optimization.

## 3 Problem Formulation

Since a number of alternatives may be available to perform an activity in the construction project, it is necessary to allocate a suitable alternative for each activity of the construction project that also meets the desired projects' objectives.

The input parameters for proposed TCEIT model are taken as activity completion time (ACT), activity completion cost (ACC) and activity environmental impact (AEI). Since each alternative requires different amount of labor, materials and equipment resources, therefore numerical values of individual ACT, ACC and AEI are determined based on activity and choice of alternative. The objective of this paper is to optimize the project completion time (PCT), project completion cost (PCC) and project environmental impact (PEI) simultaneously, which are the functions of ACT, ACC and AEI, respectively. The key assumptions made in this study are as follows: (i) An activity, if once started, must be completed without interruption. (ii) There are precedence relationships between activities, which can be represented in the form of an activity on node (AON) network. So now, these three objective optimization problems have been formulated in the following manner.

***Project Completion Time (PCT)***

PCT is the sum of ACT of all the activities in the critical path of construction project.

Objective 1: Minimize PCT

$$\text{PCT} = \sum_{A \in CP} \text{ACT}_A \tag{1}$$

where PCT is the project completion time and $\text{ACT}_\text{A}$ is the completion time of critical path activity (A).

***Project Completion Cost (PCC)***

PCC is the algebraic sum of ACC of all the activities of construction project, whereas ACC comprises direct and indirect cost of activity.

Objective 2: Minimize PCC

$$\text{PCC} = \sum_{A} \text{D.C} + \text{ I.C per day } \times \text{ PCT in days} \tag{2}$$

where PCC is the project completion cost, $\sum_A$ D.C is the sum of direct cost of each activity, and indirect cost of project can be estimated by multiplying the PCT and indirect cost per day.

***Project Environment Impact (PEI)***

In this paper, the activity environmental impact (AEI) is measured in terms of $CO_{2\text{-equivalent}}$ greenhouse gas emissions ($GHG_{CO2\text{-eq}}$). PEI is the sum of AEI of all the activities of project.

Objective 3: Minimize PEI

$$\text{PEI} = \sum_{A=1}^{n} \text{AEI}_A \tag{3}$$

where PEI is the total environmental impact of project, $\text{AEI}_A$ is the environmental impact of activity A and n is total number of activities in project.

The key constraints considered in this study are as follows: (i) Each activity should be completed for the successful completion of project; (ii) Only one out of available alternative should be initiated to complete the activity; (iii) Alternatives of each activity, which are the decision variables, must be positive integers and subject to lower and upper bound and (iv) Precedence relationships between activities should be maintained during optimization process.

# 4 Proposed NSGA-III-Based TCEIT Model

Basically, the NSGA-III is an extension of NSGA-II [16] with only difference in their selection mechanism in non-dominated front. Deb [7] used reference point-based non-dominated sorting approach in which predefined or undefined reference points are placed on a normalized hyper-plane as suggested by Das and Dennis [17]. This reference point-based approach helps the algorithm to distribute the population in higher-dimensional non-dominated fronts. Besides the selection mechanism in non-dominated front, rest of the process of NSGA-III is the same as NSGA-II.

In NSGA-III, each solution of optimization problem can be represented in the form of chromosome. A chromosome consists of number of cells (genes) equal to the total number of decision variables or activities in construction project. Thus, each gene of chromosomes contains a value (allele) of decision variable, which is the selected alternative through which activity can be executed. Each possible chromosome of given TCEIT problem is a possible solution of project. For understanding the chromosome solution, suppose a construction project consisting of nine activities and each activity can be executed either one of its three available alternate options. A chromosome solution shown in Fig. 1 can be a possible solution for delivering the project of 1–9 activities with alternatives 3, 1, 2, 3, 2, 1, 2, 3, 1, respectively. As per permutation rules there will be $3^9$ possible project delivering solutions.

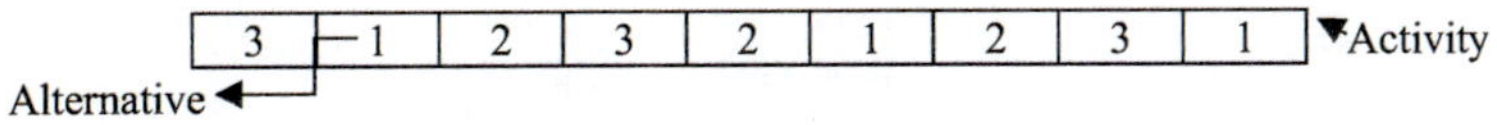


**Fig. 1** Chromosome solution for given TCEIT problem

To begin the process of NSGA-III, it is required to generate a random population of size $N$ and a set of widely distributed M-dimensional reference points $H$ on a normalized hyper-plane. Detailed description of NSGA-III procedure for TCEIT optimization is explained as follows.

***Generation and Ranking of Initial Population ($P_t$) of Size N***

At first, an initial population (parent population $P_t$) of size $N$ (i.e. $N$ chromosomes) is generated randomly. The fitness value of each population member is calculated using the equations (i)–(iii). Then, $P_t$ undergoes non-dominated sorting (*NDS*) to be distributed in non-dominated fronts. Next, the ranking of solutions ($P_t$) is done in each non-dominated front using the crowding distance concept.

***Selection (also Known as Reproduction Stage)***

In selection stage, pair-wise tournaments are carried out between $N$ pairs of randomly selected solutions and the winner solution is decided on the basis of ranking produced by NDS and crowding distance concept. Thus, a mating pool of size $N$ is determined after tournament selection.

***Simulated Binary Crossover (SBX)***

In crossover, two randomly selected population members from mating pool ($P_t$ of size $N$) undergo simulated binary crossover [18]. Two offspring solutions $C_1$ and $C_2$ from two parent solutions $P_1$ and $P_2$ are calculated as follows:

$$C_1 = 0.5\,[(1 + \beta_i)\,P_1 + (1 - \beta_i)P_2] \tag{4}$$

$$C_2 = 0.5\,[(1 - \beta_i)\,P_1 + (1 + \beta_i)\,P_2] \tag{5}$$

where $\beta_i$ is calculated as follows:

$$\beta_i = \begin{cases} (2u_i)^{1/(n_c+1)}, & \text{if } u_i < 0.5 \\ [1/2(1 - u_i)]^{1/(n_c+1)}, & \text{otherwise} \end{cases} \tag{6}$$

where $u_i$ is a random number in range [0, 1] and $n_c$ is a distribution index which is a non-negative real number.

***Polynomial Mutation (PM)***

Population generated from crossover operation undergoes polynomial mutation (PM) to generate offspring population ($O_t$) or child population [19]. PM is applied to a gene value as follows:

$$C = \mathrm{P} + (x_i^{(u)} - x_i^{(l)})\delta_i \tag{7}$$

where $P$ is parent gene value, $C$ is child/offspring gene value, $(x_i^{(u)} - x_i^{(l)})$ is the difference in upper and lower bound of gene value and $\delta_i$ is the mutation value. $\delta_i$ is calculated as follows:

$$\delta_i = \begin{cases} (2r_i)^{1/(n_m+1)} - 1, & \text{if } r_i < 0.5 \\ 1 - [2(1 - r_i)]^{1/(n_m+1)}, & \text{if } r_i \geq 0.5 \end{cases} \tag{8}$$

where $r_i$ is a random number in the range [0,1] and $n_m$ is a distribution constant of mutation.

***Non-dominated Sorting of Combined Population ($R_t$: $P_t$ U $Q_t$)***

Offspring population ($O_t$) generated after SBX and PM is combined with parent population ($P_t$), which gives the combined population ($R_t$) of size 2 $N$. These 2 $N$ solutions undergo non-dominated sorting to be distributed in non-dominated fronts ($F_1, F_2 \ldots F_l \ldots F_n$).

***Generation of New Population***

A new population from non-dominated fronts ($F_1, F_2 \ldots F_l \ldots F_n$) is generated for next generations. For this purpose, an intermediate population $S_t$ is set at zero initially and then population members are preserved from non-dominated fronts ($F_1, F_2 \ldots F_n$) until $S_t \geq N$. If $S_t$ is equal to $N$ up to front $F_l$, then $S_t$ turns into new generation ($P_{t+1}$). If $S_t > N$, at first $p_{t+1}$ population members are preserved from first $F_{l-1}$ non-dominated fronts and remaining $K$ ($N - p_{t+1}$) population members are selected from front $F_l$ based on maximum diversity.

To select $K$ members from front $F_l$, at first each objective value of population members $S_t$ (up to front $F_l$) is normalized and then $H$ number of structured reference points ($Z_S$) are generated on a normalized hyper-plane. If in a problem, there are $M$ objectives and $p$ number of divisions on each objective axis in the normalized hyper-plane, then $H$ is calculated as follows:

$$H = \binom{M + p - 1}{p} \tag{9}$$

These $H$ reference points are used to place on normalized hyper-plane having ($p$ + 1) points along each boundary. Afterward, reference line of each reference point is computed by joining each reference point to origin. Perpendicular distance from each $S_t$ member to each reference line is then calculated. A reference point of reference line is said to be associated to population member, if perpendicular distance between this population member and that reference point is minimum with respect to others.

For niche counting of reference points, firstly, a reference point with minimum number of associated population members is selected, say $j$th reference point. Now, there are two possible cases: (i) if *j*th reference point has no associated member up to front $F_{l-1}$ but has some associated population members in front $F_l$, then a population member of $F_l$ with minimum perpendicular distance to *j*th reference point is added to $p_{t+1}$ and (ii) if *j*th reference point has at least one or more associated population members up to front $F_{l-1}$ and also in front $F_l$, then a random associated population member is added to $p_{t+1}$. Otherwise, the *j*th reference point is not considered in current iteration. Since one population member is added to $p_{t+1}$ in one time, this procedure is repeated for $K$ times or until all vacant population slots of $P_{t+1}$ are filled.

Figure 2 shows the movement starting with one generation then onto next in NSGA-III. Since it is an iterative procedure, the process continues until the stopping criterion met. General stopping criteria are the maximum number of generation ($G_{max}$) or the attained convergence of solutions, or both. When optimization process terminates, final set of optimal solution is provided to the user as Pareto-optimal solutions.

# 5 Case Study and Discussion

A three-storied building construction project is taken for implementation of the proposed TCEIT model. This project consisted of 13 activities. Activities' name, successors, alternatives and values of objectives (time, cost and environmental impact) are given in Table 1. Based on available alternatives of projects' activities, there were 55,296 different ways to deliver the project. The proposed model is implemented to determine the optimal combinations of activities alternatives for successfully delivering the project.

The proposed NSGA-III-based TCEIT model is practically implemented using MATLAB R2014a. For finalizing the values of NSGA-III parameters, numerous trials were executed with varying values of NSGA-III parameters. The best possible combination of NSGA-III parameters was adopted and the values are shown in Table 2.

A total of 22 exclusive optimal combinations of activities alternatives is obtained that fulfilled the desired project objectives. The values of PCT, PCC and PEI are determined for each 22 delivering ways of project. The PCT values vary from 170 to 230 days; the PCC values vary from 487984.90 to 722083.90 and the PEI values vary from 435285.0 to 683054.8. Table 3 shows all 22 obtained Pareto-optimal solutions. However, the time–cost trade-off plot and time–environmental impact trade-off plot are shown in Figs. 3 and 4, respectively. The whole set of the obtained Pareto-optimal solutions is portrayed in three-dimensional plot, as shown in Fig. 5. In order to evaluate the impact of various activities alternatives on project performance, these two- and three-dimensional representation of obtained Pareto-optimal solutions can be used to visualize the trade-offs among PCT, PCC and PEI.

Selecting a solution from the obtained Pareto-optimal set of solutions is also a requirement of project team. For this purpose, a study [20] presented numerous

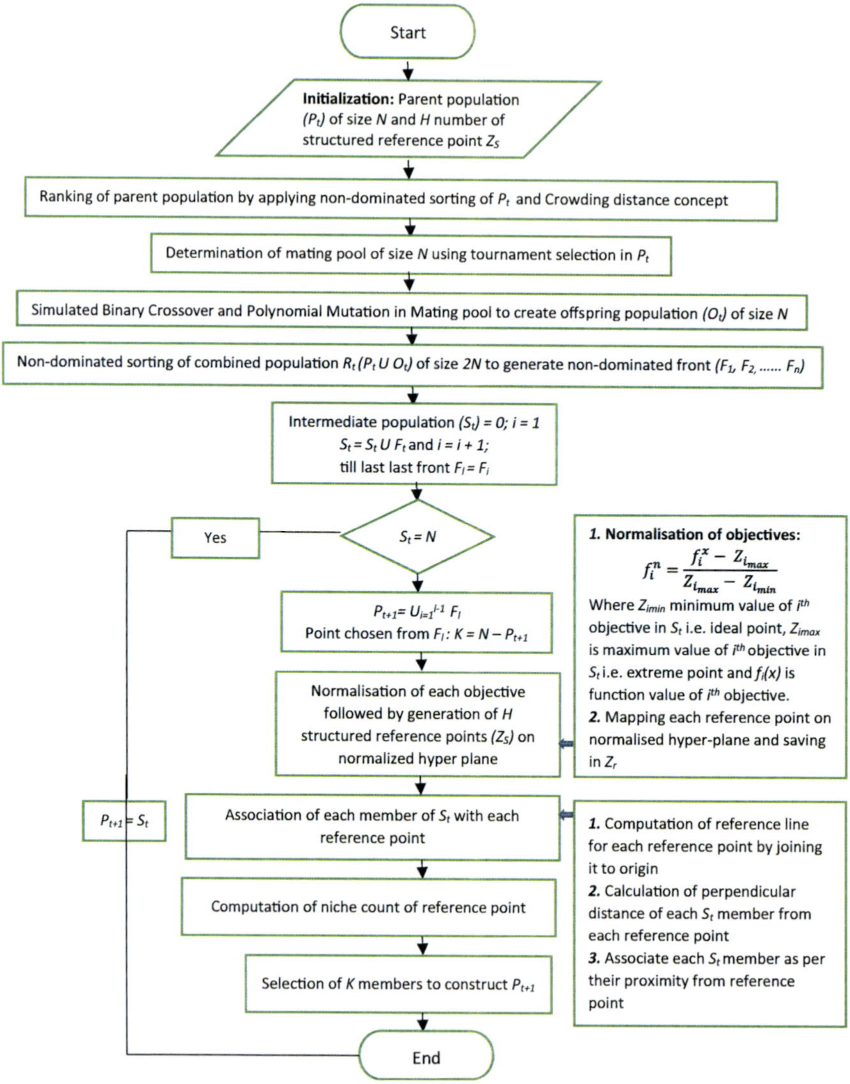


**Fig. 2** Movement starting with one generation then onto the next in NSGA-III

methodologies to select solutions from Pareto-optimal set. One of its methodologies is weighted sum method, through which a solution can be selected from Pareto-optimal set by using the following equations:

$$\text{Min} \overset{j=1}{\underset{n}{\vee}} \sum_{i=1}^{m} w_i x_{ij} \tag{10}$$

**Table 1** Details of the case study project

| S. no. | Activity name | Successors | Alt | Time (days) (ACT) | Cost (ACC in US $) | Environmental impact (AEI in $CO_{2\text{-eq}}$) |
|---|---|---|---|---|---|---|
| 1 | Ground-works | 2 | 1 | 8 | 10039.42 | 3428.86 |
| | | | 2 | 8 | 9849.86 | 5845.97 |
| 2 | Excavation | 3 | 1 | 6 | 1082.13 | 634.67 |
| | | | 2 | 6 | 891.05 | 785.83 |
| 3 | Footing | 4 | 1 | 12 | 15545.67 | 16785.56 |
| | | | | 10 | 17039.34 | 17234.65 |
| 4 | Formwork | 5 | 1 | 5 | 562.13 | 18536.34 |
| | | | 2 | 4 | 590.32 | 19537.88 |
| 5 | Retaining wall | 6 | 1 | 26 | 15834.49 | 28321.54 |
| | | | 2 | 16 | 17274.94 | 29652.76 |
| 6 | Basement | 7 | 1 | 32 | 74124.65 | 170322.45 |
| | | | 2 | 29 | 76345.78 | 198223.43 |
| | | | 3 | 23 | 84312.34 | 233671.23 |
| 7 | Slab | 8 | 1 | 22 | 32646.05 | 18346.26 |
| | | | 2 | 11 | 29759.59 | 19312.87 |
| 8 | Exterior wall | 9 | 1 | 18 | 65959.52 | 17554.67 |
| | | | 2 | 29 | 105296.94 | 70231.87 |
| | | | 3 | 11 | 157433.42 | 71232.65 |
| 9 | Interior wall | 13 | 1 | 37 | 58570.35 | 8345.76 |
| | | | 2 | 21 | 59999.39 | 8212.32 |
| | | | 3 | 32 | 57668.29 | 23864.32 |
| | | | 4 | 17 | 63321.11 | 67231.74 |
| 10 | Flooring | – | 1 | 34 | 38411.50 | 50435.65 |
| | | | 2 | 17 | 65326.48 | 65342.43 |
| | | | 3 | 12 | 50214.22 | 54235.12 |
| 11 | Exterior finish | – | 1 | 9 | 12216.23 | 34572.45 |
| | | | 2 | 12 | 3846.23 | 30235.32 |
| 12 | Interior finish | – | 1 | 41 | 90219.78 | 5323.51 |
| | | | 2 | 31 | 233034.50 | 2045.23 |
| 13 | Roof | 10–12 | 1 | 23 | 127641.84 | 134971.84 |
| | | | 2 | 24 | 81323.17 | 69273.46 |

**Table 2** Adopted values of NSGA-III parameters

| NSGA-III parameters | Value |
|---|---|
| Population size | 100 |
| Number of generation | 150 |
| SBX probability | 1 |
| SBX distribution index | 20 |
| PM probability | 1 |
| PM rate | 1/13 |
| PM distribution index | 20 |

$$\text{s.to} \quad \sum_{i=1}^{m} w_i = 1 \tag{11}$$

where $w_i$ represents the weight of $i$th objective and $x_{ij}$ represents the $j$th solution of $i$th objective. For instance, if a project team considers that all objectives are of same importance, then the weight of each objective will be 1/3. Then, from Eqs. (10) and (11), calculated solution from obtained Pareto-optimal set will be 2, 1, 2, 2, 2, 1, 2, 1, 4, 1, 2, 1, 2. PCT, PCC and PEI corresponding to this solution are 187 days, 492801.80 US dollars and 502596, respectively.

### *Validation and Comparison of Model*

In order to validate the developed model, the results obtained from the developed model and an existing model are compared with the same algorithm parameters (i.e. 150 generations, 100 population size and 20 crossover and mutation distribution index). The above presented case study project is solved using model of Ozcan-Deniz [2], which is developed as non-dominated sorting genetic algorithm II. As shown in Table 4, Pareto-optimal solutions obtained from developed model are either similar to or better than existing model that validate the applicability of the developed model.

In order to compare the performance of developed model and the model of Ozcan-Deniz [2], hypervolume of the Pareto-optimal solutions obtained from both model is compared, where hypervolume (*HV*) measures the volume covered by Pareto-optimal front solutions in a minimization problem. For instance, shaded portion shown in Fig. 6 represents the hypervolume of Pareto-optimal solutions in two objectives ($f_1$, $f_2$) minimization problem.

Higher value of hypervolume shows superiority of optimization model in searching more converge and diverse solution. The hypervolume of Pareto-optimal solutions obtained from developed model and model of Ozcan-Deniz [2] was 0.78 and 0.72, respectively, which justify that developed model is superior than the existing model.

**Table 3** Obtained pareto-optimal solutions of case study project

| S. no. | A1 | A2 | A3 | A4 | A5 | A6 | A7 | A8 | A9 | A10 | A11 | A12 | A13 | PCT | PCC | PEI |
|---|---|---|---|---|---|---|---|---|---|---|---|---|---|---|---|---|
| 1 | 1 | 2 | 2 | 2 | 2 | 3 | 2 | 3 | 4 | 1 | 2 | 1 | 1 | 170 | 640780.50 | 683054.8 |
| 2 | 2 | 2 | 2 | 1 | 2 | 3 | 2 | 1 | 4 | 1 | 2 | 2 | 2 | 172 | 645584.90 | 561815.7 |
| 3 | 1 | 2 | 1 | 2 | 2 | 3 | 2 | 1 | 4 | 1 | 2 | 2 | 2 | 173 | 644309.00 | 559951.1 |
| 4 | 1 | 2 | 2 | 2 | 2 | 3 | 2 | 3 | 2 | 1 | 2 | 1 | 2 | 175 | 591140.10 | 558337.0 |
| 5 | 1 | 2 | 2 | 2 | 2 | 3 | 2 | 1 | 4 | 1 | 2 | 1 | 1 | 177 | 549306.60 | 629376.8 |
| 6 | 1 | 2 | 2 | 2 | 2 | 3 | 2 | 1 | 4 | 1 | 2 | 1 | 2 | 178 | 502987.90 | 563678.4 |
| 7 | 2 | 2 | 1 | 2 | 2 | 1 | 2 | 3 | 2 | 1 | 2 | 2 | 2 | 179 | 722083.90 | 493678.0 |
| 8 | 1 | 2 | 2 | 2 | 2 | 1 | 2 | 3 | 4 | 1 | 2 | 1 | 2 | 180 | 584274.10 | 554007.6 |
| 9 | 2 | 1 | 2 | 2 | 2 | 2 | 2 | 3 | 2 | 1 | 2 | 1 | 2 | 181 | 583175.10 | 525155.1 |
| 10 | 2 | 2 | 1 | 2 | 2 | 1 | 2 | 3 | 4 | 1 | 2 | 1 | 2 | 182 | 582590.90 | 555975.7 |
| 11 | 2 | 2 | 1 | 2 | 2 | 2 | 2 | 1 | 2 | 1 | 2 | 2 | 2 | 183 | 632831.20 | 467901.0 |
| 12 | 1 | 2 | 2 | 2 | 2 | 2 | 2 | 1 | 4 | 1 | 2 | 1 | 2 | 184 | 495021.40 | 528230.6 |
| 13 | 1 | 2 | 2 | 1 | 2 | 2 | 2 | 1 | 4 | 1 | 2 | 1 | 2 | 185 | 494993.20 | 527229.1 |
| 14 | 2 | 2 | 1 | 2 | 2 | 1 | 2 | 1 | 2 | 1 | 2 | 2 | 2 | 186 | 630610.00 | 440000.0 |
| 15 | 1 | 2 | 2 | 2 | 2 | 1 | 2 | 1 | 4 | 1 | 2 | 1 | 2 | 187 | 492800.20 | 500329.7 |
| 16 | 1 | 2 | 2 | 1 | 2 | 1 | 2 | 1 | 4 | 1 | 2 | 1 | 2 | 188 | 492772.00 | 499328.1 |
| 17 | 1 | 2 | 1 | 2 | 2 | 1 | 2 | 1 | 2 | 1 | 2 | 1 | 2 | 193 | 487984.90 | 440861.1 |
| 18 | 2 | 2 | 1 | 1 | 2 | 1 | 2 | 1 | 1 | 1 | 2 | 2 | 2 | 203 | 629152.80 | 439131.9 |
| 19 | 1 | 2 | 1 | 2 | 1 | 1 | 1 | 1 | 2 | 1 | 2 | 2 | 2 | 207 | 632246.00 | 435285.0 |
| 20 | 1 | 2 | 1 | 2 | 1 | 1 | 1 | 1 | 2 | 1 | 2 | 1 | 2 | 214 | 489431.20 | 438563.3 |
| 21 | 1 | 2 | 1 | 2 | 1 | 1 | 1 | 1 | 1 | 1 | 2 | 2 | 2 | 223 | 630816.90 | 435418.5 |
| 22 | 1 | 2 | 1 | 2 | 1 | 1 | 1 | 1 | 1 | 1 | 2 | 1 | 2 | 230 | 488002.20 | 438696.8 |

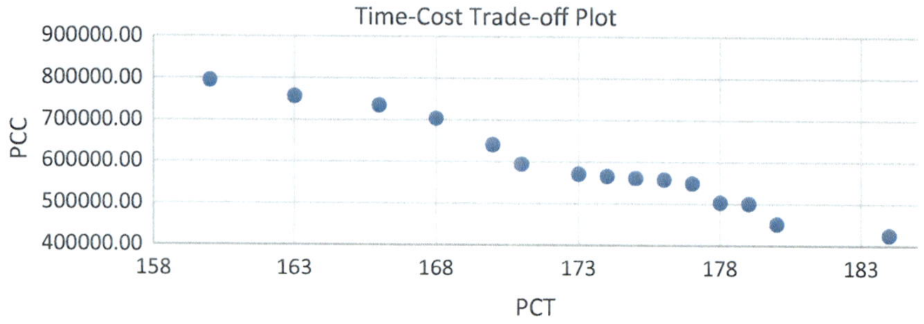


**Fig. 3** Time–cost trade-off plot

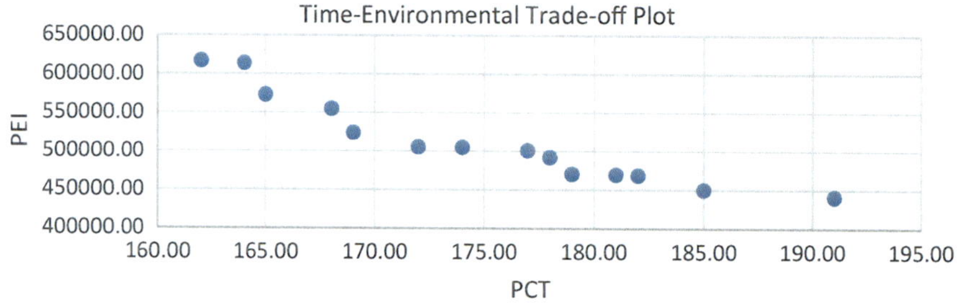


**Fig. 4** Time–environmental impact trade-off plot

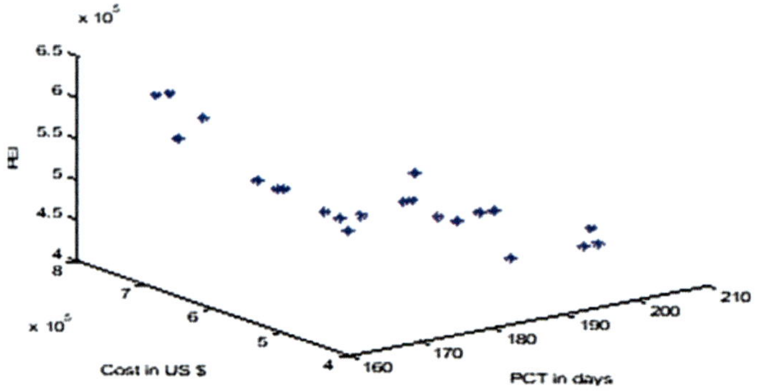


**Fig. 5** Time–cost–environmental impact trade-off plot

## 6 Conclusion

It is becoming necessary to found the optimal trade-off between project objectives due to the rapid development of technologies and requirements of project stakeholders. The main purpose of this paper is to extend the time–cost trade-off up to the time–cost–environmental impact trade-off. For this purpose, environmental impact of each

**Table 4** Comparison of results for TCEIT optimization

| Study | Ozcan-Deniz [12] | | | Developed model | | |
|---|---|---|---|---|---|---|
| Optimization algorithm | Non-dominated sorting Genetic Algorithm II | | | Non-dominated sorting Genetic Algorithm III | | |
| Objectives | Time (days) | Cost (US $) | Environment impact | Time (days) | Cost (US $) | Environment impact |
| Solutions | | | | | | |
| 1 | | | | 170 | 640780.50 | 683054.8 |
| 2 | 180 | 584274.10 | 554007.6 | 180 | 584274.10 | 554007.6 |
| 3 | 193 | 487984.90 | 440861.1 | 193 | 487984.90 | 440861.1 |
| 4 | 214 | 523365.30 | 482373.5 | 214 | 489431.20 | 438563.3 |
| 5 | 230 | 488002.20 | 438696.8 | 230 | 488002.20 | 438696.8 |

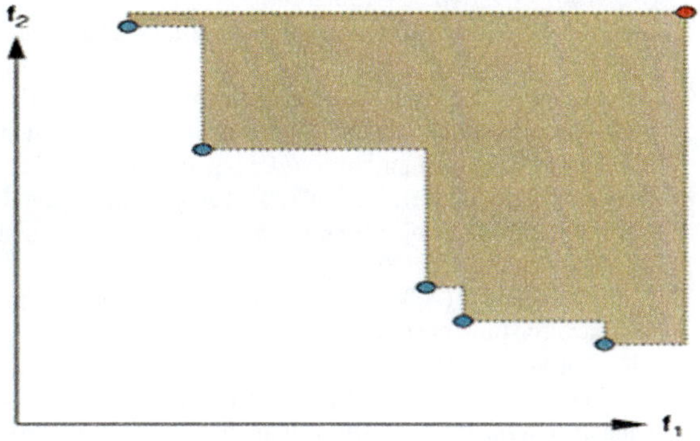

**Fig. 6** Hypervolume in case of two objectives minimization problems

alternative is estimated in terms of $CO_{2\text{-eq.}}$ greenhouse gas emissions ($GHG_{CO2\text{-eq.}}$). In order to determine the optimum alternative for each activity of construction project, non-dominated sorting genetic algorithm III is used to meet the desired projects' objectives. Validity and applicability of the proposed model is demonstrated through solving a case study project. The results of the case study reflect the potential of the proposed model. First, relationships can be established between all three objectives to expand the perspective of traditional project control mechanisms. Second, NSGA-III is found to be appropriate as well as better than NSGA-II to address the difficulties of multi-objective optimization. Third, the proposed model supports the project team and practitioner to choose one optimal solution from the obtained Pareto-optimal front. Fourth, this study and proposed model will be beneficial to project stakeholders in the perspective of maximizing profit within minimum environmental impact. Finally, this study possibly provide a better scheduling decision-making tool for project team and organization.

Although the working of proposed TCEIT optimization model has been found to be efficient, its effectiveness in large-scale projects remains to be practically tested. Further analysis is also necessary, taking into account the uncertainty in determining time, cost and environmental impact of project.

## References


1. Zahraie B, Tavakolan M (2009) Stochastic time-cost-resource utilization optimization using nondominated sorting genetic algorithm and discrete fuzzy sets. J Constr Eng Manag. https://doi.org/10.1061/(ASCE)CO.1943-7862.0000092
2. Ozcan-Deniz G, Zhu Y, Ceron V (2012) Time, cost, and environmental impact analysis on construction operation optimization using genetic algorithms. J Manag Eng. https://doi.org/10.1061/(ASCE)ME.1943-5479.0000098
3. Feng CW, Liu L, Burns SA (1997) Using genetic algorithms to solve construction time-cost trade-off problems. J Comput Civil Eng. https://doi.org/10.1061/(ASCE)0887-3801(1997)11:3(184)
4. Tiwari S, Johari S (2015) Project scheduling by integration of time cost trade-off and constrained resource scheduling. J Inst Eng (India): Ser A. https://doi.org/10.1007/s40030-014-0099-2
5. Zheng DXM, Ng ST, Kumaraswamy MM (2004) Applying a genetic algorithm-based multi-objective approach for time-cost optimization. J Constr Eng Manag. https://doi.org/10.1061/(ASCE)0733-9364(2004)130:2(168)
6. Singh G, Ernst AT (2011) Resource constraint scheduling with a fractional shared resource. Oper Res Lett. Elsevier B.V., 39(5):363–368. https://doi.org/10.1016/j.orl.2011.06.003
7. Deb K, Jain H (2014) An evolutionary many-objective optimization algorithm using reference-point-based nondominated sorting approach. Part I: Solving problems with box constraints. IEEE Trans Evol Comput. https://doi.org/10.1109/TEVC.2013.2281535
8. Panwar A, Tripathi KK, Jha KN (2019) A qualitative framework for selection of optimization algorithm for multi-objective trade-off problem in construction projects. Eng Constr Architect Manag. https://doi.org/10.1108/ECAM-06-2018-0246
9. Xiong Y, Kuang Y (2008) Applying an ant colony optimization algorithm-based multiobjective approach for time-cost trade-off. J Constr Eng Manag. https://doi.org/10.1061/(ASCE)0733-9364(2008)134:2(153)
10. Van Den Bergh F, Engelbrecht AP (2010) A convergence proof for the particle swarm optimiser. Fundamenta Informaticae. https://doi.org/10.3233/FI-2010-370
11. Fallah-Mehdipour E et al (2012) Extraction of decision alternatives in construction management projects: application and adaptation of NSGA-II and MOPSO. Expert Syst Appl. https://doi.org/10.1016/j.eswa.2011.08.139
12. Shahriari M (2016) Multi-objective optimization of discrete time–cost tradeoff problem in project networks using non-dominated sorting genetic algorithm. J Indus Eng Int 12(2):159–169. https://doi.org/10.1007/s40092-016-0148-8
13. Köppen M, Yoshida K (2007) Many-objective particle swarm optimization by gradual leader selection. In: Lecture notes in computer science (including subseries Lecture notes in artificial intelligence and Lecture notes in bioinformatics). https://doi.org/10.1007/978-3-540-71618-1_36
14. Marzouk M et al (2008) Handling construction pollutions using multi-objective optimization. Constr Manag Econ. https://doi.org/10.1080/01446190802400779
15. Cheng MY, Tran DH (2015) Opposition-based multiple-objective differential evolution to solve the time-cost-environment impact trade-off problem in construction projects. J Comput Civil Eng. https://doi.org/10.1061/(ASCE)CP.1943-5487.0000386
16. Deb K et al (2002) A fast and elitist multiobjective genetic algorithm: NSGA-II. IEEE Trans Evol Comput. https://doi.org/10.1109/4235.996017
17. Das I, Dennis JE (1998) Normal-boundary intersection: a new method for generating the Pareto surface in nonlinear multicriteria optimization problems. SIAM J Optim. https://doi.org/10.1137/S1052623496307510

18. Deb K, Agrawal RB (1994) Simulated binary crossover for continuous search space. Comp Syst. 10.1.1.26.8485Cached
19. Deb K, Goyal M (1996) A combined genetic adaptive search (GeneAS) for engineering design. Comput Sci Inform 26(4):30–45. https://doi.org/10.1.1.27.767&rep=rep1&type=pdf
20. Ferreira JC, Fonseca CM, Gaspar-Cunha A (2007) Methodology to select solutions from the pareto-optimal set: a comparative study. In: Proceedings of GECCO 2007: genetic and evolutionary computation conference. https://doi.org/10.1145/1276958.1277117

# Chapter 3
# Wideband Microstrip Stepped Two-Way Power Divider for 3G/4G Applications

**Anshul Agarwal, P. K. Singhal, and V. V. Thakare**

## 1 Introduction

Power dividers are a component of the wireless communication system which is used to split the powers in various ratios (equal or unequal) to another component of the system. The power division of a power divider is the number of outputs of the device, such as 2-way, 3-way, 4-way, 5-way, and so on.

The major application of power divider is in the distribution of RF signal to form an antenna array. Various types of power dividers are: T-junction and Wilkinson power divider [1], in which Wilkinson power divider is mostly used due to its ease in calculation and designing, but a lot of power wastage is found across the resistance. Microstrip stepped power divider is in demand in the modern communication system because the total input power is distributed to the output ports proportionally as per the design and requirements [2]. Wideband Wilkinson power divider using the Villard voltage doubler method for RF harvesting from 800 MHz to 3.11 GHz is designed [3] but some power losses across the resistor occurred during designing. The ultra-wideband power divider is designed using a multi-mode resonator technique which shows the in-phase power division from 3.1 to 10.7 GHz [4].

In the Wilkinson power divider, two comb-shaped resonators are used instead of a quarter transmission line to improve the harmonics suppression [5]. The number of output ports and frequency band of 4–10.5 GHz is increased by two separate cavity rings in a multi-way power divider [6]. Various techniques are used in designing a wide impedance power divider, but lot of power is wasted across the resistance to provide good isolation between the output ports.

In the present work, a wide impedance power divider is designed without using the resistance which saves the power and provides the equally 50% power divider in both the output ports with good isolation.

A. Agarwal (✉) · P. K. Singhal · V. V. Thakare
Department of Electronics, Madhav Institute of Technology and Science, Gwalior, MP, India

© The Author(s), under exclusive license to Springer Nature Singapore Pte Ltd. 2022 
H. M. Dubey et al. (eds.), *Artificial Intelligence and Sustainable Computing*, Algorithms for Intelligent Systems, https://doi.org/10.1007/978-981-16-1220-6_3

## 2 Power Divider Design and Parametric Analysis

Stepped power divider is designed using a three-section quarter-wave transformer of different impedances to achieve a wide frequency band with a VSWR at 1.5 for better performance. The 50 Ω input port is connected centrally as a combination of three quarter-wave transmission lines of different widths and lengths through a 64 Ω transmission line to reduce the losses along the mismatching of the impedance of the transmission line. The three quarter-wave transformers generate the three resonant frequencies of 1.58, 2.06 and 2.78 GHz simultaneously, which enhanced the bandwidth of the power divider.

The proposed two-way power divider is designed on an FR-4 substrate (thickness 1.6 mm, dielectric constant 4.3) and simulated by using CST Software Version 2018 [7]. The two-way microstrip stepped power divider with its impedance value is shown in Fig. 1.

The parameters which majorly affect the response of the power divider are also analyzed. With the help of parametric analysis, the optimum values of the parameters of power divider are obtained.

The first parameter is the width of the first section of the quarter-wave transmission line of 100 Ω says 'W1'. The impedance bandwidth reduces by increasing the value of W1. The effect on $S_{11}$ by W1 is shown in Fig. 2.

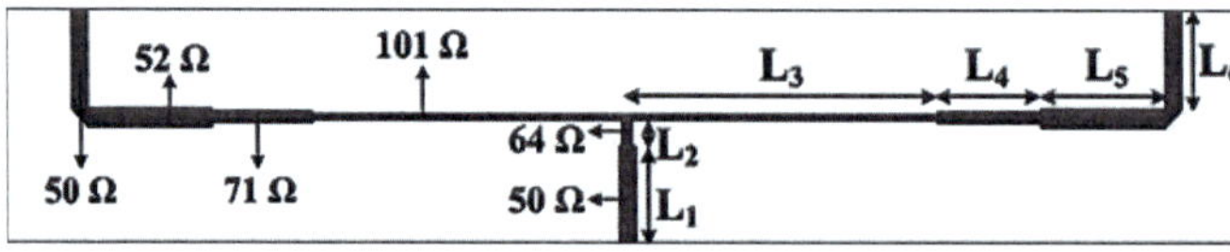


Fig. 1 Geometry of two-way power divider

Fig. 2 Simulated return loss for W1

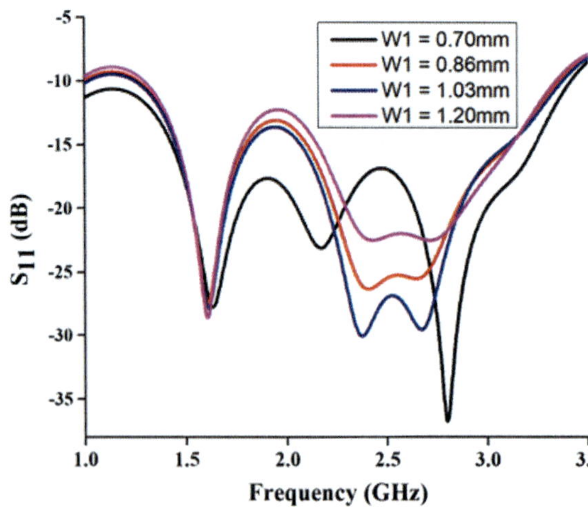

The second parameter is the length of the transmission line of 101 Ω 'L4'. The frequency band is shifted toward the lower side by increasing the value of L4. The effect of L4 on $S_{11}$ is shown in Fig. 3. L4 of 20 mm was chosen for 3G/4G applications.

The third parameter is the width of the transmission line of 64 Ω says 'W2', which connects the input port to a multi-section quarter-wave transmission line. Wide impedance bandwidth was converted into a multi-impedance band by increasing the value of W2. The effect of W2 on $S_{11}$ is shown in Fig. 4.

After parametric analysis by varying one parameter once at a time and remaining parameters are constant, the optimized length of the different quarter-wave transmission lines is listed in Table 1.

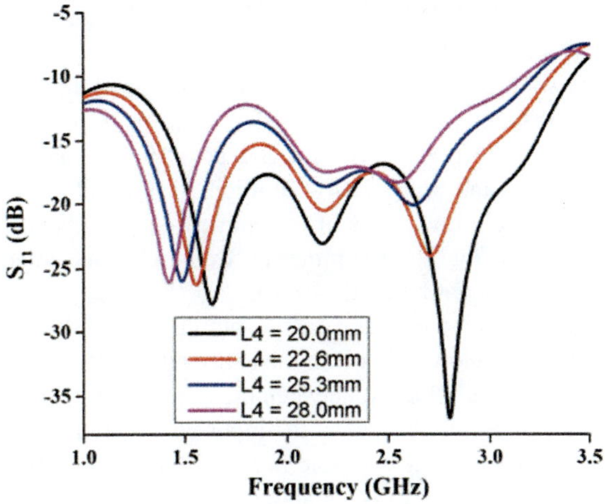


**Fig. 3** Simulated return loss for L3

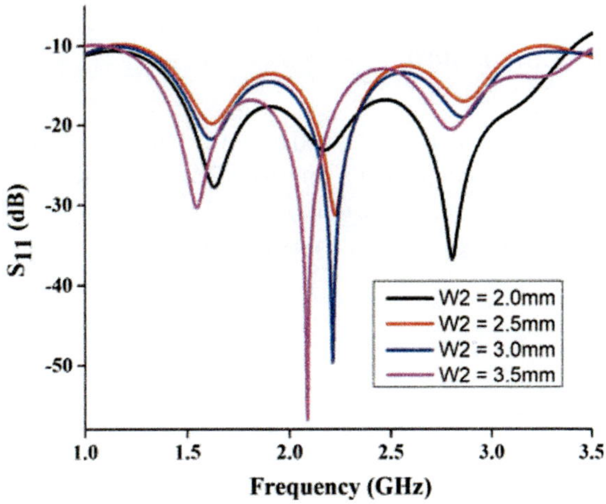


**Fig. 4** Simulated return loss for W2

**Table 1** Lengths of different quarter wave transmission lines of power divider

| Length | Value (mm) |
|---|---|
| L1 | 15 |
| L2 | 4 |
| L3 | 60 |
| L4 | 20 |
| L5 | 24 |
| L6 | 15 |

## 3 Simulated and Measured Results

The simulated and fabricated two-way power divider is shown in Fig. 5(a–b) and their response is shown in Fig. 5c.

The reflection coefficient $S_{11}$ at −15 dB having a frequency band of 1.39–3.16 GHz and transmission coefficients $S_{21}$ and $S_{31}$ achieved a 3 dB equal powers at both the output ports in the entire frequency band. The response of $S_{21}$ and $S_{31}$ is the same due to its symmetric structure. It is also noted that the phase difference between the output ports is zero.

The isolation between the two output ports is shown in Fig. 6. It is shown in Fig. 6 that $S_{23}$ is below −30 dB in the entire operating frequency band, which shows good isolation was achieved between the outputs ports of the power divider.

The input impedance at the input port (port 1) and output port (ports 2 and 3) is shown in Fig. 7. It is visible in Fig. 7 that the resistive part is nearly equal to 50 Ω and the reactive part is equal to zero.

## 4 Conclusion

Wideband microstrip stepped two-way equal power divider is designed for the 3G/4G frequency band. Using a multi-section of different widths of the transmission line, wide impedance bandwidth of 1.77 GHz is achieved. This two-way power divider is used for designing an antenna array to enhance the antenna gain and beamwidth.

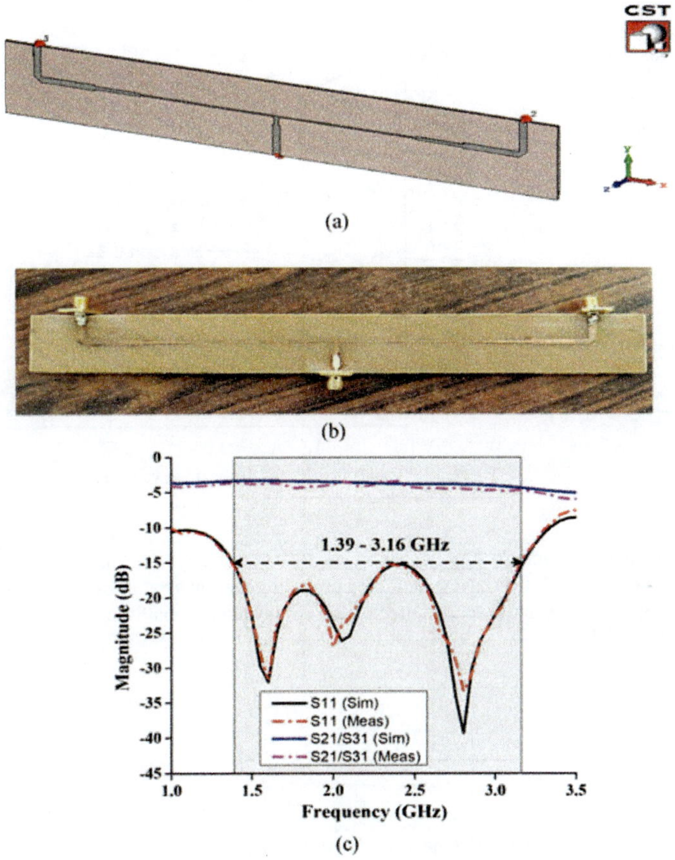


**Fig. 5** **a** Simulated design, **b** fabricated design, **c** simulated and measured reflection and transmission coefficients of the two-way power divider

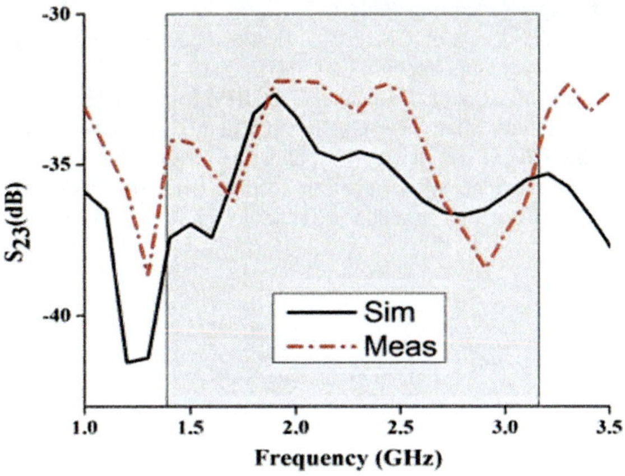


**Fig. 6** Isolation of output port of two-way power divider

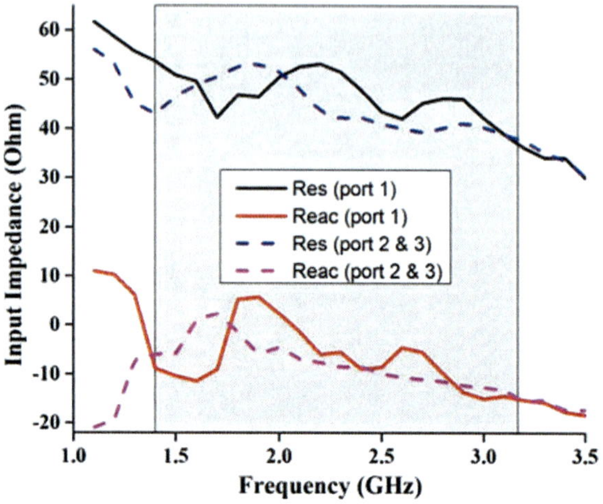


**Fig. 7** Impedance at all ports of the two-way power divider

**Acknowledgments** The authors highly acknowledge and are thankful to the Ministry of Human Resource Development (MHRD), New Delhi for providing the financial assistance under TEQIP-III. The authors are also thankful to the Director, MITS, Gwalior, India for providing the facilities for this research work.

## References

1. Pozar DM (2005) Microwave Engineering, 3rd edn. Wiley
2. Gustrau F (2012) RF and Microwave Engineering: Fundamentals of Wireless Communication. Wiley
3. Ozturk B, Coskun O (2019) Design of wideband microstrip power divider/combiner with input and output impedance matching for RF energy harvesting applications. Int J RF Microwave Comput Aided Eng 29(1):1–7
4. Mondal P, Parui SK (2016) Multi-mode resonator-based ultra-wideband power divider. Microwave Opt Technol Lett 58(10):2419–2422
5. Imani MA, Shama F, Roshani GH (2019) Miniaturized Wilkinson power divider with suppressed harmonics. Microwave Opt Technol Lett 62(4):1526–1532
6. Zakerifar H, Khaleghi A, Aliakbarian H (2017) A compact multiple-way double ring-cavity power divider. Microwave Opt Technol Lett 58(8):2006–2012
7. CST (Computer Simulation Technology) Microwave Studio (Version 2018)

# Chapter 4
# Simulated Annealing-Based Time–Cost Trade-Off Model for Construction Projects

**Amit Dhawan, Kamal Sharma, and Manoj Kumar Trivedi**

## 1 Introduction

Time and cost are the most focused objectives for construction project planning and success [1]. However, time and cost of project vary due to variation in the amount of resources utilization [2]. Utilization of modern resources and technologies increases the project cost but reduces the project time. On the other hand, traditional resources and technologies decrease the project cost but increase the project time. Time and cost are therefore competing and conflicting objectives of construction project. Also, there is a hidden trade-off (TCT) between time and cost of project. Accordingly, various researchers made effort to develop the time–cost trade-off models. Feng et al. [3] developed a TCT model based on genetic algorithm and Pareto-optimal front approach. Kalhor et al. [4] applied ant colony optimization method to propose TCT model. Zhang and Li [5] proposed a multi-objective Pareto-front-based TCT model. Anagnostopoulos and Kotsikas [6] experimentally evaluated the simulated annealing in solving TCT problems.

A construction project consists of several interconnected activities [7]. An activity has several alternatives to be executed. However, each alternative is different in completion time and resource utilization [8]. Therefore, it is required to allocate one suitable alternative for each activity of the project. Many delivering methods of a project are possible based on possible combinations of available activities alternatives. In this study, the research problem is time–cost trade-off problem and an effort is made to find optimal way or ways to deliver the project with simultaneously optimizing the time and cost of project. This problem is a NP-hard problem because complexities increase as number of contradictory objectives increases [9].

A. Dhawan · K. Sharma (✉) · M. K. Trivedi
Department of Civil Engineering, MITS, Gwalior, India

M. K. Trivedi
e-mail: manojtrivedi@mitsgwalior.in

© The Author(s), under exclusive license to Springer Nature Singapore Pte Ltd. 2022 
H. M. Dubey et al. (eds.), *Artificial Intelligence and Sustainable Computing*,
Algorithms for Intelligent Systems, https://doi.org/10.1007/978-981-16-1220-6_4

Simulated annealing (SA) is a meta-heuristic approach, which has been found successful in combinatorial optimization problem. Simulated annealing (SA) mimics the physical annealing process. This process is very useful for situations where there are a lot of local minima.

Based on the above discussion, this paper presents a time–cost trade-off model. In the next section, literature review is presented followed by SA-based proposed TCT model. Subsequently, the applicability of proposed TCT model is demonstrated by using a case study project and result comparisons. In the final section of this paper, conclusions and scope for future work is presented.

## 2 Literature Review

Literature study shows that mainly three methods are used for scheduling, i.e., deterministic methods, heuristic methods and meta-heuristic methods [10]. Deterministic methods provide exact solutions and include CPM, integer programming (IP), linear programming (LP) and dynamic programming (DP) methods for activities scheduling. Start with history, Kelley and Walker [11] proposed a critical path method for scheduling the activities of projects. Thereafter, several deterministic methods-based scheduling models were developed using CPM [12], IP/LP [13] and DP [14].

Heuristic methods are based on historical practice of problem-solving [15]. Fondhahl's approach [16] and structural stiffness [17] are prevalent heuristics methods of scheduling. Furthermore, Zhang et al. [18] and Di et al. [19] also made effort to develop heuristic methods-based scheduling models.

Meta-heuristic methods of scheduling were developed to solve two or more than two objectives optimization problems. For large-scale optimization problem, meta-heuristic methods have demonstrated the efficiency in searching relative good solutions rather than exact [4]. Feng [3] and Zheng et al. [20] developed a time–cost trade off (TCT) model based on genetic algorithm and Pareto-front approach. Yang [21] and Zhang and Li [5] proposed a TCT model based on the multi-objective particle swarm optimization (MOPSO). A study [6] also experimentally evaluated the simulated annealing algorithm in solving the TCT problem. Li and Li [22] presented multi-objective ant colony optimization-based Pareto-front approach for solving the TCT problems. Fallah-Mehdipour et al. [23] concluded that NSGA II outperform the MOPSO in determining the Pareto-front for TCT and TCQT optimization. Shahriari [24] applied NSGA II in discrete TCT optimization problem. Afruzi et al. [25] solved multi-mode resource-constrained discrete time–cost trade-off problem using adjusted fuzzy dominance genetic algorithm.

In this paper, the SA process is adopted to develop a new methodology to solve the TCT problems. Also, this study aims to develop a SA-based construction TCT model for helping the construction industries and planners to search the optimum trade-off or balance between time and cost of project.

## 3 Proposed Simulated Annealing-Based TCT Model

A construction project has a number of activities to be finished to complete a construction project. An activity (A) should be executed through one of its available alternatives. It is required to allocate a suitable alternative for each activity of construction project. Each alternative of an activity has different completion time and completion cost due to variation in amount of resources associated to each alternative.

In this paper, the input parameters are taken as activity completion time (ACT) and activity completion cost (ACC). Assume a construction project having activities $A-1, A-2 \ldots A-n$ and these can be executed through various alternatives denoted as Alt$-1$, Alt$-2$ ... Alt$-m$. Each alternative is associated with a specific amount of labor, material and equipment resources which govern the ACT and ACC. Numerical values of ACT and ACC are calculated based on activity and choice of alternative for each construction activity. The objective of this paper is to optimize the project completion time (PCT) and project completion cost (PCC) simultaneously, which are the functions of ACT and ACC, respectively. The key assumptions made in this study are as follows: (i) An activity, if once started, must be completed without interruption. (ii) There are precedence relationships between activities. These relationships may be start-to-start, start-to-finish, finish-to-finish and finish-to-start. These relationships can be represented in the form of network.

So now, this two-objective optimization problem can be defined as follows: (i) Minimization of PCT and (ii) minimization of PCC. These two construction project objectives are formulated in the following manner.

***Project Completion Time (PCT)***

PCT is an important parameter of construction project, which greatly influences the success of construction project. In this paper, precedence diagramming method (PDM) is employed to calculate PCT. PDM is based on critical path (CP) of activity on node (AON) project networks. PCT is the sum of ACT of all the activities in the critical path of construction project.

Objective 1: Minimize PCT

$$\mathrm{PCT} = \sum_{A \in CP} \mathrm{ACT}_A \tag{1}$$

where PCT is the project completion time and $\mathrm{ACT}_A$ is the completion time of critical path activity (A).

***Project Completion Cost (PCC)***

PCC is one of the key factors affecting the success of construction projects. PCC is simply the algebraic sum of each activity completion cost (ACC). ACC is the sum of direct cost and indirect cost of activity. Direct cost (D.C) includes mainly the cost of labor, material and equipment, whereas indirect cost (I.C) includes overhead expenses and outage losses.

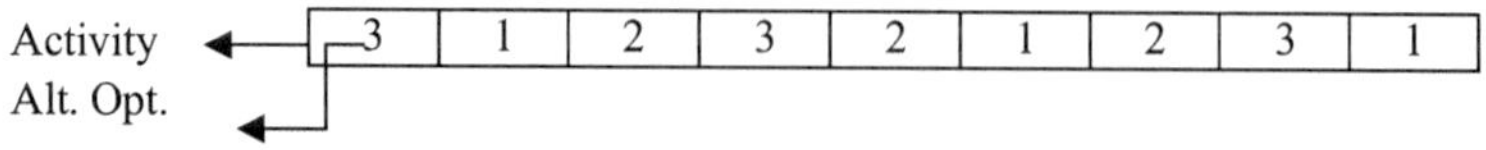


**Fig. 1** Chromosome solution of given TCT problem

Objective 2: Minimize PCC

$$\mathrm{PCC} = \sum_{A} \mathrm{D.C} + \mathrm{I.C\ per\ day} \times \mathrm{PCT\ in\ days} \tag{2}$$

where PCC is the project completion cost, $\sum_A$ D.C is the sum of direct cost of each activity. Indirect cost of construction project is simply estimated by multiplying the PCT and indirect cost per day.

As mentioned earlier, an activity can be executed through one of its available alternatives. For each activity, a particular alternative is required to choose as decision variable. And each combination of decision variables of project activities is the possible solution of this TCT problem. In SA, each solution of optimization problem can be represented in the form of chromosome. A chromosome consists of number of cells (genes) equal to total number of decision variables or activities in construction project. Thus, each gene of chromosomes contains a value (allele) of decision variable, which is the selected alternative through which that activity can be executed. Each possible chromosome of given TCT problem is a possible solution of TCT problem. For understanding the chromosome solutions of given TCT problem, suppose a construction project consisting of nine activities and each activity can be executed either one of its three available alternate options. A chromosome solution showing in Fig. 1 can be a possible solution for delivering the project of 1–9 activities with alternatives 3, 1, 2, 3, 2, 1, 2, 3, 1, respectively. As per permutation rules there will be $3^9$ possible project delivering solutions.

Simulated annealing process to solve the TCT problem starts with setting a high initial temperature ($T_i$) and a cooling schedule. Then, a chromosome solution ($\mathrm{CS}_1$), as shown in Fig. 1, is randomly generated for the given TCT problem. $\mathrm{PCT}_1$ and total $\mathrm{PCC}_1$ of $\mathrm{CS}_1$ are calculated by using the objective functions of PCT and PCC. Afterwards, another chromosome solution ($\mathrm{CS}_\mathrm{n}$) is randomly generated, where $\mathrm{CS}_\mathrm{n}$ is the nth chromosome solution of total possible chromosome solutions ($N$). Thus, $\mathrm{CS}_\mathrm{N}$ is the last chromosome solution. Then, a chromosome solution ($\mathrm{CS}_2$), between $\mathrm{CS}_n$ and $\mathrm{CS}_\mathrm{N}$, which is the closest neighbor of $\mathrm{CS}_1$ in respect of $\mathrm{PCT}_1$, is generated randomly. For $\mathrm{CS}_2$, $\mathrm{PCT}_2$ and $\mathrm{PCC}_2$ are calculated by using the objective functions. Now, there will two possible cases: (1) If $\mathrm{PCC}_2 < \mathrm{PCC}_1$ then accept the $\mathrm{CS}_2$; otherwise, (2) if $\mathrm{PCC}_2 > \mathrm{PCC}_1$ then metropolis criterion is applied, in which probability of accepting the $\mathrm{CS}_2$ is calculated by using the following formula:

$$\Delta = \exp\left(\frac{PCC_1 - PCC_2}{T_i}\right) \quad (3)$$

A random number in the range [0, 1] is then generated and compared to $\Delta$. If $\Delta$ is greater than the random number between 0 and 1, then $CS_2$ is accepted. Thus, a chromosome solution is accepted as project delivering solution. Afterward, temperature is reduced through cooling schedule for generating the next acceptable chromosome solution. After generating a list of acceptable chromosomes, that chromosome is selected as solution of construction project scheduling whose total cost is minimum at desired duration of project. If list of acceptable chromosome consists of more than one chromosomes of desired project duration, then chromosome of minimum total cost is selected and proposed as solution of construction project scheduling.

## 4 Case Study and Discussion

A three-storey under construction building construction project was taken for applying the proposed SA-based TCT model. This construction project was facing so many problems in selecting optimal combination of alternatives to deliver the project successfully. This project consisted of 13 activities. Name, successor, number of available alternate options of each activity and corresponding values of objectives (time and cost) are given in Table 1, while activity on node diagram of case study project is shown in Fig. 2.

For the above case study project, based on permutation theory, there will be 55,296 ways to deliver the project. To determine the optimal way to deliver the project look manually impossible. So, the proposed methodology is coded in MATLAB 2014a software. Then, the proposed simulated annealing base time–cost trade-off model was practically implemented on above case study project. The project desired duration was considered as 185 days. At 185 days, minimum project cost was found as 533239.80 US$. However, at different–different project durations, minimum project cost is found and are given in Table 2.

The pseudo-code for SA-based TCT model for this case study project is explained in next section. Based on time–cost values obtained in Table 2, the time–cost trade-off curve is plotted and shown in Fig. 3.

**Pseudo Code for SA Based TCT Model**

(1) $T_{\text{initial}} = 1000$
(2) Generate a random chromosome solution ($CS_1$)
(3) Calculate the $PCT_1$ and $PCC_1$ of $CS_1$ by using objective function. Repeat—
(4) Generate a random chromosome solution $CSn$

**Table 1** Details of the case study project

| S. no. | Activity name | Successors | Alternatives | Time (days) | Cost (US $) |
|---|---|---|---|---|---|
| 1 | Site clearance | 2 | 1 | 8 | 10039.42 |
| | | | 2 | 8 | 9849.86 |
| 2 | Excavation | 3 | 1 | 6 | 1082.13 |
| | | | 2 | 6 | 891.05 |
| 3 | Footing | 4 | 1 | 12 | 15545.67 |
| | | | | 10 | 17039.34 |
| 4 | Formwork | 5 | 1 | 5 | 562.13 |
| | | | 2 | 4 | 590.32 |
| 5 | Retaining wall | 6 | 1 | 26 | 15834.49 |
| | | | 2 | 16 | 17274.94 |
| 6 | Basement | 7 | 1 | 32 | 74124.65 |
| | | | 2 | 29 | 76345.78 |
| | | | 3 | 23 | 84312.34 |
| 7 | Slab | 8 | 1 | 22 | 32646.05 |
| | | | 2 | 11 | 29759.59 |
| 8 | Exterior wall | 9 | 1 | 18 | 65959.52 |
| | | | 2 | 29 | 105296.94 |
| | | | 3 | 11 | 157433.42 |
| 9 | Interior wall | 13 | 1 | 37 | 58570.35 |
| | | | 2 | 21 | 59999.39 |
| | | | 3 | 32 | 57668.29 |
| | | | 4 | 17 | 63321.11 |
| 10 | Flooring | – | 1 | 34 | 38411.50 |
| | | | 2 | 17 | 65326.48 |
| | | | 3 | 12 | 50214.22 |
| 11 | Exterior finish | – | 1 | 9 | 12216.23 |
| | | | 2 | 12 | 3846.23 |
| 12 | Interior finish | – | 1 | 41 | 90219.78 |
| | | | 2 | 31 | 233034.50 |
| 13 | Roof | 10–12 | 1 | 23 | 127641.84 |
| | | | 2 | 24 | 81323.17 |

(5) Generate a random chromosome solution $CS_2$ between $CS_n$ and $CS_N$
(6) Calculated $PCT_2$ and $PCC_2$ of $CS_2$
(7) If $PCC_2 < PCC_1$ then accept the $CS_2$, otherwise, generate a random number in the range [0, 1], and compute the $\Delta$ by using the following formula:

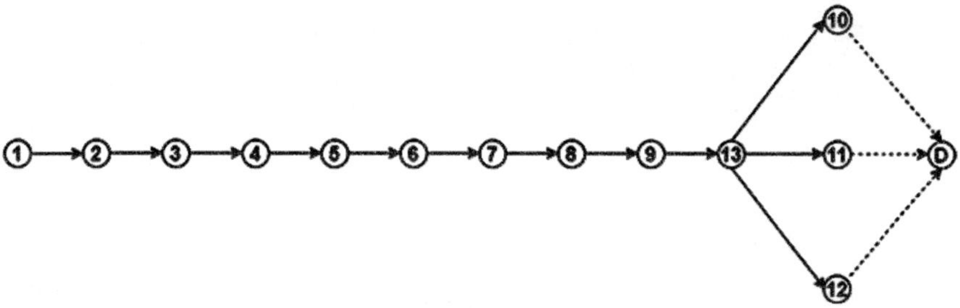


**Fig. 2** Activity on node diagram of case study project

$$\Delta = \exp\left(\frac{TC2 - TC1}{t}\right)$$

Again, if $\Delta >$ rand {0 1}, then accept the chromosome.

Reduce the temperature by 1 at each repetition. And repetitions are terminated when temperature reaches to pre-specified limit. For this case study, temperature is reduced up to 1. Therefore, 1000 repetitions were allowed. Arguably, the initial temperature and temperature reduction rate were set out based on information available in the literature.

***Comparison with Genetic Algorithm***

Results obtained from the proposed model were compared with a genetic algorithm-based (GA) TCT model as proposed by Feng et al. [5]. As shown in Table 2, at 185 days, which is the desired duration of project, minimum project cost obtained from proposed model and GA model was 533239.80 US$ and 650458.90 US$, respectively. Project cost calculated from the proposed model is 18.02% less than that of GA model. Therefore, Table 3 demonstrates that the proposed SA model performed better than the GA model in finding optimum project cost at desired project duration.

## 5 Conclusion

In recent years, a significant amount of attention is given to complete the construction project in minimum time and cost. Also, due to technological advancement and based on requirements of project stakeholders, it become necessary to find the optimal ways to deliver the project successfully when there are number of available alternatives for construction activity. In this regard, this paper proposed a systematically developed and experimentally proved time–cost trade-off model using the simulated annealing

**Table 2** Minimum project cost at different–different project durations

| S.no | A1 | A2 | A3 | A4 | A5 | A6 | A7 | A8 | A9 | A10 | A11 | A12 | A13 | PCT | PCC |
|---|---|---|---|---|---|---|---|---|---|---|---|---|---|---|---|
| 1 | 1 | 1 | 2 | 2 | 2 | 3 | 2 | 3 | 4 | 1 | 2 | 2 | 2 | 164 | 737467.60 |
| 2 | 2 | 2 | 1 | 2 | 2 | 3 | 2 | 3 | 2 | 2 | 2 | 2 | 2 | 167 | 759186.60 |
| 3 | 1 | 2 | 1 | 2 | 2 | 3 | 2 | 1 | 4 | 2 | 2 | 2 | 2 | 170 | 671224.00 |
| 4 | 1 | 2 | 2 | 2 | 2 | 3 | 2 | 3 | 4 | 1 | 2 | 1 | 2 | 171 | 594461.80 |
| 5 | 1 | 2 | 1 | 2 | 2 | 3 | 2 | 3 | 4 | 1 | 2 | 1 | 1 | 172 | 639286.80 |
| 6 | 1 | 1 | 1 | 1 | 2 | 1 | 2 | 3 | 4 | 3 | 2 | 2 | 2 | 173 | 737560.80 |
| 7 | 1 | 1 | 1 | 1 | 2 | 3 | 2 | 1 | 2 | 3 | 1 | 2 | 1 | 174 | 707641.50 |
| 8 | 1 | 2 | 2 | 2 | 2 | 3 | 2 | 3 | 2 | 1 | 2 | 1 | 2 | 175 | 591140.10 |
| 9 | 1 | 2 | 1 | 2 | 2 | 3 | 2 | 1 | 2 | 1 | 2 | 2 | 2 | 177 | 640987.30 |
| 10 | 2 | 2 | 2 | 2 | 1 | 1 | 1 | 3 | 3 | 3 | 1 | 2 | 2 | 178 | 641150.20 |
| 11 | 1 | 2 | 1 | 2 | 2 | 3 | 2 | 1 | 4 | 2 | 2 | 1 | 2 | 180 | 528409.30 |
| 12 | 2 | 2 | 2 | 2 | 2 | 3 | 2 | 1 | 2 | 1 | 2 | 1 | 2 | 182 | 499476.60 |
| 13 | 1 | 2 | 1 | 2 | 2 | 3 | 2 | 1 | 2 | 3 | 1 | 1 | 1 | 183 | 564663.90 |
| 14 | 1 | 2 | 1 | 2 | 2 | 3 | 2 | 1 | 2 | 3 | 1 | 1 | 2 | 184 | 518345.30 |
| 15 | 1 | 2 | 1 | 2 | 2 | 3 | 2 | 1 | 3 | 3 | 2 | 2 | 2 | 185 | 533239.80 |
| 16 | 1 | 1 | 1 | 1 | 2 | 2 | 2 | 1 | 4 | 1 | 2 | 1 | 2 | 187 | 493690.60 |
| 17 | 1 | 1 | 1 | 1 | 1 | 3 | 2 | 1 | 2 | 1 | 2 | 2 | 2 | 188 | 639710.10 |
| 18 | 1 | 1 | 1 | 1 | 2 | 1 | 2 | 1 | 4 | 1 | 2 | 1 | 1 | 189 | 537788.10 |
| 19 | 1 | 2 | 1 | 2 | 2 | 3 | 2 | 1 | 1 | 3 | 2 | 2 | 2 | 190 | 651360.90 |
| 20 | 1 | 1 | 1 | 1 | 1 | 1 | 2 | 3 | 4 | 2 | 2 | 1 | 1 | 192 | 654736.90 |
| 21 | 1 | 1 | 1 | 2 | 1 | 3 | 2 | 1 | 2 | 2 | 2 | 1 | 1 | 193 | 570157.20 |
| 22 | 1 | 1 | 1 | 2 | 1 | 3 | 2 | 1 | 2 | 2 | 1 | 1 | 2 | 194 | 532208.50 |
| 23 | 1 | 2 | 1 | 1 | 1 | 3 | 2 | 1 | 2 | 1 | 2 | 1 | 2 | 195 | 496704.30 |
| 24 | 1 | 1 | 1 | 1 | 1 | 3 | 1 | 3 | 2 | 1 | 2 | 1 | 1 | 198 | 637574.40 |
| 25 | 1 | 1 | 1 | 1 | 2 | 2 | 2 | 1 | 1 | 1 | 2 | 2 | 1 | 199 | 678073.20 |
| 26 | 2 | 2 | 1 | 2 | 2 | 3 | 2 | 1 | 1 | 1 | 2 | 1 | 2 | 200 | 496553.90 |
| 27 | 1 | 2 | 1 | 1 | 1 | 3 | 2 | 1 | 1 | 2 | 2 | 2 | 2 | 201 | 665005.00 |
| 28 | 1 | 1 | 1 | 1 | 2 | 2 | 2 | 1 | 3 | 1 | 2 | 1 | 2 | 202 | 488037.80 |
| 29 | 1 | 1 | 1 | 2 | 1 | 3 | 2 | 1 | 3 | 2 | 2 | 1 | 2 | 205 | 521507.40 |
| 30 | 1 | 1 | 1 | 1 | 1 | 3 | 1 | 1 | 3 | 2 | 2 | 2 | 1 | 206 | 713499.10 |
| 31 | 1 | 1 | 1 | 1 | 1 | 3 | 1 | 1 | 3 | 1 | 2 | 2 | 1 | 209 | 686584.10 |
| 32 | 1 | 2 | 1 | 2 | 2 | 3 | 1 | 1 | 1 | 3 | 2 | 1 | 2 | 211 | 511432.70 |
| 33 | 1 | 1 | 1 | 1 | 1 | 3 | 1 | 3 | 1 | 1 | 1 | 1 | 2 | 215 | 598196.70 |
| 34 | 2 | 2 | 1 | 2 | 1 | 1 | 2 | 1 | 1 | 1 | 2 | 1 | 2 | 219 | 484926.20 |
| 35 | 1 | 1 | 1 | 1 | 2 | 1 | 1 | 2 | 3 | 3 | 2 | 1 | 1 | 226 | 586161.19 |
| 36 | 1 | 1 | 1 | 1 | 1 | 3 | 1 | 2 | 3 | 3 | 1 | 1 | 2 | 228 | 556960.90 |

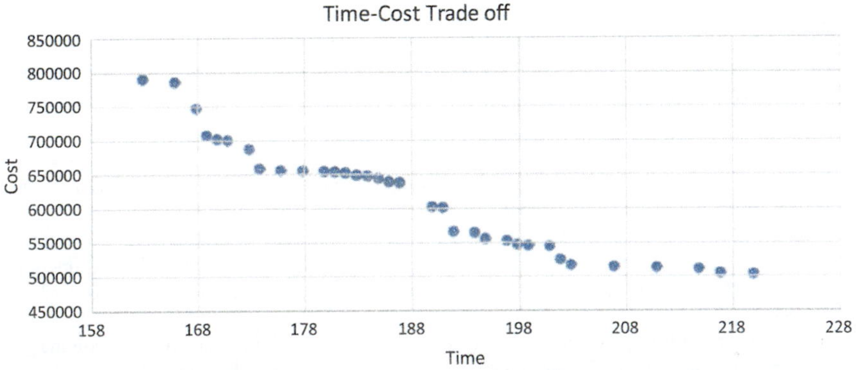


**Fig. 3** Time–cost trade-off curve

**Table 3** Results comparisons for case study project

| Algorithm | Desired project duration (days) | A1 | A2 | A3 | A4 | A5 | A6 | A7 | A8 | A9 | A10 | A11 | A12 | A13 | Minimum project cost (US $) |
|---|---|---|---|---|---|---|---|---|---|---|---|---|---|---|---|
| GA | 185 | 2 | 2 | 1 | 1 | 2 | 3 | 2 | 1 | 2 | 2 | 1 | 1 | 2 | 650458.9 |
| SA | 185 | 1 | 2 | 1 | 2 | 2 | 3 | 2 | 1 | 3 | 3 | 2 | 2 | 2 | 533239.8 |

approach. Applicability of the proposed model is demonstrated through solving a case study project. Results of the case study illustrate the potential of proposed TCT model. First, relationship between time and cost of project can be established by the simulated annealing. Second, simulated annealing is found satisfactory in discrete search for time–cost trade-off solution. Third, simulated annealing performed better than the genetic algorithm to find minimum cost desired project duration. Fourth, this study and proposed model will be beneficial to project's participants in perspective of making profit and project completion time minimization. Finally, this study possibly provides a better scheduling decision-making tool for project team and organization.

The main challenge in this work was to set a cooling schedule for algorithm. However, numerous trials were performed on changing the cooling schedule. And the best possible cooling schedule was adopted for final analysis.

Globally, even though the proposed model is found capable of finding optimal ways to deliver the project, its effectiveness in large scale and multi-construction projects are yet to be tested practically. Further study on taking uncertain time and cost is also required. Further analysis is also essential after adding one or more project's objectives, such as quality, safety, risk and others.

## References


1. Zahraie B, Tavakolan M (2009) Stochastic time-cost-resource utilization optimization using non-dominated sorting genetic algorithm and discrete fuzzy sets. J Constr Eng Manag. https://doi.org/10.1061/(ASCE)CO.1943-7862.0000092
2. Luong DL, Tran DH, Nguyen PT (2018) Optimizing multi-mode time-cost-quality trade-off of construction project using opposition multiple objective difference evolution. Int J Constr Manag. Taylor & Francis, 21(3), pp 1–13. doi:https://doi.org/10.1080/15623599.2018.1526630
3. Feng CW, Liu L, Burns SA (1997) Using genetic algorithms to solve construction time-cost trade-off problems. J Comput Civil En. https://doi.org/10.1061/(ASCE)0887-3801(1997)11:3(184)
4. Kalhor E et al (2011) Stochastic time-cost optimization using non-dominated archiving ant colony approach. Autom Constr. https://doi.org/10.1016/j.autcon.2011.05.003
5. Zhang H, Li H (2010) Multi-objective particle swarm optimization for construction time-cost tradeoff problems. Constr Manag Econ. https://doi.org/10.1080/01446190903406170
6. Anagnostopoulos KP, Kotsikas L (2010) Experimental evaluation of simulated annealing algorithms for the time-cost trade-off problem. Appl Math Comput. https://doi.org/10.1016/j.amc.2010.05.056
7. Singh G, Ernst AT (2011) Resource constraint scheduling with a fractional shared resource. Oper Res Lett. Elsevier B.V., 39(5):363–368. https://doi.org/10.1016/j.orl.2011.06.003
8. Hartmann S, Briskorn D (2010) A survey of variants and extensions of the resource-constrained project scheduling problem. Eur J Oper Res. https://doi.org/10.1016/j.ejor.2009.11.005
9. Khalili-Damghani K et al (2015) Solving multi-mode time-cost-quality trade-off problems under generalized precedence relations. Optim Methods Softw. https://doi.org/10.1080/10556788.2015.1005838
10. Panwar A, Tripathi KK, Jha KN (2019) A qualitative framework for selection of optimization algorithm for multi-objective trade-off problem in construction projects. Eng Constr Architect Manag. https://doi.org/10.1108/ECAM-06-2018-0246
11. Kelley JE, Walker MR (1959) Critical-path planning and scheduling. In: Proceedings of the eastern joint computer conference, IRE-AIEE-ACM 1959. https://doi.org/10.1145/1460299.1460318
12. Kallantzis A, Lambropoulos S (2004) Critical path determination by incorporating minimum and maximum time and distance constraints into linear scheduling. Eng Constr Architect Manag. https://doi.org/10.1108/09699980410535813
13. Ipsilandis PG (2007) Multiobjective linear programming model for scheduling linear repetitive projects. J Constr Eng Manag. https://doi.org/10.1061/(ASCE)0733-9364(2007)133:6(417)
14. Moselhi O, Ei-Rayes K (1993) Scheduling of repetitive projects with cost optimization. J Constr Eng Manag. https://doi.org/10.1061/(ASCE)0733-9364(1993)119:4(681)
15. Zhou J et al. (2013)'A review of methods and algorithms for optimizing construction scheduling. J Oper Res Soc 64(8):1091–1105. https://doi.org/10.1057/jors.2012.174
16. Fondahl JW (1962) A non-computer approach to The Critical Path Meghod for the Construction Industry. Technical Report No. 9, Stanford University, pp 1–163
17. Moselhi O (1993) Schedule compression using the direct stiffness method. Can J Civ Eng. https://doi.org/10.1139/l93-007
18. Zhang H, Li H, Tam CM (2006) Heuristic scheduling of resource-condlxstrained, multiple-mode and repetitive projects. Constr Manag Econ. https://doi.org/10.1080/01446190500184311
19. Di C et al (2013) Priority rule based heuristics for project scheduling problems with multi-skilled workforce constraints. In 2013 25th Chinese control and decision conference, CCDC 2013. https://doi.org/10.1109/CCDC.2013.6561233
20. Zheng DXM, Ng ST, Kumaraswamy MM (2004) Applying a genetic algorithm-based multi-objective approach for time-cost optimization. J Constr Eng Manag 130(2): 168–176. https://doi.org/10.1061/(ASCE)0733-9364

21. Yang IT (2007) Using elitist particle swarm optimization to facilitate bicriterion time-cost trade-off analysis. J Constr Eng Manag. https://doi.org/10.1061/(ASCE)0733-9364(2007)133:7(498)
22. Li H, Li P (2013) Self-adaptive ant colony optimization for construction time-cost optimization. Kybernetes. https://doi.org/10.1108/K-03-2013-0063
23. Fallah-Mehdipour E et al (2012) Extraction of decision alternatives in construction management projects: Application and adaptation of NSGA-II and MOPSO. Expert Syst Appl. https://doi.org/10.1016/j.eswa.2011.08.139
24. Shahriari M (2016) Multi-objective optimization of discrete time–cost tradeoff problem in project networks using non-dominated sorting genetic algorithm. J Indus Eng Int 12(2):159–169. https://doi.org/10.1007/s40092-016-0148-8
25. Afruzi EN et al (2013) A multi-mode resource-constrained discrete time-cost tradeoff problem solving using an adjusted fuzzy dominance genetic algorithm. Scientia Iranica. https://doi.org/10.1016/j.scient.2012.12.024. Application and adaptation of NSGA-II and MOPSO. Expert Syst Appl. https://doi.org/10.1016/j.eswa.2011.08.139

# Chapter 5
# AHP and NSGA-II-Based Time–Cost–Quality Trade-Off Optimization Model for Construction Projects

**Kamal Sharma and Manoj Kumar Trivedi**

## 1 Introduction

Construction projects include enormous investment as far as time, money and resources. Construction managers are accordingly required to oversee and schedule the projects carefully and in the best way. Time and cost are two principal concerns in construction projects [1]. Use of modern expensive resources reduces both the completion time of activity and project, but increases the completion cost of activity and project. On the other hand, reduction in either time or cost of activity may adversely affect the quality of overall project [2]. Hence, the relationship among these scheduling objectives is conflicting in nature and a hidden trade-off exists among them. Also, the present competitive state of the construction business makes it important to complete the project within the minimum time and cost with achieving the maximum possible quality.

A construction project consists of many interconnected activities and each activity has different alternatives through which it can be executed [3]. Based on the permutation principle, several methods are possible to complete the project based on the different combinations of alternatives of activities. Since each alternative is accompanied by different time, cost and impact on entire project quality, it is therefore required to allocate an optimum alternative to each activity that satisfies the desired projects' objectives. The allocation of optimum alternative to each activity of project provides an optimal method to complete the project. Consequently, the purpose of this paper is to find the optimal combinations of activities alternatives while optimizing the time, cost and quality of project simultaneously.

K. Sharma (✉) · M. K. Trivedi
Department of Civil Engineering, Madhav Institute of Technology and Science, Gwalior, India

M. K. Trivedi
e-mail: manojtrivedi@mitsgwalior.in

© The Author(s), under exclusive license to Springer Nature Singapore Pte Ltd. 2022 
H. M. Dubey et al. (eds.), *Artificial Intelligence and Sustainable Computing*,
Algorithms for Intelligent Systems, https://doi.org/10.1007/978-981-16-1220-6_5

Non-dominated sorting genetic algorithm II (NSGA-II) was first developed in 2002 by Deb and his co-authors [4]. NSGA-II is now one of the most used multi-objective optimization algorithm. NSGA-II is an iterative procedure to solve the multi-objective optimization problems. The one iteration of standard NSGA-II includes the generation of random initial population, selection, crossover, mutation, non-dominated sorting of combined population and generation of new population for next iteration. In the presented paper, NSGA-II is used to optimize the time, cost and entire project quality simultaneously. Besides, pairwise comparison-based analytical hierarchy process (AHP) is used to determine the relative weight of activities and quality indicators of project.

Based on the abovementioned discussion, this paper provides an AHP and NSGA-II-based time–cost–quality trade-off optimization model for construction projects. The reminder of this paper is organized as follows: the next section provides brief details of the related works; problem formulation is then presented followed by proposed AHP and NSGA-II-based TCQT model. Subsequently, the working of proposed model is demonstrated through solving a case study project. The final section of this paper presents the conclusion and suggestions for future research.

## 2 Related Works

A recent study [5] showed that mainly three types of algorithms are used for project scheduling, that is, deterministic, heuristic and meta-heuristic algorithms. Start with history, at first, Kelley and Walker [6] proposed deterministic algorithm-based critical path method of project scheduling that is broadly used to determine the minimum time required to complete the project. Since then, a number of scheduling methods have been developed using the above three types of algorithms [7–9]. In the above three algorithms, meta-heuristic algorithms were found to be more satisfying multi-objective optimization algorithms. Meta-heuristic algorithms of project scheduling mainly include the genetic algorithm (GA), particle swarm optimization (PSO), simulated annealing (SA), ant colony optimization (ACO) and non-dominated sorting genetic algorithm. Accordingly, several meta-heuristic algorithms-based time–cost and time–cost–quality trade-off models have been developed. In this order, at first, a study [10] proposed GA and Pareto-front-based approach to solve time–cost trade-off (TCT) problem; however, this study considered only the direct cost of project. To fill this gap, Xiong and Kuang [11] developed the multi-objective ACO-based time–cost trade-off model that considered both the projects' direct and indirect cost. Also, Zhang and Li [12] proposed multi-objective particle swarm optimization-based TCT model. Nevertheless, the original PSO still needs some adjustments for guaranteed convergence to a local optimum [13]. Fallah-Mehdipour et al. [14] has shown the effectiveness of NSGA-II over MOPSO in solving the TCT problem. Then, a recent study [15] proposed a NSGA-II-based discrete TCT model. The abovementioned studies were restricted to single or bi-objective optimization because probability of discovering the non-dominated solutions decreases exponentially as the number of conflicting objectives increases [16]. Given the importance of construction quality, many researchers have tried to extend the time–cost trade-off to time–cost–quality trade-off (TCQT). At first, Babu and Suresh [17] introduced the quality parameter in the study of time–cost trade-off and used three inter-related mathematical models to solve the TCQT problems. Since then only a few researchers addressed the TCQT problems. El-Rayes and Kandil [18] developed a TCQT model using the Pareto-front-based multi-objective genetic algorithm. Besides, a recent study [2] also proposed a time–cost–quality trade-off model using the multi-objective differential evolution. However, by using the relative weight of activities and quality indicators these studies were able to determine the quality in each alternative of activities, but none of these studies explained the weight assignment process. Therefore, the presented study employs the pairwise comparison-based analytical hierarchy process (AHP) to determine the relative weight of activities in project and the relative weight of quality indicators in activities.

## 3 Problem Formulation

An activity can be completed through one of its available alternatives accompanied by different execution time and cost, as well as impact on entire project quality. It is therefore necessary to allocate an appropriate alternative to each activity of the project which also meets the desired objectives of the project. However, the purpose of this paper is to assign an alternative for each activity that minimizes the time and cost of project and provide the maximum possible project quality.

Activity time (AT), activity cost (AC) and activity quality (AQ) are taken as input for the proposed TCQT model. Since it is worth mentioning that each alternative involves different amounts of labor, materials and equipment resources, the numerical values of AT, AC and AQ are determined on the basis of activity and the chosen alternative. This paper aims to optimize project time (PT), project cost (PC) and project quality index (PQI) simultaneously, which are functions of AT, AC and AQ, respectively. Some assumptions form the basis of the study presented, which are as follows: (i) Each activity must be completed without any disturbance. (ii) Precedence relationships exist between activities that can be represented as activity on node (AON) networks. (iii) Each activity has labor, material and equipment quality indicators.

Based on the above, this three-objective optimization problem can be formulated in the following manner.

**Objective 1: Minimize Project Time (PT)**

PT is an important parameter of construction project which greatly influences the success of construction project. In this paper, the precedence diagramming method (PDM) is employed to calculate PT. PDM is based on critical path (CP) of activity on node (AON) project networks. PT is the sum of AT of all the activities in the critical path of construction project.

$$\text{PT} = \sum_{A \in CP} AT_A \tag{1}$$

where PT is the project completion time and $AT_A$ is the completion time of critical path activity ($A$).

**Objective 2: Minimize Project Cost (PC)**

PC is one of the key factors affecting the success of construction projects. PC is simply the algebraic sum of each activity cost (AC). However, AC is the sum of direct cost and indirect cost of activity. Direct cost (D.C) includes mainly the cost of labor, material and equipment, while indirect cost (I.C) includes overhead expenses and outage losses.

$$\text{PC} = \sum_{A} \text{D.C} + \text{I.C per day} \times \text{PT in days} \tag{2}$$

where PC is project completion cost, $\sum_A$ D.C is sum of direct cost of each activity. Indirect cost of construction project is simply estimated by multiplying the PT and indirect cost of project per day.

**Objective 3: Maximize Project Quality Index (PQI) or Minimize 1/PQI**

Since the project quality (PQ) is of paramount importance, the objective of construction project is to measure the PQ quantitatively, which is a very challenging task. In this paper, PQ is measured in terms of project quality index (PQI), which is estimated by taking the weighted sum of each activity quality (AQ). On the other hand, AQ depends upon three quality indicators, i.e. labor, material and equipment. Therefore, AQ can be measured by taking the weighted sum of quality indicators of activity.

$$AQ = wt_A \cdot \sum_{k=1}^{k} wt_{A,k} \cdot Q^m_{A,k} \tag{3}$$

$$\text{PQI} = \sum_{A=1}^{n} wt_A \cdot \sum_{k=1}^{k} wt_{A,k} \cdot Q^m_{A,k} \tag{4}$$

where $Q^m_{A,k}$ is performance of quality indicator ($k$) in activity ($A$) using alternative ($m$); $wt_{A,k}$ is weight of quality indicator ($k$) compared to other indicators in activity ($A$); and $wt_A$ is weight of activity ($A$) compared to other activities in the project. Relative weight of activities and quality indicators can be estimated by pairwise comparison-based AHP as demonstrated in case study and discussion section. For instance, quality performance of labor, material and equipment are 85, 90 and 70, respectively. The corresponding weights of quality indicators are 0.50, 0.20 and

0.30, respectively. If the weight of activity in total project is 0.15 (15%) then AQ is calculated as [(0.85 * 0.50 + 0.90 * 0.20 + 0.70 * 0.30)] * 0.15 = 0.122.

The key constraints considered in this study are as follows: (i) Each activity should be completed for the successful completion of project. (ii) Only one out of available alternatives should be initiated to complete the activity. (iii) Alternatives of each activity are the decision variables, which are positive integers and subject to lower and upper bound. (iv) Precedence relationships between activities should be maintained during optimization process.

## 4 Research Methodology: Proposed TCQT Model

Over the years, several evolutionary multi-objective algorithms (EMOAs) have been systematically developed and they have been demonstrated for their usefulness in solving two or more objectives optimization problem. In this order, Deb and his authors [4] proposed a multi-objective optimization algorithm NSGA-II for the constraint and unconstraint optimization problems. The population in non-dominated fronts increases exponentially as the number of objective functions in the optimization problem increases. Therefore, it becomes difficult to select population members from non-dominated fronts for next generation. To alleviate, Deb and his authors [4] used maximum crowding distance concept to select individual from non-dominated front. The larger distance one individual has, the higher priority to select.

As discussed in problem formulation, an activity can be executed through one of its available alternatives. For each activity, a particular alternative is required to choose as decision variable. And each combination of decision variables of project activities is the possible solution of this TCQT problem. In NSGA-II, each solution of optimization problem can be represented in the form of chromosome. A chromosome consists of number of cells (genes) equal to total number of decision variables or activities in construction project. Thus, each gene of chromosomes contains a value (allele) of decision variable which is the selected alternative through which that activity can be executed. Each possible chromosome of given TCQT problem is a possible solution of that TCQT problem. For understanding the chromosome solutions of given TCQT problem, suppose a construction project consists of nine activities and each activity can be executed with either one of its three available alternatives. A chromosome solution shown in Fig. 1 can be a possible solution for project of 1-9 activities with alternatives 3, 1, 2, 3, 2, 1, 2, 3, 1, respectively. As per permutation rules there will be $3^9$ such possible project solution.

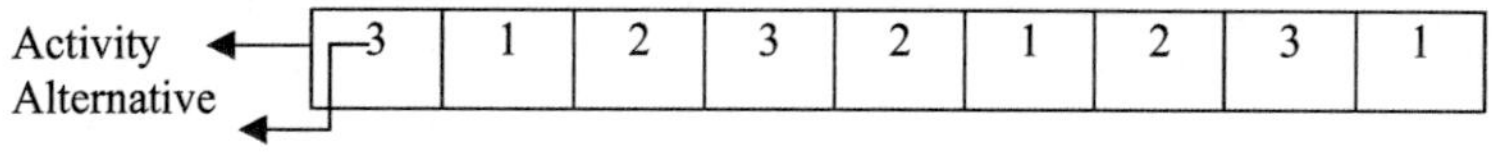


**Fig. 1** Chromosome solution of TCQT problem

NSGA-II starts with generating the random parent population ($P_t$) of size $N$. The detailed description of NSGA-II to solve the TCQT problem is explained as follows.

***Generation and Ranking of Initial Population ($P_t$) of Size N***

At first, an initial population (parent population $P_t$) of size $N$ (i.e. $N$ chromosomes) is generated randomly. The fitness value of each objective with respect to each chromosome is calculated by using equations (i)–(iii). Then, $P_t$ undergoes non-dominated sorting *(NDS)* to be distributed in non-dominated fronts ($F_1, F_2, F_3, \ldots, F_n$), where 1, 2, 3,…, $n$ represents the ranking of non-dominated fronts. Crowding distance of each solution of each non-dominated front is also calculated. The process of crowding distance calculation is described in the section of generation of new population.

***Selection (also Known as Reproduction Stage)***

In the selection stage, the parent population ($P_t$) undergoes tournament selection in which $N$ pairs of ($P_t$) are randomly selected. Then, a tournament is carried out between $N$ pairs of solutions and the winner solution is decided on the basis of ranking produced by NDS. In case, if the solutions of a pair are of the same rank, the solution with a higher crowding distance is chosen as the winner. Thus, a mating pool of size $N$ can be obtained after tournament selection.

***Crossover***

In crossover operation, two randomly selected population members from mating pool undergo simulated binary crossover (SBX). SBX was proposed by Deb and Agarwal [19]. SBX is structured as single-point crossover properties in binary-coded GA. Two offspring solutions $C_1$ and $C_2$ from two parent solutions $P_1$ and $P_2$ are calculated as follows:

$$C_1 = 0.5\,[(1 + \beta_i)\,P_1 + (1 - \beta_i)P_2] \tag{5}$$

$$C_2 = 0.5\,[(1 - \beta_i)\,P_1 + (1 + \beta_i)\,P_2] \tag{6}$$

while $\beta_i$ is calculated as follows:

$$\beta_i = \begin{cases} (2u_i)^{1/(n_c+1)}, & \text{if } u_i < 0.5 \\ [1/2(1 - u_i)]^{1/(n_c+1)}, & \text{otherwise} \end{cases} \tag{7}$$

where $u_i$ is a random number in the range [0,1] and $n_c$ is distribution index which is a non-negative real number. Usually, to preserve some good chromosomes selected in reproduction stage, crossover operation is performed over mating pool with a crossover probability ($c_p$). If $\mathrm{c_p}$ is $x\%$, it means $x\%$ chromosomes from mating pool will undergo crossover operation and the remaining $(100-x)\%$ will simply go into new population.

***Mutation***

To keep diversity in the population, a chromosome is mutated to another chromosome by a small random tweak in chromosome. Population generated from crossover operation undergo polynomial mutation (PM) to generate offspring population ($O_t$) or child population. PM reflects the bit-flip mutation (in binary coded GA) properties in real-valued decision variable [3]. PM is applied on a gene value as follows:

$$C = \mathrm{P} + (x_i^{(u)} - x_i^{(l)})\delta_i \tag{8}$$

where $P$ is parent gene value, $C$ is child/offspring gene value, $(x_i^{(u)} - x_i^{(l)})$ is difference in upper and lower bound of gene value and $\delta_i$ is mutation value. $\delta_i$ is calculated as follows:

$$\delta_i = \begin{cases} (2r_i)^{1/(n_m+1)} - 1, & \text{if } r_i < 0.5 \\ 1 - [2(1 - r_i)]^{1/(n_m+1)}, & \text{if } r_i \geq 0.5 \end{cases} \tag{9}$$

where $r_i$ is a random number in the range [0,1] and $n_m$ is distribution constant of mutation. Generally, mutation operation over population is performed with low mutation rate. In this paper, $1/n$ mutation rate is used as suggested by Deb and Goyal [20], where n is number of genes in chromosome. Higher values of $n_c$ and $n_m$ mean that there is a higher probability of generation of near-parent solution.

***Non-dominated Sorting of Combined Population ($R_t$: $P_t$ U $Q_t$)***

Generated offspring population ($O_t$) after SBX and PM is combined with parent population ($P_t$), which gives the combined population ($R_t$) of size 2 $N$. These 2 $N$ solutions then undergo non-dominated sorting to be distributed in non-dominated fronts ($F_1$, $F_2$ ...$F_1$...$F_n$).

***Generation of New Population***

A new population from non-dominated fronts ($F_1$, $F_2$ ...$F_1$... $F_n$) is generated for next generations. For this purpose, an intermediate population $S_t$ is set at zero initially and then the population members are preserved from non-dominated fronts ($F_1$, $F_2$ ...$F_n$) until $S_t \geq N$. If $S_t$ is equal to $N$ up to front $F_l$, then $S_t$ turns into new generation ($P_{t+1}$). If $S_t > N$, at first $p_{t+1}$ population members are preserved from first $F_{l-1}$ non-dominated fronts and the remaining $K$ ($N - p_{t+1}$) population members are selected from front $F_l$ based on maximum crowding distance concept. In order to select $K$ population members from $F_l$, solutions of front $F_l$ are required to sort in descending order using their crowding distance values. Then, first $K$ solutions are selected as best solutions to complete the population of size $N$ for next generation. To calculate the crowding distance of solutions of a non-dominated front, at first, all solutions are sorted in each objective domain. Then, crowding distance of $i$th solution, i.e. $d_i$ is calculated using the following formula:

**Fig. 2** *i*th, (*i* + 1)th and (*i* – 1)th solutions

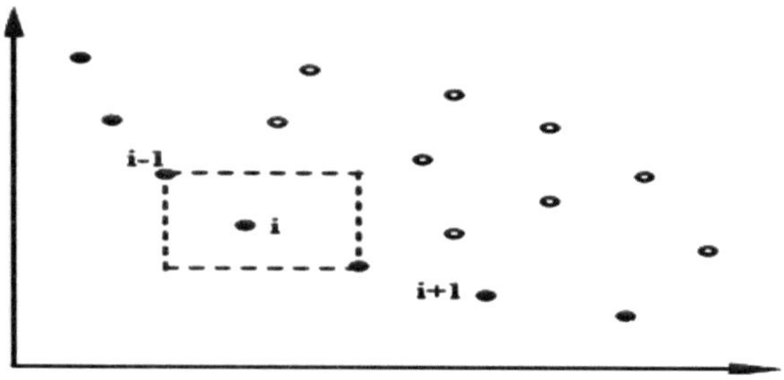


$$d_i = \sum_{m=1}^{M} \frac{f_{i+1}^{m} - f_{i-1}^{m}}{f_m^{\max} - f_m^{\min}} \tag{10}$$

where $f_{i+1}$ and $f_{i-1}$ are the objective values of *m*th objective for (*i* + 1)th and (*i*–1)th solutions, whereas $f_m^{\max}$ and $f_m^{\min}$ are the maximum and minimum objective values of *m*th objective. Figure 2 shows the position of *i*th, (*i* + 1)th and (*i* –1)th solutions in a non-dominated front.

Figure 3 shows the movement starting with one generation then onto the next in NSGA-II. Since it is an iterative procedure, the process is continued till the stopping criterion is met. General stopping criteria are the maximum number of generation or the attained convergence of solutions, or both.

## 5 Case Study and Discussion

The proposed model has been applied to several case study projects. In the presented paper, the working of the proposed TCQT model is demonstrated through solving one case study project. As shown in Table 1, the presented case study project included 13 activities and each activity has several possible alternatives. These alternatives are accompanied by different amount of time, cost and performance and weight of quality indicators. The time and cost of each alternative can be estimated using the amount of resources associated with each alternative. However, relative weight of activities in project and relative weight of quality indicators in an activity can be estimated using the pairwise comparison-based AHP.

As shown in Table 3, at first, the activities were compared pairwise with the help of project team and according to the scale given in Table 2. Then, a priority vector is determined by using the following steps: (1) sum of each column; (2) divide each element of matrix by the sum of its column; (3) priority vector is then determined by averaging across rows. As shown in Table 4, the values obtained from priority vector are simply assigned as weights to the activities in project.

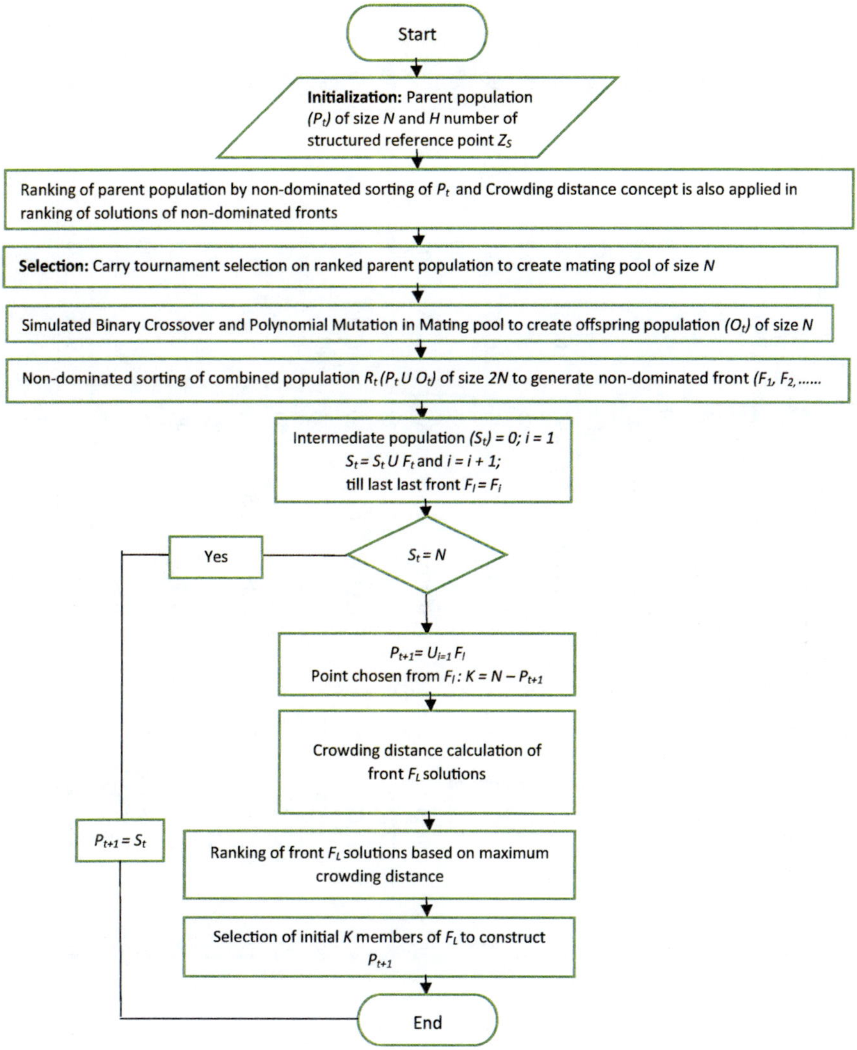


**Fig. 3** Movement starting with one generation then onto the next in NSGA-II

Similarly, the weight of quality indicators is determined for each activity. For instance, Table 5 shows the pairwise comparison matrix for quality indicators of slab activity, whereas the determined priority vector is shown in Table 6.

Similarly, the priority vector can be determined for the quality indicators of each activity. Priority vectors are assigned as weights to the quality indicators of activities as given in the case study Table 1.

**Table 1** Details of the case study project

| | | | | | | | | QI $k = 1$ | | QI $k = 2$ | | QI $k = 3$ | |
|---|---|---|---|---|---|---|---|---|---|---|---|---|---|
| S. no. | Activity name | Successors | AOs | Time (days) | Cost (US$) | Resources (units) | Act. weight $wt_i$ | IW ($wt_{i,k}$) | QP ($Q_{i,k}$) | IW ($wt_{i,k}$) | QP ($Q_{i,k}$) | IW ($wt_{i,k}$) | QP ($Q_{i,k}$) |
| 1 | Ground-works | 2 | 1 | 8 | 10039.42 | 8 | 0.160 | 0.45 | 97 | 0.30 | 99 | 0.25 | 93 |
| | | | 2 | 8 | 9849.86 | 11 | | | 70 | | 73 | | 71 |
| 2 | Excavation | 3 | 1 | 6 | 1082.13 | 12 | 0.164 | 0.20 | 94 | 0.35 | 91 | 0.45 | 89 |
| | | | 2 | 6 | 891.05 | 12 | | | 83 | | 88 | | 84 |
| 3 | Footing | 4 | 1 | 12 | 15545.67 | 38 | 0.190 | 0.30 | 95 | 0.45 | 94 | 0.25 | 95 |
| | | | | 10 | 17039.34 | 56 | | | 85 | | 87 | | 89 |
| 4 | Formwork | 5 | 1 | 5 | 562.13 | 11 | 0.133 | 0.35 | 89 | 0.35 | 92 | 0.30 | 94 |
| | | | 2 | 4 | 590.32 | 10 | | | 83 | | 83 | | 87 |
| 5 | Retaining wall | 6 | 1 | 26 | 15834.49 | 46 | 0.070 | 0.30 | 85 | 0.35 | 92 | 0.35 | 95 |
| | | | 2 | 16 | 17274.94 | 52 | | | 78 | | 95 | | 89 |
| 6 | Basement | 7 | 1 | 32 | 74124.65 | 27 | 0.057 | 0.40 | 99 | 0.30 | 78 | 0.30 | 94 |
| | | | 2 | 29 | 76345.78 | 32 | | | 90 | | 85 | | 88 |
| | | | 3 | 23 | 84312.34 | 23 | | | 85 | | 90 | | 86 |
| 7 | Slab | 8 | 1 | 22 | 32646.05 | 32 | 0.059 | 0.50 | 94 | 0.40 | 93 | 0.10 | 87 |
| | | | 2 | 11 | 29759.59 | 35 | | | 80 | | 85 | | 96 |
| 8 | Exterior wall | 9 | 1 | 18 | 65959.52 | 38 | 0.043 | 0.40 | 90 | 0.50 | 78 | 0.10 | 73 |
| | | | 2 | 29 | 105296.94 | 44 | | | 85 | | 90 | | 94 |
| | | | 3 | 11 | 157433.42 | 49 | | | 90 | | 95 | | 95 |
| 9 | Interior wall | 13 | 1 | 37 | 58570.35 | 12 | 0.025 | 0.40 | 97 | 0.50 | 98 | 0.10 | 97 |

(continued)

**Table 1** (continued)

| | | | | | | | | QI $k = 1$ | | QI $k = 2$ | | QI $k = 3$ | |
|---|---|---|---|---|---|---|---|---|---|---|---|---|---|
| S. no. | Activity name | Successors | AOs | Time (days) | Cost (US$) | Resources (units) | Act. weight $wt_i$ | IW ($wt_{i,k}$) | QP ($Q_{i,k}$) | IW ($wt_{i,k}$) | QP ($Q_{i,k}$) | IW ($wt_{i,k}$) | QP ($Q_{i,k}$) |
| | | | 2 | 21 | 59999.39 | 22 | | | 89 | | 96 | | 67 |
| | | | 3 | 32 | 57668.29 | 21 | | | 81 | | 85 | | 63 |
| | | | 4 | 17 | 63321.11 | 31 | | | 95 | | 84 | | 90 |
| 10 | Flooring | – | 1 | 34 | 38411.50 | 15 | 0.020 | 0.45 | 85 | 0.45 | 90 | 0.10 | 78 |
| | | | 2 | 17 | 65326.48 | 12 | | | 88 | | 92 | | 90 |
| | | | 3 | 12 | 50214.22 | 12 | | | 78 | | 84 | | 93 |
| 11 | Exterior finish | – | 1 | 9 | 12216.23 | 18 | 0.024 | 0.40 | 70 | 0.50 | 85 | 0.10 | 90 |
| | | | 2 | 12 | 3846.23 | 20 | | | 95 | | 88 | | 85 |
| 12 | Interior finish | – | 1 | 41 | 90219.78 | 17 | 0.022 | 0.40 | 70 | 0.50 | 85 | 0.10 | 90 |
| | | | 2 | 31 | 233034.50 | 13 | | | 95 | | 88 | | 85 |
| 13 | Roof | 10, 11, 12 | 1 | 23 | 127641.84 | 13 | 0.025 | 0.30 | 98 | 0.40 | 95 | 0.30 | 99 |
| | | | 2 | 24 | 81323.17 | 9 | | | 87 | | 86 | | 83 |

**Table 2** 1–9 scale of pairwise comparison matrix

| Intensity of importance | Definition |
|---|---|
| 1 | Equal important |
| 2 | Slight important |
| 3 | Moderate important |
| 4 | Moderate plus |
| 5 | Strong important |
| 6 | Strong plus |
| 7 | Very strong |
| 8 | Very, very strong |
| 9 | Extremely strong |

The proposed NSGA-II-based TCQT is practically implemented on above case study project using MATLAB R2014a. For finalizing the values of NSGA-II parameters, numerous trials were executed with varying values of these parameters. The best possible combination of NSGA-III parameters was adopted and the values are shown in Table 7.

A total of 25 exclusive optimal combinations of activity alternatives were obtained that met the desired project's objectives. PT, PC and PQI were determined for every 25 project completion method. PT values vary from 164 to 211 days, PC values vary from 493,529.20 to 783377.50 US$, and 1/PQI values vary from 1.096 to 1.205. All 25 obtained Pareto-optimal solutions are presented in Table 8. The three-dimensional plot for obtained PT, PC and PQI values is shown in Fig. 4.

Selecting a solution from obtained Pareto-optimal set of solutions is also a requirement of project team. For this purpose, a study [21] presented numerous methodologies to select a solution from Pareto-optimal set. One of its methodologies is weighted sum method, through which a solution can be selected from Pareto-optimal set by using the following equations:

$$\text{Min} \vee_{n}^{j=1} \sum_{i=1}^{m} w_i x_{ij} \tag{11}$$

$$\text{s.to} \quad \sum_{i=1}^{m} w_i = 1 \tag{12}$$

where $w_i$ represents the weight of $i$th objective and $x_{ij}$ represents the $j$th solution of $i$th objective. For instance, if a project team considers that all objectives are of same importance, then weight of each objective will be 1/3. Then, from Eqs. (11) and (12), the calculated solution from obtained Pareto-optimal set will be 1, 1, 1, 1, 2, 3, 2, 1, 4, 2, 2, 1, 2. PT, PC and 1/PQI corresponding to this solution are 181 days, 528572.10 US$ and 1.110, respectively.

**Table 3** Pairwise comparison matrix for project's activities

| Activities → ↓ | Groundworks | Excavation | Footing | Formwork | Retaining wall | Basement | Slab | Exterior wall | Interior wall | Flooring | Exterior finish | Interior finish | Roof |
|---|---|---|---|---|---|---|---|---|---|---|---|---|---|
| Groundworks | 1 | 2 | 3 | 3 | 3 | 2 | 3 | 6 | 7 | 6 | 6 | 5 | 2 |
| Excavation | 0.500 | 1 | 3 | 3 | 3 | 5 | 7 | 7 | 8 | 4 | 8 | 5 | 6 |
| Footing | 0.333 | 0.333 | 1 | 7 | 8 | 8 | 8 | 8 | 8 | 8 | 8 | 6 | 6 |
| Formwork | 0.333 | 0.333 | 0.142 | 1 | 5 | 6 | 7 | 9 | 9 | 8 | 5 | 6 | 6 |
| Retaining wall | 0.333 | 0.333 | 0.125 | 0.200 | 1 | 7 | 3 | 3 | 4 | 3 | 6 | 3 | 1 |
| Basement | 0.500 | 0.200 | 0.125 | 0.166 | 0.142 | 1 | 4 | 4 | 5 | 4 | 4 | 3 | 1 |
| Slab | 0.333 | 0.142 | 0.125 | 0.142 | 0.333 | 0.250 | 1 | 6 | 7 | 5 | 5 | 2 | 4 |
| Exterior wall | 0.166 | 0.142 | 0.125 | 0.111 | 0.333 | 0.250 | 0.166 | 1 | 5 | 6 | 5 | 2 | 3 |
| Interior wall | 0.142 | 0.125 | 0.125 | 0.111 | 0.250 | 0.200 | 0.142 | 0.200 | 1 | 4 | 2 | 3 | 1 |
| Flooring | 0.166 | 0.250 | 0.125 | 0.125 | 0.333 | 0.250 | 0.200 | 0.166 | 0.250 | 1 | 3 | 1 | 1 |
| Exterior finish | 0.166 | 0.125 | 0.125 | 0.200 | 0.166 | 0.250 | 0.200 | 0.200 | 0.500 | 0.333 | 1 | 4 | 3 |
| Interior finish | 0.200 | 0.200 | 0.166 | 0.166 | 0.333 | 0.333 | 0.500 | 0.500 | 0.333 | 1.000 | 0.250 | 1 | 3 |
| Roof | 0.500 | 0.166 | 0.166 | 0.166 | 1.000 | 1.000 | 0.250 | 0.333 | 1.000 | 1.000 | 0.333 | 0.333 | 1 |

**Table 4** Priority vector showing weight of each activity in project

| Activity | Weight |
|---|---|
| Groundworks | 0.160 |
| Excavation | 0.164 |
| Footing | 0.190 |
| Formwork | 0.133 |
| Retaining wall | 0.070 |
| Basement | 0.057 |
| Slab | 0.059 |
| Exterior wall | 0.043 |
| Interior wall | 0.025 |
| Flooring | 0.020 |
| Exterior finish | 0.024 |
| Interior finish | 0.022 |
| Roof | 0.025 |
| Total sum | **1.000** |

**Table 5** Pairwise comparison matrix for quality indicators of option of slab

| Quality Indicators→ ↓ | Labor | Material | Equipment |
|---|---|---|---|
| Labor | 1 | 2 | 3 |
| Material | 0.500 | 1 | 6 |
| Equipment | 0.330 | 0.166 | 1 |

**Table 6** Priority vector showing weight of quality indicators of slab activity

| Quality indicators | Weight |
|---|---|
| Labor | 0.500 |
| Material | 0.400 |
| Equipment | 0.100 |
| Total sum | **1.000** |

**Table 7** Adopted values of NSGA-III parameters

| NSGA-III parameters | Value |
|---|---|
| Population size | 100 |
| Number of generation | 150 |
| SBX probability | 1 |
| SBX distribution index | 20 |
| PM probability | 1 |
| PM rate | 1/13 |
| PM distribution index | 20 |

**Table 8** Obtained pareto-optimal front solutions

| S. no. | A1 | A2 | A3 | A4 | A5 | A6 | A7 | A8 | A9 | A10 | A11 | A12 | A13 | PT (days) | PC (US $) (dollars) | 1/PQI (1/PQI) |
|---|---|---|---|---|---|---|---|---|---|---|---|---|---|---|---|---|
| 1 | 2 | 2 | 2 | 1 | 2 | 3 | 2 | 3 | 4 | 1 | 2 | 2 | 1 | 164 | 783377.50 | 1.205 |
| 2 | 1 | 1 | 2 | 1 | 2 | 3 | 2 | 3 | 4 | 1 | 2 | 2 | 2 | 165 | 737439.40 | 1.141 |
| 3 | 1 | 1 | 1 | 2 | 2 | 3 | 2 | 3 | 4 | 1 | 2 | 2 | 2 | 166 | 735974.00 | 1.114 |
| 4 | 1 | 2 | 1 | 1 | 2 | 3 | 2 | 3 | 4 | 1 | 2 | 2 | 2 | 167 | 735754.70 | 1.113 |
| 5 | 1 | 1 | 2 | 1 | 2 | 3 | 2 | 1 | 4 | 2 | 2 | 2 | 2 | 169 | 672880.50 | 1.146 |
| 6 | 1 | 2 | 2 | 2 | 2 | 3 | 2 | 3 | 4 | 1 | 2 | 1 | 2 | 171 | 594461.80 | 1.169 |
| 7 | 1 | 1 | 1 | 2 | 2 | 3 | 2 | 3 | 4 | 2 | 2 | 1 | 1 | 172 | 666392.90 | 1.112 |
| 8 | 1 | 2 | 1 | 2 | 2 | 3 | 2 | 3 | 4 | 1 | 2 | 1 | 2 | 173 | 592968.20 | 1.128 |
| 9 | 2 | 2 | 1 | 1 | 2 | 3 | 2 | 3 | 4 | 1 | 2 | 1 | 2 | 174 | 592750.40 | 1.169 |
| 10 | 1 | 1 | 1 | 1 | 2 | 3 | 1 | 3 | 4 | 2 | 2 | 2 | 2 | 175 | 765747.20 | 1.094 |
| 11 | 1 | 1 | 2 | 1 | 2 | 3 | 2 | 3 | 2 | 1 | 2 | 1 | 2 | 176 | 591303.00 | 1.144 |
| 12 | 1 | 1 | 2 | 1 | 2 | 2 | 2 | 3 | 4 | 1 | 2 | 1 | 2 | 178 | 586658.20 | 1.143 |
| 13 | 1 | 1 | 1 | 2 | 2 | 3 | 2 | 1 | 4 | 2 | 2 | 1 | 1 | 179 | 574919.00 | 1.118 |
| 14 | 2 | 1 | 1 | 2 | 2 | 3 | 2 | 1 | 4 | 1 | 2 | 1 | 2 | 180 | 501495.80 | 1.176 |
| 15 | 1 | 1 | 1 | 1 | 2 | 3 | 2 | 1 | 4 | 2 | 2 | 1 | 2 | 181 | 528572.10 | 1.110 |
| 16 | 1 | 1 | 1 | 1 | 1 | 3 | 2 | 3 | 4 | 3 | 2 | 1 | 2 | 184 | 603493.70 | 1.103 |
| 17 | 2 | 1 | 1 | 2 | 2 | 2 | 2 | 1 | 4 | 1 | 2 | 1 | 2 | 186 | 493529.20 | 1.176 |
| 18 | 2 | 1 | 1 | 2 | 1 | 3 | 2 | 1 | 4 | 1 | 2 | 1 | 2 | 190 | 500055.70 | 1.173 |
| 19 | 1 | 1 | 1 | 1 | 2 | 3 | 1 | 1 | 4 | 2 | 2 | 1 | 2 | 192 | 531458.60 | 1.103 |
| 20 | 1 | 1 | 1 | 1 | 2 | 3 | 1 | 2 | 4 | 2 | 2 | 2 | 2 | 193 | 713610.70 | 1.097 |
| 21 | 1 | 1 | 1 | 1 | 2 | 3 | 2 | 1 | 3 | 1 | 2 | 1 | 2 | 196 | 496004.30 | 1.113 |

(continued)

**Table 8** (continued)

| S. no. | A1 | A2 | A3 | A4 | A5 | A6 | A7 | A8 | A9 | A10 | A11 | A12 | A13 | PT (days) | PC (US $) (dollars) | 1/PQI (1/PQI) |
|---|---|---|---|---|---|---|---|---|---|---|---|---|---|---|---|---|
| 22 | 1 | 1 | 1 | 1 | 2 | 2 | 1 | 1 | 4 | 2 | 2 | 1 | 2 | 198 | 523492.00 | 1.102 |
| 23 | 1 | 1 | 1 | 1 | 2 | 2 | 1 | 2 | 4 | 2 | 2 | 2 | 2 | 199 | 705644.20 | 1.096 |
| 24 | 1 | 1 | 1 | 1 | 1 | 1 | 2 | 1 | 4 | 3 | 2 | 1 | 2 | 200 | 501832.10 | 1.106 |
| 25 | 1 | 1 | 1 | 1 | 1 | 1 | 1 | 1 | 4 | 3 | 2 | 1 | 2 | 211 | 504718.60 | 1.099 |

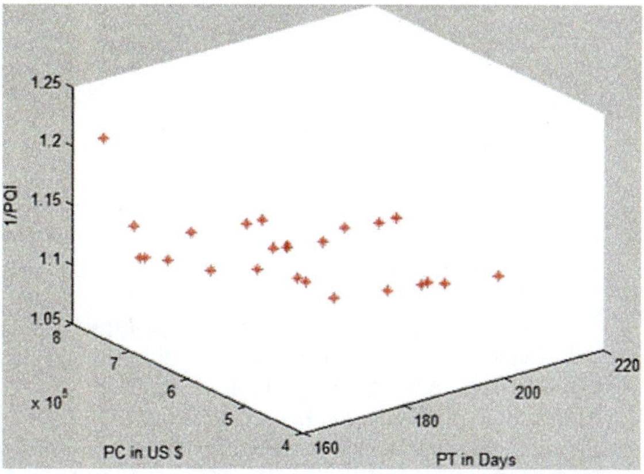


**Fig. 4** Time–cost–quality trade-off plot

## 6 Conclusion

To date, many methodologies have been systematically developed and algorithmically established to optimize the time–cost and time–cost–quality trade-off. These methodologies ranged from deterministic methods to heuristic and meta-heuristic methods. The presented paper provides a meta-heuristic NSGA-II-based time–cost–quality trade-off optimization model that employs the AHP technique to assign the weight to the activities and quality indicators. The advantages of proposed model are demonstrated through analyzing a case study project that illuminates the capabilities of the proposed model. First, the time, cost and quality of the project can interact with each other which makes it important to optimize them together. Second, NSGA-II is found suitable in order to solve the multi-objective optimization problems. Third, the proposed model is found effective in generating satisfactory and quality Pareto-optimal solutions. Fourth, the proposed model also assists in selecting the one solution from Pareto-optimal solutions. Finally, this study feasibly provides a sound tool to construction organizations for worthy decision making in project scheduling.

The main challenge in this work was to quantify the project quality quantitatively. However, in this paper, the project quality is estimated in terms of PQI that uses the AHP technique to determine the weight for activities and quality indicators.

Though the proposed TCQT model is systematically established to determine the quality Pareto-optimal solutions, its usefulness and applications in multi-construction projects are however to be verified. Additional revision is also requisite with taking the time, cost and other project's objective as uncertain. Further study is also necessary after adding one or more projects' objectives such as safety and others.

## References


1. Zahraie B, Tavakolan M (2009) Stochastic time-cost-resource utilization optimization using nondominated sorting genetic algorithm and discrete fuzzy sets. J Constr Eng Manag. https://doi.org/10.1061/(ASCE)CO.1943-7862.0000092
2. Luong DL, Tran DH, Nguyen PT (2018) Optimizing multi-mode time-cost-quality trade-off of construction project using opposition multiple objective difference evolution. Int J Constr Manag. Taylor & Francis 21(3):1–13. https://doi.org/10.1080/15623599.2018.1526630
3. Singh G, Ernst AT (2011) Resource constraint scheduling with a fractional shared resource. Oper Res Lett. Elsevier B.V., 39(5):363–368. https://doi.org/10.1016/j.orl.2011.06.003
4. Deb K et al (2002) A fast and elitist multiobjective genetic algorithm: NSGA-II. IEEE Trans Evol Comput. https://doi.org/10.1109/4235.996017
5. Panwar A, Tripathi KK, Jha KN (2019) A qualitative framework for selection of optimization algorithm for multi-objective trade-off problem in construction projects. Eng Constr Architect Manag. https://doi.org/10.1108/ECAM-06-2018-0246
6. Kelley JE, Walker MR (1959) Critical-path planning and scheduling. In: Proceedings of the eastern joint computer conference, IRE-AIEE-ACM 1959. https://doi.org/10.1145/1460299.1460318
7. Elazouni A (2009) Heuristic method for multi-project finance-based scheduling. Constr Manag Econ. https://doi.org/10.1080/01446190802673110
8. Ipsilandis PG (2007) Multiobjective linear programming model for scheduling linear repetitive projects. J Constr Eng Manag. https://doi.org/10.1061/(ASCE)0733-9364(2007)133:6(417)
9. Zheng DXM, Ng ST, Kumaraswamy MM (2004) Applying a genetic algorithm-based multi-objective approach for time-cost optimization. J Constr Eng Manag. https://doi.org/10.1061/(ASCE)0733-9364(2004)130:2(168)
10. Feng CW, Liu L, Burns SA (1997) Using genetic algorithms to solve construction time-cost trade-off problems. J Comput Civil Eng. https://doi.org/10.1061/(ASCE)0887-3801(1997)11:3(184)
11. Xiong Y, Kuang Y (2008) Applying an ant colony optimization algorithm-based multiobjective approach for time-cost trade-off. J Constr Eng Manag. https://doi.org/10.1061/(ASCE)0733-9364(2008)134:2(153)
12. Zhang H, Li H (2010) Multi-objective particle swarm optimization for construction time-cost tradeoff problems. Constr Manag Econ. https://doi.org/10.1080/01446190903406170
13. Van Den Bergh F, Engelbrecht AP (2010) A convergence proof for the particle swarm optimiser. Fundamenta Informaticae. https://doi.org/10.3233/FI-2010-370
14. Fallah-Mehdipour E et al (2012) Extraction of decision alternatives in construction management projects: application and adaptation of NSGA-II and MOPSO. Expert Syst Appl. https://doi.org/10.1016/j.eswa.2011.08.139
15. Shahriari M (2016) Multi-objective optimization of discrete time–cost tradeoff problem in project networks using non-dominated sorting genetic algorithm. J Indus Eng Int 12(2):159–169. https://doi.org/10.1007/s40092-016-0148-8
16. Köppen M, Yoshida K (2007) Many-objective particle swarm optimization by gradual leader selection. In: Lecture notes in computer science (including subseries Lecture notes in artificial intelligence and lecture notes in bioinformatics). https://doi.org/10.1007/978-3-540-71618-1_36
17. Babu AJG, Suresh N (1996) Project management with time, cost, and quality considerations. Eur J Oper Res. https://doi.org/10.1016/0377-2217(94)00202-9
18. El-Rayes K, Kandil A (2005) Time-cost-quality trade-off analysis for highway construction. J Constr Eng and Manag. https://doi.org/10.1061/(ASCE)0733-9364(2005)131:4(477)

19. Deb K, Agrawal RB (1994) Simulated binary crossover for continuous search space. Compl Syst. http://citeseerx.ist.psu.edu/viewdoc/download?doi=10.1.1.27.767&rep=rep1&type=pdf
20. Deb K, Goyal M (1996) A combined genetic adaptive search (GeneAS) for engineering design. Comput Sci Inform. citeulike-article-id:9625478
21. Ferreira JC, Fonseca CM, Gaspar-Cunha A (2007) Methodology to select solutions from the pareto-optimal set: a comparative study. In: Proceedings of GECCO 2007: genetic and evolutionary computation conference. https://doi.org/10.1145/1276958.1277117

# Chapter 6
# A Wavelet-Based Segmentation Technique for Medical Images

**Rinisha Bagaria, Sulochana Wadhwani, and A. K. Wadhwani**

## 1 Introduction

In the area of medical science, new applications such as analysis, recognition, classification and diagnosis of a different medical image are emerging day-by-day in MIP, especially in the realm of image segmentation [1, 2]. But segmentation of monochrome images is one of the complex tasks in the area of image processing (IP). The method segmentation is a root step in many newly developed methods of multi-dimensional signal processing. Image segmentation (IS) is the basic preprocessing stage which is used to process the identification of the surface and finding the regions of the digital image which corresponds for understanding all computer vision [3]. Segmentation helps to lessen the relevant information for effortless analysis and is extremely useful in analysis, compression and enhancement of an image [4, 5]. The target of the segmentation is to differentiate the object and the background in an image. However, the segmentation process has become a new challenge today because of its complicated surround and varying brightness on the image. Poor quality of the segmented image may lead to inaccurate and unsuccessful outcomes or results. All methods are plotted in the MATLAB environment. In the first section, there is a short introduction to the work related to image processing and image segmentation. In the second section, there is a full description of the images used in medical science and its details. But we only choose X-ray images to work on our proposed method. In the third section, various kinds of segmentation methods are named, but we work only on wavelet technique which is our proposed method. In this section, a complete

R. Bagaria (✉) · S. Wadhwani · A. K. Wadhwani
Electrical Engineering Department, MITS Gwalior, Gwalior, India
e-mail: rinisha.bagaria19@mitsgwalior.in

S. Wadhwani
e-mail: sulochana_wadhwani@mitsgwalior.in

A. K. Wadhwani
e-mail: akwadhwani@mitsgwalior.in

© The Author(s), under exclusive license to Springer Nature Singapore Pte Ltd. 2022 
H. M. Dubey et al. (eds.), *Artificial Intelligence and Sustainable Computing*, Algorithms for Intelligent Systems, https://doi.org/10.1007/978-981-16-1220-6_6

description is given about wavelets and his family. Our fourth section is experimental results and discussions. It has five parts, namely, data acquisition, evaluation index, experimental data, display of image and visual display of method. In the last section, we conclude our research work.

## 2 Techniques for Medical Images

Imaging usage is increasing rapidly because of the development of image processing-based techniques, including image analysis, image recognition and image enhancement. There are various kind of medical images used in the field of imaging, such as X-rays, ultrasound, CT scans, MRI and many more [6]. In this research paper, only X-ray images are taken for processing and discussion in detail.

### *2.1 X-Rays*

An X-ray (also known as X-radiation) is a waveform of high energy or it can be said that X-rays are waves of electromagnetic radiation. An X-ray has a wavelength which ranges from 10 pm to 10 nm. X-rays can recognize bone structures and hence they have been created for utilization in clinical imaging. Radiology is a particular field of medical science that utilizes radiography and different procedures for indicative imaging. X-rays are particularly valuable in the location of pathology of the skeletal framework; they are also helpful for distinguishing some illness measures in delicate tissue. These strategies incorporate X-rays, computed tomography (CT) examines and magnetic resonance imaging (MRI) tests. There is next to no hazard on having one X-ray test. Nonetheless, with rehashed tests there is a danger that the X-rays may harm a few cells in the body, potentially prompting disease later on. The portion of X-rays radiation is constantly kept to the base and expected to get a decent image of the specific body part being checked. Likewise, radiographers who take the X-ray pictures consistently wear lead covers or go behind a defensive screen when the X-rays are terminated to keep away from rehashed presentation to X-rays [7–10]. Pregnant ladies, if conceivable, ought not to have the X-ray test as there is a little danger that X-rays may make a variation from the norm unborn kid. This is the reason ladies are asked before having the X-rays in the event that they are, or may be, pregnant.

## 3 Image Segmentation Methods

### 3.1 Segmentation Techniques

Segmentation is a usually utilized procedure in advanced processing of an image and examination to parcel a picture into different parts or locales, which frequently is dependent on the attributes of the pixels in the image [11]. Image segmentation could include isolating forepart from background, or grouping regions of pixels dependent on similitude in color or shape. Generally, segmentation is focused on the two main properties, such as discontinuity of an image and similarity in an image [12]. There are various kinds of image segmentation techniques, namely, edge detection method, thresholding based, region growing algorithm, feature-based clustering, texture-based, watershed algorithm and wavelet transform. But the focus is on one segmentation technique, i.e. wavelet transform, which is considered as our proposed method.

### 3.2 Wavelet Transform in Image Processing

Wavelet is used for calculating the image transform, shift coefficients for the transformation and for calculating the inverse transformation. Because scale in the wavelet domain is analogous to the frequency in the Fourier domain, most of the Fourier-based filtering techniques have an equivalent "wavelet domain" counterpart. They are effective instruments for smoothing or blurring images. It works as horizontal, vertical and diagonal directionality of the single-scale wavelet transform [13]. Fourier transform provides result in sine wave. Wavelet has scaling (shrinking or expanding) and shifting (dealing with a long wave) properties [14]. Wavelet transform can be bifurcated into two parts: continuous wavelet transform (CWT) and discrete wavelet transform (DWT). There are many applications of wavelet transform such as filtering and time domain analysis for CWT and de-noising and compression for DWT [15, 16]. In the DWT, an image signal can be dissected by going it through an analysis filter bank followed by an abolish activity [17]. It comprises a low-pass and a high-pass filter at every decomposition level, and is normally utilized in image compression and image recognition [18, 19]. At the point when an input signal applied to these filters, it is parted into two groups. The low-pass filter compares to an averaging activity and separates the unwanted data of the signal. The high-pass filter compares to a differencing activity and extricates the detail data of the signal. The tasks of the filtering are then demolished by two.

A 2-D transform can be cultivated by performing two separate 1-D transforms. To start with, the image is separated along the x-measurement utilizing low pass and high pass investigation channels and devastated by two. Coefficients of low-pass filters are put away on the left aspect of the matrix and high-pass filtered on the right. As a result of demolition, the complete structure of the transformed image is similar to the original image [20]. At that point, it is trailed by separating the sub-image along the y-measurement and obliterated by two. At long last, it parts the image into four groups signified by LL, HL, LH and HH after single-level decomposition. This deals with the properties of wavelets which is a numerical tool [21]. This transform is intended to give great time goal and helpless recurrence goal at high frequencies and great recurrence goal and helpless time goal at low frequencies. Wavelet implies wave which is an oscillating periodic function in terms of time having limited energy [22]. Interestingly, wavelets are restricted waves and have their vitality moved in time or space and are fit to the examination of transient signals.

Moreover, Fourier transform utilizes waves to examine signals [23]. The key bit of leeway of wavelet strategies is the assortment of wavelet capacities accessible, and in this way permitting the most proper to be picked for the signal under scrutiny. This is rather a Fourier investigation limited to one component morphology: the sinusoidal. There is an assortment of wavelets, for example, Daubechies, Symmlet or Coiflet, which produce symmetrical wavelet bases. Here, we choose Haar method from the wavelet family because it is already experimented for X-ray images, and Haar wavelet is best in use. Wavelet decomposition is described as a multi-scale and multi-level wavelet decomposition which is used to perform on the images after applying the process of averaging [24]. The determination of number of levels of wavelet decomposition affects the subsequent de-noising effect. If the level of decomposition is very small, then the de-noising effect will be unsatisfactory. To extract low-frequency components and eliminate high-frequency noise, the decomposition levels will be expanded to some extent, and if the levels are too high, the error will be considered as large; moreover, it mainly focuses on the reflection of the characteristics of the wavelet basis rather than the input signal being examined on its own, which results in false frequencies and loss of information. To obtain more wavelet coefficients, the level of decomposition can be suitably increased.

## 4 Experimental Results and Discussion

### *4.1 Data Acquisition*

To assess the ability of the wavelet process in the segmentation field [25, 26], the proposed Haar wavelet was evaluated on three datasets. Three X-ray images are shown in Fig. 1 and named as image A, image B and image C, respectively, in Fig. 1 A–C. These X-rays images are derived from the Vidhya Imaging Centre, Gwalior with the help of Dr. Anil Kumar. And we are also able to take X-rays in our research lab.

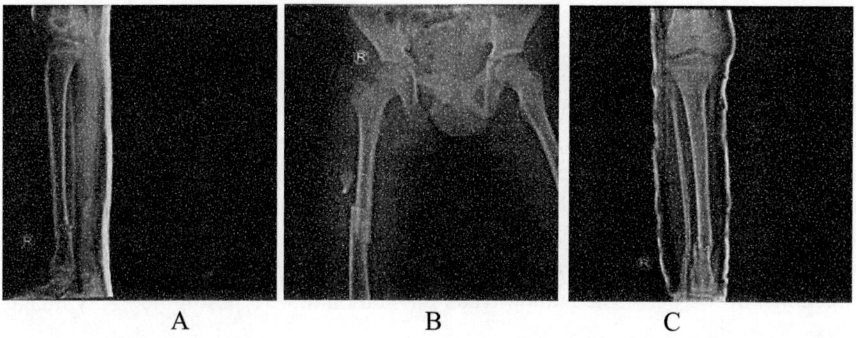

Fig. 1 Experimental data sets

## 4.2 Evaluation Index

Noise is an irregular variety of brightness or color information in images and is typically a part of electronic noise. The adaptive noise detector is utilized to recognize the kind of noise, for example, Gaussian commotion, Salt and Paper and so forth. Image distortion is a major issue in image processing. Image is distorted because of different sorts of commotion, such as Gaussian noise, Poisson noise, Speckle noise, Salt and Paper noise and a lot more key noise types, in the event of advanced images.

Generally, the SNR ratio is explained as the quality of the signal or an image [27]. Mathematically, it is shown in Eq. (1).

$$SNR = 10log\frac{S_{original}}{S_{noise}} \tag{1}$$

$S_{original}$ : $Original\ signal\ without\ noise$
$S_{noise}$ : $Noisy\ signal$

It is utilized in imaging to portray an image quality. The affectability of an (advanced or film) imaging framework is commonly portrayed in the provisions of the signal level that yields a limit level of SNR.

Entropy: It is a factual proportion of irregularity that can be utilized to portray the texture of the input image [28] and is shown in Eq. (1).

$$E = -\sum P. * log^2(p) \tag{2}$$

where p contains the normalized histogram counts returned from imhist (MATLAB function).

### 4.3 Display of Image

Taking Fig. 1A–C for instance, we can acquire the five levels of decomposition diagram by utilizing the proposed wavelet transform technique. For decomposed image, there are approximate component (AC), vertical detail component (VDC), horizontal detail component (HDC) and diagonal detail component (DDC). It tends to be seen from the table that the wavelet-based segmentation technique can successfully hold the subtleties of the original image in the horizontal, vertical and diagonal directions. So we performed this wavelet method on three types of the X-ray image which act as data sets.

In Fig. 2, the first image, i.e. image "A" is taken and processed using the MATLAB software. After applying the wavelet technique we find out its five levels of decomposition and each level of decomposition has four components such as AC, VDC, HDC and DDC. By visualization, we can say that only the third level of decomposition gives better results than the other levels of decomposition.

In Fig. 3, the second image, i.e. image "B" is taken and processed using the MATLAB software. After applying the wavelet technique we find out its five levels of decomposition and each level of decomposition has four components such as AC, VDC, HDC and DDC.

By visualization, we can say that only the third level of decomposition gives better results than the other levels of decomposition.

In this Fig. 4, the third image, i.e. "image C" is taken and processed using the MATLAB software. After applying the wavelet technique we find out its five levels of decomposition and each level of decomposition has four components such as AC, VDC, HDC and DDC. By visualization, we can say that only the third level of decomposition gives better results than the other levels of decomposition.

### 4.4 Experimental Data

The SNR and entropy are used in this comparison. To test the effectiveness of the suggested method, we have done so many tests on the three images. The procedure was carried out using MATLAB 2019 on a laptop with configuration of an Intel Core i7 8th-generation processor. And the results are visible in Table 1. This table consists of two main categories, which are SNR value and entropy value. They are further divided into four sections, namely, AC, VDC, HDC and DDC. Out of these four components, the approximate component is not our area of interest, so we can neglect it, but we need other three detail components for getting our results.

| Level of De-com-posi-tion | AC of decomposed image | VDC of decomposed image | HDC of decomposed image | DDC of decomposed image |
|---|---|---|---|---|
| 1st |  | | | |
| 2nd |  | | | |
| 3rd |  | | | |
| 4th |  | | | |
| 5th |  | | | |

**Fig. 2** Five-level decomposition of image "A"

## 4.5 *Visual Display of Method*

To see the visual effect of the suggested wavelet transform, the obtained results of the five levels of decomposition using Haar wavelet are shown in Figs. 2, 3 and 4. According to numeric values, it can be said that the SNR value should be maximum

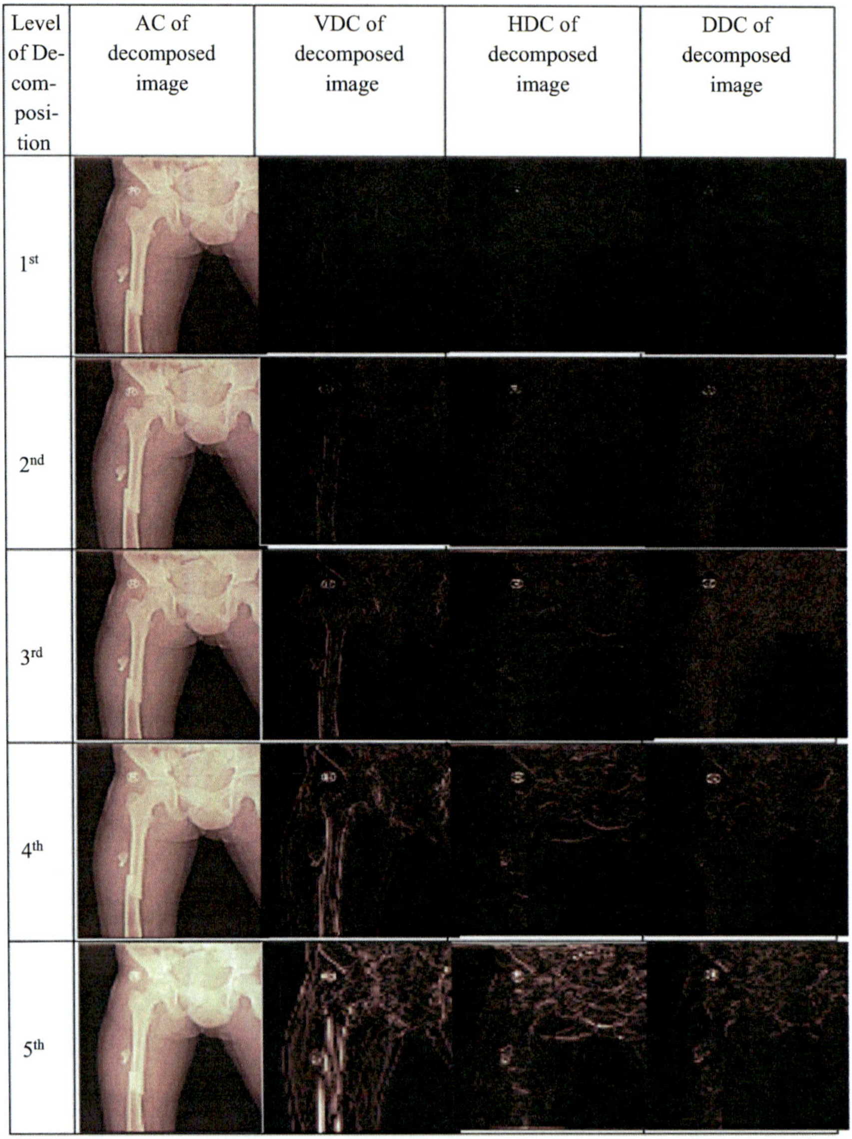


**Fig. 3** Five-level decomposition of image "B"

and the entropy value should be minimum for better segmentation. By visualization and numeric values, we can say that the third level of decomposition of Haar wavelet gives best results for segmentation of an X-ray image. In Fig. 5 we take graph between SNR value of all images with all three detail components of third level of decomposition. So in Fig. 5, it is visible that the vertical detail component has the highest value in all three images (A, B and C).

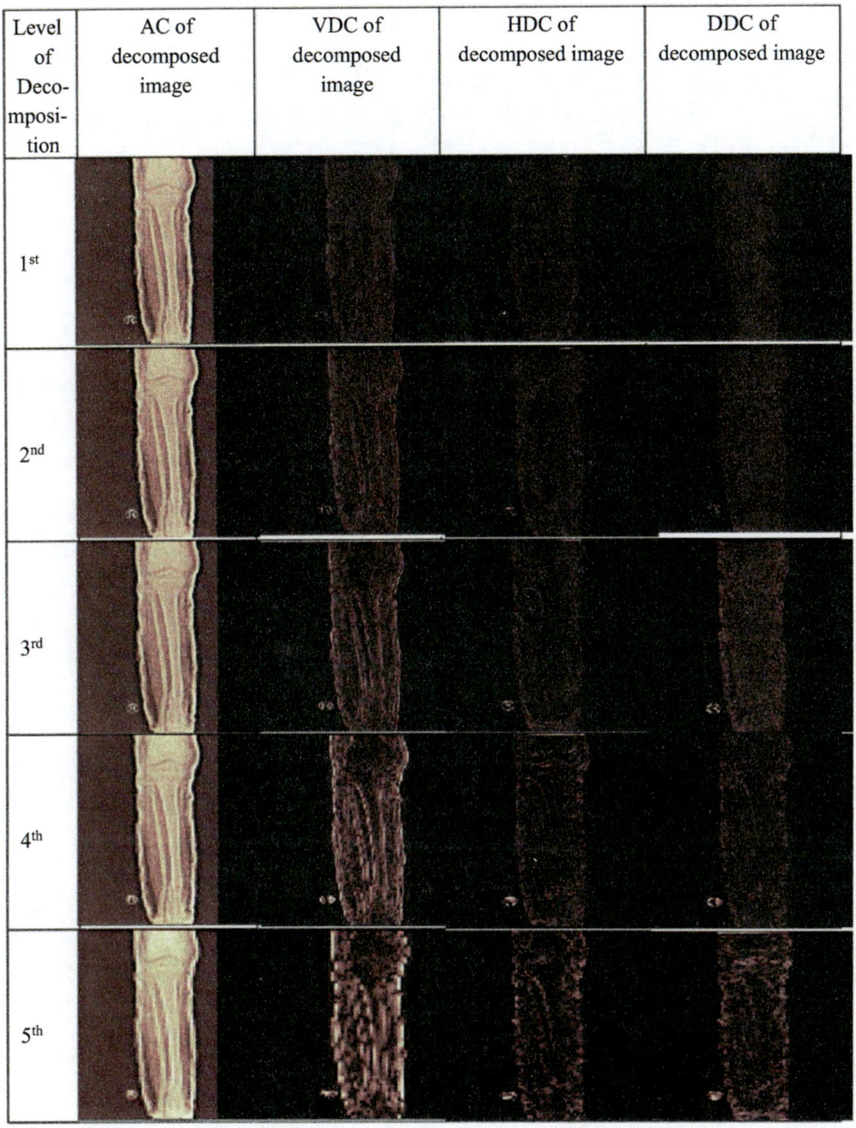


**Fig. 4** Five-level decomposition of image "C"

In Fig. 6 we take graph between entropy value of all images with all three detail components of third level of decomposition. So in Fig. 6, it is visible that the vertical detail component has the lowest values of all three images (A, B and C).

**Table 1** The results of five levels of decompositions on three images data set

| Data set | Level of decom-position | SNR value of AC | SNR value of VDC | SNR value of HDC | SNR value of DDC | Entropy value of AC | Entropy value of VDC | Entropy value of HDC | Entropy value of DDC |
|---|---|---|---|---|---|---|---|---|---|
| Image "A" | 1st | 21.51 | 29.22 | 21.78 | 22.66 | 0.9856 | 0.6295 | 0.6588 | 0.6314 |
| | 2nd | 21.55 | 21.26 | 22.40 | 21.17 | 0.9851 | 0.6751 | 0.7330 | 0.7714 |
| | **3rd** | 20.36 | **21.35** | **19.45** | **19.43** | 0.9828 | **0.6655** | **0.7345** | **0.7694** |
| | 4th | 20.10 | 20.77 | 19.69 | 18.72 | 0.9825 | 0.6761 | 0.7798 | 0.8282 |
| | 5th | 18.32 | 21.13 | 17.45 | 17.47 | 0.9818 | 0.7085 | 0.7408 | 0.8396 |
| Image "B" | 1st | 11.89 | 32.69 | 22.36 | 22.60 | 0.0150 | 1.2220 | 1.2417 | 1.1525 |
| | 2nd | 11.71 | 25.96 | 21.89 | 20.06 | 0.0111 | 1.3043 | 1.3717 | 1.4882 |
| | **3rd** | 11.46 | **22.75** | **22.68** | **19.66** | 0.0077 | **1.2771** | **1.3724** | **1.4491** |
| | 4th | 11.14 | 20.38 | 21.15 | 20.09 | 0.0140 | 1.3208 | 1.4720 | 1.5648 |
| | 5th | 10.52 | 20.80 | 20.00 | 21.35 | 0.0000 | 1.3174 | 1.4378 | 1.6420 |
| Image "C" | 1st | 17.94 | 32.18 | 21.85 | 23.64 | 0.9170 | 0.8172 | 0.8150 | 0.7798 |
| | 2nd | 17.90 | 22.71 | 23.85 | 21.07 | 0.9164 | 1.1545 | 1.1602 | 1.1907 |
| | **3rd** | 17.50 | **20.08** | **19.24** | **19.58** | 0.9150 | **1.2932** | **1.3063** | **1.3515** |
| | 4th | 17.41 | 18.22 | 18.07 | 18.06 | 0.9163 | 1.5756 | 1.5948 | 1.7099 |
| | 5th | 14.90 | 18.52 | 14.39 | 17.96 | 0.9146 | 1.5930 | 1.5407 | 1.7854 |

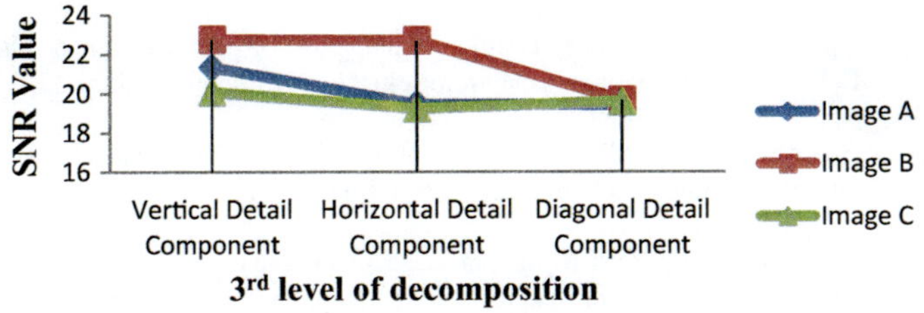


**Fig. 5** Graph plotted between the SNR value and third level of decomposition for all data sets

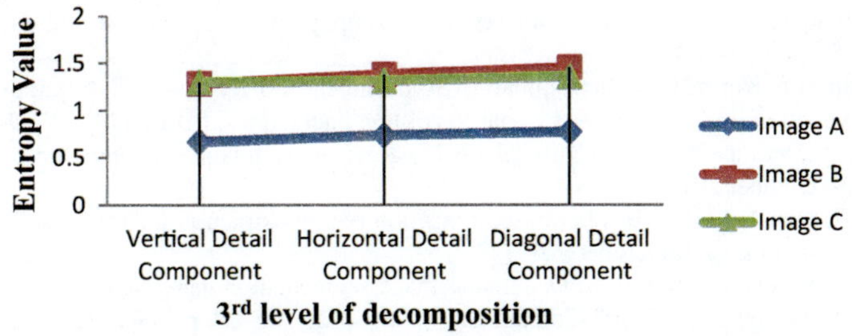


**Fig. 6** Graph plotted between entropy value and third level of decomposition for all data sets

## 5 Conclusions

In this research work, wavelet-based segmentation technique is used for better feature extraction of X-ray images. From Figs. 2, 3 and 4, it can be easily seen that this method of image segmentation not only provide the effective segmented image in terms of the required region of interest but also retain many other vital information of the image. Also from Table 1, it can be observed that the third level of decomposition gives better results for the X-ray images considered in this study.

## References

1. Yogamangalam R, Karthikeyan B (2013) Segmentation techniques comparison in image processing. Int J Eng Technol (IJET) 5(1). ISSN: 0975-4024 (Feb–Mar 2013)
2. Khan AM, Ravi S (2013) Image segmentation methods: a comparative study. Int J Soft Comput Eng (IJSCE) 3(4). ISSN: 2231-2307 (September 2013)
3. Abdullaha SLS, Jamilc N (2012) Segmentation of natural images using an improved thresholding-based technique. In: International symposium on robotics and intelligent sensors (IRIS), pp 938–944 (Hamirul'Aini Hambalia,b)
4. Maheshan MS, Harish BS, Nagadarshan N (2018) On the use of image enhancement technique towards robust sclera segmentation, pp 466–473

5. Tsai D-Y, Kojima K (2003) Wavelet coefficient mapping for enhancement of chest X-ray images. In: XVII Imeko World Congress Metrology in the 3rd Millennium, June 22–27, January 2003
6. Mahima SP, Soam SS, Tripathi SP (2019) Recognition of X-rays bones: challenges in the past, present and future. Int J Innov Technol Explor Eng (IJITEE) 8(10). ISSN: 2278–3075 (August 2019)
7. Umadevi N, Geethalakshmi SN (2012) Enhanced segmentation method for bone structure and diaphysis extraction from X-ray images. Int J Comput Appl 37(3):0975–8887
8. Wong AKO, Manske SL (2020) A comparison of peripheral imaging technologies for bone and muscle quantification: a review of segmentation techniques. J Clin Densitom Assess Manag Musculoskelet Health 23(1)
9. Dograa J, Jaina S, Sooda M (2018) Segmentation of MR images using hybrid kmean-graph cut technique. In: International conference on computational intelligence and data science (ICCIDS), pp 775–784
10. Kazeminial S, Karimil N, Mirmahboub B, Soroushmehr SMR, Samavil S, Najarian K (2015) Bone extraction in X-ray images by analysis of line fluctuations. IEEE, 978-1-4799-8339-1
11. Kannan S, Vairaprakash G, Nalini G (2014) Review on image segmentation techniques. RTRICS (October 2014)
12. Parida P, Bhoi N (2017) Wavelet based transition region extraction for image segmentation. Future Comput Inf J 2 (October 2017)
13. Wanjari MT, Dhore MP (2016) Document image segmentation using wavelet transform and gabor filter technique. IOSR J Comput Eng (IOSR-JCE), 25–29. e-ISSN: 2278-0661, ISSN: 2278-8727
14. Abid Fourati W, Bouhlel MS (2014) Trabecular bone image segmentation using wavelet and marker-controlled watershed transformation. Hindawi Publishing Corporation, Article ID 891950, p 10 (June 2014) (Chin J Eng)
15. Li J, Dai W (2009) Image quality assessment based on the correlation coefficient and the 2-D discrete wavelet transform. In: Proceedings of the IEEE international conference on automation and logistics Shenyang, China, IEEE, 978-1-4244-4795-4 (August 2009)
16. Xizhi Z (2008) The application of wavelet transform in digital image processing. In: 2008 International conference on multimedia and information technology, IEEE, 978-0-7695-3556-2
17. Larbi M, Rouini A, Messal Z, Larbi S (2019) Medical Image segmentation based on wavelet transformation and level set method. The 4th international conference on power electronics and their applications (ICPEA), IEEE, 978-1-7281-2726-2, 25-27 (September 2019)
18. Kumari A, Bisht M (2013) Optimal wavelet filter maximizes the cross correlation coefficient with an ECG signal. Int J Innov Technol Res 1(2):191–193 (February–March 2013)
19. Kaur A, Kaur L, Gupta S (2012) Image recognition using coefficient of correlation and structural similarity index in uncontrolled environment. Int J Comput Appl 59(5):0975–8887 (December 2012)
20. Angoth V, Dwith CYN, Singh A (2013) A novel wavelet based image fusion for brain tumor detection. Int J Comput Vis Signal Process 1–7 (January 2013)
21. Tsai D-Y, Lee Y (2003) A method of medical image enhancement using wavelet-coefficient mapping functions. In: IEEE conference neural networks & signal processing, Nanjing, China, 14–17 Dec 2003
22. Shigeta H, Mashita T, Kikuta J, Seno S, Takemura H, Ishii M, Matsuda H (2017) Bone marrow cavity segmentation using graph-cuts with wavelet-based texture feature. J Bioinform Comput Biol 15(5):1740004 (16 pages) (August 2017)
23. Nagendra H, Mukherjee S, Kumar V (2013) Wavelet based non linear thresholding techniques for pre processing ECG signals. Int J Biomed Adv Res (IJBAR), ISSN: 2229-3809
24. Gavlasova A, Prochazka A, Mudrova M (2006) Wavelet based image segmentation (January 2006)
25. Bagaria R, Wadhwani S, Wadhwani AK (2019) Automatic location of fracture region in a bone x-ray using MATLAB. In: IEEE national conference on advance trends and techniques in wireless communication and soft computing (Feb 2019)

26. Bagaria R, Wadhwani S, Wadhwani AK (2020) Different techniques for identification of a bone fracture in analysis of medical image. In: 9th IEEE international conference on communication systems and network technologies (CSNT), 978-1-7281-4976-9 (April 2020)
27. Gao J, Wang B, Wang Z, Wang Y, Kong F (2019) A wavelet transform-based image segmentation method. Optik Int J Light Electron Opt (December 2019)
28. Ramadan ZM (2017) Using entropy and 2-D correlation coefficient as measuring indices for impulsive noise reduction techniques. Int J Appl Eng Res 12(21):11101–11106. ISSN 0973-4562

# Chapter 7
# MVO-Based Approach for Optimal DG and Capacitor Allocation

**Kingshuk Roy, Laxmi Srivastava, and Shishir Dixit**

## 1 Introduction

Power demand is increasing rapidly day-by-day in the current scenario. Power loss in the line and deterioration in system voltage, system stability and system power quality have become quite often due to the large increment in load demand. Previously hydro, nuclear, thermal are the only energy resources utilized for electricity generation. These traditional generators which utlilizes the conventional resources for electrical power generation supply to the widely connected load through transmission and distribution lines. The major reasons for weakening of the traditional electric power stations are due to shortage of conventional resources and also environmental issues, etc. So distribution and transmission costs increased consequently. As the conventional resources have their limitations and also the power demand is increasing day-by-day, these challenges shift the attention of researchers toward load-connected distributed generation. Distributed generation placement is a modern way to solve many of these above-mentioned challenges. In simple ways, DG can be defined as dispersed generation which is not centrally dispatched. Distributed generation or DG is connected nearer to the load side and mostly uses the renewable energy resources for generation. There are basically four types of DG [1].

Type 1. DG supplies real power only (i.e. solar photovoltaic, fuel cell)

Type 2. DG supplies reactive power only (i.e. capacitors, synchronous condenser)

Type 3. DG supplies both real and reactive power. (i.e. synchronous generator)

Type 4. DG supplies real power but consumes reactive power (i.e. wind turbine)

The authors have used various sensitivity approach for optimal DG allocation [2]. Fuzzy based multi-objective PSO have been implemented in [3] for optimal DG and capacitor allocation in 33-bus radial distribution system. In this paper authors have

K. Roy (✉) · L. Srivastava · S. Dixit
Madhav Institute of Technology and Science, Gwalior, India

L. Srivastava
e-mail: srivastaval@hotmail.com

© The Author(s), under exclusive license to Springer Nature Singapore Pte Ltd. 2022 
H. M. Dubey et al. (eds.), *Artificial Intelligence and Sustainable Computing*,
Algorithms for Intelligent Systems, https://doi.org/10.1007/978-981-16-1220-6_7

taken real power loss minimization, voltage profile enhancement and balancing of current at the section of system as objective function. Combination of loss sensitivity factor and bacteria foraging optimization algorithm has been implemented in [4] for optimal capacitor and DG allocation in 33-bus system. In [5] the authors have implemented back-tracking algorithm for optimal DG allocation in 33-bus and 94-bus system. In this paper the authors have taken real power loss reduction and enhancement in system voltage profile as system objective function. Authors in [6] have taken cost minimization as the objective function of the problem. Total cost includes energy loss cost, maintenance and investment cost etc. for capacitor and DG units. GA has been implemented in this problem to calculate the sizes and locations of DG and capacitor units by minimizing objective function in this paper. Combination of moth-flame algorithm and index vector method has been implemented to calculate adequate sizing and location of only DG units [7] in 33-bus distribution system. A decomposition-based multi-objective evolutionary algorithm (MODA/G) has been used in [8] for optimal DG and Qc allocation. In this paper the authors have taken minimization of real and reactive powers loss as the objective functions. Combined GA and MSA has been implemented for optimal DG and capacitor allocation problem in 33-bus and 69-bus system [9]. The salp swarm optimization is used in [10] for reconfiguration of 33 and 69 bus system using DG and capacitor. In [11] the author developed a modified GA approach for allocation of fuel cell type DG and capacitor. In [12] author employed an newly developed Spring search algorithm for DG allocation in 33 bus radial distribution system.

## 2 Problem Formulation

The problem of allocation of DGs and capacitors are formulated as constrained optimization problem for finding optimal size and location of DGs and capacitors for minimizing active power loss while satisfying the operational constraints.

### 2.1 Objective Function

The active power loss minimization for n bus system has been taken as main optimization problem, which is formulated as

$$P_l = \sum_{i=1}^{n} \left( \left| I_i^2 \right| \times R_i \right) \tag{1}$$

$$\mathcal{F} = \min(P_l) \tag{2}$$

Here $I_i$ is the current at $ith$ branch. $R_i$ is the resistance of $ith$ branch. $P_l$ is the active power loss of the line.

## 2.2 Constraints of the System

**Power flow constraint**

$$P_{Gi} - P_{Di} = \sum_{i=1}^{N}\sum_{j=1}^{N} V_i V_j (G_{ij}\cos(\phi_I - \phi_j) + B_{ij}\sin(\phi_i - \phi_j)) \quad (3)$$

$$Q_{Gi} - Q_{Di} = \sum_{i=1}^{N}\sum_{j=1}^{N} V_i V_j (G_{ij}\sin(\phi_i - \phi_j) - B_{ij}\cos(\phi_i - \phi_j)) \quad (4)$$

$$\forall_i = 1, 2, 3 \ldots\ldots. N$$

Here $G_{ij}$, $B_{ij}$ are the conductance and susceptance of the line between bus i and j, respectively. $P_{Gi}$ and $P_{Di}$ are real power generation and demand at $i$th bus, respectively. $Q_{Gi}$ and $Q_{Di}$ are reactive power generation and demand at $i$th bus, respectively. $V_i$ and $\phi_i$ are the voltage magnitude and angle at $i$th bus, respectively

$$P_i = P_{Gi} - P_{Di} \quad (5)$$

$$Q_i = Q_{Gi} - Q_{Di} \quad (6)$$

$P_i$ and $Q_i$ are the net active and reactive power at $i$th bus, respectively.

**Voltage limit constraint**

$$V_{min} < V_i < V_{max} \quad (7)$$

$$\forall_i = 1, 2, 3 \ldots\ldots. N \quad (8)$$

**Current limit constraints**

$$I_i < I_{rated} \quad (9)$$

$$\forall_i = 1, 2, 3 \ldots N \quad (10)$$

**DG and capacitor power limit constants**

$$P_{DG}^{min} < P_{DG} < P_{DG}^{max} \tag{11}$$

$$Q_{QC}^{min} < Q_{QC} < Q_{QC}^{max} \tag{12}$$

$P_{DG}^{max}$ and $P_{DG}^{min}$ are the maximum and minimum real power generation limits of DG. $Q_{QC}^{max}$ and $Q_{QC}^{min}$ are the minimum and maximum reactive power generation limits of capacitor.

# 3 Optimization Algorithm

Alimirjalili in 2016 has developed this algorithm [13]. Big bang theory believed that a huge explosion created the whole universe. Nothing was there before the huge explosion. Multi-verse theory is also a new theory in astrophysics and cosmology. According to this theory several times the big bang has occurred, and like our universe there are many similar universe. Worm holes, white holes and black holes are used to develop the MVO algorithm. Exploration of algorithm has done using the characteristics of white/black holes. Exploitation of the algorithm has done using the characteristics of worm holes. Variable of the solution implies object of the universe. Universe corresponds to solution of the problem. Inflation rate of the universe implies the fitness value of solution. White holes are present in the universes having higher inflation rate. Black holes are present in the universes having lower inflation rate.

## 3.1 *Computational Steps to Perform MVO Algorithm*

**White/black hole tunnel**

Depending upon the inflation rate, one universe is selected at every iteration using Roulette wheel technique. White/black hole tunnel are used to transfer the object from universe having high inflation rate to universe having low inflation rate. To mathematically model this behavior

$$A = \begin{bmatrix} u_1^1 \; u_1^2 \; .. \; .. \; u_1^d \\ u_2^1 \; u_2^2 \; .. \; .. \; u_2^d \\ .. \; .. \; .. \; .. \; .. \\ .. \; .. \; .. \; .. \; .. \\ u_k^1 \; u_k^2 \; .. \; .. \; u_k^d \end{bmatrix} \tag{13}$$

$$u_k^i = u_k^i \;\; if \;\; p1 \geq N(A_i)$$

$$u_k^j \;\; if \;\; p1 < N(A_i) \tag{14}$$

Here total universes are $k$ and objects or parameters (or variable) in the universe are $d$. $u_k^i$ defined $k$th universe's $i$th object. $i$th universe is represented by $A_i$ and normalize inflation rate of $i$th universe is $N(A_i)$. Here $p1$ is a random number between [0, 1] interval. $u_k^j$ represents $k$th universe's $j$th parameter which is selected by Roulette wheel technique. In this step exploration of the optimization has been done.

**Wormhole tunnel**

Assumption is made while developing this algorithm that every universe has one wormhole. Wormhole transfers the object of one universe to the best universe get so far. This phenomenon has been developed by the following equation:

$$u_k^i = \begin{cases} \begin{bmatrix} U_j + DTR \times \left(\left(ub_j - lb_j\right) \times p4 + lb_j\right) p3 < 0.5 \\ U_j + DTR \times \left(\left(ub_j - lb_j\right) \times p4 + lb_j\right) p3 \geq 0.5 \end{bmatrix} & p2 < WP \\ u_k^i & p2 \geq WP \end{cases} \tag{15}$$

where $U_j$ is the best universe's jth variable. $u_k^i$ is the $i$th universe's $k$th parameter. $WP$ and $DTR$ are coefficients, which may be constant or may be a function of iteration.

$$DTR = 1 - \frac{n^{1/6}}{N^{1/6}} \tag{16}$$

$$WP = 0.2 + (n \times \frac{0.8}{N}) \tag{17}$$

In the above equation n denotes present equation, and maximum iteration is denoted by N. $DTR$ and $WP$ have been taken using above equation to get better computational results.

## 3.2 MVO Algorithm Overview

In the MVO algorithm, the optimization algorithm starts with creation of set of random universes. At every step through black/white hole tunnel object of higher inflation rate universe tends to move toward lower inflation rate universe. Every iteration object from each universe tends to move toward the best universe through wormholes. This process will continue until the end criteria have not been met.

# 4 Computational Results

The MVO algorithm has been computed on IEEE 33-bus radial distribution system of total connected load of 3.715 MW and 2.3 MVAr [14]. MATLAB R2013a software

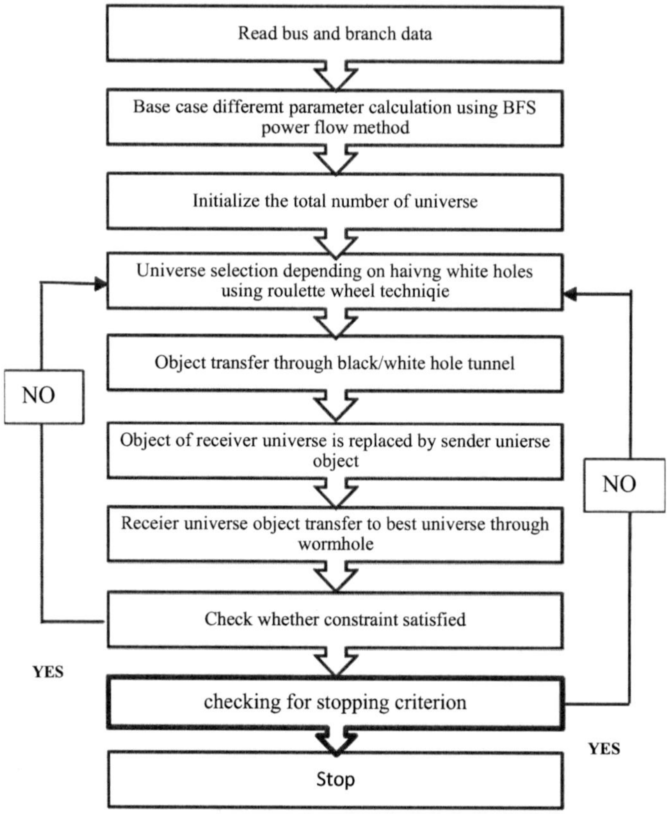


**Fig. 1** Flowchart of the computational steps

has been used for computation. Computational steps for MVO algorithm have been described in flowchart of Fig. 1.

The X/R ratio of distribution system is low compared to transmission system. Due to this reason the conventional load flow methods (i.e. Gauss–Seidel load flow method and Newton–Raphson load flow methods) converge very slowly during computation of the distribution system. The advantage of using backward/forward sweep (BFS) load flow method is fast convergence compared to other conventional load flow method. In this paper backward/forward sweep power flow method [15] has been used to compute and observe the 33-bus system profile before any DG or capacitor placement. The MVO algorithm has been implemented on 33-bus system to find optimal location of DGs and Qc units. Table 1 presents the result summary for different cases that has been obtained after implementation of MVO. It is clearly observed from Table 1 that before connecting any devices to the system real power loss was 210.9856 kW. Thereafter, placement of one DG at computed location and

**Table 1** Results summary of obtained results

| 33 Bus system | | | |
|---|---|---|---|
| Device | Parameters | | |
| | Real power loss (kW) | Minimum voltage (at bus number) | Location (bus no.) and size (in kW) |
| Without DG | 210.9856 | 0.9038 (18) | – |
| One DG | 111.1445044 | 0.9410820 (18) | 6 and 2500 |
| One Qc | 151.3809901 | 0.9164816 (18) | 30 and 1257.9973129 |
| Simultaneous DG and Qc | 58.4705230 | 0.9531906 (18) | 1. 6 and 2500<br>2. 30 and 1255.5350181 |

size as obtained with MVO implementation is shown in Table 1, and the real power loss of the system has reduced by 47.32% to that of base case. Further, placement of one capacitor into the system has reduced the real power loss to 28.241% to that of base case. Simultaneous DG and capacitor placement into the system has reduced the system power loss by 72.3% as compared to the base case. It can also be clearly observed from Table 1 that advancement in system minimum voltage is evident when simultaneously one DG and one Qc are connected to the system. The system minimum voltage was 0.9038 p.u at bus 18 before DG and Qc integration into the system. Minimum voltage of the system becomes after placement of one DG as 0.9410 p.u. at bus 18. An increase of 4.11% in minimum voltage has been observed here. After placement of one capacitor unit the system minimum voltage becomes 0.9164 p.u at bus 18. In this case minimum voltage has been increased by 1.39% in comparison of base case. Simultaneous capacitor and DG placement in the system results in rise in minimum voltage up to 5.45% of the base case. System power loss after placement of simultaneous DG and capacitor placement has been reduced the most as 70.3%. Figure 2 shows the obtained voltage profile for the different cases, i.e., without DG or capacitor, with one DG, with one capacitor, and simultaneous DG and capacitor.

Table 2 provides the comparison summary of obtained results from MVO with other recently reported results, i.e., analytical approach, PSO, hybrid PSO for 33-bus system. It is clearly observed from the table that the constraint optimization objective function is providing the best minimized value of the objective function by implementation of MVO algorithm. The highest reduction in objective function (minimization of power loss) with MVO has been found the best minimum for minimization constrained optimization problem. This minimization has been found for one DG allocation as 47.32%, whereas it has been found 28.241% for one Qc placement.

Figure 3 shows the convergence characteristics for single DG placement. These results show that it has been converged to eight iterations. Figures 4 and 5 illustrate convergence characteristics of single Qc unit placement and simultaneous DG and Qc placement, respectively. For single DG placement, the characteristics have

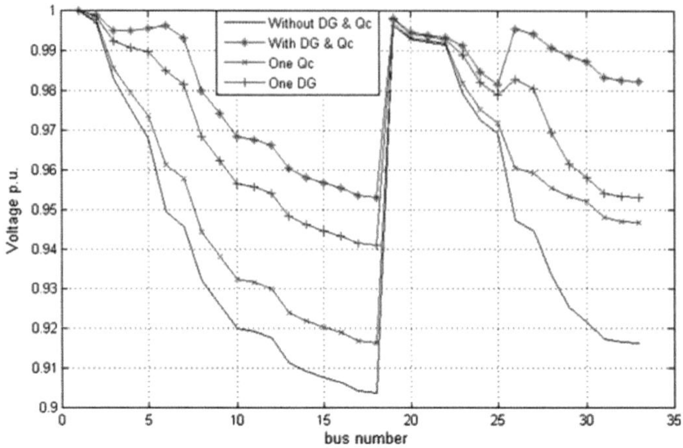


**Fig. 2** Without DGs voltage profile for 33-bus system

been converged to nine iterations, while for simultaneous DG and Qc placement the characteristics have been converged to 18 iterations.

## 5 Conclusion

In this paper, a population-based algorithm, known as MVO algorithm, has been implemented for DG and capacitor allocation in 33-bus radial distribution system. The base case power loss and other parameters are computed using BFS power flow method. After calculation of base case system parameters (i.e. real power loss, system voltage profile etc.), the MVO algorithm has been applied on load buses for optimal capacitor and DG allocation. Presence of simultaneous DG and Qc unit in distribution system causes best improvement in the voltage profile and minimization in real power losses. The comparison of obtained results from MVO has been compared with other recently reported results, i.e., analytical approach, PSO, hybrid PSO, to demonstrate the effectiveness of algorithm.

**Table 2** Comparison of obtained result

| Case | Parameters | | Analytical approach [16] | PSO [14] | Hybrid PSO [17] | MVO (Implemented) |
|---|---|---|---|---|---|---|
| No DG and Qc (Base Case) | Power loss (kW) | | 213.3 | 211 | 211 | **210.9856** |
| | Minimum voltage in p.u (at bus number) | | 0.9065 (18) | 0.9038 (18) | 0.9038 (18) | **0.9038 (18)** |
| Only DG | Location (at bus no.) and size (MW) | | (18) and 1.0 MW | (6) and 3.15 MW | (6) and 2.49 MW | **(6) and 2.5 MW** |
| | Power loss (kW) | | 142.34 | 115.29 | 111.17 | **111.1445044** |
| | Minimum voltage in p.u (at bus number) | | 0.9311(33) | 0.9506 (18) | 0.9401 (18) | **0.9410820 (18)** |
| | Loss reduction (%) | | 33.29 | 45.26 | 47.31 | **47.32** |
| Only Qc | Location (at bus no.) and size (MVAr) | | (33) and 1.0 MVAr | (30) and 1.23 MVAr | (30) and 1.23 MVAr | **(30) and 1.25799731** |
| | Power loss (kW) | | 164.6 | 151.41 | 151.41 | **151.3809901** |
| | Minimum voltage in p.u (at bus number) | | 0.91654 (18) | 0.92 (18) | 0.92 (18) | **0.9164816 (18)** |
| | Loss reduction (%) | | 22.83 | 28.24 | 28.24 | **28.241** |
| Both DG and Qc simultaneously | Locations (at bus no.) and size | DG | (18) and 1.0 MW | (6) and 2.483 MW | (6) and 2.483 MW | Both DG and Qc simultaneously |
| | | Capacitor | (33) and 1.0 MVAr | (30) and 1.223 MVAr | (30) and 1.223 MVAr | |
| | Power loss (kW) | | 96.70 | 58.51 | 58.45 | **58.4405230** |
| | Minimum Voltage in p.u (at bus number) | | 0.96003 (30) | – | 0.9570 (18) | **0.96111 (18)** |
| | Loss reduction (%) | | 54.66 | 72.27 | 72.29 | **72.30** |

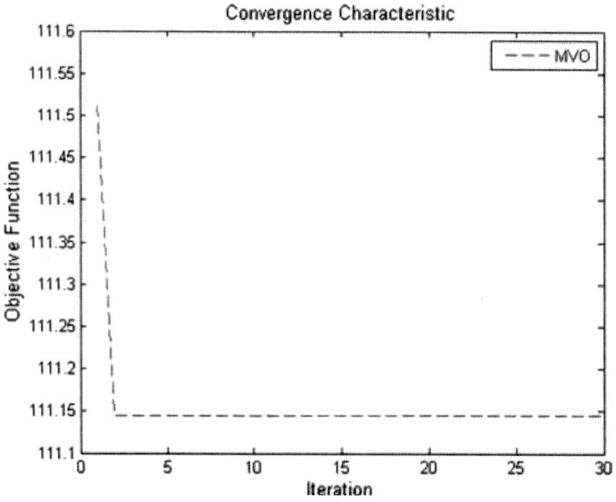


**Fig. 3** Convergence characteristics for one DG

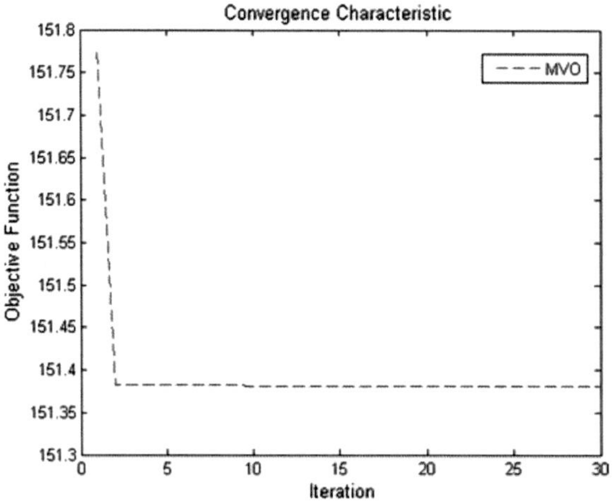


**Fig. 4** Convergence characteristics for one Qc

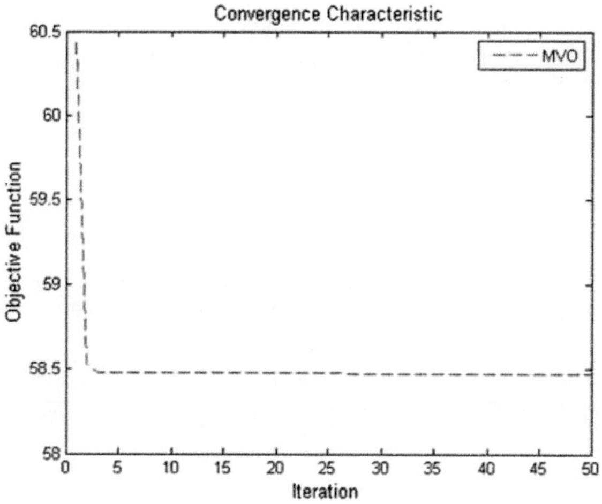


**Fig. 5** Convergence characteristics for simultaneous DG and Qc

# References


1. Yadav A, Srivastava L (2014) Optimal placement of distributed generation: an overview and key issues. In: International conference on power, signals, controls and computation, Thrissur (January, 2014)
2. Murty VVSN, Kumar A (2013) Comparison of optimal DG allocation methods in radial distribution systems based on sensitivity approaches. Int J Electr Power Energy Syst 53:246–256
3. Zeinalzadeh A, Mohammadi Y, Moradi MH (2015) Otpimal multi objective placement and sizing of multiple DGs and shunt capacitor banks simultaneously considering load uncertainty via MOPSO. Electr Power Energy Syst 67:336–349
4. Kowsalya M, Mohamed Imran A (2014) Optimal distribution generation and capacitor placement in power distribution networks for power loss minimization. In: International conference on advances in electrical engineering, Vellore
5. El-Fergany A (2015) Optimal allocation of multi-type distributed generators using baktracking search optimization algorithm. Electr Power Energy Syst 64:1197–1205
6. Rahmani-andebili M (2016) Simultaneous placement of DG and capacitor in distribution network. Electr Power Syst Res 131:1–10
7. Das A, Srivastava L (2017) Optimal placement and sizing of distribution generation units for power loss reduction using moth-flame optimization algorithm. In: International conference on intelligent computing, instrumentation and control technologies (July, 2017)
8. Biswas PP, Mallipeddi R, Suganthan PN, Amaratunga GAJ (2017) A multiobjective approach for optimal placement and sizing of distributed generation and capacitors in distribution network. Appl Soft Comput 60:268–280 (November 2017)
9. Mohamed EA, Mohamed A-AA, Mitani Y (2018) Hybrid GMSA for optimal placement and sizing of distributed generation and shunt capacitors. J Eng Sci Rev 11:55–65
10. Kola S, Jayabarathi T (2019) Optimal placement of renewable distribution generation and capacitor banks in distribution systems using salp swarm algorithm. Int J Renew Energy Res 9(1):96–107
11. Das S, Das D, Patra A (2019) Operation of distribution network with optimal placement and sizing of dispatchable DGs and shunt capacitors. Renew Sustain Energy Rev 113
12. Dehghani M, Montazeri Z, Malik OP (2020) Optimal sizing and placement of capacitor banks and distributed generation in distribution system using spring search algorithm. Int J Emerg Electr Power Syst 21(1):1–7

13. Mirjalili S, Mirjalili SM, Hatamlou A (2016) Multi-verse optimizer: a nature-inspired algorithm for global optimization. Neural Comput Appl 27(2):495–513
14. Kansal S, Kumar V, Tyagi B (2013) Optimal placement of different type of DG sources in distribution networks. Int J Electr Power Energy Syst 53:752–760
15. Ravi Teja B, Murthy VVSN, Kumar A (2016) An efficient and simple load flow approach for radial and meshed distribution networks. Int J Grid Distrib Comput 9(2):85–102
16. Gopiya Naik S, Khatod DK, Sharma MP (2013) Optimal allocation of combined DG and capacitor for real power loss minimization in distribution networks. Electr Power Energy Syst 53:967–973
17. Kansal S, Kumar V, Tyagi B (2016) Hybrid approach for optimal placement of multiple DGs of multiple types. Int J Electr Power Energy Syst 75:226–235

# Chapter 8
# Analysis of Phase Error in Rotman Lens Design with Different Beam Angles

**Deep Kishore Parsediya and Pramod Kumar Singhal**

## 1 Introduction

Microwave lenses were introduced very earlier for multiple beam-forming applications [1]. Based on their principle, Rotman lens design was introduced for planer array feed [2]. The tri-focal point-based beam-forming Rotman lens was introduced in 1963 [2]. This design has been the benchmark for all further research due to its simplicity, accuracy and reliability. Recently, Rotman lens designs have also been proposed for various 5G applications [3]. Microstrip model of Rotman lens is widely demanded due to its low profile, small size, easy fabrication and low phase error benefits [4]. Such lens are first considered for air and then dielectric region is scaled by inverse root of dielectric constant [5]. The geometry of lens includes the beam ports and array ports. Former is basically used to insert the beam, while latter mainly encompasses multiple antennas for multi-directional transmission. However, the impedance matching is generally performed by introducing dummy ports in the upper and lower side of lens geometry [6].

During the analysis of these lenses, the important design parameter that needs to be considered is phase error, which is mainly related to the array antenna aperture [7]. The lens design should be optimized according to the minimum phase errors. Many methods were proposed to minimize the phase errors, like introducing elliptical beam contour [7], non-focal design [8], by optimized on-axis, off-axis and beam angle ratio [9], and by optimized length of transmission lines [10, 11]. In this paper, an analysis of phase error has been presented at different beam angles for different values of normalized on-axis focal length (parameter 'g').

D. K. Parsediya (✉) · P. K. Singhal
Madhav Institute of Technology & Science, Gwalior, Madhya Pradesh, India
e-mail: parsediyadeep@mitsgwalior.in

© The Author(s), under exclusive license to Springer Nature Singapore Pte Ltd. 2022

H. M. Dubey et al. (eds.), *Artificial Intelligence and Sustainable Computing*,
Algorithms for Intelligent Systems, https://doi.org/10.1007/978-981-16-1220-6_8

## 2 Rotman Lens Arrangement

The geometry of Rotman lens mainly consists of beam contour and array contour. Both of these contours encompass the beam ports and array ports, respectively [6]. Former is used to energize lens design. Hence the incoming signals are fed to lens geometry via these beam ports. However, the latter are connected to different antennas for forward transmissions. Figure 1 depicts the basic arrangement required for Rotman lens design geometry.

From [2], the tri-focal point-based Rotman lens (having two off-axis focal points as $F_1$, $F_2$ and one on-axis focal point $G_1$) will have the geometry of beam and array contour as per Fig. 1. This can be achieved by following set of equations [2]:

$$F_1P + W + N\sin\alpha = F + W_o \tag{1}$$

$$G_1P + W = G + W_o \tag{2}$$

$$F_2P + W - N\sin\alpha = F + W_o \tag{3}$$

where 'P' is a point on array port; 'W' is length of transmission line; '$W_0$' is length of line passing through center point (origin), F and G are normalized off-axis and on-axis focal length; and N is lens aperture. By [2, 3], the relationship between on-axis parameter (g) and beam angle (α) can be expressed as:

$$g = 1 + \alpha^2/2 \tag{4}$$

where 'g' is the normalized on-axis focal length and 'α' is beam angle.

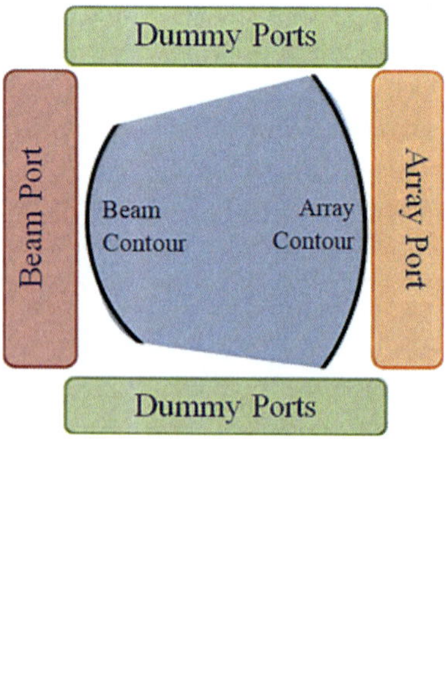


**Fig. 1** Multiple beam-forming Rotman lens arrangements

## 3 Simulation Results

The geometry of Rotman lens design for different beam angle is simulated by MATLAB software. For simulation, all the design parameters are normalized with the focal length ($f$). Figure 2 exhibits the simulated Rotman lens beams geometries at different beam angle ($\boldsymbol{\alpha} = 30°$, $35°$ and $40°$). For each beam angle, the lens is simulated for different values of parameter 'g' to get optimum phase error.

It has been observed from Fig. 2 that the increment in beam angle ($\alpha$) will result in the beam and array ports more close. For example, at ($\alpha = 30°$) the beam port arc and array port arc are slightly away as compared to the beam angle ($\alpha = 40°$). However, the increment in the value of parameter 'g' will bring the beam and array port arc closer to each other. It means Fig. 2 also depicts the Rotman lens beam-array contour geometry for different values of parameter 'g'.

Figures 3, 4 and 5 exhibit the phase error variations with respect to eta ($\pm\eta$) for different beam angles $\boldsymbol{\alpha} = 30°, 35°$ and $40°$, respectively. For better analysis of phase error with respect to these beam angles, the scan angle up to $\boldsymbol{\theta} = 35°, 40°$ and $45°$ has been considered. The graphs encompass phase error variation for $\eta = \pm 0.8$ with different values of parameter 'g', i.e., $g = 1.05, 1.10, 1.15$ and $1.20$, respectively.

Table 1 exhibits the values of phase error for beam angles $\boldsymbol{\alpha} = 30°, 35°$ and $40°$. These errors are calculated for different values of parameter 'g'. By Table 1 it can be concluded that for higher beam angle the phase error can be minimized only with slightly high value of parameter 'g' which is mainly considered during lens designing.

Table 2 exhibits different values of parameter 'g' for beam angles $\boldsymbol{\alpha} = 30°, 35°$ and $40°$ obtained from the conventional relationship between parameter 'g' and angle '$\alpha$' [2]. Optimized values of parameter 'g' for all such beam angles are also achieved by simulating the Rotman lens geometry using software MATLAB with multiple iterations. Table 2 also depicted the optimized values of parameter 'g' for different beam angles.

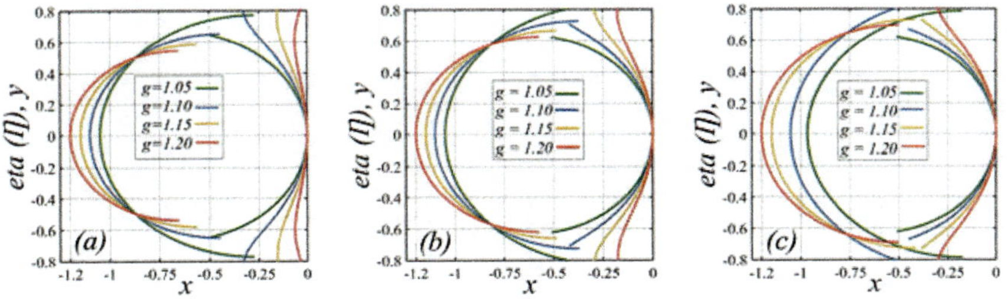


**Fig. 2** Lens geometry for different values of parameter 'g' **a** with beam angle $\alpha = 30°$, **b** with beam angle $\alpha = 35°$ and **c** with beam angle $\alpha = 40°$

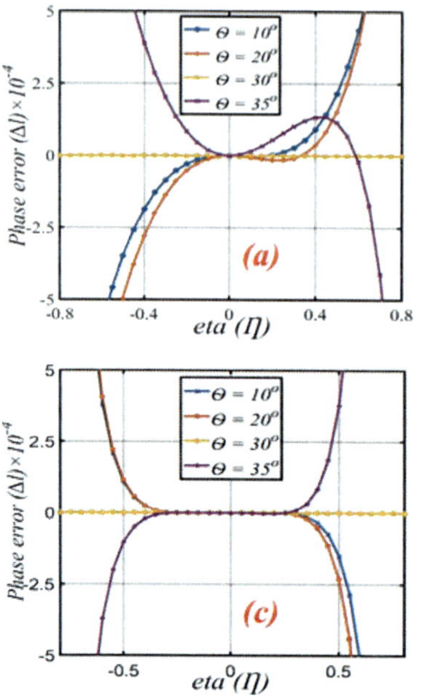


**Fig. 3** Path length error for lens with beam angle 30° at **a** g = 1.05, **b** g = 1.10, **c** g = 1.15 and **d** g = 1.20

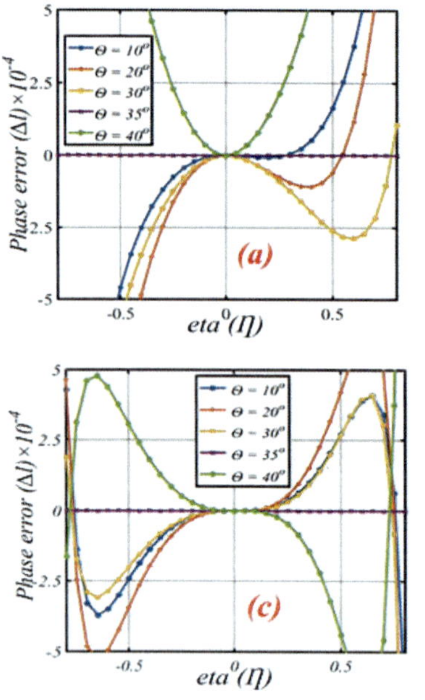


**Fig. 4** Path length error for lens with beam angle 35° at **a** g = 1.05, **b** g = 1.10, **c** g = 1.15 and **d** g = 1.20

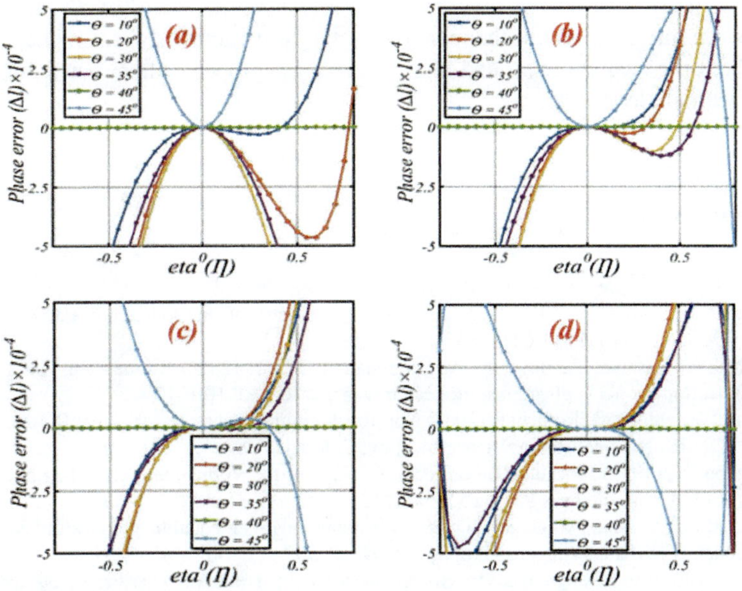


**Fig. 5** Path length error for lens with beam angle 40° at **a** g = 1.05, **b** g = 1.10, **c** g = 1.15 and **d** g = 1.20

**Table 1** Comparison of phase error at different beam angles

| Beam angle | Phase error $(\Delta l)$for $g = 1.05$ | Phase error $(\Delta l)$for $g = 1.10$ | Phase error $(\Delta l)$for $g = 1.15$ | Phase error $(\Delta l)$ for $g = 1.20$ |
|---|---|---|---|---|
| $\alpha = 30°$ | 6.6 | 3.7 | 3.8 | 25.8 |
| $\alpha = 35°$ | 9.9 | 6.6 | 4.4 | 1.6 |
| $\alpha = 40°$ | 19.3 | 10.6 | 7.2 | 5.4 |

**Table 2** Comparison of parameter 'g' at different beam angles

| | Beam angle $\alpha = 30°$ | Beam angle $\alpha = 35°$ | Beam angle $\alpha = 40°$ |
|---|---|---|---|
| Using Eq. (4) | 1.137 | 1.1864 | 1.2435 |
| Simulated | 1.139 | 1.1898 | 1.2673 |

## 4 Conclusion

In this paper an analysis of phase error for Rotman lens with three different beam angles (30°, 35° and 40°) has been presented. Different values of parameter 'g' offer different phase error in all cases. Based on simulation, it has been found that the phase error can be optimized for higher beam angle by specifically increasing the value of parameter 'g'. With the increased value of parameter 'g' for higher beam angle, both the beam contour and array contour are properly aligned which offers minimized path length error.

## References


1. Ruze J (1950) Wide-angle metal-plate optics. Proc IRE 38:53–59
2. Rotman W, Turner RF (1963) Wide-angle microwave lens for line source applications. IEEE Trans Antennas Propag 11:623–632
3. Ershadi S, Keshtkar A, Bayat A, Abdelrahman A, Xin H (2018) Rotman lens design and optimization for 5G applications. Int J Microw Wirel Technol 10:1–10
4. Dong J, Zaghloul AI, Rotman R (2008) Non-focal minimum phase-error planar Rotman lens. USNC/URSI National Radio Science Meeting, Colorado, USA
5. Gagnon DR (1989) Procedure for correct refocusing of the Rotman lens according to Snell's law. IEEE Trans Antennas Propag 37:390–392
6. Singhal PK, Sharma PC, Gupta RD (2003) Rotman lens with equal height of array and feed contours. IEEE Trans Antennas Propag 51:2048–2056
7. Hansen RC (1991) Design trades for Rotman lenses. IEEE Trans Antennas Propag 39:464–472
8. Dong J, Zaghloul AI, Rotman R (2010) Phase-error performance of multi-focal and non-focal two-dimensional Rotman lens designs. IET Microw Antennas Propag 4:2097–2103
9. Park J, Park D (2013) Phase error minimization by refocusing Rotman lens. In: Proceedings of the Asia-Pacific microwave conference, Seoul, South Korea.
10. Rajabalian M, Zakeri B (2012) An implemented non-focal Rotman lens. In: Radar conference (EuRAD), Paris, France, September 9–11.
11. Rajabalian M, Zakeri B (2015) Optimisation and implementation for a non-focal Rotman lens design. IET Microw Antennas Propag 9:982–987

# Chapter 9
# Optimal Integration of Multi-type DG in Radial Distribution Network

**Shradha Singh Parihar and Nitin Malik**

## 1 Introduction

A distributed generation (DG) is a small decentralized generation of power generally located near the load center to fulfill the local demand. The usage of DG resources in the distribution network (N/w) increases rapidly due to the economic concerns of the utilities. The optimal integration of DG enhances the voltage profile and minimizes N/w power losses. For cleaner power production, DG uses non-conventional resources like photovoltaic systems [1–3]. The optimal placement of DG enhances the network voltage magnitude and reduces the active and reactive power losses but it may produce an adverse impact if they are not integrated optimally in the N/w. Since load flow (LF) analysis is required for future planning and operation of power utilities, it becomes necessary to perform LF analysis of the distribution network including one or more distributed generators.

Authors in [4] proposed an analytical method to find DG rating operating at unity power factor (pf). Numerous numerical techniques like classical Kalman filter algorithm [5] and mixed integer nonlinear programming [6] are used to optimally integrate DG in the distribution N/w. In [7], expression for voltage stability margin has been developed to induct DG into the distribution network optimally in the presence of various load levels. An analytical technique based on active and reactive component of the line currents is proposed in [8] to find the best DG allocation for power loss minimization. In [9], a power stability index (PSI) is demonstrated for determining the optimal site of DG. Authors in [10] demonstrated the impact of different types of DGs and load growth on the performance of the distribution network. A novel voltage stability index (VSI) is developed in [11] to allocate DG units optimally in the N/w. Authors in [12] presented an analytical technique for integrating DG in the radial distribution network (RDN) for maximum benefit. Evolutionary algorithms

S. S. Parihar · N. Malik (✉)
The NorthCap University, Sector 23A, Gurugram, India

© The Author(s), under exclusive license to Springer Nature Singapore Pte Ltd. 2022

H. M. Dubey et al. (eds.), *Artificial Intelligence and Sustainable Computing*,
Algorithms for Intelligent Systems, https://doi.org/10.1007/978-981-16-1220-6_9

like genetic algorithm [13], artificial bee colony [14], teaching–learning-based optimization algorithm (TLBO), quasi-oppositional TLBO [15] and chaos-embedded SOS algorithm [16] have been implemented to solve DG allocation problem in the RDN. In [17], a bacterial foraging approach is presented to determine the location and rating of the DG. A bio-inspired computing algorithm is well explained in [18].

In this article, a new fitness index has been presented for finding the optimal location of Type I and IV DG followed by an analytical approach for finding the optimum DG rating. The selection of optimal location and rating of DG are independent of each other. The weighted-sum approach is utilized to convert the tri-objective constrained problem into a single-objective constrained problem. The network load is considered as a constant power type and examined on the 12-node network. The multi-objective function has been optimized considering various constraints. The results attained from BR matrix for without (base case) and with DG are further compared to the existing results to reveal the efficiency of the presented technique. The magnitude of voltage profile, annual cost of energy loss (ACEL), DG cost, annual energy loss savings (AELS) and N/w power losses are also analyzed.

## 2 Mathematical Modeling

The single line diagram (SLD) of a sample distribution network is displayed in Fig. 1. The electrical equivalent diagram of a branch connected between bus a and b of sample distribution N/w is given in Fig. 2.

From Fig. 2, we get

$$S_b = P_b + jQ_b = V_b * I_b^* \tag{1}$$

**Fig. 1** SLD of sample distribution N/w

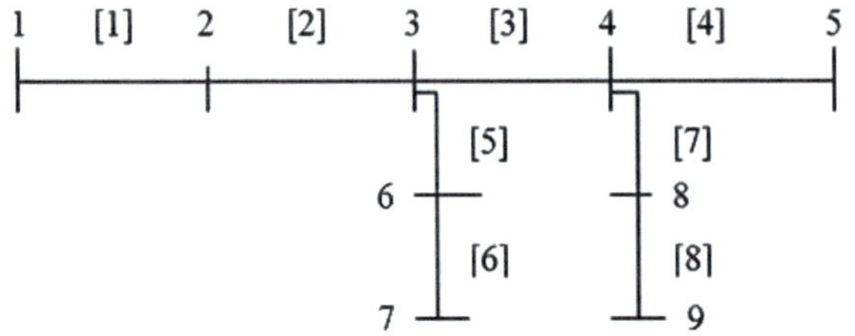


**Fig. 2** Electrical equivalent of one branch of Fig. 1

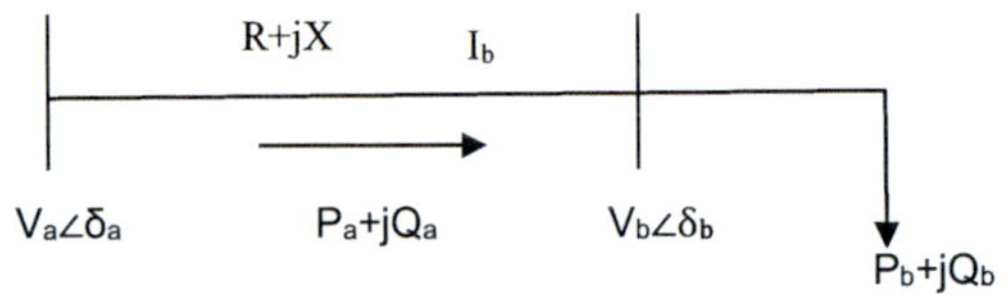

where $V_b$ and $V_a$ represent the receiving- and sending-end bus voltage, respectively. $I_b$ represents branch current. $P_b$ and $Q_b$ show the total active and reactive power load, respectively, fed through node b. The magnitude of injected current at node b is

$$I_b = \frac{P_b - jQ_b}{V_b{}^*} \tag{2}$$

Taking conjugate of both sides of (2), we get

$$S_b^* = P_b - jQ_b = V_b^* * I_b \tag{3}$$

The magnitude of $V_b$ is calculated using (1) and (3) and mentioned in (4) as

$$V_b = [[(RP_b + XQ_b - \frac{1}{2} * V_a^2)^2 - (R^2 + X^2)(P_b^2 + Q_b^2)]^{1/2} - (RP_b + XQ_b - \frac{1}{2} * V_a^2)]^{1/2} \tag{4}$$

From (2) and (3), we get

$$I_b^2 = \frac{(P_b{}^2 + Q_b^2)}{V_b{}^2} \tag{5}$$

The active ($P_{loss}$) and reactive power loss ($Q_{loss}$) of a branch linked between node a and b is

$$P_{loss} = I_b^2 * R = \frac{(P_b{}^2 + Q_b^2)}{V_b{}^2} * R \tag{6}$$

$$Q_{loss} = I_b^2 * X = \frac{(P_b{}^2 + Q_b^2)}{V_b{}^2} * X \tag{7}$$

## 3 DG Modeling

The small DG resources generally operates in constant power mode, i.e., the generation node is assumed to model as a constant negative PQ load. The integration of DG units in the RDN creates many positive impacts on operating characteristics of the network such as voltage profile enhancement, network loss reduction, annual energy savings, etc. The total net load will get reduced by the amount of power penetrated by the DG. The DG unit is taken as a generating source with constant P and Q output based on the power delivering capability of the DGs and are classified as [19]

(a) Type I (penetrates active power only, pf = 1)

Example: fuel cell, battery and photovoltaic

(b) Type II (penetrates reactive power only, pf = 0)
Example: synchronous condenser

(c) Type III DG (penetrates active power and consumes reactive power)
Example: induction generators

(d) Type IV (penetrates active and reactive power, non-unity pf)
Example: synchronous generator, wind power and cogeneration

The N/w power loss minimization and N/w bus voltage profile improvement attained from Type III DG is worst among all types of DGs as also concluded in [20]. The equivalent active and reactive load at bth node where the DG is placed can be written as (8) and (9), respectively.

$$P_{eq}(b) = P_b - P_{DGb} \tag{8}$$

$$Q_{eq}(b) = Q_b - Q_{DGb} \tag{9}$$

where $Q_{DGb}$ and $P_{DGb}$ represent the reactive and active power penetrated by the DG placed at bus b, respectively. The magnitude of reactive power injected by Type IV DG at bth bus is

$$Q_{DGb} = P_{DGb} \times tan(cos^{-1}((pf)_{DG})) \tag{10}$$

where $(pf)_{DG}$ represents the DG pf.

## 4 Solution Methodology to Site and Size DG Optimally

Step I: Read the network data at various nodes. Initialize the voltage at each node including the source node to 1.0 p.u. Set network power losses to zero.

Step II: Find the nodes beyond every branch and the total nodes number ($N_b$) by articulating BR matrix as in [21]. Compute N/w power fed at the end of the branch and determine the node's voltage using (4) starting with the node closest to the source node.

Step III: Repeat steps I and II till the convergence of the algorithm is not attained. A node voltage tolerance of 0.0001 p.u in successive iterations is considered as convergence criteria.

Step IV: Compute branch current, active and reactive power losses in every branch using (5), (6) and (7), respectively.

# 5 Fitness Index for Optimal Placement of DG

## 5.1 Voltage Stability Index

The VSI [22] of node *i* can be determined as

$$\begin{aligned} VSI = & \left\{(V_a)^4\right\} - 4 * \{P(b) * X - Q(b) * R\}^2 \\ & - 4 * \{P(b) * R + Q(b) * X\} * \left\{(V_a)^2\right\} \end{aligned} \quad (11)$$

The node having lowest VSI value is subjected to voltage collapse and is selected as the optimal node for the DG penetration.

## 5.2 Active Power Loss Index (ALI)

ALI is an amount of total active power loss (APL) occurred in an RDN and can be represented as

$$MinimizeALI = \sum_{\mathrm{b=1}}^{\mathrm{Nb}} {I_b}^2 R \quad (12)$$

## 5.3 Voltage Deviation Index (VDI)

The VDI [23] is the deviation of the node voltages from rated voltage and can be given as

$$MinimizeVDI = abs(\sum_{i=2}^{Nb} \left( \frac{V_{rated} - V_b}{V_{rated}} \right)) \quad (13)$$

where $V_{\mathrm{rated}}$ is the rated node voltage value (1 p.u).

## 5.4 Fitness Index (FI)

For the optimal penetration of DG, the three objectives are considered using a weighted-sum approach to determine FI

$$FI = (w1 * VSI) + (w2 * VDI) + (w3 * ALI) \tag{14}$$

$$\sum_{k=1}^{3} \mathrm{w(k)} = 1; \qquad \mathrm{w(k)} \in (0, 1) \tag{15}$$

where w1, w2 and w3 represent the weights for VSI, VDI and ALI, respectively. The values selected for w1 is 0.2, w2 is 0.3 and w3 is 0.5, as per the relative importance. The value of FI is determined for every line and lies between 0 and 1. If the value of the proposed FI is nearer to zero, then the network is unstable and vice-versa.

## *5.5 Operational Limits*

The following N/w limits are satisfied for the evaluation purpose:

(a) Power balance limits:

$$P_{GN} = P_D + P_{loss} \tag{16}$$

$$Q_{GN} = Q_D + Q_{loss} \tag{17}$$

where $Q_{GN}$ and $P_{GN}$ are the reactive and active power generated and $Q_D$ and $P_D$ represent the network reactive and active load demand, respectively.

(b) Voltage limits:

$$0.95p.u \leq V_b \leq 1.05p.u \tag{18}$$

(c) Thermal limits:

$$I_b \leq I_{thermal} \tag{19}$$

where $I_{thermal}$ represents the thermal limit of the $I_b$.

(d) DG power generation limits:

$$0 \leq P_{DGi} \leq \sum P_{Load} \tag{20}$$

$$0 \leq S_{DGi} \leq \sum S_{Load} \tag{21}$$

where $S_{DGi}$ represents apparent power generation. $\sum P_{Load}$ and $\sum S_{Load}$ is the network total active and apparent power load connected, respectively.

## 6 Methodology for Optimal Integration of DG

Step I: Solve the load flow problem for the base case to calculate the node voltages, line current and network power losses.

Step II: Determine VSI, ALI, VDI from (11), (12) and (13), respectively. Using weighted-sum approach the value of FI has been calculated as mentioned in (14). If branch linked to nodes a and b are found to have the lowest value of FI, then bth node is nominated as an optimal node for the DG placement. For optimal sizing, an analytical approach [9] has been applied where the active power load at an optimal bus is varied up to 100% of total N/w load at a step size of 0.01 subject to all limits from (16) to (21).

Step III: The rating of DG at which lowest total real power loss is attained is nominated as optimal DG size.

Step IV: Stop if limits are not satisfied and rerun load flow to calculate network power losses.

## 7 Cost of DG and Annual Energy Losses

The ACEL is represented in (22) as

$$ACEL = \left(\sum_{i=2}^{N_b} P_{loss} * T * E\right)\$ \tag{22}$$

where T and E are annual time duration (8760 h) and energy cost (0.06 \$/kWh), respectively.

The DG cost is expressed in terms of active power generation

$$C(P_{DG}) = a * P_{SPV}^{\,2} + b * P_{SPV} + c \quad \$/hr \tag{23}$$

Cost coefficients of DG are assumed as [24]: a = 0, b = 20 and c = 0.25.

## 8 Results and Discussion

The presented approach for the optimal penetration of Type I and Type IV DG in RDN is simulated on the 12-node network. Line and the nominal load data for 12-node RDN have been taken from [25]. The 12-node network is a radial distribution network having 12 load points with a total load of (0.435 + j0.405) MVA. The base MVA and kV for the test network are chosen as 100 and 11, respectively. The results attained from the presented approach for voltage profile, N/w power losses, ACEL and AELS are compared with the existing results to illustrate the efficiency of the approach. A program to attain results for the test network is written in MATLAB 7.1 and simulated on microprocessor @ 2.40 GHz and 4 GB RAM.

The FI value of each branch of the 12-node network is illustrated in Fig. 3. The branch connecting nodes 8 and 9 is found out to have the lowest value of 0.1564. Hence, node 9 is nominated as the optimal location to locate different types of DG in the N/w. The penetration of DG shows a noteworthy effect on network power losses. The total real power losses variation with Type I DG size is given in Fig. 4 and demonstrates a parabolic curve. Hence, the curve has only one optimal value before and above which the power losses increase. The accuracy of the selection of DG rating is dependent on the step size selected.

The optimal DG size selected for Type I and Type IV DG is 239 kW and 316.78 kVA, respectively. The results obtained after optimal DG allocation in the distribution N/w are mentioned in Table 1, which concludes that the placement of DG gives better results than the base case in all respect.

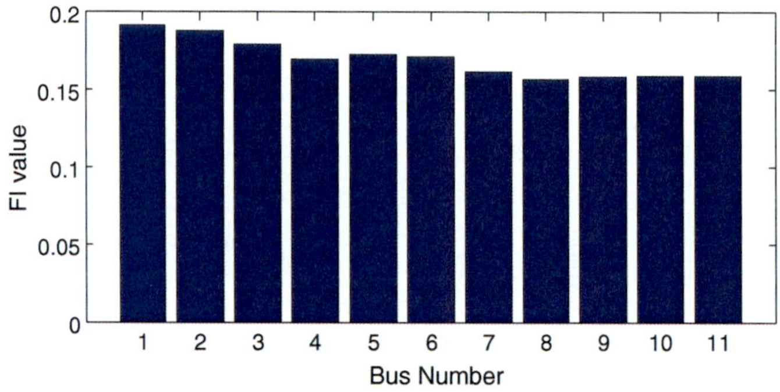


**Fig. 3** Fitness index value for each branch in 12-node network

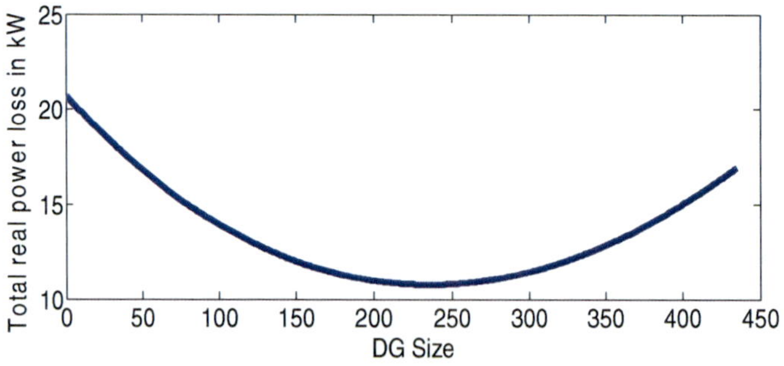


**Fig. 4** Variation in total APL with different DG size at a step size of 0.01

**Table 1** Simulated results of 12-node network for Type I and IV DG

| | Base case | Type I DG | Type IV DG |
|---|---|---|---|
| Optimal location | – | 9 | 9 |
| Optimal DG Size | – | 239 kW | 316.78 kVA |
| Minimum node voltage ($V_{min}$) in p.u @node (% Improvement in voltage) | 0.9444@12 | 0.9851@7 (4.3%) | 0.9905@7 (4.8%) |
| APL (kW) | 20.6891 | 10.5487 | 3.1915 |
| APL reduction in kW (Reduction in APL in %) | – | 10.14 (49.01%) | 17.4976 (84.57%) |
| Reactive power loss (kVAr) | 8.0319 | 4.0262 | 1.1148 |
| Cost of PDG ($/hr) | – | 5.03 | – |
| Annual cost of energy loss ($) | 10,874.19 | 5544.39 | 1766.45 |
| Annual energy loss savings ($) | – | 5329.8 | 9107.74 |

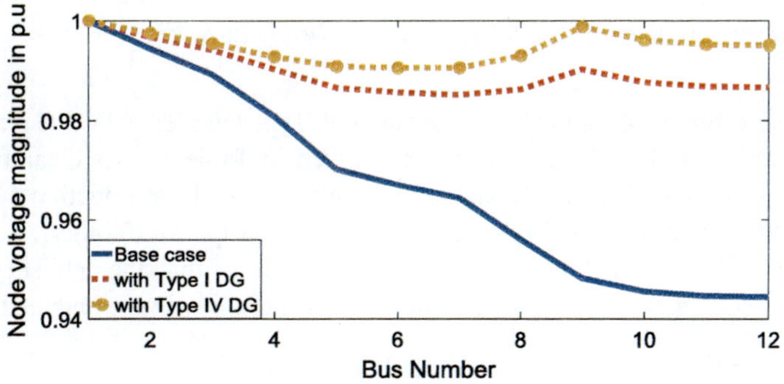


**Fig. 5** Effect of Type I and Type IV DG on voltage profile of the 12-node network

## 8.1 Impact of DG on N/w Voltage Profile

The voltage magnitude at every bus of the 12-node network without and with Type I and Type IV DG is examined and displayed in Fig. 5. The $V_{min}$ is upgraded from 0.9444 p.u at node 12 to 0.9851 at node 7 for Type I DG and 0.9905 p.u at node 7 for Type IV DG. This resulted in the % voltage enhancement of 4.3 and 4.8% for Type I DG and Type IV DG, respectively. It is also found that for both the types of DG, the approach meets all the specified limits.

## 8.2 Impact of DG on N/w Power Losses

The APL in the 12-bus N/w without placing DG was found to be 20.6891 kW. After the optimal placement of Type I DG, the N/w loss reduces to 10.5487 kW, whereas the reactive power loss of the N/w is reduced from 8.0319 kVAr to 4.0262 kVAr

(From Table 1). For Type IV DG, the active and reactive power loss minimizes to 3.1915 kW and 1.1148 kVAr, respectively (from Table 1). Hence, % reduction of 49.01 and 84.57% in APL has been achieved after penetration of Type I and Type IV DG, respectively.

## 8.3 Impact of DG on AELS

The DG placement in the distribution N/w minimizes the annual cost of energy loss from $10,874.19 to $5544.39 and $1766.45 for Type I and Type IV, respectively. Therefore, the annual energy loss savings of $5329.8 and $9107.74 is achieved considering Type I and Type IV DG, respectively (from Table 1).

## 8.4 Comparison of Results for 12-Node Network

The test results obtained for the base case and with Type I and Type IV DG placement are compared with the previous results and given in Table 2. It is clear from the obtained results that Type I DG yields maximum power loss reduction (49.01%) with an annual saving of $5329.8 and comes out to be superior to others. Similarly, the % reduction in APL achieved with Type IV DG is more than the published result. The comparison shows that the magnitude of the minimum voltage obtained at node

**Table 2** Performance comparison of 12-node network for Type I and Type IV DG

| | DG siting method | DG size | APL (kW) | % APL reduction | $V_{min}$ (p.u) | Cost of PDG ($/hr) | ACEL ($) | AELS ($) |
|---|---|---|---|---|---|---|---|---|
| Base case | Proposed method | – | 20.6891 | – | 0.9444@12 | – | 10,874.19 | – |
| | [24] | – | 20.7135 | – | 0.9434@12 | – | 10,887.03 | – |
| Type I DG (kW) | Proposed method | 239 | 10.549 | 49.01 | 0.9851@7 | 5.03 | 5544.39 | 5329.8 |
| | VSI [24] | 235 | 10.774 | 48.00 | 0.9835@7 | 4.95 | 5662.79 | 5224.2 |
| | PSI [9] | 234.9 | 10.774 | 47.98 | 0.9835@7 | 4.948 | 5662.80 | 5224.2 |
| | Power loss sensitivity [26] | 231.69 | 10.776 | 47.98 | 0.9831@7 | 4.884 | 5664.00 | 5223.0 |
| Type IV DG (kVA) | Proposed method | 316.78 | 3.1915 | 84.57 | 0.9905@7 | – | 1766.45 | 9107.74 |
| | VSI [24] | 305 | 4.4937 | 78.00 | 0.9913@6 | – | 2361.89 | 8525.10 |

7 from the presented method is better than the published results for Type I DG. The annual energy loss savings for Type I and IV DG show remarkable benefits from the presented approach.

## 9 Conclusion

In this work, a detailed strategy is projected to optimally allocate Type I and IV DG in the RDN to reduce losses and to improve the voltage profile of distribution feeders. The application of the weighted-sum method transforms the tri-objective problem as a single-objective problem by merging all three indices (VSI, ALI and VDI). The lowest value of the proposed fitness index helps in identifying the location to place DG optimally. An analytical method is applied further to determine the optimum rating of the DGs. The constant power type load model is examined. The results have been simulated for DG installation in the 12-node network and found to give notable performance. The presented method is carried out to mitigate N/w power losses, enhancement in voltage profile, ACEL and AELS. Further, the results are compared and found out to be superior to the published results in all respect. It is also found that the Type IV DG penetration provides more effective results as compared to Type I DG.

## References

1. Georgilakis PS, Hatziargyriou ND (2013) Optimal distributed generation placement in power distribution networks: models, methods and future research. IEEE Trans Power Syst 28(3):3420–3428
2. Rezaei N, Haghifam MR (2008) Protection scheme for a distribution system with distributed generation using neural networks. Int J Electr Power Energy Syst 30(4):235–241
3. Hashim TJ, Mohamed A (2013) Fuzzy logic based coordinated voltage control for distribution network with distributed generations. Int J Electr Comput Energ Electron Commun Eng 7(7):369–374
4. Acharya N, Mahat P, Mithulananthan N (2006) An analytical approach for DG allocation in primary distribution network. Int J Electr Power Energy Syst 28(10):669–678
5. Soo-Hyoung L, Jung-Wook P (2009) Selection of optimal location and size of multiple distributed generations by using Kalman Filter algorithm. IEEE Trans Power Syst 24(3):1393–1400
6. Rueda-Medina AC, Franco JF, Rider MJ, Padilha-Feltrin A, Romero R (2013) A mixed integer linear programming approach for optimal type, size and allocation of distributed generation in radial distribution system. Electr Power Syst Res 97:133–143
7. Essallah S, Khedher A, Bouallegue A (2019) Integration of distributed generation in electrical grid: Optimal placement and sizing under different load conditions. Comput Electr Eng 79:1–14. https://doi.org/10.1016/j.compeleceng.2019.106461
8. Viral R, Khatod DK (2015) An analytical approach for sizing and siting of DGs in balanced radial distribution networks for loss minimization. Electr Power Energy Syst 67:191–201
9. Aman MM, Jasmon GB, Mokhlis H, Bakar AHA (2012) Optimal placement and sizing of a DG based on a new power stability index and line losses. Electr Power Energy Syst 43(1):1296–1304

10. Parihar SS, Malik N (2020) Optimal integration of multi-type DG in RDS based on novel voltage stability index with future load growth. Evol Syst. https://doi.org/10.1007/s12530-020-09356-z
11. Parihar SS, Malik N (2020) Optimal allocation of renewable DGs in a radial distribution system based on new voltage stability index. Int Trans Electr Energy Syst 30(4):e12295. https://doi.org/10.1002/2050-7038.12295
12. Naik SNG, Khatod DK, Sharma MP (2015) Analytical approach for optimal siting and sizing of distributed generation in radial distribution networks. IET Gener Transm Distrib 9(3):209–220
13. Zhao Y, An Y, Ai Q (2014) Research on size and location of distributed generation with vulnerable node identification in the active distribution network. IET Gener Transm Distrib 8(11):1801–1809
14. Abu-Mouti FS, El-Hawary ME (2011) Optimal distributed generation allocation and sizing in distribution systems using artificial bee colony algorithm. IEEE Trans Power Deliv 26(4):2090–2101
15. Sultana S, Roy PK (2014) Multi-objective quasi-oppositional teaching learning based optimization for optimal location of distributed generator in radial distribution systems. Int J Electr Power Energy Syst 63:534–545
16. Saha S, Mukherjee V (2016) Optimal placement and sizing of DGs in RDS using chaos embedded SOS algorithm. IET Gener Transm Distrib 10(14):3671–3680
17. Imran MA, Kowsalya M (2014) Optimal size and siting of multiple distributed generators in distribution system using bacterial foraging optimization. Swarm Evolut Comput 15:58–65
18. Gedeon T (2017) Bio-inspired computing tools and application: position paper. Int J Inf Technol 9(1):7–17
19. Hung DQ, Mithulananthan N, Bansal RC (2010) Analytical expression for DG allocation in primary distribution network. IEEE Trans Energy Convers 25(3):814–820
20. Pradeepa H, Ananthapadmanabha T, Sandhya RDN, Bandhavya C (2015) Optimal allocation of combined DG and capacitor units for voltage stability enhancement. Proc Technol 21:216–223. https://doi.org/10.1016/j.protcy.2015.10.091
21. Thomas MS, Ranjan R, Malik N (2012) Deterministic load flow algorithm for balanced radial AC distribution systems. In: IEEE fifth power india conference, pp1–6, 19–22 Dec 2012
22. Murthy GVK, Sivanagaraju S, Satyanarayana S, Rao BH (2015) Voltage stability enhancement of distribution system using network reconfiguration in the presence of DG. Distrib Gener Altern Energy J 30(4):37–54
23. Injeti, S.K: A Pareto optimal approach for allocation of distributed generators in radial distribution systems using improved differential search algorithm. Journal of Electrical Systems and Information Technology 5(3), 908–927 (2018).
24. Murty VVSN, Kumar A (2015) Optimal placement of DG in radial distribution systems based on new voltage stability index under load growth. Electr Power Energy Syst 69:246–256
25. Das D, Nagi HS, Kothari DP (1994) Novel method for solving radial distribution Networks. In: IEE Proc Gener Transm Distrib 141(4):291–298.
26. Murthy VVSN, Kumar A (2013) Comparison of optimal DG allocation methods in radial distribution system based on sensitivity approaches. Electr Power Energy Syst 53:450–467

# Chapter 10
# Optimization of Congo Red Dye by Iron Oxide@AC

**Nirbhay Kumar, Vinita Khandegar, and Sanigdha Acharya**

## 1 Introduction

Nowadays, treatment of dye effluent is a major challenge in textile industries to save the environment from harmful and carcinogenic effects of dyes. In this regard, Korbahti et al. [1] reported that the majority of textile industries uses about 125–150 L water to produce 1 kg of textile fiber. Approximately 90% of 125–150 L water was used in dyeing and finishing operations only. Therefore, dyeing operation generates huge quantity of wastewater that contains auxiliary chemical as well as coloring residues. As a result, these residues can affect the aquatic life [2]. Congo red (CR) dye comes in benzidine group and carcinogenic category [3]. Therefore, removal of CR from effluent is a serious concern [4]. Various industries used different processes for removal of dyes from wastewater, which include membrane filtration, adsorption, co-precipitation, ultrafiltration, nano filtration, reverse osmosis, activated carbon, bio-adsorbents, and low-cost adsorbents [5]. But still effective and economic technique is required for removal of toxic dyes. One of the effective and economic techniques is adsorption which attracted attention due to its simple operation and regeneration of adsorbent. Activated carbon (AC) is extensively used in many processes because of its microporous structure and high surface area [6]. Combining AC with nanoparticles may make it more cost-effective, because nanosize particles are cost-effective and may be reused again and again [7]. Therefore, in the present study, AC was modified by coating the iron oxide magnetic particles for removal of CR from synthetic dye

N. Kumar · V. Khandegar (✉) · S. Acharya
University School of Chemical Technology, Guru Gobind Singh Indraprastha University, Dwarka, New Delhi, India
e-mail: vinita@ipu.ac.in

S. Acharya
e-mail: sanigdha@ipu.ac.in

© The Author(s), under exclusive license to Springer Nature Singapore Pte Ltd. 2022 
H. M. Dubey et al. (eds.), *Artificial Intelligence and Sustainable Computing*, Algorithms for Intelligent Systems, https://doi.org/10.1007/978-981-16-1220-6_10

solution. Optimization of contact time, concentration of dye, amount of adsorbent, and temperature is taken into account while evaluating the adsorption capacity of the system.

# 2 Materials and Methods

## 2.1 *Preparation of FAC*

A 5 mg/L solution of ferric chloride was prepared and 0.25 g of activated carbon was added in every 100 mL of prepared ferric chloride solution. The mixture was agitated at 150 rpm and centrifuged for 1 min at 2000 rpm. Residue was dried at 100 °C for 24 h, and activated carbon coated with iron oxide magnetic nanoparticles was obtained and stored at room temperature.

## 2.2 *Experiment*

Simulated CR solution of 10 mg/L was made by dissolving 10 mg CR dye in 1 L of distilled water. The batch adsorption kinetic study was performed at different conditions. A defined 200 mL of dye solution was used in adsorption process and kept in BOD incubator shaker at 150 rpm at 30 °C. Clear solution and adsorbent were separated using filter paper. The final concentration of dye in the clear solution was examined by UV–VIS spectrophotometer (Hitachi UV-2900) at wavelength 485 nm. The dye removal efficiency (DRE) was calculated via Eq. (1) and the equilibrium adsorption capacity (qe, mg/g) was calculated via Eq. (2), where $C_i$ and $C_f$ are initial and final dye concentrations in the solution (mg/L). $C_e$ is equilibrium concentrations of dye (mg/L), m = adsorbent dosage (g) and V is volume of solution (L) used in experiment.

$$\text{DRE (\%)} = \left(\frac{C_o - C_i}{C_o}\right) \times 100 \tag{1}$$

$$q_e(mg/g) = \left[\frac{(C_0 - C_e)}{m}\right] \times V \tag{2}$$

# 3 Results and Discussion

## *3.1 Effect of Contact Time*

Experiments were performed in order to remove CR from simulated solution at different contact time from 4 to 12 h. The DRE and $q_e$ at different contact time are shown in Fig. 1. It can be seen from Fig. 1 that the contact time has significant effect on the DRE and $q_e$. Further, it can be observed from Fig. 1 that DRE increases with increase in contact time. Since more time provides more contact of adsorbent in the dye solution, as a result removal efficiency is increased significantly (93–97.5%) and the adsorption capacity is also increasing with contact time. After 8 h the removal efficiency was marginally increasing since all of the adsorption sites were occupied in the given time, so 8 h was taken as optimum time for adsorption [8].

## *3.2 Effect of Adsorbent Dose*

The adsorbent amount used in adsorption process is a crucial parameter for application of adsorbent in industrial scale. Therefore, optimization of adsorbent amount is important and in the present study effect of adsorbent amount was investigated from 5 to 20 mg in 200 mL dye solution. Experiments were conducted at 8 h adsorption time, 150 rpm shaking speed, 7 pH on 10 mg/L dye concentration. Effect of adsorbent dose on DRE and $q_e$ are presented in Fig. 2. From Fig. 2 DRE rises with increase in

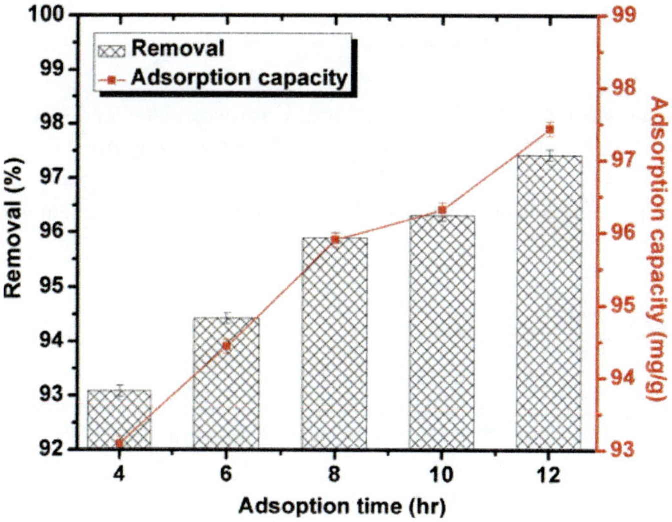


**Fig. 1** Effect of contact time

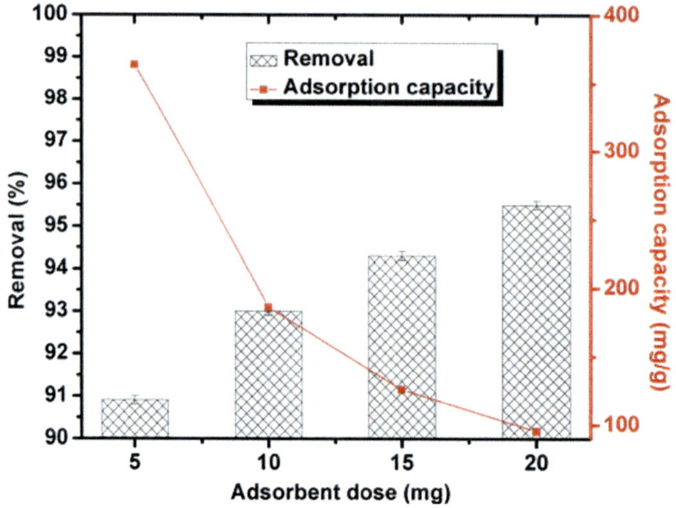


**Fig. 2** Effect of adsorbent dose

adsorbent dose (5–20 mg) and the adsorption capacity is decreasing with increase in dose since adsorption capacity is inversely proportional to the adsorbent dose, so as we are increasing the adsorbent dose reduction in adsorption capacity was observed. This may be attributed that by increasing adsorbent dose, DRE increases because the surface area of the adsorbent was increased [9].

## *3.3 Effect on Dye Concentration*

Effect of dye concentration from 2 to 10 mg/L was investigated in the present study and all other variables kept constant. In Fig. 3, DRE decreased from 99 to 94% when dye concentration increased. The possible reason may be due to that at higher concentration adsorption area is not available [7]. Our results are in agreement with [10].

## *3.4 Effect of Temperature*

Effect of temperature on DRE and $q_e$ using iron oxide @AC was investigated at different temperatures of 28–40 °C. The effect of temperature on DRE and $q_e$ was presented in Fig. 4. It can observed from Fig. 4 that no effect of temperature on DRE and qe was observed at different temperatures. This remark directed to the conclusion that the proposed adsorbent can be considered for an exothermic process. Similar findings were reported previously by some researchers for sepiolite [11], and durian seed activated carbon [12] as adsorbents.

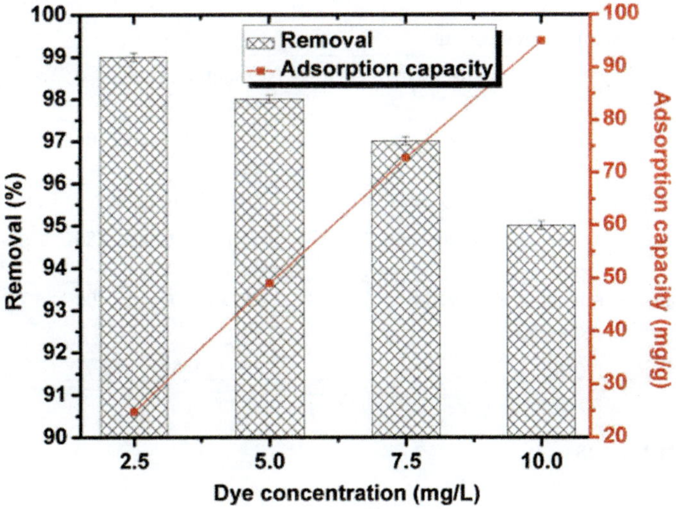


**Fig. 3** Effect of dye concentration

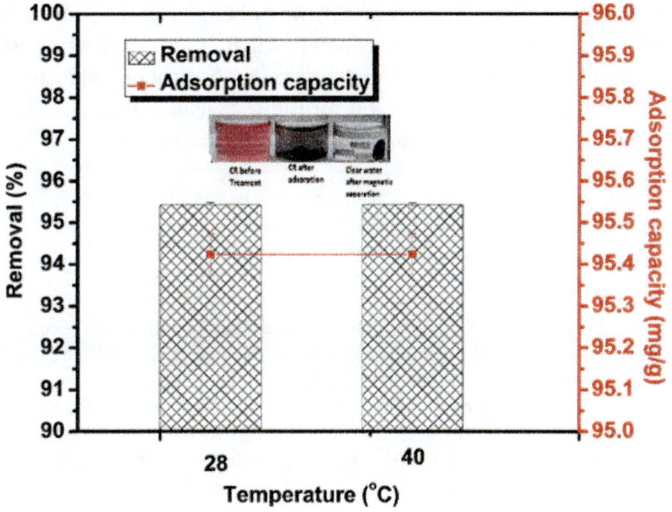


**Fig. 4** Effect of temperature

## 3.5 Comparison with the Literature

The adsorption capacity of iron oxide @AC was compared with the literature for removal of Congo red dye and presented in Table 1. It can be observed that the adsorption capacity of iron oxide @AC nanocomposite is relatively high with other nanomaterials which signifies that iron oxide @AC is active for removal of Congo red dye from aqueous solutions.

**Table 1** Comparison with previous study

| References | Adsorbent | Conditions time, concentration, dose | $q_e$ (mg/g) |
|---|---|---|---|
| Namasivayam et al. [13] | Chitosan magnetite | 60 min, 10 mg/L, 50 mg | 1.8257 |
| Jumadi et al. [14] | $Fe_3O_4$@AC | 250 min, 10 mg/L | 11.22 |
| Zhang and Kong [15] | Orange peel | 60 min, 60 mg/L, 250 mg | 22.4 |
| Namasivayam and Kavitha [16] | Coir pith | 45 min, 20 mg/L, 200 mg | 6.72 |
| Present study | FAC | 8 h, 10 mg/L, 5 mg | 97 |

## 4 Conclusions

In this study, iron oxide @AC nanoparticles were prepared and used as adsorbents for removal of CR dye from aqueous solution. It was found from the batch study that the concentration of dye, amount of adsorbent, and contact time played a significant role on DRE and $q_e$. The optimum time 8 h was obtained in the present study for removal of 10 mg/L of CR dye. It can be concluded that the prepared nanoparticles are presenting good removal efficiency. The dye removal from actual industrial effluent using prepared nanoparticles has been left for future considerations. Further work is required to establish surface functionalization and characterization.

## References

1. Körbahti BK, Tanyolaç A (2008) Electrochemical treatment of simulated textile wastewater with industrial components and Levafix Blue CA reactive dye: optimization through response surface methodology. J Hazard Mater 151:422–431
2. Barrera-Díaz C, Bilyeu B, Roa G, Bernal-Martinez L (2011) Physicochemical aspects of electrocoagulation. Sep Purif Rev 40:1–24
3. Salman JM, Hassan FM, Jouda SA (2015) Removal of Congo red dye from aqueous solution by using natural materials preparation of adsorbent effect of adsorbent dose. Mesop Environ 1:82–89
4. Lorenc-Grabowska E, Gryglewicz G (2007) Adsorption characteristics of Congo Red on coal-based mesoporous activated carbon. Dye Pigment 74:34–40
5. Ong S-T, Keng P-S, Lee W-N, Ha S-T, Hung Y-T (2011) Dye waste treatment. Water 3:157–176
6. Iqbal MJ, Ashiq MN (2007) Adsorption of dyes from aqueous solutions on activated charcoal. J Hazard Mater 139:57–66
7. Joshi S, Garg VK, Kataria N, Kadirvelu K (2019) Applications of Fe3O4@AC nanoparticles for dye removal from simulated wastewater. Chemosphere 236:124280
8. Tyagi U (2020) Adsorption of dyes using activated carbon derived from pyrolysis of vetiveria zizanioides in a fixed bed reactor. Groundw Sustain Dev 10:100303
9. Namasivayam C, Sumithra S (2005) Removal of direct red 12B and methylene blue from water by adsorption onto Fe (III)/Cr (III) hydroxide, an industrial solid waste. J Environ Manage 74:207–215

10. Kera NH, Bhaumik M, Pillay K, Ray SS, Maity A (2017) Selective removal of toxic Cr(VI) from aqueous solution by adsorption combined with reduction at a magnetic nanocomposite surface. J Colloid Interface Sci 503:214–228
11. Özdemir Y, Doğan M, Alkan M (2006) Adsorption of cationic dyes from aqueous solutions by sepiolite. Microporous Mesoporous Mater 96:419–427
12. Ahmad MA, Ahmad N, Bello OS (2015) Adsorption kinetic studies for the removal of synthetic dye using durian seed activated carbon. J Dispers Sci Technol 36:670–684
13. Namasivayam C, Muniasamy N, Gayatri K, Rani M, Ranganathan K (1996) Removal of dyes from aqueous solutions by cellulosic waste orange peel. Bioresour Technol 57:37–43
14. Jumadi J, Kamari A, Rahim NA, Wong STS, Yusoff SNM, Ishak S, Abdulrasool MM, Kumaran S (2019) Removal of methylene blue and congo red by magnetic chitosan nanocomposite: characterization and adsorption studies. J Phys Conf Ser 1397, 0–8
15. Zhang Z, Kong J (2011) Novel magnetic Fe3O4@C nanoparticles as adsorbents for removal of organic dyes from aqueous solution. J Hazard Mater 193:325–329
16. Namasivayam C, Kavitha D (2002) Removal of Congo Red from water by adsorption onto activated carbon prepared from coir pith, an agricultural solid waste. Dye Pigment 54:47–58

# Chapter 11
# Enviro-Economic Analysis of Ginger Drying in Hybrid Active Greenhouse Solar Dryer

**Pushpendra Singh and M. K. Gaur**

## 1 Introduction

Ginger is one of the most consumed crop in the world not only in kitchens for making various dishes but also for medicinal purpose. It is also a very high moisture crop so for storage it must be dried to a safe moisture level. Solar dryers are developed to dry the crops to safe moisture level by using a renewable energy source, i.e., sun. The solar dryers are not limited for agricultural purposes but it is also used in industrial purposes for large-scale drying or space heating [1, 2].

Solar energy is one of the renewable energy sources in which very faster innovations and developments are taking place. It has so many positive impacts, and also has some minor negative effect that needs to be considered for making it 100% clean source of energy [3]. The greenhouse is one of the inventions for harnessing solar energy for space heating, drying, or agricultural purposes. For making the greenhouse more efficient, it is operated mostly in active mode. In an active mode, the electric power required to operate the fans is powered by PV panels [4, 5]. By 2020, mixed-mode dryers can reduce $CO_2$ emission by 23% [6]. The use of even 40% efficient solar dryer will reduce conventional energy consumption by 27–80% [6], [7].

The hybrid dryers were developed to dry the high moisture crops as these require more energy to maintain faster drying [8]. The hybrid dryers can also be operated in off sunshine period by storing the excess energy supplied by secondary sources like biomass, LPG, or solar energy used separately in solar water heaters, etc. The use of some additional device in hybrid dryer makes it little bit costly but the extra amount invested on it can be compensated by faster drying, which means reduced drying time per batch.

P. Singh (✉) · M. K. Gaur
Department of Mechanical Engineering, Madhav Institute of Technology and Science, Gwalior 474005, India

M. K. Gaur
e-mail: gmanojkumar@rediffmail.com

© The Author(s), under exclusive license to Springer Nature Singapore Pte Ltd. 2022

H. M. Dubey et al. (eds.), *Artificial Intelligence and Sustainable Computing*, Algorithms for Intelligent Systems, https://doi.org/10.1007/978-981-16-1220-6_11

The cost and the environmental impact are some major criteria that decide the feasibility of developed product for commercial scale. Various researches had been done on the economic analysis of hybrid solar dryers [9–12]. Boonyasri et al. carried out the economic analysis of forced-mode greenhouse solar dryer for drying pork. The payback time of the dryer was 1.15 years and the capital cost was 49500. Baht [2]. Hamdani et al. carried out an economic analysis of biomass-integrated hybrid dryer. The break-even point for the developed dryer was 2.6 years with NPV of $21.091 [13]. Dhanushkodi et al. carried out the life cycle cost analysis of a hybrid dryer for drying cashew nuts. The experimentation was carried out on solar, biomass, and hybrid drying system and the payback time was 1.58, 1.32, and 1.99 years, respectively [14]. Prakash et al. developed the modified greenhouse dryer with thermal storage material on the floor of the dryer. The payback time was 1.11 years and 1.89 years, while the capital cost was Rs. 8994.50 and Rs. 12844.50 for passive and active mode [15]. Kaewkiew et al. developed the semi-cylindrical greenhouse dryer whose payback time for the dryer was about 2 years which is very less from a commercial point of view [16].

With the integration of new devices to the greenhouse dryer, the cost of dryer increases and the devices attached to it has also some impact on the environment. So it becomes necessary to check the economic viability and impact of a dryer on the environment. In this paper, the environmental and economic analysis of the hybrid active greenhouse solar dryer (HAGSD) is carried out. The dryer is made hybrid by attaching the evacuated tube solar water heater with the heat exchanger inside the greenhouse dryer.

## 2 Experimental Setup and Instrumentation

The experimentation on designed hybrid greenhouse solar dryer has been carried out at Madhav Institute of Technology and Science, Gwalior (26° 14′ N, 78° 10′ E). The frame of dryer is made of hollow square-shaped iron rods which are fixed at the required position by means of welding. The joints are properly filled by means of welding and m-seal so as to make it air tight as much as possible. The UV stabilized polythene sheet is used as a cover material. Four DC fans were provided to induce the air inside the dryer to operate it in forced mode. The PV panels are also attached in the roof of dryer for supplying electricity to operate the fans. The hot water from ETSC is circulated through copper tubes of heat exchanger placed inside the dryer by means of DC pump. The water in tube type evacuated tube solar collector was attached with the heat exchanger placed inside the dryer. The hot water from ETC is supplied inside the dryer through the copper tubes of heat exchanger bed. The experimental setup is shown in Fig. 1.

For measuring the temperature, K-type thermocouples of accuracy ±0.1 °C are placed at different location inside the dryer. Pyranometer of Central Electronics Limited with ±1 W/m$^2$ accuracy is used to measure the solar irradiance. Digital weight balance is used to measure the weight of crop at one hour interval. The

**Fig. 1** The constructed hybrid active greenhouse solar dryer

Datataker DT85 datalogger is used to record and display the reading of measured temperature and solar radiation.

The ginger is purchased from the local market and washed properly with fresh water. The ginger is cut into slices of 5 mm thickness and then spread evenly on the drying tray inside the greenhouse.

## 3 Environmental Analysis of HAGSD

### *3.1 Embodied Energy*

The total energy required to make the complete setup of the greenhouse solar dryer is termed as embodied energy of dryer ($E_{em}$) [17, 18]. The different materials used to construct the HAGSD are given in Table 1 with their corresponding embodied energy and quantity of material used.

### *3.2 Energy Payback Time (EPBT)*

EPBT is the amount of time taken by the hybrid dryer to recover the energy invested in it during its construction. Mathematically it is expressed as [17, 19].

$$\text{Energy Payback Time} = \frac{\text{Embodied Energy (kWh)}}{\text{Annual energy output (kWh/yr)}} \quad (1)$$

The annual energy output ($E_a$) is calculated using the relation (2) [20]:

**Table 1** Embodied energy of different material used to construct HAGSD

| S. No. | Materials | Embodied energy (kWh/kg) | Quantity (kg) | Total embodied energy (kWh) |
|---|---|---|---|---|
| 1. | Hollow Cast Iron Pipe for frame | 8.89 | 207 | 1840.23 |
| 2. | Hollow Cast Iron Pipe for Drying bed | 8.89 | 75 | 666.75 |
| 3. | Polycarbonate sheet | 10.1974 | 22 | 224.34 |
| 4. | Wire mesh | 9.67 | 6 | 58.02 |
| 5. | Fitting materials | | | |
| | (i) Kundi (Door lock) | 55.28 | 0.2 | 11.06 |
| | (ii) Kabja/Hinges | 55.28 | 1.5 | 82.92 |
| | (iii) Nut bolts with washer | 9.67 | 2 | 19.34 |
| 6. | Copper tube for heat exchanger | 19.61 | 42 | 823.62 |
| 7. | Aluminum Duct tape for insulation | 55.28 | 0.05 | 2.76 |
| 8. | PV Module | 739 kWh/module | 2 Nos. | 1478.00 |
| 9. | Plastic components like DC fan and DC motor casing etc. | 19.44 | 0.3 | 5.83 |
| 10. | Copper wire used for electricity supply | 19.61 | 0.1 | 1.96 |
| 11. | Paint | 25.11 | 1 | 25.11 |
| 12. | Evacuated Tube Collector | 422.5 kWh/m$^2$ | 4 m$^2$ | 1690.00 |
| 13. | Aluminum tube with T and L joints | 55.28 | 25 | 1382.00 |
| **Total Embodied Energy of the HGSD** | | | | **8311.95 kWh** |

$$E_a = \text{Daily thermal output} \times \text{No.of sunshine days in a year} \tag{2}$$

Generally, the number of sunshine days in a year is taken as about 290 days in India but it may vary from region to region and climatic condition of the place. The daily thermal output is determined by using the relation given by [20].

$$\text{Daily thermal output (kWh)} = \frac{\text{Moisture Evaporated per day (kg)} \times \text{Latent heat (J/kg)}}{3.6 \times 10^6} \tag{3}$$

The value of the latent heat of evaporation of water is taken as 2430 kJ/kg.

## 3.3 Carbon Dioxide Emitted by the HAGSD

Considering that the electricity is generated using coal then the $CO_2$ emission is taken as approximately 0.98 kg/kWh. The annual $CO_2$ emission is given by Eq. 4 [15].

$$CO_2\text{Emission per year} = \frac{\text{Embodied Energy} \times 0.98}{\text{Life of dryer}} \quad (4)$$

There are various losses related to electricity generation and transmission. If the domestic appliance losses ($L_{da}$) and transmission and distribution losses ($L_{dt}$) are considered then the Eq. 4 becomes

$$CO_2\text{ Emission per year} = \frac{1}{1 - L_{da}} \times \frac{1}{1 - L_{dt}} \times \frac{\text{Embodied Energy} \times 0.98}{\text{Life of dryer}} \quad (5)$$

Generally, the value of $L_{da}$ and $L_{dt}$ is taken as 20% and 40%, respectively, then the Eq. 5 is given as:

$$CO_2\text{ Emission per year} = \frac{\text{Embodied Energy}}{\text{Life of dryer}} \times 2.042,\ \text{kg} \quad (6)$$

## 3.4 $CO_2$ Mitigation by HAGSD

The $CO_2$ mitigation per kWh of the greenhouse dryer is given by Eq. 7 [18].

$$CO_2\text{mitigation/KWh(Y)} = \frac{1}{1 - L_a} \times \frac{1}{1 - L_{td}} \times 0.98 = 2.042\text{kg/kWh} \quad (7)$$

If the $CO_2$ mitigation is considered for the entire life of the dryer, then it is given by Eq. 8 as

$$CO_2\text{mitigation (lifetime)} = \text{Embodied Enegy} \times 2.042,\ \text{kg} \quad (8)$$

The net $CO_2$ mitigated by greenhouse during its entire lifetime is given by Eq. 9.

$$\text{Net Mitigation over lifetime} = \text{Total } CO_2 \text{ mitigation} - \text{Total } CO_2 \text{ emission}$$

$$\therefore \text{Net Mitigation over lifetime} = (E_{ao} \times L - \text{Embodied Energy}) \times 2.042,\ \text{kg} \quad (9)$$

where $E_{ao}$ is the annual energy output and L is the life of dryer which ranges from 5 to 35 years.

### 3.5 *Carbon Credit Earned by Dryer*

One ton mitigation of $CO_2$ emission is equal to one carbon credit. The carbon credit earned from the HAGSD is calculated using the relation given in Eq. 10 [21].

$$\text{Carbon Credit earned} = \text{Net}\,CO_2\ \text{mitigation} \times \text{Price per ton of}\,CO_2\ \text{mitigation} \tag{10}$$

## 4 Economic Analysis of HAGSD

The annual cost of dryer ($C_{an}$) is given as [11]

$$C_{an} = C_{acp} + C_{mt} - S_v + C_{acf} \tag{11}$$

$C_{acp}$ is the annual capital cost and $C_{acf}$ is the annual operational cost of a fan, which are calculated using Eqs. 12 and 13.

$$C_{acp} = C_{cc}F_{cp} \tag{12}$$

$$C_{acf} = N_f \times P_f \times C_{ue} \tag{13}$$

where $C_{cc}$ is the capital cost of the dryer, $N_f$ is the number of hours the fan run in a year, $P_f$ is the rated power consumed by a fan during its operation, and $C_{ue}$ is the electricity charge for one unit.

In this case the daily operation hour of fan is 8 h and taking 290 days of full sunshine days in year in this location, then the number of hours the fan operate in a year is 2320 h. The rated power of one DC fan is 6 W, so the rated power of 4 DC fans is 24 W. The charge of one unit of electricity is taken as Rs. 5/kWh.

In this case, annual maintenance cost ($C_{mt}$) and annual salvage value ($S_v$) is taken as 3% and 10% of the annual capital cost of dryer [22].

The capital recovery ($F_{cp}$) is calculated using Eq. 14 as

$$F_{cp} = \frac{d(1+d)^n}{(1+d)^n - 1} \tag{14}$$

where d is the rate of interest on the amount invested for a long time.

The total mass of dried ginger in the dryer annually ($M_{pa}$) is given by Eq. 15.

$$M_{pa} = \frac{M_{pd}D}{D_b} \tag{15}$$

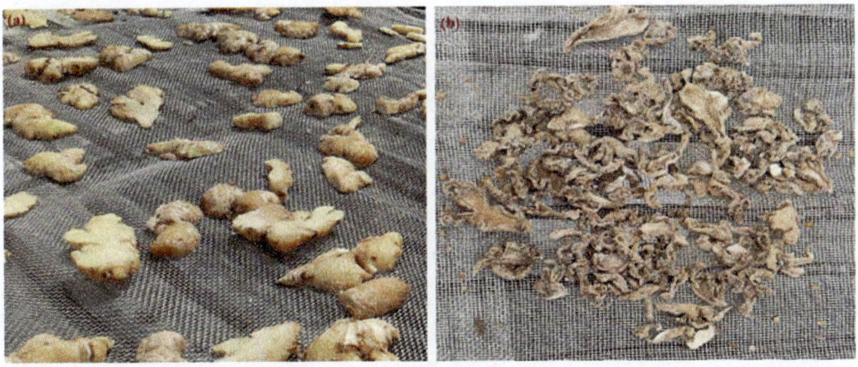


**Fig. 2** **a** Ginger before drying and **b** Ginger after drying

where $M_{pd}$ is the mass of product dried in a dryer per batch, D is the number of days for which the dryer is used for drying in a year and $D_b$ is the number of days taken for drying the material per batch.

In our case, the capacity of the dryer is 30 kg per batch and in one batch, about 3 days is taken to dry the ginger. The photograph of the ginger before and after drying is shown in Fig. 2.

Then the drying cost for one kg of material ($C_u$) is given by

$$C_u = \frac{C_{an}}{M_{pa}} \tag{16}$$

Cost of fresh product per one kg of dried product ($C_{fd}$) is shown by Eq. 17 as

$$C_{fd} = C_{fp} \times \frac{M_f}{M_{pd}} \tag{17}$$

where $M_f$ is the mass of fresh product per batch and $C_{fp}$ is the cost of one kg of fresh product.

Cost of one kg of product dried inside the dryer ($C_{ud}$) is given by:

$$C_{ud} = C_{fd} + C_u \tag{18}$$

The savings obtained from one kg of product dried inside the dryer ($S_{kg}$) is calculated using Eq. 19 as follows:

$$S_{kg} = SP_{dp} - C_{ud} \tag{19}$$

$SP_{dp}$ is the selling price of dried material for one kg.

The savings obtained from drying one batch is given as

$$S_b = S_{kg} \times M_{pd} \tag{20}$$

The savings obtained from drying of material per day is given by Eq. 21

$$S_d = \frac{S_b}{D_b} \tag{21}$$

The annual savings obtained from the greenhouse dryer in kth year is given by

$$S_k = S_d D(1 + R_{if})^{k-1} \tag{22}$$

where $R_{if}$ is the inflation rate.

Finally, the payback time ($P_b$) is calculated using the relation given by Eq. 23 as follows:

$$P_b = \frac{ln\left[1 - \frac{C_{cc}}{S_1}(d - R_{if})\right]}{ln\left(\frac{1+R_{if}}{1+d}\right)} \tag{23}$$

## 5 Result and Discussion

The developed hybrid dryer is tested in the month of September 2020 under partially cloudy sky. The variation in ambient temperature, outside relative humidity, and greenhouse room temperature during experimentation period is shown in Fig. 3.

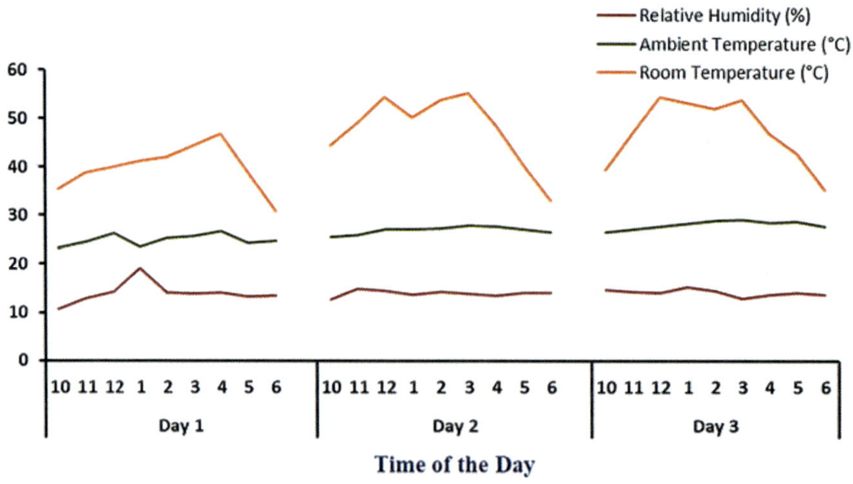


**Fig. 3** Variation in relative humidity, ambient, and room temperature

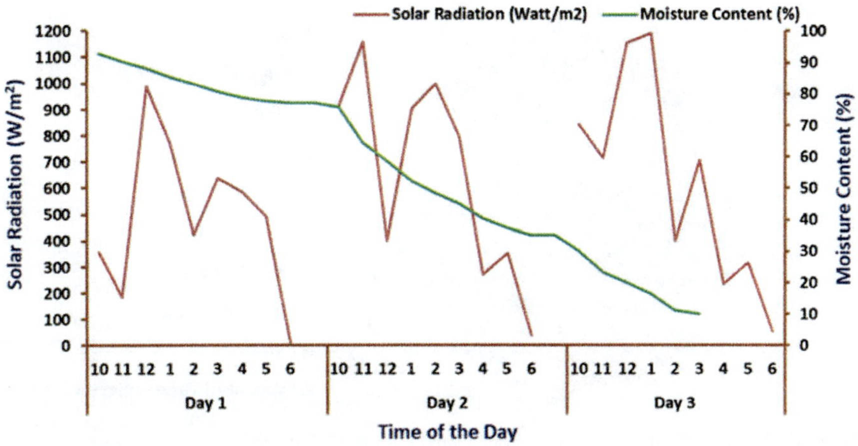


**Fig. 4** Variation in solar radiation during experimentation period

The ambient temperature ranges from 23.19 °C to 29.03 °C during experimentation time. The room air temperature varies between 30.94 and 55.33 °C. The maximum value of solar radiation during experiment was 1190.63 W/m$^2$. The fluctuations in solar irradiance as shown in Fig. 4 are due to partially cloudy weather during experimentation period.

The initial moisture content of ginger is 92.80% (wb) and it is dried to 10% (wb) moisture level in 21 h as shown in Fig. 4. Initially the moisture removal rate is faster as there is a sharp decrease in the weight of ginger from 1000 to 756 g and then 614 g in 2 h only. This is due to faster diffusion and evaporation rate of unbound moisture.

## 5.1 *Environmental Analysis*

The environmental impact of any system depends mostly on its embodied energy. The embodied energy of the materials used in the hybrid dryer is shown in Fig. 5. Cast iron and evacuated tube collector contributes 50.49% together and rest of them is by other materials used in the dryer.

The various environmental parameters calculated to show the impact of dryer on environment are shown in Table 2. From the table it is clear that the energy invested on the dryer can be earned back in very small time. i.e., in 1.54 years only. The total life of dryer is considered as 30 years and in its total lifetime, the dryer will emit 565.77 kg of $CO_2$ while it will mitigate 16,972.99 kg. This helps a lot in reducing carbon footprint and from the net mitigated $CO_2$, about Rs. 17,992.33 (taking 15 USD/ton of $CO_2$ mitigated equivalent to Rs. 1094.42/ton of $CO_2$ mitigated) can be earned from the hybrid dryer as a carbon credit.

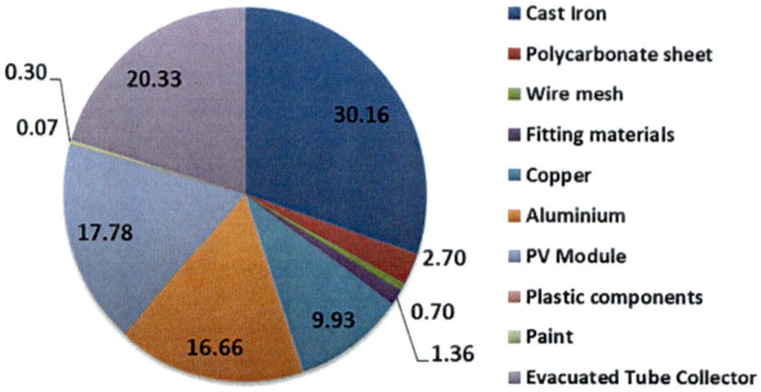


**Fig. 5** Percentage of embodied energy contributed by different material used in the dryer

**Table 2** Environmental analysis of HAGSD

| Environmental parameters | Value |
|---|---|
| EPBT | 1.54 Years |
| Moisture evaporated | 27.6 kg/batch |
| $CO_2$ emitted (lifetime) | 565.77 kg |
| $CO_2$ mitigation (lifetime) | 16972.99 kg |
| Net $CO_2$ mitigation (lifetime) | 16407.23 kg |
| Carbon credit earned | Rs. 17992.33 |

## 5.2 *Economic Analysis*

The economic analysis is one of the important factors justifying the feasibility of the developed system. The various economic parameters calculated to evaluate the economic feasibility of HAGSD is shown in Table 3. The capital cost of the dryer is very high due to inclusion of two evacuated tube solar collectors but the invested amount can be earned back in 1.15 years only, which is very less time as dryer life is 30 years. The annual maintenance is replacement of polythene sheet every year and there is no other maintenance required.

The ginger is one of the most consumed products which are used not only as spices but also for a medicinal purpose. Hence, it can be dried and stored for its easier and cheaper availability in all seasons. The developed dryer had a capacity of drying 30 kg per batch and annually it can produce 232 kg of dried ginger. The selling price of dried ginger varies between Rs. 1000 and Rs. 3000 per kg. In the calculation, it is taken as Rs. 1500/kg and this gives the annual savings of Rs. 550.70 per day.

**Table 3** Economic analysis of HAGSD

| Cost parameters | Value | Cost parameters | Value |
|---|---|---|---|
| Capital Cost of Dryer | Rs. 1,69,692 | Quantity of dried Ginger annually | 232 kg |
| Capital Recovery Factor | 9% | Drying Cost of 1 kg of Ginger | Rs. 61.62 |
| Annual Capital Cost | Rs. 15073.30 | Cost of 1 kg fresh ginger | Rs. 60.00 |
| Annual Maintenance Cost | Rs. 452.20 | Mass of fresh product per batch (kg) | 30 kg |
| Annual Salvage Value | Rs. 1507.33 | Cost of fresh product per kg of dried product | Rs. 750.00 |
| Annual Operational cost of fan | Rs. 278.40 | Cost of 1 kg ginger dried inside the dryer | Rs. 811.62 |
| Annual Cost of Dryer | Rs. 14296.57 | Selling price of 1 kg dried ginger | Rs. 1500 |
| Life of Dryer | 30 Years | Saving per Kg | Rs. 688.38 |
| Interest Rate | 8% | Saving per batch | Rs. 1652.10 |
| Inflation Rate | 6% | Saving per day | Rs. 550.70 |
| Payback Time | 1.15 Years | Saving after 1 year | Rs. 1,59,703.43 |
| Mass of Ginger dried per batch | 2.40 kg | | |

## 6 Conclusions

The developed HAGSD is suitable for drying high moisture crops and it is also economically and environmentally feasible. Although the capital cost is very large for small-scale purpose and for its use in rural areas, but once the amount is invested it can be earned back in very less time of only 1.15 years. The maximum temperature difference between ambient and inside room air is 27.52 °C, which is good for thin layer drying of high moisture crop. The dryer also mitigates about 16407.23 kg of $CO_2$ in its lifetime, which will help in reducing the carbon footprints. This makes the dryer a sustainable solution for the drying purpose as it only emits 565.77 kg of $CO_2$ in its lifetime of 30 years.

## References

1. Abdulmalek SH, Assadi MK, Al-Kayiem HH, Gitan AA (2018) A comparative analysis on the uniformity enhancement methods of solar thermal drying. Energy 148:1103–1115. https://doi.org/10.1016/j.energy.2018.01.060
2. Boonyasri M, Lertsatitthanakorn C, Wiset L, Poomsa N (2011) Performance analysis and economic evaluation of a greenhouse dryer for pork drying. KKU Eng J 38:433–443
3. Gunerhan H, Hepbasli A, Giresunlu U (2009) Environmental impacts from the solar energy systems. Energy Sources, Part A Recover Util Environ Eff 31:1131–1138. https://doi.org/10.1080/15567030701512733

4. Eltawil MA, Azam MM, Alghannam AO (2018) Solar PV powered mixed-mode tunnel dryer for drying potato chips. Renew Energy 116:594–605. https://doi.org/10.1016/j.renene.2017.10.007
5. Chauhan PS, Kumar A, Nuntadusit C (2018) Heat transfer analysis of PV integrated modified greenhouse dryer. Renew Energy 121:53–65. https://doi.org/10.1016/j.renene.2018.01.017
6. Tripathy PP (2015) Investigation into solar drying of potato: effect of sample geometry on drying kinetics and $CO_2$ emissions mitigation. J Food Sci Technol 52:1383–1393. https://doi.org/10.1007/s13197-013-1170-0
7. Arata A, Sharma V, Spagna G (1993) Performance evaluation of solar assisted dryers for low temperature drying application-II Experimental results. Energy Convers Manag 34:417–426
8. Shaikh TB, Kolekar AB (2015) Review of Hybrid Solar Dryers. Int J Innov Eng Res Technol [IJIERT] 2:1–7
9. Mustapha MK, Salako AF, Ademola SK, Adefila IA (2014) Qualitative performance and economic analysis of low cost solar fish driers in Sub-Saharan Africa. J Fish 2:64. https://doi.org/10.17017/jfish.v2i1.2014.23
10. Ndukwu MC, Simo-Tagne M, Abam FI, Onwuka OS, Prince S, Bennamoun L (2020) Exergetic sustainability and economic analysis of hybrid solar-biomass dryer integrated with copper tubing as heat exchanger. Heliyon 6: https://doi.org/10.1016/j.heliyon.2020.e03401
11. ELkhadraoui A, Kooli S, Hamdi I, Farhat A (2015) Experimental investigation and economic evaluation of a new mixed-mode solar greenhouse dryer for drying of red pepper and grape. Renew Energy 77:1–8. https://doi.org/10.1016/j.renene.2014.11.090
12. Selvanayaki S, Sampathkumar K (2017) Techno-economic analysis of solar dryers. In: Prakash O, Kumar A (eds) Green energy technol. Springer, p. 463–93 (2017). https://doi.org/10.1007/978-981-10-3833-4_16
13. Hamdani, Rizal TA, Muhammad Z (2018) Fabrication and testing of hybrid solar-biomass dryer for drying fish. Case Stud Therm Eng 12:489–96. https://doi.org/10.1016/j.csite.2018.06.008
14. Dhanushkodi S, Wilson VH, Sudhakar K (2015) Life cycle cost of solar biomass hybrid dryer systems for cashew drying of nuts in India. Environ Clim Technol 15:22–33. https://doi.org/10.1515/rtuect-2015-0003
15. Prakash O, Kumar A, Laguri V (2016) Performance of modified greenhouse dryer with thermal energy storage. Energy Reports 2:155–162. https://doi.org/10.1016/j.egyr.2016.06.003
16. Kaewkiew J, Nabnean S, Janjai S (2012) Experimental investigation of the performance of a large-scale greenhouse type solar dryer for drying chilli in Thailand. Procedia Eng 32:433–439. https://doi.org/10.1016/j.proeng.2012.01.1290
17. Eltawil MA, Azam MM, Alghannam AO (2018) Energy analysis of hybrid solar tunnel dryer with PV system and solar collector for drying mint (MenthaViridis). J Clean Prod 181:352–364. https://doi.org/10.1016/j.jclepro.2018.01.229
18. Prakash O, Kumar A (2014) Environomical analysis and mathematical modelling for tomato flakes drying in a modified greenhouse dryer under active mode. Int J Food Eng 10:1–13. https://doi.org/10.1515/ijfe-2013-0063
19. Kesavan S, Arjunan TV, Vijayan S (2019) Thermodynamic analysis of a triple-pass solar dryer for drying potato slices. J Therm Anal Calorim 136:159–171. https://doi.org/10.1007/s10973-018-7747-0
20. Vijayan S, Arjunan TV, Kumar A (2020) Exergo-environmental analysis of an indirect forced convection solar dryer for drying bitter gourd slices. Renew Energy 146:2210–2223. https://doi.org/10.1016/j.renene.2019.08.066
21. Tiwari S, Tiwari GN (2016) Exergoeconomic analysis of photovoltaic-thermal (PVT) mixed mode greenhouse solar dryer. Energy 114:155–164. https://doi.org/10.1016/j.energy.2016.07.132
22. Sreekumar A, Manikantan PE, Vijayakumar KP (2008) Performance of indirect solar cabinet dryer. Energy Convers Manag 49:1388–1395. https://doi.org/10.1016/j.enconman.2008.01.005

# Chapter 12
# A Comparative Study of Meta-Heuristic-Based Task Scheduling in Cloud Computing

**Ravi Kumar and Jai Bhagwan**

## 1 Introduction

In the real world environment, it can be seen that clouds are formed by the association of water vapors. Similarly in cloud computing, the cloud is formed by the association of networks and different kinds of devices. This cloud computing environment establishes big-sized data centers and delivers virtual resources like computational power, storage space and applications etc. as on-demand self-service over the internet [1]. Cloud computing relies on two entities, i.e., infrastructure and service providers. Cloud computing is based on the architecture "Everything as a Service" (with short-form EaaS or XaaS) and structured into three models as "Infrastructure-as-a-Service (IaaS), Platform-as-a-Service (PaaS) and Software-as-a-Service (SaaS)" [2, 3]. These service models are based on a "pay-as-you-go" policy which attract the clients toward cloud computing and hence many sectors have started using the cloud services and traffic of clients over cloud service providers (CSP) which are being increased day by day [4].

The reliability of cloud services depends upon many factors; one of them is task scheduling. The principle aim of task scheduling is to increase the availability of resources and lower downs the execution time/computation cost of tasks. The tasks can be a query entered by a user or a request to access the memory or storage space. To achieve the objective of the best possible allotment of resources, the tasks must be executed in a good way using a task scheduling algorithm. The task scheduling algorithm focuses on performance metrics such as response time, makespan, resource utilization and cost [5]. The meta-heuristic algorithms give solutions near to optimal within the effective time duration and worthy for computation of the infinite and complex problems. That is why we are more interested to use these kinds of algorithms to enhance task scheduling in cloud computing.

R. Kumar (✉) · J. Bhagwan (✉)
Department of Computer Science & Engineering, Guru Jambheshwar University of Science & Technology, Hisar, India

© The Author(s), under exclusive license to Springer Nature Singapore Pte Ltd. 2022 
H. M. Dubey et al. (eds.), *Artificial Intelligence and Sustainable Computing*, Algorithms for Intelligent Systems, https://doi.org/10.1007/978-981-16-1220-6_12

In Sect. 2, task scheduling using meta-heuristic techniques (ACO, PSO, GWO, WOA and FPA) are investigated. In Sects. 3 and 4, literature observations and research gaps are discussed, respectively. Section 5 includes the conclusion of this research paper.

## 2 Meta-Heuristic-Based Task Scheduling

### *2.1 Ant Colony Optimization (ACO) Algorithm*

ACO algorithm was first introduced by Marco Dorigo in 1990 and published it as an ant system. ACO algorithms are very helpful in finding a path toward their goals. It is capable of solving "Travelling Salesman Problem (TSP), Graph coloring problem, Traffic routing, and various scheduling problems" [6]. The solution to the task scheduling problem can be obtained by mapping tasks with the available resources as the same way ants approach to find the food in the ant system.

A sequence of tasks (Xi) with a sequence of resources (Yj) is carried out using the ant system, and the probabilistic function is stated as

$$P_{ij}^{k}(t)\begin{cases} \frac{[\tau_{ij}(t)^{\alpha}][\eta_{ij}(t)^{\beta}]}{\sum_{k\in \text{allowed}}[\tau_{ik}(t)^{\alpha}][\eta_{ik}(t)^{\beta}]} & \text{if j} \in \text{allowed}_k \\ 0 & \text{otherwise} \end{cases} \quad (1)$$

where $\tau_{ij}$ indicates pheromone strength found on the path. $\eta_{ij}$ indicates the heuristic function. This function can be formulated as $\eta_{ij} = \frac{1}{d_{ij}}$, where $d_{ij}$ is the distance between $(X_i, Y_j)$.

This algorithm drives the objective function to lower down the makespan while the allocation of tasks on virtual machines. A heuristic function is also taken into account as a time function, which gives the allocation path of tasks ($T_i$) on a virtual machine ($VM_j$).

A comparative analysis of ACO centric task scheduling has been done and shown above in Table 1.

### *2.2 Particle Swarm Optimization (PSO) Algorithm*

PSO was introduced by Kennedy and Eberhart in 1995. PSO combines features like local search and global search in a single algorithm [19].

Let the position of *i*th particle at iteration t, and at iteration (t + 1) the *i*th particle updates its position with velocity. The Eq. (2) is given as

$$\vec{X_i}(t+1) = \vec{X_i}(t) + \vec{V_i}(t+1) \quad (2)$$

**Table 1** A variety of ACO-aided task scheduling algorithms

| Work citation | Improved methodology | Improved performance metrics | Nature of tasks | Simulation tools |
|---|---|---|---|---|
| Tawfeek et al. [7] | Standard ACO | Makespan | Independent | Cloudsim |
| Li et al. [8] | Pheromone update (LBACO) | Makespan and resource allocation | Independent | Cloudsim |
| Ming et al. [9] | Polymorphic ACO | Resource allocation and completion time | Independent | Cloudsim |
| Gao et al. [10] | Multi-objective ACO | Energy intake and resource misuse | Independent | Java |
| Wen et al. [11] | ACO combined with PSO | Makespan and resource operation | Workflow | Cloudsim, MATLAB |
| Hu et al. [12] | ACO combined with GA | Makespan | Independent | MATLAB |
| Fang and Li [13] | Improved VM-ACO | Task latency and Degree of imbalance | Independent | Cloudsim |
| Nie and Li [14] | Improved ACO | Execution time and task cost | Independent | Cloudsim |
| Nasr et al. [15] | Hybrid of CRO with ACO | Makespan and cost | Workflow | Cloudsim |
| Wei [16] | Improved ACO | Convergence rate and resource utilization | Independent | Cloudsim |
| Li [17] | Hybrid with symbiotic algorithm (ACO-SOS) | Completion time and cost consumption | Independent | Cloudsim |
| He et al. [18] | Original ACO | Makespan and resource consumption | Independent | Cloudsim |

The velocity $\overrightarrow{V_i}(t)$ updates after particle information exchanged is

$$\overrightarrow{V_i}(t+1) = \omega * \overrightarrow{V_i}(t) + C_1 r_1 \left(\overrightarrow{X_i^{best}} - \overrightarrow{X_i}(t)\right) + C_2 r_2 \left(\overrightarrow{Z_i^{best}} - \overrightarrow{X_i}(t)\right) \quad (3)$$

where $\omega$ is the inertial weight. $r_1$ and $r_2$ are random numbers. $\overrightarrow{X_i^{best}}$ is the local best position of particle i. $\overrightarrow{Z_i^{best}}$ is the global best position of particle i. $C_1$ and $C_2$ are learning parameter constants. To apply PSO in scheduling problems, particles are assumed as tasks and a set of particles covers the set of tasks.

A comparative analysis of the improvement of PSO centric task scheduling is shown in Table 2.

## 2.3 Gray Wolf Optimization (GWO) Algorithm

This is a swarm intelligence technique introduced by Seyedali Mirjalili. The social hierarchy of wolfs acts as a significant part in designing the algorithm for an optimal solution [35].

To find out an optimal solution by this technique, the gray wolves are denoted as alpha (α), beta (β), delta (δ) and omega (ω). Equations (4) and (5) are formulated in the way the wolves surround the prey.

$$\vec{D} = \left|\vec{C}.\vec{X}_p(t) - \vec{X}(t)\right| \tag{4}$$

$$\vec{X}(t+1) = \vec{X}_p(t) - \vec{A}.\vec{D} \tag{5}$$

where $\vec{X}(t)$ and $\vec{X}_p(t)$ are the position vectors of wolf and prey, respectively, at iteration t. $\vec{A}$ and $\vec{C}$ are the coefficient vectors and evaluated as

$$\vec{A} = 2.\vec{a}.\vec{r_1} - \vec{a} \tag{6}$$

$$\vec{C} = 2.\vec{r_2} \tag{7}$$

where $\vec{r_1}$ and $\vec{r_2}$ are random vectors ranges [0, 1] and $\vec{a}$ is a vector whose values drop from 2 to zero. According to the hierarchy, the position of wolves (α, β, δ) can be formulated as in Eq. (8). The wolf which has the best awareness about the position of prey will be considered as alpha (α) and treated as the best or optimum solution. The remaining wolves update their position as follows:

$$\vec{D_\alpha} = \left|\vec{C_1}.\vec{X_\alpha} - \vec{X}\right|, \vec{D_\beta} = \left|\vec{C_1}.\vec{X_\beta} - \vec{X}\right|, \vec{D_\delta} = \left|\vec{C_1}.\vec{X_\delta} - \vec{X}\right| \tag{8}$$

$$\vec{X_1} = \vec{X_\alpha} - \vec{A}.\vec{D_\alpha}, \vec{X_2} = \vec{X_\beta} - \vec{A}.\vec{D_\beta}, \vec{X_3} = \vec{X_\delta} - \vec{A}.\vec{D_\delta} \tag{9}$$

$$\vec{X}(t+1) = \frac{\vec{X_1} + \vec{X_2} + \vec{X_3}}{3} \tag{10}$$

To perform task scheduling using GWO, the position of gray wolves updated with respect to their prey, i.e., the mapping of the tasks (Ti) onto the resources (Rj) or virtual machines is processed by updating their positions.

**Table 2** A variety of PSO-aided task scheduling algorithms

| Work citation | Improved methodology | Improved performance metrics | Nature of tasks | Simulation tools |
|---|---|---|---|---|
| Pandey et al. [20] | PSO combined with heuristic | Communication cost and time | Independent | Cloudsim |
| Xu et al. [21] | QoS-PSO with Berger Model | Execution time and efficiency | Independent | Cloudsim |
| Zarei et al. [22] | Mutation and hierarchical PSO | Convergence rate and computation time | Independent | MATLAB |
| Wu et al. [23] | Revise discrete PSO | Communication and execution cost | Workflow | Cloudsim |
| Yassa et al. [24] | DVFS combined with PSO | Makespan, cost and energy consumption | Workflow | Cloudsim |
| Liu and Wang [25] | Self-adapting Inertial PSO | Resource utilization and makespan | Independent | MATLAB |
| Ramezani et al. [26] | Improved PSO | VM setup time and load balance | Independent | Cloudsim |
| Pan and Chen [27] | Improved PSO | Resource utilization and load balance | Independent | Cloudsim |
| Sidhu et al. [28] | Load rebalance PSO | Makespan and load balance | Independent | Homo/Heterogeneous |
| Yingqiu et al. [29] | Applied mutation in PSO | Makespan | Workflow | Cloudsim |
| Wu [30] | Iterative select. operator PSO | Execution time | Independent | Cloudsim |
| Al-Maamari and Omara [31] | Modified PSO | Makespan and resource utilization | Independent | Cloudsim |
| Pradhan and Bisoy [32] | LBMPSO | Makespan and resource utilization | Independent | Cloudsim |
| Richa and Keshavamurthy [33] | Improved PSO | Degree of imbalance | Independent | Cloudsim |
| Alsaidy et al. [34] | Hybrid of PSO LJFP, MCT | Makespan, energy consumption | Independent | MATLAB |

**Table 3** A variety of GWO-aided task scheduling algorithms

| Work citation | Improved methodology | Improved performance metrics | Nature of tasks | Simulation tools |
|---|---|---|---|---|
| Gupta et al. [36] | Learning based QoS GWO | Execution time and resource utilization | Independent | Cloudsim |
| Khalili and Babamir [37] | Pareto based multi-objective GWO | Makespan and Cost | Workflow | Workflowsim |
| Natesan and Arun [38] | GA combined with GWO | Execution time/cost and energy intake | Workflow | Cloudsim |
| Natesha et al. [39] | Multi-objective GWO | Makespan, energy and resource usage | Independent | Python Simulator |
| Kumar [40] | Hybrid of GELS with GWO | Energy consumption and Makespan | Independent | Cloudsim |
| Alzaqebah [41] | Modified multi-objective GWO | Makespan and cost | Independent | Cloudsim |
| Kumar et al. [42] | Democratic GWO | Throughput and resource utilization | Independent | Cloudsim |
| Kumari et al. [43] | Standard GWO in DTSP | Execution time | Independent | Not declared |
| Bacanin et al. [44] | Standard GWO | Makespan | Independent | Cloudsim |
| Natesan and Chokkalingam [45] | Improved GWO | Completion time and cost | Independent | Cloudsim |
| Bansal and Singh [46] | Standard GWO | Total cost | Independent | Cloudsim |
| Mohammadzadeh et al. [47] | Chaotic binary GWO (HCGWO) | Makespan, cost and energy consumption | Workflow | Cloudsim |

A comparative analysis of improvement studied in GWO centric task scheduling is shown in Table 3.

## 2.4 *Whale Optimization Algorithm (WOA)*

Whale optimization algorithm (WOA) is also a nature-inspired meta-heuristic algorithm proposed by Mirjalili and Lewis [48].

The helical shape movement of whale toward prey is formulated by Eq. 11.

$$\vec{X}(t+1) = \vec{X_p}(t) + \vec{D'}.e^{jk}.\cos(2\pi k) \tag{11}$$

where j is a constant which defines spiral shape in logarithmic value, k is random value that lies between $[-1, 1]$, dot (.) is dot product and $\vec{D'}$ is evaluated by $\left|\vec{X_p}(t) - \vec{X}(t)\right|$. For optimization, this phenomenon is formulated as follows:

$$\vec{X}(t+1) = \begin{cases} \vec{X_p}(t) - \vec{A}.\vec{D} & \text{if } p < 0.5 \\ \vec{X_p}(t) + \vec{D'}.e^{jk}.cos(2\pi k) & \text{if } p \geq 0.5 \end{cases} \tag{12}$$

where p is probability of choosing shrinking encircling/spiral method that lies between [0, 1]. The mathematical equation for exploration is given as:

$$\vec{D} = \left|\vec{C}.\overrightarrow{X_{rand}} - \vec{X}\right| \tag{13}$$

$$\vec{X}(t+1) = \overrightarrow{X_{rand}} - \vec{A} \cdot \vec{D} \tag{14}$$

where $\overrightarrow{X_{rand}}$ is a random position vector selected for the current population.

A comparative analysis of improvement studied in WOA centric task scheduling is shown in Table 4.

**Table 4** A variety of WOA-aided task scheduling algorithms

| Work citation | Improved methodology | Improved performance metrics | Nature of tasks | Simulation tools |
|---|---|---|---|---|
| Strumberger et al. [49] | Hybrid of ABC and exploration FS | Makespan and cost | Artificial and NASA iPSC | Cloudsim |
| Saravanan and Kumaravel [50] | Modified WOA | Makespan and energy consumption | Independent | Cloudsim |
| Premalatha and Ramakrishnan [51] | Combined ABC with WOA | Completion time and power consumption | NASA dataset | Mobile devices (Wi Fi enabled) |
| Sanaj et al. [52] | Original WOA | Resource utilization and cost | Independent | Cloudsim |
| Subalakshmi and Jeyakarthic [53] | WOA with tumbling effect (WOA-TRA) | Energy consumption and resource utilization | Independent | Cloudsim |

## 2.5 Flower Pollination Algorithm (FPA)

FPA was presented by X. Yang in 2012. This technique was based on pollination exercise among flowers. The flower pollination is of two types: self-pollination and cross-pollination. Both kinds of pollination gives birth to the fittest solution and represented as $R_*$ [54]. This flower constancy is modeled as a mathematical equation given by

$$f_i^{t+1} = f_i^t + L\left(f_i^t - R_*\right) \tag{15}$$

where $f_i^t$ is pollen i at iteration t, $R_*$ is the current best fit solution among all solutions at the current step and L is the influence of pollination.

However, a transfer agent can fly in global space in different distance steps, and for modeling this exercise, Levy flight is introduced [54]. A Levy distribution is drawn from L >0 and the equation is given as

$$L \sim \frac{\lambda \Gamma(\lambda) \sin(\pi \lambda/2)}{\pi} \frac{1}{s^{1+\lambda}}, (s \gg s_0 > 0) \tag{16}$$

where $\Gamma(\lambda)$ is a function of standard gamma. This distribution is applicable only for steps, i.e., s>0.

Flower constancy for local pollination is modeled as:

$$f_i^{t+1} = f_i^t + \in \left(f_j^t - f_k^t\right) \tag{17}$$

where $f_j^t$ and $f_k^t$ are pollen of different flowers belonging to the same plant species.

A comparative analysis of improvement studied in FPA centric task scheduling is shown in Table 5.

# 3 Literature Observations

By a deep study of the literature of ACO, PSO, GWO, WOA and FPA, we observed the following:

(1) The improvement is achieved by making integration of two or more algorithms. The algorithm trapped in a local search for a solution is avoided by an algorithm that is better in a global search solution. Also, other techniques were combined with standard meta-heuristic techniques like mutation which improved the convergence rate and avoid local optima, Tabu search that provided solution very close to optimal to handle complex problems, pareto that maximized coverage and spread ratio, fuzzy technique used to handle the complex and vast problem, Taguchi experiment that integrated to obtain best parameter values,

**Table 5** A variety of FPA-aided task scheduling algorithms

| Work citation | Improved methodology | Improved performance metrics | Nature of tasks | Simulation tools |
|---|---|---|---|---|
| Gupta et al. [55] | Standard FPA | Makespan, resource utilization | Synthetic dataset | MATLAB |
| Kaur and Sidhu [56] | Standard FPA | Makespan | Independent | Cloudsim |
| Khurana and Singh [57] | Modified FPA | Completion time and VM idle time | Workflow | Not declared |
| Usman et al. [58] | Energy efficient FPA | Resource allocation and energy consumption | Real data (PlanetLab, CoMon) | MultiRecCloudsim |
| Gokuldhev et al. [59] | Hybrid of GWO and FPA | Energy consumption and makespan | Uniform/ non-uniform heterogeneous environment | MATLAB |
| Khurana and Singh [60] | Hybrid of GWO and FPA | Improved reliability (cost/ time ratio) | Workflow | Not declared |

tumbling effect for good exploration ability and DVFS that minimized the energy consumption for solving complex problems.

(2) The advances were also achieved by modifying the operators like evaporation of trials of pheromone in ACO, polymorphism form of ants (single set of solution update by ants), reward and punishment coefficient to update pheromone strategy, inertia (ω) and learning factor in PSO, the addition of mean value in the equations of (hunting and encircling) in GWO and tumbling effect in WOA.

(3) The improvements in load balance were achieved by placements, migrations and consolidation of VMs, using DVFS, detection of overloaded and underloaded nodes.

## 4 Research Gaps

Till now, researchers have worked very hard to make effective and efficient scheduling. But still, there are open issues which can be taken into account for further research work and those are as follows:

- The researchers have done many practices to combine one algorithm with other and applied them in task scheduling. Still, there are algorithms whose fusion needs to consider like hybridization of two or more soft-computing techniques at different levels.

- Artificial intelligence, machine learning and neural network can be applied to produce learning-based algorithms and hence may reduce the number of iterations and computation required for an optimum solution.
- With a single objective, these algorithms worked fine but further on, more effective multi-objective algorithms can be considered which ensure QoS with VM placement/consolidation, load balancing/rebalancing with cost-saving and tackle resource wastage with deadline meeting constraints.
- A variety of tasks can be included like left/right-skewed tasks, workflow and real-world tasks problems. Also, their scheduling can be performed on a vast heterogeneous environment and offloading on smartphone devices and laptop devices etc.
- Security and privacy are the biggest concern for CSP and clients. Algorithms must be designed in such a way that ensures confidentiality and secrecy during user request for service and resources/task scheduling.

## 5 Conclusion

This paper critically examined the ACO, PSO, GWO, WOA and FPA-based task scheduling algorithms and their modifications, hybridization and simulation results. We conclude that PBACO has given better performance as compared to other modifications of ACO. In this technique, PSO is summed up into ACO which dominated PBACO results. GWO, WOA and FPA dominated PSO and PBACO results. But GWO, WOA and FPA performance is dangling with stagnation in local optima. So to extend GWO, WOA and FPA, a technique is required that can avoid them from falling into local optima with better convergence rate and QoS-awareness. So the suggested approach is to develop a hybrid of meta-heuristic techniques with multi-objectives or integration of an independent mathematical formula.

With consecutive research works, task scheduling is attaining better appearance in terms of makespan, resource consumption, energy intake and running cost. So, to gear up the scheduling in a vast cloud environment and real-world problems, machine learning and neural network concepts can also be incorporated with existing algorithms in the forthcoming time.

## References

1. Er-raji N, Benabbou F, Eddaoui A (2016) Task scheduling algorithms in the cloud computing environment: survey and solutions. Int J Adv Res Comput Sci Softw Eng 6(1):604–608
2. Bokhari MU, Shallal QM, Tamandani YK (2013) Cloud computing service models: a comparative study. In: 3rd international conference on computing for sustainable global development, pp 890–895

3. Goyal S (2014) Public vs private vs hybrid vs community—cloud computing: A critical review. Int J Comput Netw Inform Secur 6(3):20–29
4. Bhagwan J, Kumar S (2016) An intense review of task scheduling algorithms in cloud computing. Int J Adv Res Comput Commun Eng 5(11):605–611
5. Kalra M, Singh S (2015) A review of metaheuristic scheduling techniques in cloud computing. Egyptian Inform J 16(3):275–295
6. Dorigo M, Birattari M, Stutzle T (2006) Ant colony optimization. IEEE Comput Intell Mag 1(4):28–39
7. Tawfeek M, El-sisi A, Keshk A, Torkey F (2015) Cloud task scheduling based on ant colony optimization. Int Arab J Inform Technol 12(2):129–137
8. Li K, Xu G, Zhao G, Dong Y, Wang D (2011) Cloud task scheduling based on load balancing ant colony optimization. In: Proceedings of 6th annual ChinaGrid conference, pp 3–9
9. Ming W, Chunyan Z, Feng Q, Yu C, Qiangqiang S, Wanbing D (2015) Resources allocation method on cloud computing. In: Proceedings of international conference on service science, pp 199–201
10. Gao Y, Guan H, Qi Z, Hou Y, Liu L (2013) A multi-objective ant colony system algorithm for virtual machine placement in cloud computing. J Comput Syst Sci 79(8):1230–1242
11. Wen X, Huang M, Shi J (2012) Study on resources scheduling based on ACO allgorithm and PSO algorithm in cloud computing. In: Proceedings of 11th international symposium on distributed computing and applications to business, engineering and science, vol 1, no 6, pp 219–222
12. Hu W, Li K, Xu J, Bao Q (2015) Cloud-computing-based resource allocation research on the perspective of improved ant colony algorithm. In: Proceedings of international conference on computer science and mechanical automation, pp 76–80
13. Fang Y, Li X (2017) Task scheduling strategy for cloud computing based on the improvement of ant colony algorithm. In: Proceedings of international conference on computer technology, electronics and communication, pp 571–574
14. Nie Q, Li P (2016) An improved ant colony optimization algorithm for improving cloud resource utilization. In: Proceedings of international conference on cyber-enabled distributed computing and knowledge discovery, pp 311–314
15. Nasr AA, El-Bahnasawy NA, Attiya G, El-Sayed A (2019) Cost-effective algorithm for workflow scheduling in cloud computing under deadline constraint. Arabian J Sci Eng 44(4):3765–3780
16. Wei X (2020) Task scheduling optimization strategy using improved ant colony optimization algorithm in cloud computing. J Ambient Intell Humanized Comput
17. Li Y (2019) ACO-SOS-based task scheduling in cloud computing. Int J Performab Eng 15(9):2534–2543
18. He Z, Dong J, Li Z, Guo W (2020) Research on task scheduling strategy optimization based on aco in cloud computing environment. In: IEEE 5th Information technology and Mechatronics engineering conference, pp 1615–1619
19. Kennedy J, Eberhart R (1995) Particle swarm optimization. In: Proceedings of ICNN'95 international conference on neural networks, Australia, vol 4, pp 1942–1948
20. Pandey S, Wu L, Guru SM, Buyya R (2010) A particle swarm optimization-based heuristic for scheduling workflow applications in cloud computing environments. In: Proceedings of international conference on advanced information networking and applications, pp 400–407
21. Xu A, Yang Y, Mi Z, Xiong Z (2015) Task scheduling algorithm based on PSO in cloud environment. In Proceedings of IEEE 15th international conference on scalable computing and communications, pp 1055–1061
22. Zarei B, Ghanbarzadeh R, Khodabande P, Toofani H (2011) MHPSO: a new method to enhance the particle swarm optimizer. In: 6th international conference on digital information management, pp 305–309
23. Wu Z, Ni Z, Gu L, Liu X (2010) A revised discrete particle swarm optimization for cloud workflow scheduling. In: Proceedings of international conference on computational intelligence and security, pp 184–188

24. Yassa S, Chelouah R, Kadima H, Granado B (2013) Multi-objective approach for energy-aware workflow scheduling in cloud computing environments. Scient World J
25. Liu Z, Wang X (2012) A PSO-based algorithm for load balancing in virtual machines of cloud computing environment. Lecture Notes Comput Sci (including subseries Lecture Notes in Artificial Intelligence and Lecture Notes in Bioinformatics) 7331(1):142–147
26. Ramezani F, Lu J, Hussain FK (2014) Task-based system load balancing in cloud computing using particle swarm optimization. Int J Parallel Prog 42(5):739–754
27. Pan K, Chen J (2015) Load balancing in cloud computing environment based on an improved particle swarm optimization. In: Proceedings of the IEEE international conference on software engineering and service sciences, pp 595–5982
28. Sidhu MS, Thulasiraman P, Thulasiram RK (2013) A load-rebalance PSO heuristic for task matching in heterogeneous computing systems. In: Proceedings of the IEEE symposium on swarm intelligence, IEEE symposium series on computational intelligence, pp 180–187
29. Yingqiu L, Shuhua L, Shoubo G (2016) Cloud task scheduling based on chaotic particle swarm optimization algorithm. In: Proceedings of international conference on intelligent transportation, big data and smart city, pp 493–496
30. Wu D (2018) Cloud computing task scheduling policy based on improved particle swarm optimization. In: Proceedings of international conference on virtual reality and intelligent systems, pp 99–101
31. Al-Maamari A, Omara FA (2015) Task scheduling using PSO algorithm in cloud computing environments. Int J Grid Distributed Comput 8(5):245–256
32. Pradhan A, Bisoy SK (2020) A novel load balancing technique for cloud computing platform based on PSO. J King Saud University—Comput Inform Sci
33. Richa, Keshavamurthy BN (2020) Improved PSO for task scheduling in cloud computing. In: Frontiers in intelligent computing: theory and applications, pp 467–477
34. Alsaidy SA, Abbood AD, Sahib MA (2020) Heuristic initialization of PSO task scheduling algorithm in cloud computing. J King Saud University—Comput Inform Sci
35. Mirjalili S, Mirjalili SM, Lewis A (2014) Grey wolf optimizer. Adv Eng Softw 69:46–61
36. Gupta P, Ghrera SP, Goyal M (2018) QoS aware grey wolf optimization for task allocation in cloud infrastructure. In: Proceedings of first international conference on smart system, innovations and computing, vol 79, no 1
37. Khalili A, Babamir SM (2017) Optimal scheduling workflows in cloud computing environment using pareto-based grey wolf optimizer. Concurr Comput 29(11):1–11
38. Gobalakrishnan N, Arun C (2018) A new multi-objective optimal programming model for task scheduling using genetic gray wolf optimization in cloud computing. Computer Journal 61(10):1–14
39. Natesha BV, Sharma NK, Domanal S, Guddeti RMR (2018) GWOTS: Grey wolf optimization based task scheduling at the green cloud data center. In: Proceedings of 14th international conference on semantics, knowledge and grids, pp 181–187
40. Kumar KP (2018) gravitational emulation-grey wolf optimization technique for load balancing in cloud computing. In: Proceedings of the 2nd international conference on green computing and internet of things, pp 177–184
41. Alzaqebah A (2019) Optimizer in cloud computing environment. In: 2nd International conference on new trends in computing sciences, pp 1–6
42. Pani AK, Dixit B, Patidar K (2019) Resource allocation using democratic grey wolf optimization in cloud computing environment. Int J Intell Eng Syst 12(4):358–366
43. Nayak SK, Panda CS, Padhy SK (2019) Dynamic task scheduling problem based on grey wolf optimization algorithm. In: 2nd international conference on advanced computational and communication paradigms, pp 1–5
44. Bacanin N, Bezdan T, Tuba E, Strumberger I, Tuba M, Zivkovic M (2019) Task scheduling in cloud computing environment by grey wolf optimizer. In: 27th telecommunications forum
45. Natesan G, Chokkalingam A (2020) An improved grey wolf optimization algorithm based task scheduling in cloud computing environment. Int Arab J Inform Technol 17(1):73–81

46. Bansal N, Singh AK (2018) Grey wolf optimized task scheduling algorithm in cloud computing. In: Proceedings of the 7th international conference on FICTA, pp 137–145
47. Mohammadzadeh A, Masdari M, Gharehchopogh FS, Jafarian A (2020) Improved chaotic binary grey wolf optimization algorithm for workflow scheduling in green cloud computing. Evolut Intell
48. Mirjalili S, Lewis A (2016) The whale optimization algorithm. Adv Eng Softw 95:51–67
49. Strumberger I, Bacanin N, Tuba M, Tuba E (2019) Resource scheduling in cloud computing based on a hybridized whale optimization algorithm. Appl Sci 9(22)
50. Saravanan N, Kumaravel T (2019) An efficient task scheduling algorithm using modified whale optimization algorithm in cloud computing. Int J Eng Adv Technol 9(2):2533–2537
51. Premalatha M, Ramakrishnan B (2019) Hybrid whale-bee optimization (HWBO) based optimal task offloading scheme in MCC. Int J Innov Technol Exploring Eng 8(4):281–292
52. Sanaj MS, Joe PPM, Valanto A (2020) Profit maximization based task scheduling in hybrid clouds using whale optimization technique. Inform Secur J: Global Perspect 29(4):155–168
53. Subalakshmi N, Jeyakarthic M (2020) Optimal whale optimization algorithm based energy efficient resource allocation in cloud computing environment. IIOAB J 11(2):92–102
54. Yang XS (2012) Flower pollination algorithm for global optimization. Unconventional Comput Natural Comput 7445:240–249
55. Gupta I, Kaswan A, Jana PK (2017) A flower pollination algorithm based task scheduling in cloud computing. In: International conference on computational intelligence, communications, and business analytics, pp 97–107
56. Kaur J, Sidhu BK (2017) A new flower pollination based task scheduling algorithm in cloud environment. In: 4th international conference on signal processing, computing and control (ISPCC), pp 457–462
57. Khurana S, Singh RK (2020) Modified flower pollination based task scheduling in cloud environment using virtual machine migration. Int J Innov Technol Exploring Eng (IJITEE) 8(9):856–1860
58. Usman MJ, Ismail AS, Chizari H et al (2019) Energy-efficient virtual machine allocation technique using flower pollination algorithm in cloud datacenter: a panacea to green computing. J Bionic Eng, 354–366
59. Gokuldhev M, Singaravel G, Mohan NRR (2020) Multi-objective local pollination-based gray wolf optimizer for task scheduling heterogeneous cloud environment. J Circuit Syst Comput 29(7)
60. Khurana S, Singh RK (2020) Workflow scheduling and reliability improvement by hybrid intelligence optimization approach with task ranking. EAI Endorsed Trans Scalable Inform Syst 7(27)

# Chapter 13
# Electromagnetic Modelling and Parameters Extraction of Metal Contact and Capacitive Type RF MEMS Switch

**Manish Gaikwad, Nitin Deshmukh, and Vishram Sawant**

## 1 Introduction

The trending radio frequency micro electro mechanical system (RF MEMS) devices are beneficial in numerous fields, like defence, medical, telecommunication and space applications, and IoT. The increasing demand of high-speed technologies to operate at very high frequencies and at high-speed internet has given tremendous scope for development of RF MEMS switches [1, 2]. The advanced wireless criteria are coming up which necessitate reconfigurable devices to decrease the number of mechanisms and the cost of the equipment. However, system performance turns out to be more critical when large number of switches are used [3]. The classification of RF MEMS switches is done on the basis of its actuation mechanism (thermal, electrostatic, piezoelectric, and magnetostatic), type of contact (metal to metal and capacitive), actuation movement (lateral and vertical), and circuit configuration (series and shunt) [4, 5]. As the device size scale comes under microns, it offers several advantages like high linearity, low power consumption, low cost, high reliability, low insertion loss, low operating voltage, and high isolation; therefore, it also gives an opportunity to fabricate the device in large number of scale [6, 7].

The high actuation voltage is undesirable while designing and fabricating the device. The actuation voltage of switch can be reduced by using dielectric material over contact electrode and by applying stoppers of high capacitance. The challenges involved in DC contact switch are surface degradation, less lifetime, high actuation voltage, and contact heating. However, in capacitive RF MEMS switch, the dielectric charging is a major drawback which increases release time and self-actuation of switch. The stiction in capacitive contact switches occurs due to charging of dielectric material [8].

The purpose of RF MEMS is to produce devices with low cost, light weight, and consumes low power. The MEMS is a mechanized expertise used to produce

M. Gaikwad (✉) · N. Deshmukh · V. Sawant
Rajiv Gandhi Institute of Technology, Andheri, Mumbai 400053, Maharashtra, India

© The Author(s), under exclusive license to Springer Nature Singapore Pte Ltd. 2022

H. M. Dubey et al. (eds.), *Artificial Intelligence and Sustainable Computing*,
Algorithms for Intelligent Systems, https://doi.org/10.1007/978-981-16-1220-6_13

tiny mechanical devices with incorporated electronics by integrated circuit batch processing techniques. The MEMS are geometrically categorized into cantilever type and fixed ends support structures which sense, actuate, and control the system parameters. The micro machining emerging out of MEMS technology has shown great potential to enhance the accuracy and efficiency of industrial products to improve the quality of consumer needs [9].

In this paper finite element analysis (FEA) is performed using Ansys software to analyse the mechanical stability of switch and stresses involved in switch membrane during actuation. The main objective of this work is to design a switch for low actuation voltage and for understanding the effect of RF frequency on the proposed design by using electromagnetic analysis. The design, simulation, and analysis approach is used to realize the electrostatic and electromagnetic performance parameters of switch. The comparison of results of DC contact and capacitive contact switch used in this paper is shown for better understanding of its performance parameters.

## 2 Design of Switch

The proposed design of RF MEMS switch using fixed end supports is shown in Fig. 1. The lower layer of switch gives a platform to entire switch assembly which is made of glass and it is expressed as a substrate. Above the substrate three parallel metal layers are deposited to form a transmission line which conducts the RF signal. For ensuring higher thermal stability of switch, the coplanar waveguides are made of gold material because of its high thermal conductivity. Likewise, the actuating membrane and supporting anchors are designed by assigning gold metallic properties. The impedance characteristics for lossless signal transmission are applied as 50 Ω, and the lesser the impedance, the better will be its performance. This value of characteristic impedance is greatly reliant on the frequency of the applied signal. The four anchors are positioned on two micro grids placed on substrate which are used to clamp the actuating membrane and for maintaining the air gap between two contact points.

(a) (b)

**Fig. 1** Proposed geometrical topologies of RF MEMS switch. **a** Capacitive; **b** metal to metal contact

**Table 1** Design specification of switch

| Parameter | Length (μm) | Width (μm) | Height (μm) |
|---|---|---|---|
| Substrate thickness | 500 | 140 | 500 |
| CPW thickness | 140 | 120 | 0.15 |
| Actuating membrane | 70 | 70 | 0.5 |
| Dielectric layer | 140 | 120 | 0.15 |
| Air gap | | | 2 |
| Perforation | 10 | 10 | 0.5 |

The perforations of 10 μm × 10 μm are applied over membrane to increase squeeze film damping and lowering the pull-in voltage. A very thin insulating layer of silicon nitride is placed above the transmission line to improve capacitance in the middle of the transmission line and top membrane. The important specifications of the RF MEMS switch are shown in Table 1.

The working principle of designed switch comes from Newton's spring mass system, where initially spring possesses mechanical energy to lift the membrane steadily without any displacement. To displace the mass of membrane, the pulling voltage acts as an external force that enables it to make contact with signal line. In the presented design, the meanders work as spring, and the membrane mass is lifted up and down during its actuation and off state [10].

## 3 Performance Analysis of Switch

The performance of proposed RF MEMS switch can be extracted and analysed by simulation tools such as Ansys and high-frequency structure simulator (HFSS). In this work for the electromechanical analysis of switch Ansys is used and electromagnetic performance of switch is observed by using HFSS.

### *3.1 Deflection and Stresses*

In Ansys the 20-node solid 186 element is used for generating 3D geometry, meshing, and structural analysis of the switch because it supports higher deflection, stress hardening, and high elasticity values. The actuation force is applied on membrane to realize the desired deflection of beam and mechanical stresses generated during this movement are recorded. The maximum deflection obtained by performing the simulation is 2 μm. Further, to understand the mechanical strength of switch, von

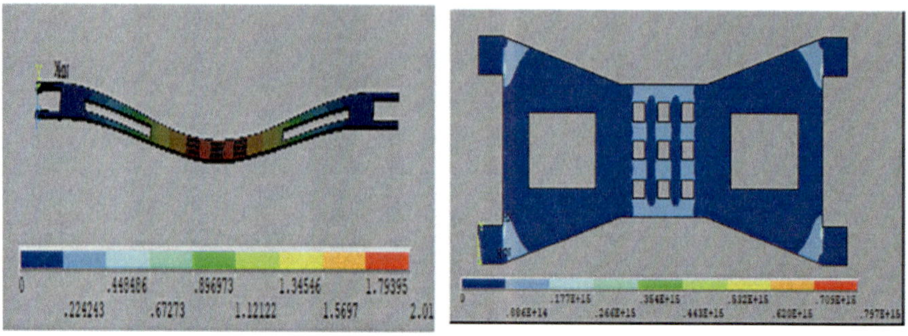


(a) (b)

**Fig. 2** Performance outcomes of switch during actuation. **a** Deflection of switch; **b** mechanical stresses induced on membrane

Mises stresses are distinguished and characterized on the beam by way of contact area, minimum and maximum stresses as shown in Fig. 2. The 775 μN force enables the membrane to make contact with bottom electrode which produces maximum mechanical distortion stress as 797 kPa near the corner points of the anchor.

### 3.2 *Natural Frequency of Switch*

The natural frequency is one at which a system has a tendency to oscillate in the nonexistence of any driving force. While an object oscillates at a frequency parallel to its natural frequency, its vibration generosity grows substantially which could create long-lasting damage to the system. The structural behaviour of RF MEMS switch on various modes of natural frequency is shown in Fig. 3. The modal simulation is performed by fixing all degrees of freedom and considering four modes for natural frequency. As the number of modes increases then frequency also increases.

The analytical method is also used to calculate the natural frequency of switch with respect to the increasing values of beam thickness. When thickness of beam increases the natural frequency of beam also starts to increase. The mathematical expression of natural frequency is given by [1],

$$\mathrm{fn} = \frac{1}{2\Pi}\sqrt{\frac{k}{m}} \tag{1}$$

where k is the beam stiffness in N/m and m is mass of beam.

Figure 4 presents the natural frequency and release time of switch for capacitive and metal to metal contact RF MEMS switch. At the lowest possible thickness higher values of release time are obtained, but such thin metal membrane has a chance of stiction problem because of low stiffness. Therefore, there is a need to optimize

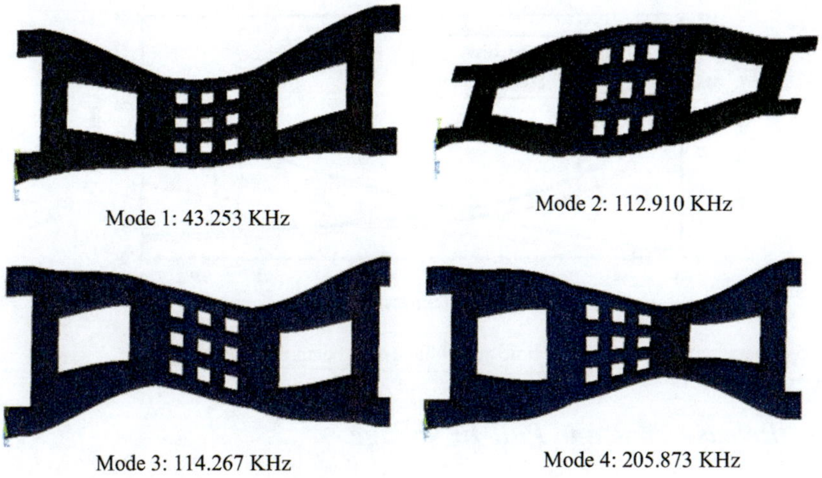


**Fig. 3** Modal analysis of switch

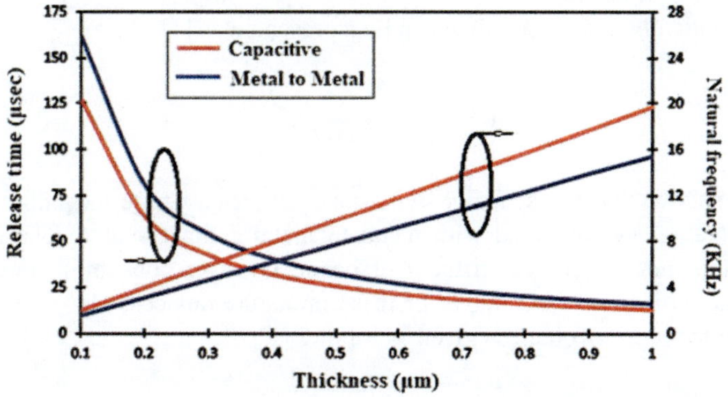


**Fig. 4** The comparison of natural frequency and release time on varying beam thickness

the switch specifications by analysing all possible performance characteristics for ensuring better stability and reliability of switch. The calculated natural frequency at optimized beam thickness of 0.5 μm is 9.83 kHz for capacitive design and 7.72 kHz for metal to metal contact switch. Therefore, the capacitive design exhibits more natural oscillations than DC contact switch.

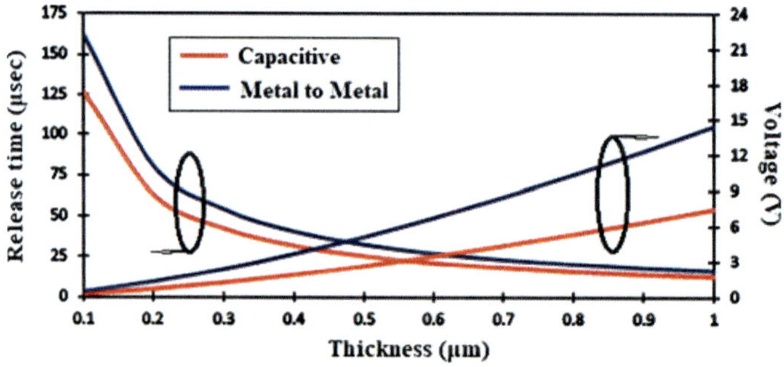


**Fig. 5** Characterisation of release time and voltage over beam thickness

## *3.3 Release Time and Pull-In Voltage*

The time required for membrane to separate contact with bottom electrode and come back to its original position is called release time of switch. It greatly depends on the thickness value of beam.

Mathematically, the release time can be represented as [1]

$$t_r = \frac{1}{4fn} = \frac{1}{2\Pi}\sqrt{\frac{K}{m}} \tag{2}$$

Figure 5 shows that the capacitive switch takes less release time and pull-in voltage during actuation as compared with metal to metal contact switch. The increase in beam thickness improves stiffness of beam, however consumes more power and voltage which is not desirable for these miniature devices. The mathematical expression for pull-in voltage is given as follows [1]:

$$V_{pi} = \sqrt{\frac{8Kg^3}{27A\,\epsilon_0}} \tag{3}$$

where A is capacitive contact area between electrodes in $\mu m^2$, g is air gap in µm. The release time of switch at 0.5 µm beam thickness is 25.4 µs and 32.4 µs for capacitive and DC contact switch. Likewise, the actuation voltage is 2.64 V and 5.1 V for capacitive and metal to metal contact switch, respectively.

## *3.4 Pull-In Voltage and Deflection*

The FEA method is used to realize the effect of electrostatic force on deflection and pull-in voltage of switch. The increasing deflection steps of membrane cause raise in pull-in voltage and this phenomenon is represented in Fig. 6. The capacitive design

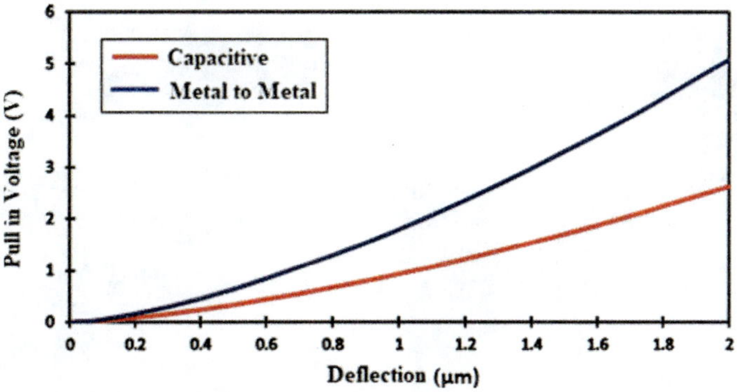


**Fig. 6** Pull-in voltage over stepwise deflection of switch

of switch is highly sensitive to applied voltage and requires less voltage to make contact with bottom electrode. However, the metal contact switch is consuming high voltage and requires high electrostatic force to travel from top to bottom electrode.

## 4 Electromagnetic Analysis of Switch

The electromagnetic waves are responsible for transmitting energy in the form of UV light, microwaves, gamma rays, wireless signals, and infrared radiations. The electric field investigation is executed to understand the electric charge intensity on transmission line as well as on actuating membrane. For this analysis the electrically conducted material and wave port are selected wisely, and the radiation pattern is applied to ensure the radiation of switch only in specified atmosphere. Figure 7 shows the electric field effect on capacitive and DC contact RF MEMS switch.

The energy of electromagnetic waves is transmitted through pulsations of electric and magnetic fields. The maximum E-field effect measured on Au transmission line is $2.094 \times 10^6$ V/m and for membrane is $2.20 \times 10^6$ V/m. Further, maximum E-field effect on silicon nitride is $1.234 \times 10^6$ V/m and on membrane is $2.53 \times 10^6$ V/m. The electromagnetic analysis is performed to realize the changes in designed topologies when RF power is supplied to it. The designed switch has tendency to reflect, absorb, and transmit the RF power through transmission line, and based on this the scattering parameters are obtained [11].

The current advancements in technologies have made it possible to manufacture devices that has virtually no mass and consumes almost negligible DC power. The linear step frequency setup for interpolating sweep type analysis is enabled from analysis setup. The step size for frequency setup is fixed at 0.1 GHz by applying 0.5% error tolerance.

The insertion loss of the switch is dominated by reflection coefficient and the loss of transmission line up to 15 GHz to keep it below −0.1 dB. Figure 8 shows the

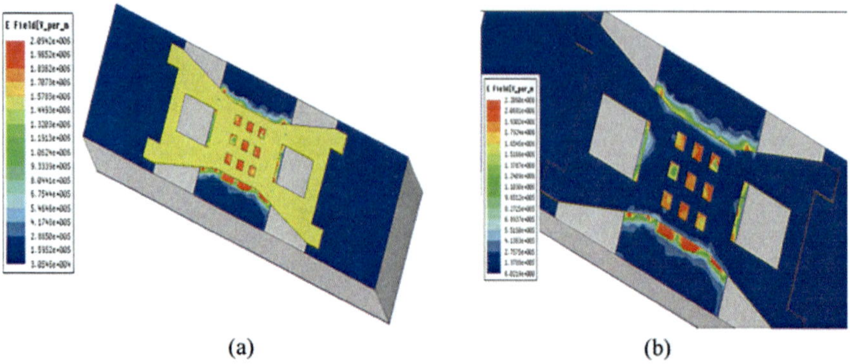


**Fig. 7** E-field effect on proposed RF MEMS switch. **a** Electric field effect on Au transmission line, **b** E-field effect on DC contact switch membrane, **c** E-field effect on dielectric layer, **d** E-field effect on capacitive switch membrane

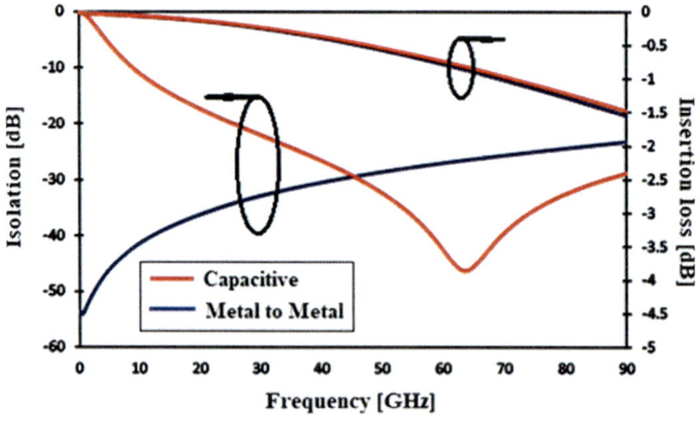


**Fig. 8** Analysis of insertion loss and isolation

insertion loss as −1.47 dB and −1.54 dB at 90 GHz and the measured isolation is −28.73 dB and −23.18 dB at 90 GHz for capacitive and metal to metal contact switch, respectively. The highest isolation is obtained for metal to metal contact switch and is only up to 2.4 GHz frequency which starts to decrease by increasing radio frequency of switch. However, capacitive switch design shows very high isolation at frequency more than 60 GHz.

The input rf power reflections due to mechanical imperfections and impedance mismatch are measured in terms of return loss of switch. The measured return loss of switch in upstate is −5.66 dB and −5.84 dB at 90 GHz for metal to metal contact and capacitive contact switch, respectively, and it is shown in Fig. 9.

Figure 10 shows the relation between isolation and return loss of switch in actuated and unactuated state. The return loss of switch in down state is −0.20 dB and −0.20 dB at 90 GHz, respectively, for metal to metal contact and capacitive

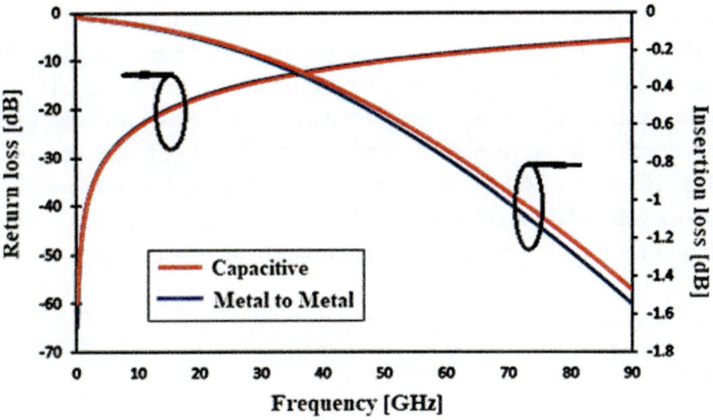


**Fig. 9** Analysis of return loss and insertion loss

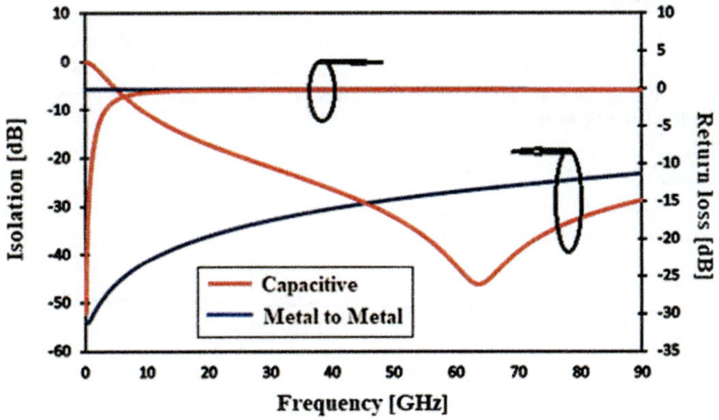


**Fig. 10** Analysis of isolation and return loss

contact switch. Therefore, S matrix significantly permits to correctly describe the behaviour of complex networks more effortlessly and helps to understand the S parameters on various frequency bands.

## 5 Conclusion

In this paper the novel RF MEMS metal to metal contact and capacitive contact switch is presented and the performance characteristics of both designs are obtained by using analytical and simulation method. The proposed design is modelled by applying appropriate perforations to minimize actuation voltage of switch. The achieved actuation voltage for capacitive switch is 2.64 V and 5.1 V for metal to metal contact switch. In terms of release time, the switch capacitive design shows 27% more release speed than metal to metal contact design of switch. The electric field analysis is performed using Ansys to understand the electrical surroundings of structure of charges. The maximum electric field obtained at 90 GHz frequency suggests high electrical charges on metal transmission line as compared to dielectric layer. The recorded RF performance switch at 90 GHz frequency shows insertion loss as −1.47 dB and −1.54 dB for capacitive contact and metal to metal contact switch, respectively. The maximum isolation for capacitive switch is −46.28 dB at 63.66 GHz and for metal to metal contact is more than −50 dB up to 2.4 GHz.

## References

1. Rebeiz GM (2013) RF MEMS theory, design, and technology. Wiley New York
2. Srinivasa Rao K, Vasantha K, Girija Sravani K (2020) Design and simulation of capacitive SPDT RF MEMS switch to improve its isolation. Microsyst Technol. https://doi.org/10.1007/s00542-020-04795-9
3. Angira M, Rangra K (2014) A low insertion loss, multi-band, fixed central capacitor based RF-MEMS switch. Microsyst Technol (2014). https://doi.org/10.1007/s00542-014-2378-2
4. Yuan X-L, Huang Q-A, Liao X-P (2008) Analysis of electromagnetic interference of a capacitive RF MEMS switch during switching. Microsyst Technol 14:349–360
5. Jaafar H, Beh KS, Yunus NAM, Hasan WZW, Shafie S, Sidek O (2014) A comprehensive study on RF MEMS switch. Microsyst Technol 20:2109–2121
6. Kurmendra, RK (2020) A review on RF micro-electro-mechanical-systems (MEMS) switch for radio frequency applications. Microsystem Technologies. https://doi.org/10.1007/s00542-020-05025-y
7. Ravirala AK, Bethapudi LK, Kommareddy J, Thommandru BS, Jasti S, Gorantla PR, Puli A, Karumuri GS, Karumuri SR (2017) Design and performance analysis of uniform meander structured RF MEMS capacitive shunt switch along with perforations. Microsyst Technol. https://doi.org/10.1007/s00542-017-3403-z
8. Peroulis D, Capacitive MEMS switches. Encyclopedia of Nanotechnology. https://doi.org/10.1007/978-90-481-9751-4
9. Srinivasa Rao K, Ashok Kumar P, Guha K, Sailaja BVS, Vineetha KV, Baishnab KL, Girija Sravani K (2018) Design and simulation of fixed–fixed flexure type RF MEMS switch

for reconfigurable antenna. Microsystem Technologies. https://doi.org/10.1007/s00542-018-3983-2

10. Shanthi G, Srinivasa Rao K, Girija Sravani K (2020) Design and analysis of a RF MEMS shunt switch using U-shaped meanders for low actuation voltage. Microsyst Technol. https://doi.org/10.1007/s00542-020-04864-z
11. Giridhar MS, Prasobhkumar PP, Jambhalikar A, John J, Islam R, Vedaprakash K, Viswanathan M (2015) Estimation of power handling limits of ohmic contact rf mems switches due to joule heating. ISSS national conference on MEMS smart materials, structures and systems, Cochin

# Chapter 14
# Artificial Neural Networks as a Tool for Thermal Comfort Prediction in Built Environment

**Rishika Shah, R. K. Pandit, and M. K. Gaur**

## 1 Introduction

Over the past few years, a growing trend has been observed interlinking environmental wellbeing, quality of life, and urban microclimate. The alarming rise in temperature levels worldwide has also been a concern for city planners and urban designers [1]. Ensuing global climate change has aggravated the discomfort levels of city dwellers which often suffer from heat stress issues, leading to increasing fatality rate, especially when urbanization is at its peak and cities are adding up tons of built up areas to already existing ones. It must be noted that comfortable cityscapes are significantly related to mental and physical health of city dwellers. For this reason, the idea of outdoor thermal comfort assessment is being deemed important and is a matter of research that should be involved in urban planning practices and developing potential smart cities [2]. For many years, researchers in city planning have worked over urban heat island issues to contest the rising temperatures. The progression in this field has brought the introduction of advanced tools in evaluation of thermal comfort. In accordance with this, a key goal in the city building industry is achieving satisfactory thermal comfort levels through urban design elements and policy implementation while making functional spaces and building models that emerge as paradigm for sustainable development goals [3].

Conventionally, thermal comfort in built environments has been examined using various thermal comfort indices which came into being since the 1970s, such as physiologically equivalent temperature (PET) and predicted mean vote (PMV) [4]. The most recent thermal comfort evaluation scale is UTCI which is designed to be used in universal context. These variants use statistical methods to analyze the survey results and meteorological parameters. Though these methods have brought an initiation in thermal comfort evaluation, these do not find application that can

R. Shah (✉) · R. K. Pandit · M. K. Gaur
Madhav Institute of Technology and Science, Gwalior 474005, India

© The Author(s), under exclusive license to Springer Nature Singapore Pte Ltd. 2022

H. M. Dubey et al. (eds.), *Artificial Intelligence and Sustainable Computing*, Algorithms for Intelligent Systems, https://doi.org/10.1007/978-981-16-1220-6_14

be applied in dynamic environment with complex narratives. Consequently, artificial intelligence presents leading edge technology toward analytical techniques of modeling the complex and nonlinear nature of the interaction between inhabitants and their built environments. Machine learning, and more so artificial neural network (ANN), is being widely employed to meet this purpose [5].

It is observed that, however, an extensive amount of critical review is available in the literature on application of ANN for indoor built environments, but there is still a lack of review studies on use of ANN for prediction of outdoor thermal comfort in built environment. On that account, a short review was conducted on the same. Out of 135 studies shortlisted on the basis of keywords search, like "thermal comfort", "PET", "PMV", "ANN", "urban microclimate", and combination of these, only 17 studies were found to have direct connection with the use of ANN as a tool in analyzing outdoor thermal comfort. Moreover, these studies range from 2011 to 2020. Thus out of the few studies shortlisted, the aim of this work is to (i) present the studies published over the past 10 years on ANN-based thermal comfort prediction; (ii) analyze the models and framework of these studies; and (iii) highlight potential future research directions.

## 2 ANN and Thermal Comfort in Built Environment

Over recent years, application of artificial neural networks (ANNs) has enhanced in various aspects of built environment studies. Regardless of this and advantages of the approach, in particular nonlinearity, and wide variety of data sets there is, overall, a limited number of literatures available. It cannot be denied that building physics and urban physics is closely intertwined while working on temperature levels, urban heat island, thermal comfort, and energy load of an urban area.

But with better employment of ANN for indoor built environment, the attention needs to be directed toward outdoor built environment. While the outdoor built environment, which is governed by many influencing factors (i.e., changing built forms, street networks, green area percentage, etc.), is more dynamic in comparison to indoors, often enough it has been unquestioningly presumed that the same model can be applied for outdoor thermal comfort assessment as that of indoor thermal comfort evaluation.

In view of this fundamental contrast, prediction of outdoor thermal comfort cannot be attained by solely relying on the supposition of linearity between thermal sensation and thermal comfort. The ANN models, which are biologically inspired computational models, have been used to portray complicated nonlinear interrelationships among a multitude of such factors. Though attempts have long begun toward utilization of artificial neural network (ANN) for various aspects of outdoor built environment, there is still a need for scientific involvement in assessment of outdoor thermal comfort.

## 3 ANN-Based Outdoor Thermal Comfort Assessment

On the basis of type of outdoor built environment taken as the study area, the studies can be divided into three main classifiers: built form geometry, urban parks, and street canyons. The brief overview on studies that have used ANN as a tool to predict outdoor thermal comfort has been composed in Table 1. The shortlisted research articles have been categorized into seven categories:

(1) Year of research,
(2) Geographic location,
(3) Climate—on the basis of Köppen climate classification,
(4) Model
(5) Focus parameter—the parameters for which the study was conducted,
(6) Input parameters—the parameters which were monitored to perform the study,
(7) Period of study—for recording and simulation.

The above categories were used to determine how the recent research is and to trace the research trends in the proposed field (year of research), to map the location and microclimate where studies have been conducted and which locations and climate zone have remained unexplored (geographic location and climate), to discover the state-of-the-art in the proposed field (model), to note different parameters considered by different researchers for their objectives and what results were produced because of these parameters or lack thereof (focus parameter and input parameters), and to comprehend the time period considered by researchers (period of study). The accumulation of this categorized information helped in creating the understanding of various findings and research gaps.

## 4 Training Algorithms and Validation

From the above studies, it was learned that the functional process of ANN depends upon the huge historic database for training, the choice of input parameters, and training algorithms, and MLP is the most common type of ANN used. Its structure consists of neurons that are organized in layers, which starts with an input layer followed by hidden layer and output layer. The input layer distributes the input signals to the first hidden layer which transmits the signal with each neuron of the next hidden layer, in correspondence with a weight factor. In this way each neuron gets processed through input and hidden layer in addition to activation function and finally the outcome reaches the output layer. The actual outcome is then compared for its closeness with the desired outcome which determines the accuracy of the ANN model. The training weights thus play a role in reducing the error rate as they are adjusted, and the whole process is repeated until the error reaches an acceptable range.

This matrix-based tool contains patterns and variables represented by rows and columns. The data set which is used in a particular study is known as sample. To

**Table 1** Brief overview of studies on ANN-based outdoor thermal comfort assessment

| Authors | Location | Climate | Model | Focus parameter | Input parameter | Period of study |
|---|---|---|---|---|---|---|
| Bozorgi et al. [6] | Isfahan, Iran | BWk | MLP, Feed forward | Ts | NDBI map, urban pixels, the percentage of green cover pixels within neighborhood size of 25 pixels, and NDVI index within 30 pixels buffer zone | 30 days, June 2015 |
| Chan and Chau [7] | Hong Kong | Cwa | MLP | PMV | Ta, Tmrt, RH, WS, met, clo | 11 am to 5 pm Dec 2014 and Mar 2015, between Jun and Aug 2015 and between Oct 2015 and Feb 2016 |
| Chronopoulos et al. [8] | Greece | Csa | MLP, back propagation | Ta | Ta, RH | 12 Jun 2003 (15:00 h) to 16 Aug 2003 (04:00 h) |
| Dombaycı and Golcu [9] | Turkey | Csa | MLP | Ta | Month of the year and day of the month and also the Ta value of the previous day | Year 2006 |
| Gobakis et al. [10] | Greece | Csa | Elman, Feed-Forward and Cascade | Ta | Date, Time, Ta, SR | 06/04/2009 to 07/09/2009 |
| Moustris et al. [11] | Greece | Csa | MLP, back propagation | PET | Number of month (1–12), the number of day (1–31), the hour of the day (1–24), and Ta | May–Sep and Oct–Apr |

(continued)

**Table 1** (continued)

| Authors | Location | Climate | Model | Focus parameter | Input parameter | Period of study |
|---|---|---|---|---|---|---|
| Weerasuriyaa et al. [11] | Hong Kong & Netherlands | Cwa & Cfb | MLP, feed forward | UTCI | Ta, Tmrt, RH, WS | |
| Vouterakos et al. [12] | Greece | Csa | MCA, feed forward | DI | Station number, consecutive hours during the day when DI ≥ 24°C for each one of the next 3 days, month number, Ta, RH, DI for each one of the six previous days | May–Sept 2009 |
| Protic et al. [13] | Serbia | Cfa | MLP | PET | Ta, VP, SR, WS | Jul and Aug 2014 |
| Papantoniou and Kolokotsa [14] | Ancona—Italy, Chania—Greece, Granada—Spain and Mollet—Spain | | Feed forward, Elman, Cascade | Ta | Time, Ta, SR, WS | First 20 days of Jan |
| Moustris et al. [15] | Greece | Csa | MLP | CP, LD | Ta, VP, SR, WS, RH | 2015 to 2017 |
| Moustris et al. [16] | Greece | Csa | MLP | CP, DI | | May–September |
| Moustris et al. [17] | Greece | Csa | MLP | PET | Ta, SR, WS, RH | 15/06/2005–31/12/2011 |
| Kamoutsis et al. [17] | Greece | Csa | MLP | THI | Ta, RH | 23 June and 28 August 2007 |
| Chronopoulos et al. [18] | Greece | Csa | MLP | THI | Ta, RH | 23 June and 28 August 2007 |

(continued)

**Table 1** (continued)

| Authors | Location | Climate | Model | Focus parameter | Input parameter | Period of study |
|---|---|---|---|---|---|---|
| Lee et al. [19] | Seoul, Korea | Dwa | MLP | UHI | Ta, Precipitation, RH, WD and WS albedo, water area, green area, road area, building area | 01:00 January 1st 2012 to 24:00 December 31st 2012 |
| Ketterer and Matzarakis [20] | Stuttgart, Germany | | MLP, Feed forward | PET | Altitude, Slope, H/W, cold air flow, WS, land use types, build-up area, SVF, thermal image, building volume and number of trees | July 3 and 4, 2014, every 10 s |

develop the predictive model of high accuracy, the obtained sample is divided into three "subsets" for training, validating, and testing the model. The training subset uses the data to "train-educate the network" to achieve best fit of nonlinear function. The validation subset helps determining the adjustment of weights or evasion of over training through error evaluation, while the development of neural network is under process. And the test subset, the unknown data set for network, is used to test the network's efficiency.

This development process can be described mathematically as follows [21]:

$$u = \sum_{i=1}^{n} w_i x_i \tag{1}$$

$$\mathrm{y} = \mathrm{f}\,(u - \theta) \tag{2}$$

where $w_i$ is the weights, $x_i$ is the input parameters for the developed ANN model, y is the output parameter that is to be predicted, u is the activation potential of neuron, θ is the threshold value, and *f ()* is the activation function.

Due to its simplicity, sigmoid function (Eq. 3) is most popularly used in development of artificial neural networks.

$$f(x) = \frac{1}{\left[1 + \exp(-x)\right]} \tag{3}$$

In the above-mentioned studies, four types of neural network were found: feed forward, back propagation, cascade, and Elman. When the flow of transmission of input signals is from one end (layer) to another (layer) in forward manner, this type of network is known as feed forward neural network. Another type of most applied neural network is back propagation neural network, which has a forward pass and backward pass as steps of programming. The weight of the network is fixed in the course of forward pass. On the contrary, the weights are adjusted in backward pass to work upon the error. The cascade neural network also consists of the same structure of layers: input layer, hidden layers, and output layers. In cascade neural network the first hidden layer receives from input layer, but the next hidden layer receives weight from input layer as well as from previous hidden layer, which have biases. This goes on till the last hidden layer after which there is an output layer. The Elman neural network is the same as feed forward neural network, except for the context layer and its additional input parameters.

Algorithms such as scaled conjugate gradient (SCG), BFGS quasi-Newton, and Levenberg–Marquardt (LM) are faster algorithms than the other algorithms and use standard numerical optimization techniques [9]. The Levenberg–Marquardt algorithm was used for training of networks.

In order to validate the predictive accuracy of the developed models, the most common statistical evaluation indices found in the literature are coefficient of determination ($R^2$), root mean square error (RMSE), mean bias error (MBE), and index of agreement (IA), which are used generally to evaluate forecasting models. The coefficient of determination gives the idea about the percentage of variance and are checked by values ranging from 0 to 1. The model's accuracy is directly proportional to the closeness of value toward 1. Root mean square error calculates the models accuracy on how well the predicted value matches with the observed value. The lower the value of RMSE, the better the predictive ability of the model. Similarly, lower the MBE better the accuracy. Since the principal concern is forecast values of thermal comfort with respect to their corresponding thermal classes, the statistical indices used were true predicted rate (TPR), which represents the ratio of predicted thermal comfort values that were predicted correct, false alarm rate (FAR), which represents the ratio of predicted thermal comfort values that do not fall in the thermal classes as that of observed values, and success index, which validates the success of developed models by checking whether the predicted values fall in the particular thermal class or not.

## 5 Results

As stated earlier, a total of 135 studies were shortlisted, out of which only 10% of studies had direct link with investigation of outdoor thermal comfort assessment using ANN (Fig. 1). In this short review, it is observed that out of the few studies available in the context of ANN and assessment of outdoor thermal comfort, most are based on built form geometry (48%) followed by urban parks (26%) and street canyons (13%) (Fig. 2). As is evident from Table 1, air temperature is the most common parameter serving as both input parameter and focus parameter for prediction of thermal comfort. The most common input parameters are meteorological variables—relative humidity, solar radiation, and wind speed. Only few studies have taken into account of personal factors, such as clothing insulation and metabolic rate. PET has been determined as most popular output variable to assess thermal comfort followed by PMV, THI, and UTCI (Fig. 3). Most studies have used day-time data set mainly

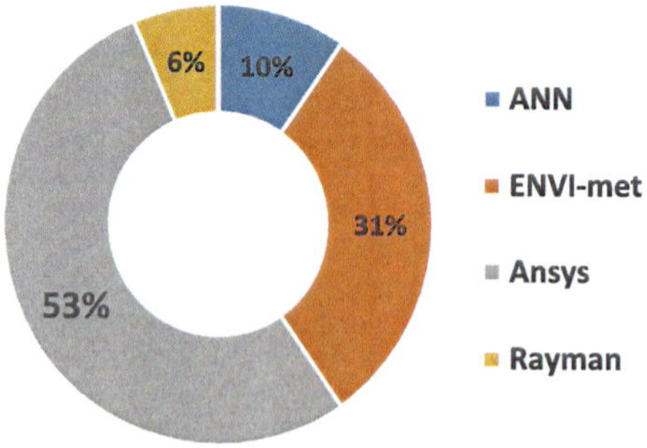


**Fig. 1** Graph showing percentage of ANN studies among other tools for investigating outdoor thermal comfort

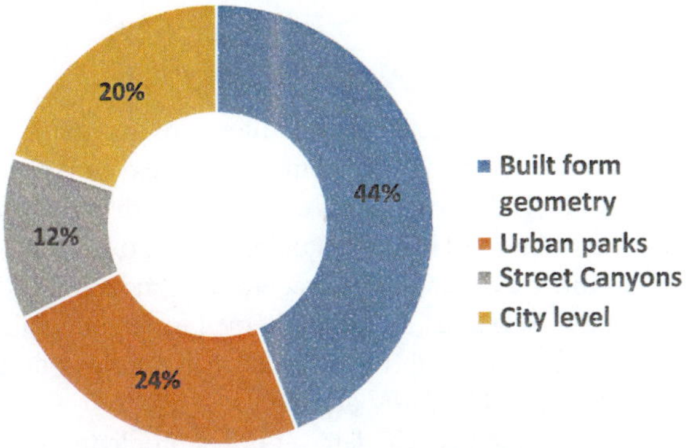


**Fig. 2** Percentage share of ANN-based studies on the basis of selected built environment setting

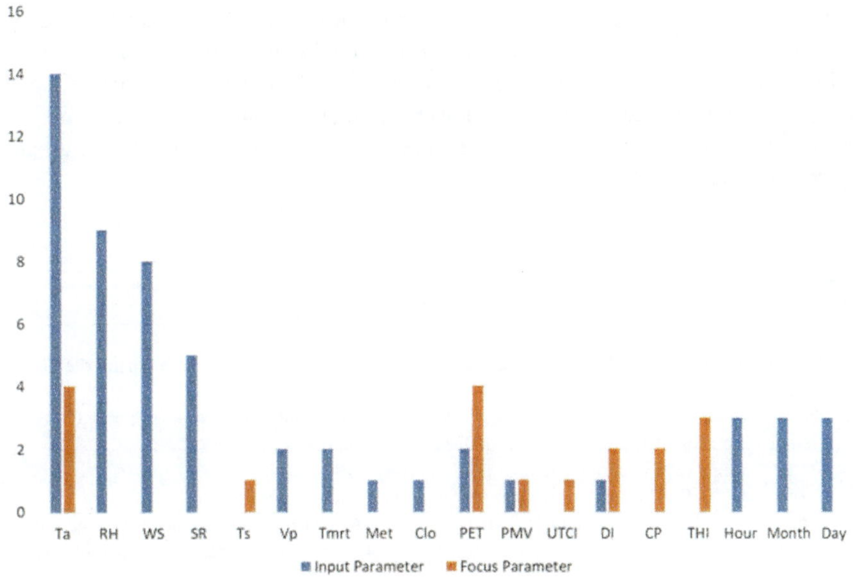


**Fig. 3** Share of input and focus parameters in outdoor thermal comfort analysis using ANN

during the warm and humid period, either for a span of 4–5 months or 2–3 days. In case of time period spanning for months, hourly recording of data has been considered, and in case of few days, data has been monitored in an interval of seconds or 15 min. With such low number of cases it is hard to comment on the dearth of geographic locations and climate zones studies.

## 6 Conclusion

It is apparent from the research articles that the literature in regard of evaluation of outdoor thermal comfort in the built environment is devoid of ANN as a potential tool. The thermal comfort predictive models concerned with just air temperature as their output variable were observed in initial studies. With more advanced monitoring devices it became possible for researchers to record solar radiation and mean radiant temperature. These output variables, however, helped to map the urban heat island (UHI) which has direct linkages with outdoor thermal comfort but did not predict direct thermal comfort indices like PMV and PET.

The availability of this wide historic data set poses problem, making the studies a tough task in the stage of data collection. Another shortcoming that has arrived from these studies is that the ANN models developed in these studies cannot be used for the same scenarios in different locations. Lack of generalization of these models can be a possible future direction to be explored. Similarly, these studies have been observed to work only on supervised learning method, and future studies in built environment can employ more complex neural network architecture. Future studies can also explore the study of thermal comfort in street canyons on the basis of urban physics parameters to create a generalized model for street canyons of selected urban physics configuration.

## References


1. Basha G et al (2017) Historical and projected surface temperature over india during the 20 th and 21 st century, 1–10. https://doi.org/10.1038/s41598-017-02130-3
2. Peng C, Elwan A (2014) An outdoor-indoor coupled simulation framework for Climate Change–conscious Urban Neighborhood Design. Simulation 90:874–891
3. United Nations (2017) Global status report 2017
4. Hoof J van (2010) Thermal comfort: research and practice. Front Biosci 15, 765
5. Yang S, Pun M, Chen W, Feng B, Dubey S (2020) Model predictive control with adaptive machine-learning-based model for building energy efficiency and comfort optimization. Appl Energy 271:115147
6. Bozorgi M, Nejadkoorki F (2018) Land surface temperature estimating in urbanized landscapes using artificial neural networks. Environ Monit Assess. https://doi.org/10.1007/s10661-018-6618-2
7. Chan SY, Chau CK (2019) Development of arti fi cial neural network models for predicting thermal comfort evaluation in urban parks in summer and winter. Build Environ 164:106364
8. Chronopoulos KI, Tsiros IX, Dimopoulos IF, Alvertos N (2008) An application of artificial neural network models to estimate air temperature data in areas with sparse network of meteorological stations. J Environ Sci Heal Part A Toxic/Hazard. Subst. Environ, 1752–1757. https://doi.org/10.1080/10934520802507621
9. Dombaycı OA, Golcu M (2009) Daily means ambient temperature prediction using artificial neural network method: a case study of Turkey. Renew Energy 34:1158–1161
10. Gobakis K et al (2011) Development of a model for urban heat island prediction using neural network techniques. Sustain Cities Soc 1:104–115

11. Moustris K, Tsiros IX, Tseliou A, Nastos P (2018) Development and application of artificial neural network models to estimate values of a complex human thermal comfort index associated with urban heat and cool island patterns using air temperature data from a standard meteorological station. Int J Biometeorol. https://doi.org/10.1007/s00484-018-1531-5
12. Vouterakos PA et al (2012) Forecasting the discomfort levels within the greater Athens area, Greece using artificial neural networks and multiple criteria analysis. Theor Appl Climatol 110:329–343
13. Ivana S, Bogdanovic P, Vukadinovic AV, Jasmina M, Radosavljevic MA, Mitkovic MP (2016) Forecasting of outdoor thermal comfort index in urban open spaces The Nis Fortress Case Study. Therm Sci 20:1531–1540
14. Papantoniou S, Kolokotsa D (2015) Prediction of outdoor air temperature using Neural Networks; application in 4 European cities. Energy Build. https://doi.org/10.1016/j.enbuild.2015.06.054
15. Moustris K, Kavadias KA, Za D, Kaldellis JK (2020) Medium, short and very short-term prognosis of load demand for the Greek Island of Tilos using artificial neural networks and human thermal comfort-discomfort biometeorological data 147:100–109
16. Moustris KP, Tsiros IX, Ziomas IC, Paliatsos AG (2010) Artificial neural network models as a useful tool to forecast human thermal comfort using microclimatic and bioclimatic data in the great Athens area (G. J. Environ. Sci. Heal., Part A Toxic/Hazard Subst Environ 45:447–453
17. Moustris KP, Nastos PT, Paliatsos AG (2013) One-day prediction of biometeorological conditions in a mediterranean urban environment using artificial neural networks modeling. Adv Meteorol 2013
18. Chronopoulos K, Kamoutsis A, Matsoukis A, Manoli E (2012) An artificial neural network model application for the estimation of thermal comfort conditions in mountainous regions. Greece Atmósfera 25(2):171–181
19. Lee YY, Kim JT, Yun GY (2016) The neural network predictive model for heat island intensity in Seoul. Energy Build 110:353–361
20. Ketterer C, Matzarakis A (2016) Mapping the Physiologically Equivalent Temperature in urban areas using artificial neural network. Landsc Urban Plan 150:1–9
21. Viotti P, Liuti G, Genova PDi (2002) Atmospheric urban pollution: applications of an artificial neural network (ANN) to the city of Perugia. Ecol Modell 148:27–46

# Chapter 15
# Impact of Channel-Hot-Carrier Damage in Dopingless Devices at Elevated Temperature

**Meena Panchore and Jawar Singh**

## 1 Introduction

In the recent past, junctionless field-effect transistor (JLFET) has attracted remarkable attention due to its enhanced current driving capability and superior gate controllability over the channel that minimizes the short-channel effects (SCEs) [1]. Although the performance of these devices is superior in many fonts, but the requirement of high doping concentration makes it vulnerable to process variation issues caused by the random dopant fluctuation (RDF) [2, 3]. Hence, in literature, the problem of RDF is addressed by keeping the channel almost intrinsic (lightly doped) along with source/drain electrode workfunction engineering, and the device is referred to as the dopingless (DL) JLFETs [4, 5], which offer enough fabrication margins to mitigate variability issues caused by RDF. However, these new device structures suffer from hot-carrier reliability issues when scaled to deep sub-micron regime. Especially when supply voltage does not scale at the same pace with the device dimensions, then the impact of channel-hot-carrier (CHC) [6–8] becomes more pronounced due to the increased electric field with elevated temperature. Therefore, performance evaluation against hot-carrier reliability at high temperature becomes extremely important for these emerging devices. Recently, a few works have been dealt with hot-carrier reliability of JLFETs with different stress conditions. The CHC effect in DL-JLFETs and JLFETs has been studied [9, 10] at room temperature. It was found that the performance of JLFET was degraded under CHC stress

M. Panchore (✉)
Department of Electronics and Communication Engineering, National Institute of Technology, Patna 800005, Bihar, India
e-mail: meenap.ec@nitp.ac.in

J. Singh
Department of Electrical Engineering, Indian Institute of Technology Patna, Patna 801106, Bihar, India
e-mail: jawar@iitp.ac.in

© The Author(s), under exclusive license to Springer Nature Singapore Pte Ltd. 2022 
H. M. Dubey et al. (eds.), *Artificial Intelligence and Sustainable Computing*,
Algorithms for Intelligent Systems, https://doi.org/10.1007/978-981-16-1220-6_15

and deteriorates the overall circuit performance significantly than its counterpart DL-JLFET.

Further, in recent past, the study has suggested that the effect of localized charges in JLFETs is higher at room temperature and reduces at high temperature [11, 12]. But no study is available in the literature yet concerning the high temperature operation of DL and JLFETs due to CHC stress. Hence, in this paper, we investigate the CHC phenomena at high temperature in both dopingless and junctionless devices and circuits. We find that deterioration of drain current in JLFETs is enhanced at elevated temperature due to CHC damage. Especially, the heavily doped JLFET is more vulnerable to CHC damage at elevated temperature as compared to DL devices. Hence, the electrical parameters of JLFET, namely, transconductance, threshold voltage and OFF-state current are more severely affected with this damage than DL-JLFET. The degradation of drain current causes significant impact on delay time of JLFET-based digital circuits than its dopingless counterpart.

## 2 Device Description and Simulation Setup

The cross-sectional view of devices is shown in Fig. 1a, b. The uniform device dimensions are considered for both DL and JLFETs excluding doping concentrations and work functions. The device parameters employed in our simulation are [4]: Si layer thickness ($Tsi$) = 10 nm, oxide thickness = 1 nm, gate length ($L_g$) = 15 nm, and extension for source and drain regions ($L_{ext}$) = 15 nm. The workfunction of gate electrode is taken as 5.5 eV for nJLFET with same doping concentration across all the regions of silicon body ($10^{19}$ cm$^{-3}$). However, in nDL-JLFET, the workfunction of gate is taken as 4.73 eV with intrinsic silicon body ($10^{15}$ cm$^{-3}$). Using charge-plasma (CP) concept, S/D regions are formed and therefore, the range of workfunctions of metal electrodes has been considered [5] as $\phi_m < \chi_{Si} + (E_G/2q)$, where $E_G$ is the bulk silicon bandgap, $\chi_{Si}$ is electron affinity of bulk silicon ($\chi_{Si} = 4.17$) and q is elementary charge. Here, we employed the Hafnium (workfunction = 3.9 eV) to form S/D region on Si layer having thickness of 10 nm. For both p-type DL and JLFET, we have taken the gate electrode workfunctions as 4.86 eV and 3.96 eV,

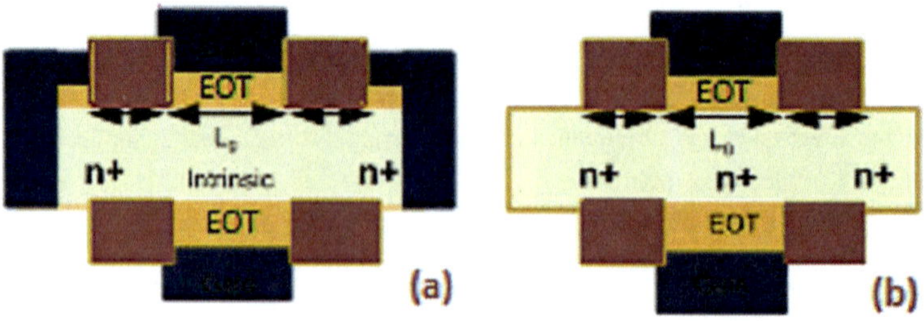


**Fig. 1** Schematic cross-sectional view of **a** nDL and **b** nJLFET

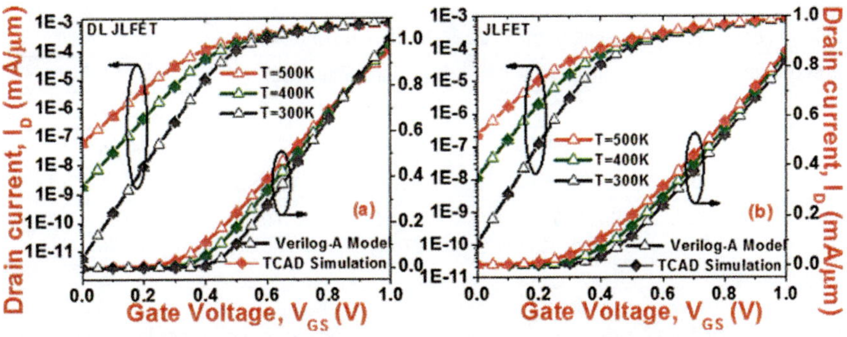


**Fig. 2** Drain current characteristics of **a** DL, and **b** JLFETs at various temperatures

respectively. The simulations for DL and JLFETs were conducted through TCAD software (Silvaco ATLAS) with default parameters for Si [13]. The device-level models, such as band-gap narrowing (BGN), Lombardi mobility model (CVT) and Shockley Read Hall (SRH), were enabled during simulations. To analyse the channel-hot-carrier (CHC) effect in devices, we incorporated hot-carrier injection model (HEI) and energy balance transport (EBT).

In order to examine the CHC damage at elevated temperature, we have chosen n-type device with an effective channel length of 15 nm. The standard measure-stress-measure (MSM) [14] approach is employed in worst-case scenario ($V_D = V_G = 1.9$) for evaluation of CHC stress at various temperatures. In view of this, the higher stress voltage condition is employed than nominal voltage condition, and drain current variation is monitored for pre- and post-CHC stressed devices by varying gate operating voltage from 0 to 1 V while keeping the drain operating voltage constant (1 V). The measurements were conducted at various temperatures with CHC stress. In order to simulate circuit-level performance, the compact model has been developed for devices and validated with Silvaco TCAD results. The drain current characteristics of DL and JLFETs are verified from our previously published paper [4], as shown in Fig. 2a, b. It is noted that the Verilog-A models developed for nDL and nJLFET are matched with TCAD simulation results.

## 3 Results and Discussion

The drain current reduction is monitored by comparing fresh and stressed device characteristics. It can be observed from Table 1 that the drain current is reduced by 20% in DL-JLFET due to CHC stress of 2000s at 500 K. However, reduction in drain current of JLFET is 33.7% due to the same stress and temperature. From this, it is confirmed that drain current deterioration increases due to increase of temperature against CHC stress. However, the high drain current deterioration in JLFET is because of high doping concentration in channel region that leads to injection of large number

**Table 1** Drain current degradation (in $mA/\mu m$) in DL and JLFETs

| Stress conditions | Devices | Temperature (K) | | |
|---|---|---|---|---|
| | | 300 k | 400 k | 500 k |
| Without stress | JLFET | 0.82 | 0.85 | 0.86 |
| | DL JLFET | 1.01 | 0.98 | 0.95 |
| With stress (2000s) | JLFET | 0.65 | 0.62 | 0.57 |
| | DL JLFET | 0.89 | 0.80 | 0.76 |

of charge carriers in oxide near the drain end; as a consequence, considerable change in $V_{TH}$ occurs. However, in DL-JLFET, the drain current deterioration is relatively low with increasing temperature against CHC than conventional JLFET. It is because of lightly doped channel in DL-JLFET that causes low electric field near drain side, and hence assured high reliability due to CHC effect.

The reduction in drain current is due to significant shift in $V_{th}$ at high temperature. This comes from the dominant effect of the mobility reduction and that can be observed from the transconductance ($g_m$) estimation. Therefore, it is important to estimate the high temperature CHC effect on electrical parameters of devices. Hence, here, we have investigated the impact of CHC at different temperatures on various performance metrics of JLFETs. Figure 3a–d indicates the temperature dependence CHC effect on different device parameters, such as $I_{on}$, $I_{off}$, $V_{th}$ and $g_m$, respectively. It is observed that deterioration in device parameters of JLFETs increases with increase

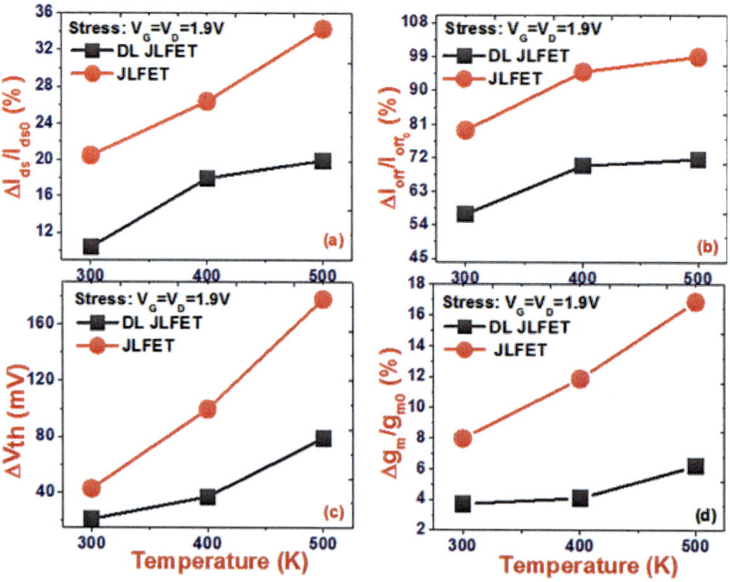


**Fig. 3** Deterioration of **a** ON-state current $I_{on}$, **b** OFF-state current $I_{off}$, **c** $V_{th}$, and **d** $G_m$, DL and JLFETs under CHC stress of 2000s at various temperatures

**Table 2** Operating frequency (in GHz) of Ring Oscillator (ROs) based on DL and JLFET

| Devices | | Stress conditions | Supply voltage (V) | | |
|---|---|---|---|---|---|
| | | | 0.5 V | 0.7 V | 0.9 V |
| DL JLFET | 300 K | Fresh<br>2000s | 5.9<br>4.59 | 10.9<br>9.97 | 15.7<br>14.6 |
| | 400 K<br>500 K | Fresh<br>2000s<br>Fresh<br>2000s | 9.1<br>6.98<br>11.01<br>8.27 | 13.9<br>12.4<br>17.23<br>14.2 | 18.1<br>16.3<br>20.8<br>17.5 |
| JLFET | 300 K<br>400 K<br>500 K | Fresh<br>2000s<br>Fresh<br>2000s<br>Fresh<br>2000s | 2.25<br>0.703<br>5.12<br>0.701<br>5.54<br>0.688 | 5.27<br>1.94<br>8.01<br>1.9<br>9.19<br>1.75 | 7.75<br>3.14<br>10.14<br>3.01<br>11.9<br>2.74 |

in temperature due to CHC stress of 2000s. The ON- and OFF-state current of JLFET deteriorated by 1.7 and 1.4 times as compared to DL due to CHC stress with temperature range of 500 K. Likewise, the $V_{th}$ and $g_m$ of JLFET deteriorated by 2.5 and 2.7 times in comparison with DL due to same stress and temperature conditions. It is because of the density of interface charges increases with increase in $V_{th}$ of device, and as a result, mobility reduction in the channel occurs, which degrades the performance of conventional JLFET. However, the $V_{th}$ deterioration in DL is comparatively lower than the JLFET because of fewer amount of interface charges created at interface due to lower electric field at drain side, which results in less mobility reduction in channel.

In order to estimate CHC damage with temperature at circuit level, the stressed and fresh device models are incorporated. This model is more efficient in large system design and quickly predict device transfer characteristics. For circuit level simulation, we have considered an 11-stage ring oscillator (RO) realized with JLFETs. Table 2 shows the performance of RO based on conventional and DL-JLFETs due to pre- and post-CHC stress of 2000s at various temperatures. The operating frequency of RO realized with conventional JLFET is degraded by 68.8%, 86.3% and 87.59% than pre-CHC stress at temperature range of 300 K, 400 K and 500 K, respectively, whereas DL-JLFET-based RO has negligible (22%, 23.24% and 24.8%) impact of CHC at same temperature and stress conditions.

Moreover, the performance of standard SRAM cell designed with DL and JLFETs is also analysed at elevated temperature under CHC stress. For this, we have studied only dynamic performance metrics, which includes assessment of write and read delay as these parameters are more sensitive with CHC stress. From Fig. 4a, b, we can see that the read delay of DL-based SRAM cell is declined by 1.5, 1.6 and 1.95 times as compared to fresh device, while, in JLFET, the read delay is declined by 1.7, 3.1 and 6.48 times as compared to fresh device due to CHC stress of 2000s and operating voltage of 0.5 V in the temperature range from 300 to 500 K, respectively.

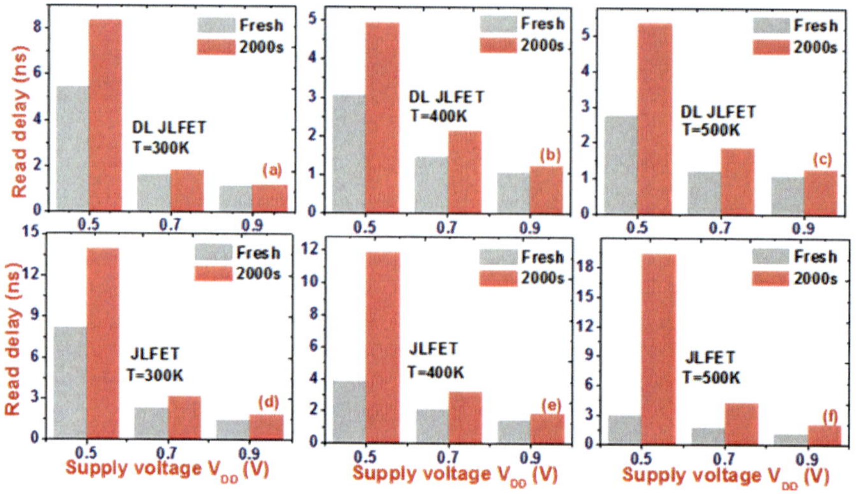


**Fig. 4** Read delay of DL and JLFET SRAM cells for various operating voltages and CHC stress at different temperatures

Likewise, for the same stress and supply voltage, the write delay of DL-JLFET SRAM cell is declined by 12.8%, 14.8% and 17.4% and for JLFET SRAM cell, write delay is declined by 20.6%, 43.3% and 87.9%, respectively, as shown in Fig. 5a, b. The delay time of JLFET SRAM cell is affected more due to shift in $V_{th}$ increases with CHC stress at elevated increasing temperature.

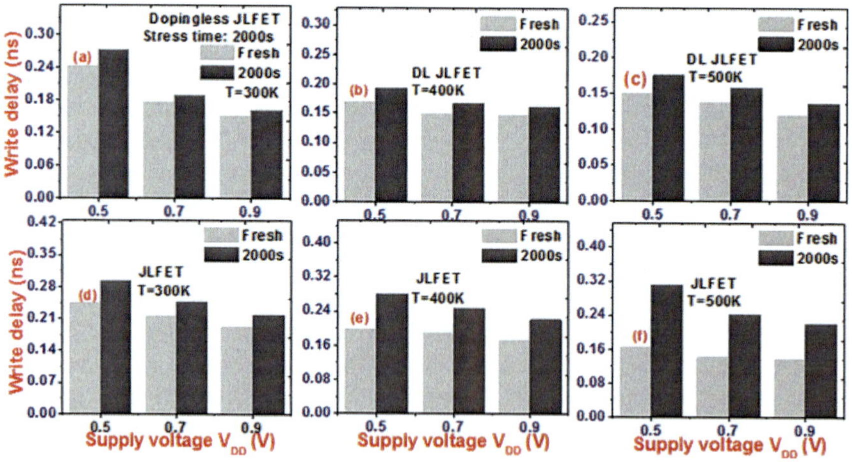


**Fig. 5** Write delay of DL and JLFET SRAM cells for various operating voltages and CHC stress at different temperatures

## 4 Conclusion

The performance of DL devices has been analysed under CHC stress at high temperature and compared with its conventional counterpart JLFET. For this, the electrical parameters such as threshold voltage, transconductance and ON–OFF current characteristics of JLFETs are examined under CHC stress at various temperatures. We found that the threshold voltage and drain current of JLFET are more susceptible to CHC effect at elevated temperature than DL-JLFET. Similarly, the performance of RO and SRAM cell-based JLFETs is more sensitive to CHC stress at elevated temperature in contrast to its DL counterpart. Hence, at elevated temperature, the CHC damage causes significant impact on performance of heavily doped deep sub-micrometre devices and their circuits in future.

## References

1. Gnani E, Gnudi A, Reggiani S, Baccarani G (2011) Theory of the Junctionless Nanowire FET. IEEE Trans Electron Devices 58(9):2903–2910
2. Gnudi A, Reggiani S, Gnani E, Baccarani G (2012) Analysis of threshold voltage variability due to random dopant fluctuations in junctionless fets. IEEE Electron Device Lett 33(3):336–338
3. Leung G, Chui CO (2012) Variability impact of random dopant fluctuation on nanoscale junctionless finfets. IEEE Electron Device Lett 33(6):767–769
4. Sahu C, Singh J (2014) Charge-plasma based process variation immune junctionless transistor. IEEE Electron Device Lett 35(3):411–413
5. Sahu C, Singh J (2015) Potential benefits and sensitivity analysis of dopingless transistor for low power applications. IEEE Trans Electron Devices 62(3):729–735
6. Procel LM, Crupi F, Trojman L, Franco J, Kaczer B (2016) A defect-centric analysis of the temperature dependence of the channel hot carrier degradation in nMOSFETs. IEEE Trans Device Mater Reliab 16(1):98–100
7. Sutaria KB, Mohanty A, Wang R, Huang R, Cao Y (2015) Accelerated aging in analog and digital circuits with feedback. IEEE Trans Device Mater Reliab 15(3):384–393
8. Park CH, Ko MD, Kim KH, Lee SH, Yoon JS, Lee JS, Jeong YH (2012) Investigation of low-frequency noise behavior after hot-carrier stress in an n-channel Junctionless Nanowire MOSFET. IEEE Electron Device Lett 33(11):1538–1540
9. Panchore M, Singh J, Mohanty SP (2016) Impact of channel hot carrier effect in junction-and doping-free devices and circuits. IEEE Trans Electron Devices 63(12):5068–5071
10. Panchore M, Singh J, Mohanty SP, Kougianos E (2016) Compact behavioral modeling and time dependent performance degradation analysis of junction and doping free transistors. In: 2016 IEEE international symposium on Nanoelectronic and information systems (iNIS), pp 194–199
11. Pratap Y, Haldar S, Gupta RS, Gupta M (2014) Performance evaluation and reliability issues of junctionless csg mosfet for rfic design. IEEE Trans Device Mater Reliab 14(1):418–425
12. Pratap Y, Haldar S, Gupta RS, Gupta M (2015) Localized charge-dependent threshold voltage analysis of gate material- engineered junctionless nanowire transistor. IEEE Trans Electron Devices 62(8):2598–2605
13. ATLAS Device Simulator Software (2012) Silvaco. Santa Clara, CA, USA
14. Magnone P, Crupi F, Wils N, Tuinhout HP, Fiegna C (2012) Characterization and modeling of hot carrier-induced variability in subthreshold region. IEEE Trans Electron Devices 59(8):2093–2099

# Chapter 16
# An Optimum Multi-objective Dynamic Scheduling Strategy for a Hybrid Microgrid System Using Fuzzy Rank-Based Modified Differential Evolution Algorithm

**Sunita Shukla and Manjaree Pandit**

## 1 Introduction

The present challenge for power system engineer is to develop a power grid network that is sustainable and offers socio-economic benefits. This need has paved path for microgrids (MGs). Large-scale integration of various distributed energy sources (DER) via microgrids has helped meet the increasing load demand and mitigate fossil fuel emissions. Also, during instances of grid instability or extreme climatic conditions, microgrids can be operated in an islanded mode mitigating fluctuation while maintaining stability [1–4]. However, the uncertain and intermittent output characteristic of renewable energy resources (RER) may cause imbalances in the power generation-demand balance resulting in grid instability. The energy storage system (ESS) incorporated with RER can help to mitigate these uncertainties, help increase the level of RER penetration into the grid and make RER more grid viable.

In recent years, a lot of research with emphasis on storage to overcome the intermittency and uncertainty drawbacks of renewable sources has been done [5–9]. In [5], the authors solved a MG problem in real time to minimize RER costs while considering hidden costs incurred due to battery charging/discharging. In [8], the authors proposed a two-stage stochastic battery incorporated in multi-MG DER optimization model to minimize costs with effective reserve scheduling while maintaining frequency. Thus, ESS incorporation can enable optimal participation of RER in the economic dispatch and help in minimizing spinning reserve and adjustment costs incurred due to RER uncertainties [10]. ESS can charge and discharge according to load demand and help maintain the generation-demand balance. Figure 1 shows an illustration of a modern utility grid system.

S. Shukla (✉) · M. Pandit
Department of Electrical Engineering, Madhav Institute of Technology and Science, Gwalior, India

© The Author(s), under exclusive license to Springer Nature Singapore Pte Ltd. 2022 
H. M. Dubey et al. (eds.), *Artificial Intelligence and Sustainable Computing*, Algorithms for Intelligent Systems, https://doi.org/10.1007/978-981-16-1220-6_16

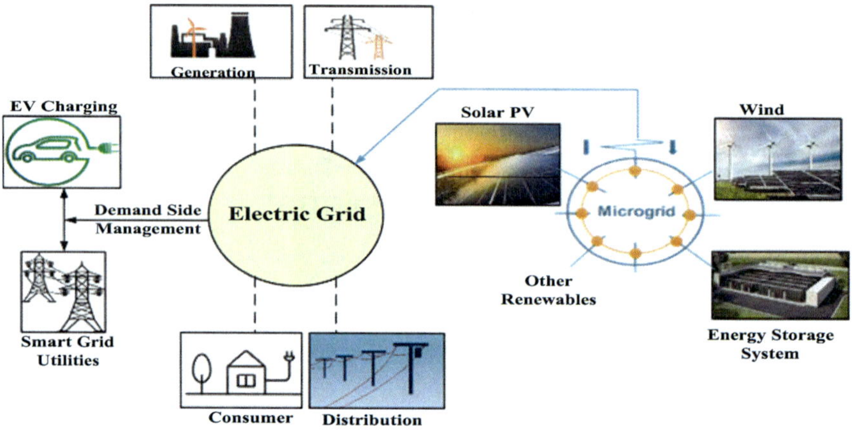


**Fig. 1** Interconnection of components in a modern electric grid network

In this paper, an optimal scheduling strategy for a battery incorporated multi-objective optimization (MOO) MG problem is proposed. The contribution of this paper can be summarized as follows:

(i) A fuzzy rank method-based modified differential evolution algorithm is proposed to minimize three objectives, viz., cost, emission and loss simultaneously.
(ii) The proposed scheduling strategy takes into consideration the uncertainties in wind power and the impact of solar irradiation forecast on solar PV power generation.
(iii) A small-scale battery charges and discharges to optimize participation of RER in the generator scheduling strategy.
(iv) Fuzzy rank method (FRM) is used to maximize the rank of solutions by finding a best comprised solution between the MOO functions such that no objective dominates the other.
(v) The results are validated with the weighted sum approach (WSA) method.

## 2 Problem Formulation

Large-scale deployment of renewable energy sources is globally seen as an optimum alternative source to fossil fuel for power generation. Charging and discharging of the battery storage can help mitigate the RER uncertainties and improve quality and efficiency of RER. In this paper, a multi-objective battery incorporated micro-grid problem with high renewable energy is developed to optimize three objectives, i.e. cost, emission and loss simultaneously under dynamic conditions. Battery has been incorporated to enable full exploitation of RER contribution. It can charge in case of excess power availability and discharges during RER power deficit and help maintain the generation-load demand balance. The mathematical model of the MOO scheduling problem is formulated using Sects. 2.1–2.3.

## 2.1 Optimal Cost Objective Function

The objective of the cost function ($C_f$) is the minimization of the total cost calculated based on the participation level of fossil fuel, wind and solar PV generating units as shown below.

$$C_f = \sum_{k=1}^{T}\sum_{k=1}^{N_g} fcost_p + WPcost_p + \sum_{k=1}^{T}\sum_{k=1}^{N_s} Scost_p \tag{1}$$

where $fcost_p$ is the fuel cost in \$/h; $WPcost_p$ and $Scost_p$ are the wind and solar power generation cost in \$/h, respectively, for the pth unit at the kth time interval. $N_g$ and $N_s$ are the number of thermal and solar power generating units, while T is the total time in hours.

### 2.1.1 Fuel Cost Function

The fuel cost function $fcost_p$ for the pth power unit, $P_{Gp}^k$ at the kth time interval is defined using the quadratic cost coefficients $p_p, q_p, r_\mathrm{p}$ as follows:

$$\sum_{k=1}^{T}\sum_{p=1}^{N_g} fcost_p(P_{Gp}^k) = \sum_{k=1}^{T}\sum_{\mathrm{p}=1}^{N_g} p_p + q_p P_{Gp}^k + r_p (P_{Gp}^k)^2 \tag{2}$$

### 2.1.2 Wind Cost Function

The wind power generation cost, $WPcost$ can be calculated as follows [11]:

$$WPcost = Dcost_p + Pcost_p + Rcost_p \tag{3}$$

where

$$Dcost_p = d_{cp} \times w_p \tag{4}$$

$$PCost_p = v_u \int_{w_{sch}}^{w_R} \left( (w - w_p) \times f(w) \right) d_{cp} \tag{5}$$

$$RCost_p = v_{ov} \int_{w_R}^{w_{sch}} \left(\left(w_p - w\right) \times f(w)\right) dw \tag{6}$$

where $d_{cp}$ is the direct cost coefficient; $w_p$ is the wind speed in miles/s of the pth wind power unit; $w_{sch}$ and $w_R$ are the scheduled and rated wind power in miles/s; $Dcost_p$ is the direct cost, $Pcost_p$ is the over estimation penalty cost and $Rcost_p$ is the reserve scheduling penalty cost in \$/h for the pth wind power unit, respectively.

In this paper, the uncertain wind power (WP) random variable $w$ has been portrayed using the Weibull probabilistic density function $(pdf)$. The reserve scheduling penalty, $v_u$ and overestimated cost coefficients, $v_{ov}$ are set to 0.1. The WP random variable $w$ is transformed into continuous and discrete distributions based on regions defined by Eq. (7). In this paper, the rated value of wind speed is selected as 15 miles/s with 5 miles/s cut in speed and 25 miles/s as cut out wind speed.

$$\begin{cases} w = 0 & for\ v < v_{wp}\ and\ v > v_{wR} \\ w = w_R \times \frac{v_{wo} - v_{wp}}{v_{wR} - v_{wp}} & for\ v_{wp} \le v \le v_{wR} \\ w = w_R & for\ v_{wR} \le v \le v_{wo} \end{cases} \tag{7}$$

### 2.1.3 Solar PV Cost Function

The cost of generating power using solar PV units, $Scost$ is expressed as [12]

$$Scost = \sum_{k=1}^{T} \sum_{p=1}^{N_s} P_{sp}^{k} \times I_{sp}^{k} \times PVrate_p \tag{8}$$

where

$$P_{sp}^{k} = P_{rp} \times \left(1 + \left(t_o - t_{ref}\right) \times t_{coeff}\right) \times \left(S_{irr}(k) \div 1000\right) \tag{9}$$

In Eqs. (8) and (9), $I_{sp}^{k}$ is the on/off variable of the pth solar unit; $PVrate_p$ is the per unit solar cost in \$/h; $S_{irr}(k)$ is the solar incident irradiance at kth time interval; $P_{rp}^{k}$ is the rated power value and $P_{sp}^{k}$ is the output power value of the pth solar power unit in KW, while $t_o$ is the ambient temperature in °C, $t_{ref}$ is the reference temperature in °C and $t_{coeff}$ is the temperature coefficient in per °C, respectively.

In this paper, per unit $t_{ref}$ is set at 25 °C; $t_{coeff}$ at −0.005 per °C; solar cost at 0.08 \$/KWh for the rated solar power units of 100 KW and 30 KW, respectively.

### 2.2 Emission Function

The emission function $f_{EM}$ over the time horizon T can be computed using the emission coefficient (α, β and γ) of the pth generating unit as

$$f_{EM} = \sum_{k=1}^{T}\sum_{p=1}^{N_g} E_p\left(P_{gp}^k\right) = \alpha_p + \beta_p P_{gp}^k + \gamma_p (P_{gp}^k)^2 \tag{10}$$

### 2.3 Loss Function

The loss function $f_L^k$ for the kth interval using transmission loss coefficients $U_{pq,} U_{0p}, U_{00}$ is defined as

$$f_L^k = \sum_{\text{p}=1}^{N_T}\sum_{q=1}^{N_T} P_{Gp}^k U_{pq} P_{Gq}^k + \sum_{p=1}^{N_T} U_{0\text{p}} P_{Gp}^k + U_{00} \tag{11}$$

where $\text{N}_\text{T}$ is the sum of all the DERs such that $\text{N}_\text{T} = \text{N}_\text{G} + \text{N}_\text{WP} + \text{N}_\text{S}$; $P_{Gp}^k$ and $P_{Gq}^k$ are the generated power of the pth and qth generating power unit at the kth time interval.

### 2.4 Battery

In this paper, battery is charged or discharged at the kth hour to accommodate the fluctuations in RER power generation using (12) and (13), respectively.

$$P_{ch}^k = P_B^{k-1} + \left(P_{Gs}^k + P_{Gw}^k - P_D^k\right) \times \eta_{ch} / \eta_{inv} \tag{12}$$

$$P_{dis}^k = P_B^{k-1} + \frac{1}{\eta_{dis}} \times \left(P_{Gs}^k + P_{Gw}^k - P_D^k\right) / \eta_{inv} \tag{13}$$

where $P_B^{k-1}$ is the battery power capacity at the start of the charge or discharge time period; $P_{ch}^k$ and $P_{dis}^k$ are the battery power charging and discharging capacity at the kth time interval; $P_{Gs}^k$, $P_{Gw}^k and P_D^k$ are the generated solar power, generated wind power and load demand at the kth time interval, respectively, while $\eta_{ch}/\eta_{dis}/\eta_{inv}$ are the charging/discharging/inverter efficiencies, respectively.

## 2.5 Multi-objective Function

The MOO function for the battery incorporated microgrid with high RER contribution comprises the three objectives defined in Sects. 2.1–2.3 and can be formulated as

$$MinF(f_c, f_{EM}, f_L) \tag{14}$$

subject to the following constraints.

### 2.5.1 Generating Limit Constraint

The inequality constraint for the pth power generator for the kth time step can be written as:

$$P_{Gpmin}^{\mathrm{k}} \leq P_{Gp}^{k} \leq P_{Gpmax}^{\mathrm{k}} \tag{15}$$

where $P_{Gpmin}^{\mathrm{k}}$ and $P_{Gpmax}^{\mathrm{k}}$ are the minimum and maximum generator operating limits for the kth time interval.

### 2.5.2 Equality Constraint for the Load-Supply Balance

The algebraic sum of power generated by all the DERs and the charging or discharging battery power must be equal to the total sum of the load connected and losses incurred at the kth time interval and is given by (16).

$$\sum_{p=1}^{N_G}(P_{Gp}^k) + \sum_{w=1}^{N_{WP}}(P_{Gw}^k) + \sum_{s=1}^{N_S}(P_{Gs}^k) + P_{dis}^k - P_{ch}^k = P_D^k + f_L^k \tag{16}$$

### 2.5.3 Ramping Constraints for Generating Units

The generator unit output during the dynamic dispatch is limited by the up/down $(R_{up}/R_{dp})$ ramping limits during two consecutive intervals.

$$R_{dp} \leq P_{Gp}^k - P_{Gp}^{k-1} \leq R_{up} \tag{17}$$

### 2.5.4 Battery State of Charge Constraints

To maintain optimum battery operational life, its state of charge should be well within the limits. The equality and inequality constraints governing the optimal operation of the battery can be expressed as:

$$P_B^k = P_B^{k-1} + P_{ch}^k \eta_{ch} - P_{dis}^k / \eta_{dis} \tag{18}$$

$$P_{Bmin}^k \le P_B^k \le P_{Bmax}^k \tag{19}$$

where $P_{Bmin}^k = 0.2 \times C_{BESS}$ and $P_{Bmax}^k = 0.8 \times C_{BESS}$

In this paper, the operating voltage is assumed to be constant, state of charge is assumed to be 80% taking into consideration the standby losses and aging effect with depth of discharge set at 20%. The battery is selected to charge and discharge at a rate ≤20% of its rated capacity ($C_{BESS}$) over the 24-hour time horizon.

## 3 Proposed Solution Methodology

In this paper, a hybrid fuzzy rank method-based modified differential evolution is used to find the most optimal MG schedule that satisfies three objectives, viz., total costs, emission and loss simultaneously. The proposed optimal scheduling strategy aims to reduce the penalty and reserve scheduling costs arising due to forecast errors as a result of the stochastic and uncertain behavior of RER. The flowchart of the proposed solution approach is shown in Fig. 2.

In the first stage a modified mixed integer nonlinear DE algorithm [13] is applied to evaluate the values of cost, emission and loss. At this stage population sample is selected. Crossover is performed to search, optimize and create new candidates from the population sample size. Lastly mutation is done to keep candidates with best fits as the optimal solutions. An advantage of DE algorithm for MOO is that it gives a large number of non-dominated solutions in a single run. In the next stage, fuzzy rank sets are generated for each objective. The fuzzy sets then assign a rank between 0 and 1 to the large non-dominated solutions with a rank of zero assigned to the worst solution and one to best solution for each objective using (20). The rank of the pth solution for the nth objective, $\mu(f_{in})$ can be computed as:

$$\mu(f_{p,n}) = \begin{cases} 1; & f_{p,n} \le f_{p,n}^{min} \\ \frac{f_{p,n}^{max} - f_{p,n}}{f_{p,n}^{max} - f_{p,n}^{min}}; & f_{p,n}^{min} < f_{p,n} < f_{p,n}^{max} \\ 0; & f_{p,n} \ge f_{p,n}^{max} \end{cases} \tag{20}$$

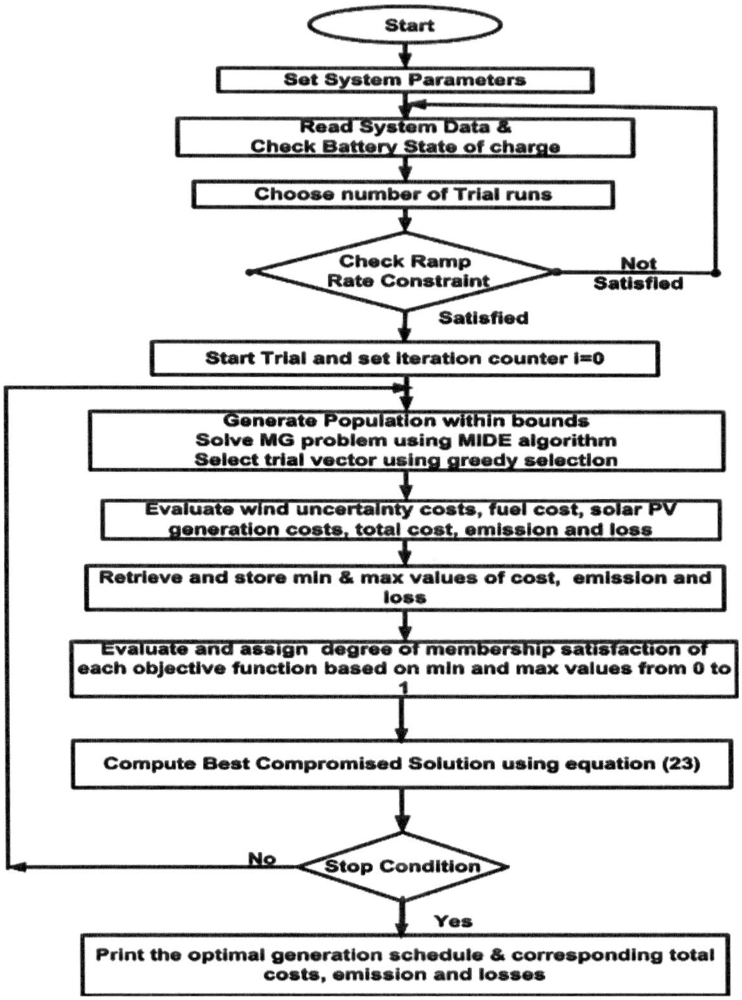


**Fig. 2** Flowchart of the proposed fuzzy rank method-based modified DE algorithm

where $f_{p.n}^{min}$ and $f_{p,n}^{max}$ are the worst and best values of the nth objective for the minimization problem. The fuzzy ranked sets of the three objectives can then be normalized collectively to obtain a best compromised solution using (21)–(23) such that no objective function dominates the other.

$$MinF_T = \left(\frac{1}{1 + rank_p}\right) \tag{21}$$

$$rank_i = \min\left[\mu\left(f_{p,n}\right)\right] \quad \text{For} \forall \text{ n} \tag{22}$$

$$BCS = \max\left(rank_p\right) \quad \text{For} \, \forall \, \text{p} \tag{23}$$

The proposed methodology aims to maximize the level of attainment for all objectives.

# 4 Numerical Analysis, Discussion and Results

The optimal scheduling problem of a five DER battery integrated microgrid with significant renewable generation is investigated in this paper. The MOO microgrid problem is solved considering minimization of total generation cost (including penalty and reserve cost), emission content and losses using a fuzzy rank method-based modified DE algorithm. Fuzzy rank method assigns a fuzzy rank to each solution based on the fulfillment level of each objective. The aim is to find an optimal schedule renewable generating units which would maximize this rank. A 10 KW battery is incorporated to maximize the contribution of RER in the optimal schedule and accommodate the fluctuations in RER by charging or discharging to maintain the generation-load balance requirement.

The results obtained using the proposed optimal scheduling strategy are compared with the weighted sum approach to validate the proposed algorithm. The developed approach is tested over a time horizon of 24 h for dynamic operating conditions with and without battery incorporated. Coding of all the algorithms has been carried out using MATLAB 2017 on a 2.4 GHz, core i5-8265U processor with 8 GB RAM.

## *4.1 Structure of the Microgrid Test System*

A five DER microgrid connected to a 14-bus configuration is taken from [14] as a test system. The microgrid model has been modified to comprise a wind turbine, two solar PV units, one microturbine and a diesel generator unit representing the utility grid. A small-scale battery system is also incorporated to enhance the socio-economic benefits of RER in the microgrid scheduling. Loss coefficients are from [14], and the ambient temperature and solar radiation are from [12].

## *4.2 Contribution of DERs and Battery in the Optimal Dynamic Scheduling of the Microgrid*

Hourly variation of the contribution of power and operational costs of DERs with load is shown in Figs. 3 and 4, respectively. Figure 5 shows the hourly charging and discharging power capacity of the battery. From Figs. 3 and 4, it can be deduced

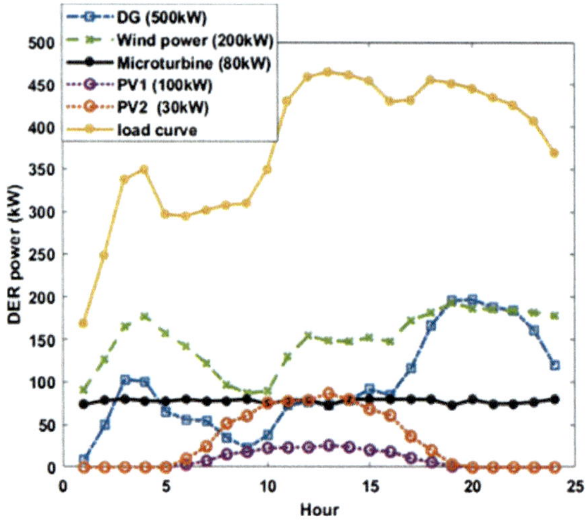


**Fig. 3** Hourly DER power variation using MIDE

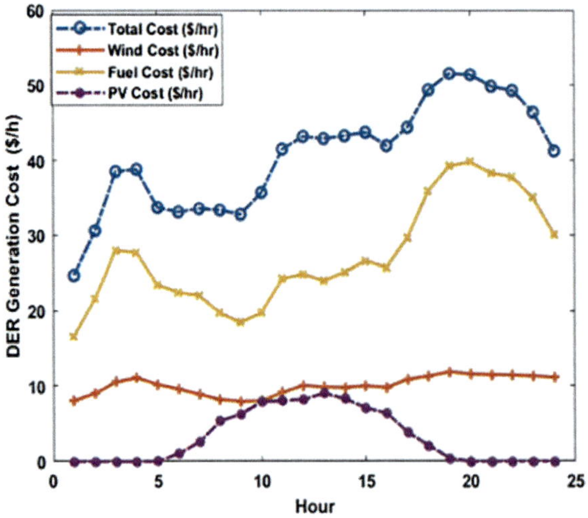


**Fig. 4** Hourly cost variation of the DERs

that DG combination is more costly among the other DERs and observed to follow the cost curve. Wind power is observed to contribute most power in the optimal schedule strategy with microturbine contributing near its full capacity throughout the day. Solar power is the cheapest among all sources with its contribution limited to a few hours only subject to its availability. DG share can be observed to fluctuate and follow the load curve pattern for the day. Hence, it can be concluded that DG is contributing as a reserve unit responsible for keeping up with the load demand and maintaining the power balance. Lastly, the operational costs of the battery are assumed to be negligible as its contribution capacity is negligible compared to other

Fig. 5 Hourly charging/discharging of battery power

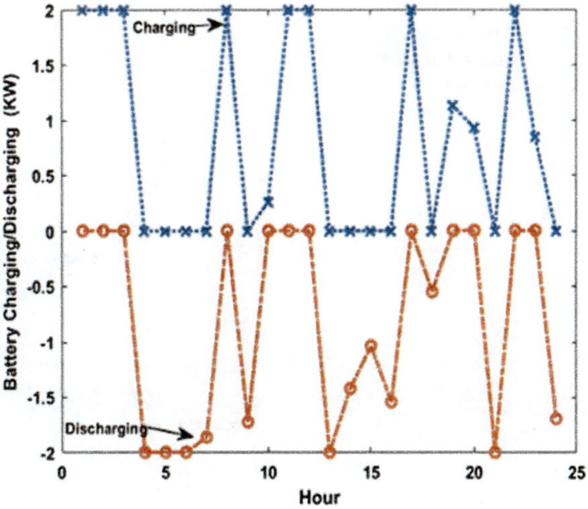


DERs. However, during charging durations, battery is considered as a load and the charging costs are included in the overall costs of DG units.

## *4.3 Comparison of Pareto Fronts for Three Objective Functions*

Pareto fronts can be plotted to judge the merit of optimality and obtain a trade-off between multi-objective functions from a large number of non-dominant solutions. To show the full range of Pareto optimal solutions, Pareto fronts showing best compromised solution for the MOO function are plotted in Figs. 6, 7 and 8 for low (load = 169 KW), medium (load = 350 KW) and high (load = 462 KW) loading conditions.

Fig. 6 Three objective Pareto fronts for load = 169 KW

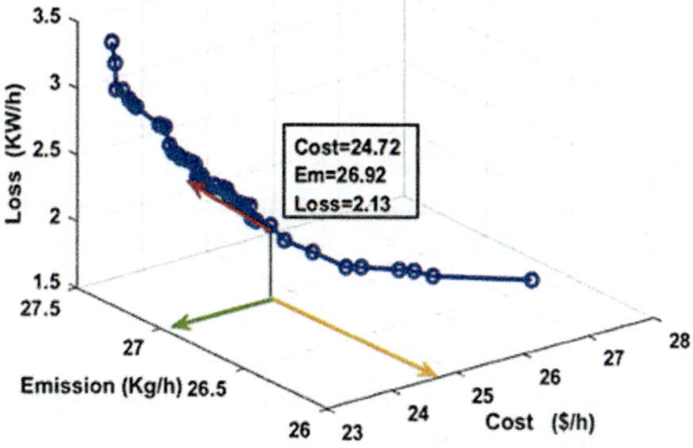

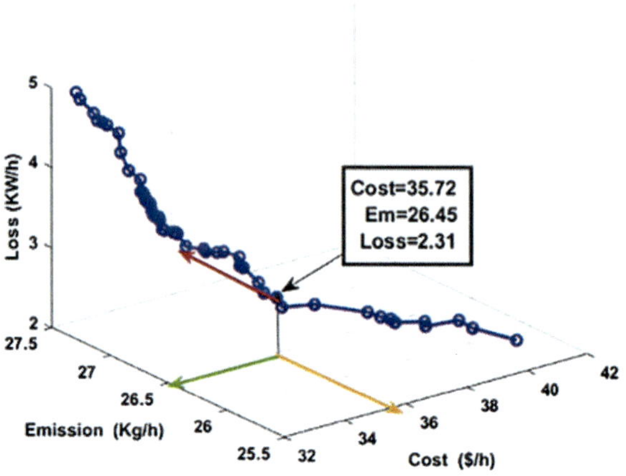


**Fig. 7** Three objective Pareto fronts for load = 350 KW

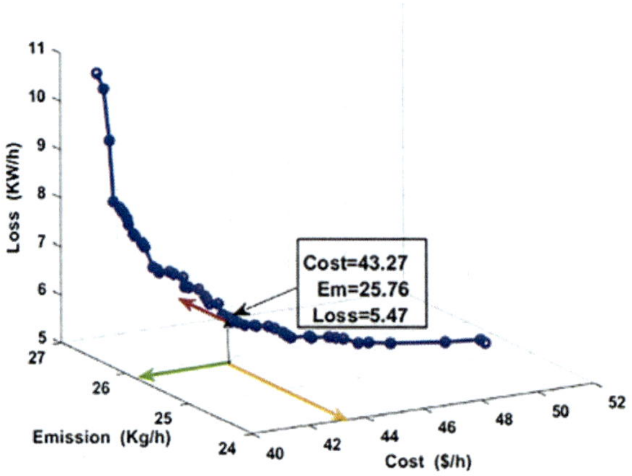


**Fig. 8** Three objective Pareto front for load = 462 KW

## *4.4 Comparison of the Proposed Strategy with Weighted Sum Approach for the Multi-objective Dynamic Optimal Scheduling Problem*

For the dynamic operating conditions, generator ramp rate limit constraint given by (17) has to be satisfied between two consecutive optimal scheduling hours. The performance of the WSA and FRM using the modified DE algorithm is compared for different loading conditions in Table 1. Table 2 shows the comparison between WSA and FRM for the simultaneous optimization of cost–emission–loss objectives for dynamic operating conditions with and without battery. All the solutions obtained are observed to satisfy the nonlinear as well as the stochastic wind and discrete solar PV constraints.

From Table 1, the FRM method is observed to give better ranking solution with higher minimum attainment level computed over the three objectives. From

**Table 1** Comparison between WSA and FRM under dynamic operating conditions for few loading conditions

| Load (kW) | Method | Total Cost ($/h) | Emission (Kg/h) | Loss (KW/h) | M.V. (cost) ($\mu_c$) | M.V (Em) ($\mu_{Em}$) | M.V. (Loss) ($\mu_{Loss}$) | Rank of solution ($Rank_i$) |
|---|---|---|---|---|---|---|---|---|
| 169 | WSA | 24.73 | 26.90 | 2.14 | 0.54 | 0.70 | 0.49 | 0.49 |
| | FRM | 24.71 | 26.92 | 2.13 | 0.56 | 0.65 | 0.52 | 0.52 |
| 295 | WSA | 33.13 | 26.11 | 4.54 | 0.57 | 0.67 | 0.80 | 0.57 |
| | FRM | 33.12 | 26.11 | 4.54 | 0.57 | 0.67 | 0.80 | 0.57 |
| 350 | WSA | 35.96 | 26.61 | 2.39 | 0.74 | 0.40 | 0.83 | 0.40 |
| | FRM | 35.74 | 26.44 | 2.31 | 0.79 | 0.55 | 0.91 | 0.55 |
| 430 | WSA | 41.60 | 26.06 | 4.53 | 0.76 | 0.32 | 0.93 | 0.32 |
| | FRM | 41.49 | 25.86 | 4.59 | 0.81 | 0.52 | 0.88 | 0.52 |
| 465 | WSA | 42.99 | 25.93 | 5.05 | 0.58 | 0.51 | 0.93 | 0.51 |
| | FRM | 42.90 | 25.92 | 5.00 | 0.63 | 0.52 | 0.97 | 0.52 |

**Table 2** Comparison of WSA and FRM for multi-objective dynamic scheduling

| Method | Total cost/day ($) | Emission/day (kg) | Loss/day (kW) | Fuel cost/day ($) | Wind cost/day ($) | PVCost/day ($) |
|---|---|---|---|---|---|---|
| WSA* | 1057.50 | 797.88 | 152.93 | 680.28 | 287.54 | 89.68 |
| FRM* | 1050.02 | 807.32 | 140.15 | 675.93 | 284.41 | 89.68 |
| WSA** | 979.92 | 615.39 | 178.23 | 655.07 | 248.33 | 76.52 |
| FRM** | 974.92 | 612.62 | 175.84 | 657.16 | 241.23 | 76.52 |

*Without battery inclusion. **With battery included

Table 2, the emission and cost objectives are observed to reduce significantly in both cases when battery is incorporated. Both WSA and FRM are observed to give non-dominating solutions. However, the FRM method produces a better result in terms of cost and emission objectives.

## 5 Conclusion

This paper proposes a hybrid fuzzy rank method-based modified DE algorithm to solve a multi-objective optimization problem for a utility connected microgrid with high renewable energy contribution. A small-scale battery is also incorporated to accommodate the variability of RER. The target is to optimize cost, emission and losses simultaneously and achieve maximum attainment level of each objective. Three objective Pareto fronts have also been plotted for different loading conditions showing the best compromised multi-objective solution. Some of the results of the proposed algorithm have also been compared with WSA for the five DER microgrid system with and without battery. Due to better RER variability accommodation and less scheduling of the DG reserve capacity, fossil fuel emission over the day is observed to reduce significantly while using battery. In both the cases, the RER uncertainty costs are also observed to reduce significantly when battery is included in the optimal schedule. However, the proposed algorithm is found to show slightly superior results for at least two objectives as compared to WSA algorithm. The proposed optimal scheduling strategy is also found to produce a solution which has a greater satisfaction level for the cost–emission–loss objectives.

## References

1. Gholami A, Shekari T, Aminifar F, Shahidehpour M (2016) Microgrid scheduling with uncertainty: the quest for resilience. IEEE Trans Smart Grid 7:2849–2858
2. Ding T, Lin Y, Bie Z, Chen C (2017) A resilient microgrid formation strategy for load restoration considering master-slave distributed generators and topology reconfiguration. Appl Energy 199:205–216
3. Fang X, Yang Q, Dong W (2018) Fuzzy decision based energy dispatch in offshore industrial microgrid with desalination process and multi-type DGs. Energy 148(C):744–755
4. Sadeghi Sarcheshmeh M, Taher SA, Mazidi MA (2020) stochastic frequency security constrained optimization model for day-ahead energy and reserve scheduling of islanded multi-microgrids systems. Int Trans Electric Energy Syst 30
5. Zhang Z, Wang J, Ding T (2017) A two-layer model for microgrid real—time dispatch based on energy storage system charging/discharging hidden costs. IEEE Trans Sustain Energy 8:33–42
6. Gil-González W, Montoya OD, Holguín E, Garces A, Grisales-Noreña LF (2019) Economic dispatch of energy storage systems in dc microgrids employing a semidefinite programming model. J Energy Storage 21:1–8
7. Montoya OD, Gil-González W, Grisales-noreña LF, Orozco-Henao C, Serra F (2019) Economic dispatch of BESS and renewable generators in DC microgrids using voltage-dependent load models. Energies 12, undefined
8. Sarcheshmeh MS, Taher SA M, A MM (2020) A stochastic frequency security constrained optimization model for day-ahead energy and reserve scheduling of islanded multi-microgrids systems. Int Trans Electr Energy Syst 30(6): e12386. https://doi.org/10.1002/2050-7038.12386
9. Liu Z, Zhang H, Dong J, Yu H (2020) A prediction-based optimization strategy to balance the use of diesel generator and emergency battery in the microgrid. Int J Energy Res 44:5425–5440
10. Reddy SS, Bijwe PR, Abhyankar AR (2015) Optimum day-ahead clearing of energy and reserve markets with wind power generation using anticipated real-time adjustment costs. Int J Electr Power Energy Syst 71:242–253
11. Hetzer J, Yu DC, Bhattarai K (2008) An economic dispatch model incorporating wind power. IEEE Trans Energy Convers 23:603–611
12. Khan NA et al (2015) Combined emission economic dispatch of power system including solar photo voltaic generation. Energy Convers Manag 92:82–91
13. Shukla S, Pandit M (2020) Mixed-integer differential evolution algorithm for optimal static/dynamic scheduling of a microgrid with mixed generation. In: Nature inspired optimization for electrical power system, algorithms for intelligent systems. https://doi.org/10.1007/978-981-15-4004-2_7, pp 83–99
14. Basu AK, Bhattacharya A, Chowdhury S, Chowdhury SP (2012) Planned scheduling for economic power sharing in a CHP-based micro-grid. IEEE Trans Power Syst 27(1):30–38. https://doi.org/10.1109/TPWRS.2011.2162754

# Chapter 17
# Performance Analysis of Different Tilt Angles-Based Solar Still

**Vikas Kumar Thakur, M. K. Gaur, and M. K. Sagar**

## 1 Introduction

Solar still is completely dependent upon the renewable energy, which is able to convert the brackish and impure water into potable water. It is an eco-friendly and also economic-friendly device. The fabrication cost of solar still is very low and its design is also very simple, hence requirement of person advanced skillset is not needed to operate the device [1]. A drawback of solar still is that its productivity and efficiency are very low. Increasing the productivity of solar still remains a big challenge. To increase the performance of solar stills, it is necessary to increase the evaporation rate of the basin water and this is possible only when the basin water temperature increases [2].

In this paper, the effect of solar still on the performance due to different tilt angles of the glass cover has been studied. The performance of solar still has been affected by many parameters. (i) Natural parameters: wind speed, atmospheric temperature, climate condition and solar radiation intensity, etc. (ii) Construction parameters: body materials, insulation materials, inclination angle of the glass cover, and glass cover thickness and (iii) operational parameters such as the orientation of solar stills, basin water quality and water depth [4].

Cherraye et al. [5] constructed six different angles-based (10°, 15°, 20°, 30°, 35° and 45°) solar still and studied their performance in cold and hot climates. The highest productivity was achieved at 20° in summer and 30° in cold climates. At lower angle the slope was not sufficient enough to get water droplets dripping on the surface of the condenser, and at higher angle the large shadow of the side walls and the important distance gap resulting in productivity decrease at low and high tilt angle.

---

V. K. Thakur (✉) · M. K. Gaur · M. K. Sagar
Department of Mechanical Engineering, Madhav Institute of Technology and Science, Gwalior 474005, India

© The Author(s), under exclusive license to Springer Nature Singapore Pte Ltd. 2022 
H. M. Dubey et al. (eds.), *Artificial Intelligence and Sustainable Computing*, Algorithms for Intelligent Systems, https://doi.org/10.1007/978-981-16-1220-6_17

Kumar et al. [6] analyzed solar still with two different glass cover angles (23° and 30°) in Roorkee (latitude: 29.8543°N), which exhibit extreme cold climate. A system condensing cover angle was created without the latitude and another was fabricated with latitude of the site. From the result it was found that the latitude-based solar still had higher productivity than other stills. 30° tilt angle is near to the latitude of place, so the incident radiation is normal to the glass cover, thus the maximum amount of radiation is reached at the basin water, which increases the evaporation rate of basin water.

Azooz and Younis [7] constructed 10 simple basin-type solar still at 10 different tilt angles (10°, 15°, 20°, 25°, 30°, 35°, 40°,45°, 50° and 55°). The higher productivity achieved at 25°, due to less cavity gap and sufficient angle for gliding the water droplets.

In the current experimental attempt, different tilt angles of glass cover (11°, 26°, 26° and 41°) were investigated. Variations in glass cover angle are affected by the radiation falling on the glass surface. The inclination angle of the glass cover is one of the important reasons for increasing and decreasing the internal heat transfer of the solar still. Also, slide of water droplets at internal surface of condensing cover is dependent upon the tilt angle of glass cover.

## 2 Experimental Setup

Figure 1A and B shows the schematic diagram of the setup and conventional setup, respectively. The experimental setup can be observed in Fig. 2 showing the inclination angle (11°, 26° and 41°) of the glass cover for different solar stills. The angle of solar still is chosen based on the latitude of Gwalior city (26.2183°N). The glass cover angle of the experimental setup has been made 15° lower and 15° higher than the latitude of Gwalior city. The experimental setup is installed at the campus of Madhav Institute of Technology and Science, Gwalior, where the latitude and longitude is

**Fig. 1** Experimental setup with different tilt angles of the glass cover

**Fig. 2** Calibrated temperature indicator

26.2183°N and 78.1828°E, and the height from the sea level is 211 m. The area of each experimental setup is 1 $m^2$, which is placed south-facing. The lower height of all solar stills is 0.20 m, and the higher height is 0.44 m, 0.74 m and 1.12 m, for 11°, 26° and 41° inclination angle of glass cover, respectively.

The inside body of solar still is made of 2 mm zinc-coated stainless steel sheet, and 20 mm thermocol sheet and 20 mm plywood have been used to prevent the heat loss. Normal stainless steel sheets have been used to protect the plywood from sunlight and rain water. In all solar stills, a 5 mm toughest glass with high light transmission capacity has been installed for the condensing cover.

To make the setup airtight m-seal and silicone gel has been used and the glass cover is also sealed with silicone gel to make setup airtight. A tank has been installed in all the setup through which water is filled in the basin and the tank is kept at the proper height for gravity effect. All the tanks are fitted with valves that can be opened and closed as required. For the removal of condensed water, a distilled water collector tray has been provided in all the stills, due to which the distilled water is collected in the outside bottle. Tables 1, 2 and 3 show the hourly performance of 11°, 26° and 41° inclined solar still.

A calibrated temperature indicator has been used to measure the temperature of various parts of the solar still as shown in Fig. 2. A salorimeter has been used to measure hourly solar radiation, which is shown in Fig. 3.

## 3 Mathematical Modeling

The following assumptions have been considered for the energy balance of various parts of the experimental setup:

1. The heat capacity of the glass cover and insulating material is considered negligible.
2. The experimental setup is completely airtight; there is no any kind of leakage from setup.

**Table 1** Hourly observation taken for passive solar still with inclination angle of 11°, on a typical day in October, 2020

| Time (h) | $T_{go}$ | $T_{gi}$ | $T_v$ | $T_w$ | $T_{bl}$ | $T_a$ | I(t) | Hourly yield (ml) |
|---|---|---|---|---|---|---|---|---|
| 9 am | 35.8 | 44.3 | 39.1 | 30.2 | 30.1 | 34 | 240 | 0.00 |
| 10 am | 36.9 | 46.1 | 40.6 | 32.6 | 32.4 | 34.8 | 430 | 2 ml |
| 11 am | 39.3 | 49.2 | 43.2 | 39.9 | 39.2 | 35.7 | 496 | 8 ml |
| 12 am | 43.6 | 53.8 | 47.9 | 48 | 47.4 | 35.7 | 501 | 40 ml |
| 1 pm | 48.9 | 54.2 | 52.8 | 54 | 53.2 | 36.5 | 610 | 60 ml |
| 2 pm | 50.2 | 56.9 | 53.5 | 56.9 | 55.9 | 36.4 | 582 | 100 ml |
| 3 pm | 51.2 | 57.4 | 57.3 | 57.4 | 56.4 | 35.6 | 480 | 290 ml |
| 4 pm | 48.9 | 55.2 | 55.1 | 55.4 | 54.4 | 34.9 | 310 | 160 ml |
| 5 pm | 44.8 | 51.1 | 50.1 | 52.0 | 51.1 | 33.7 | 165 | 110 ml |
| 6 pm | 40.4 | 50.1 | 49.1 | 47.8 | 47.1 | 32.3 | 70 | 110 ml |
| 7 pm | 37.9 | 43.2 | 42.2 | 45.2 | 44.5 | 31.6 | 0 | 90 ml |
| Total daily distillate collection | | | | | | | | **970 ml** |

**Table 2** Hourly observation taken for passive solar still with inclination angle of 26°, on a typical day in October, 2020

| Time (h) | $T_{go}$ | $T_{gi}$ | $T_v$ | $T_w$ | $T_{bl}$ | $T_a$ | I(t) | Hourly yield (ml) |
|---|---|---|---|---|---|---|---|---|
| 9 am | 35.8 | 44.3 | 39.1 | 30.2 | 30.1 | 34 | 240 | 0.00 |
| 10 am | 38.2 | 42.6 | 48.5 | 32.0 | 33.1 | 34.8 | 430 | 0.00 |
| 11 am | 40.6 | 44.6 | 50.5.2 | 38.6 | 40.2 | 35.7 | 496 | 10 ml |
| 12 am | 43.9 | 48.1 | 54.2 | 47.2 | 49.2 | 35.7 | 501 | 95 ml |
| 1 pm | 48.9 | 54.2 | 52.8 | 54 | 53.2 | 36.5 | 610 | 105 ml |
| 2 pm | 50.4 | 53.9 | 58.1 | 57.2 | 59.8 | 36.4 | 582 | 250 ml |
| 3 pm | 51.6 | 54.2 | 58.9 | 58.1 | 60.4 | 35.6 | 480 | 240 ml |
| 4 pm | 48.8 | 51.0 | 57.8 | 56.1 | 58.3 | 34.9 | 310 | 180 ml |
| 5 pm | 45.3 | 46.7 | 55.2 | 52.8 | 54.8 | 33.7 | 165 | 170 ml |
| 6 pm | 41.0 | 42.1 | 49.1 | 48.7 | 50.3 | 32.3 | 70 | 110 ml |
| 7 pm | 38.5 | 39.5 | 42.1 | 46.0 | 47.5 | 31.6 | 0 | 98 ml |
| Total daily distillate collection | | | | | | | | **1258 ml** |

3. The base fluid inside the basin has been considered at constant level.
4. The water vapor inside the basin behaves like an ideal gas.
5. The setup is placed in the south-facing.

The energy balance of various parts of the experimental setup is given below [8]:
Outer glass cover:

$$\frac{K_g}{L_g}\left(T_{gi} - T_{go}\right) = h_{t,go-a}\left(T_{go} - T_a\right) \tag{1}$$

**Table 3** Hourly observation taken for passive solar still with inclination angle of 41°, on a typical day in October, 2020

| Time (h) | $T_{go}$ | $T_{gi}$ | $T_v$ | $T_w$ | $T_{bl}$ | $T_a$ | I(t) | Hourly yield (ml) |
|---|---|---|---|---|---|---|---|---|
| 9 am | 35.8 | 44.3 | 39.1 | 30.2 | 30.1 | 34 | 240 | 0.00 |
| 10 am | 40.5 | 44.5 | 51.0 | 31.6 | 31.6 | 34.8 | 430 | 0.00 |
| 11 am | 42.8 | 46.5 | 53.0 | 38.8 | 38.4 | 35.7 | 496 | 10 ml |
| 12 am | 46.2 | 50.2 | 56.5 | 47.6 | 47.4 | 35.7 | 501 | 30 ml |
| 1 pm | 51.1 | 53.9 | 61.3 | 54.4 | 54.2 | 36.5 | 610 | 50 ml |
| 2 pm | 52.4 | 54.8 | 61.4 | 57.6 | 57.2 | 36.4 | 582 | 250 ml |
| 3 pm | 53.3 | 55.2 | 60.9 | 58.1 | 57.5 | 35.6 | 480 | 100 ml |
| 4 pm | 50.0 | 52.3 | 57.8 | 55.6 | 55.4 | 34.9 | 310 | 140 ml |
| 5 pm | 45.6 | 47.4 | 55.1 | 52.2 | 51.8 | 33.7 | 165 | 210 ml |
| 6 pm | 41.3 | 42.1 | 48.2 | 47.9 | 47.5 | 32.3 | 70 | 150 ml |
| 7 pm | 38.6 | 39.2 | 42.4 | 44.9 | 44.8 | 31.6 | 0 | 100 ml |
| Total Daily Distillate Collection | | | | | | | | **1040 ml** |

**Fig. 3** Solar radiation measuring device, pyranometer and solarimeter

Inner glass cover:

$$I(t)A_g\alpha_g\tau_g + h_{t,w-gi}\left(T_w - T_{gi}\right)A_b = \frac{K_g}{L_g}\left(T_{gi} - T_{go}\right) \quad (2)$$

Basin water:

$$m_w c_{pw}\frac{dT_w}{dt} + h_{t,w-gi}\left(T_w - T_{gi}\right)A_b = I(t)A_g\alpha_w\tau_w + h_{Cn,bl-w}(T_{bl} - T_w)A_b \quad (3)$$

And

Basin liner:

$$I(t)A_g\alpha_{bl}\tau_{bl} = h_{Cn,bl-w}(T_{bl} - T_w)A_b + h_{Cd,bl-a}(T_{bl} - T_a)A_b \tag{4}$$

In case of variation of glass cover angle, area of glass is calculated through the following equations [9]:

$$A_g = A_b Sec\beta$$

where $\beta$ is the inclination angle of the glass cover.

After rearranging the Eqs. (4) and (3), one gets

$$\frac{dT_w}{dt} + aT_w = \frac{A_b}{M_w C_w} f(t) \tag{5}$$

where

$$a = \frac{A_b(U_t + U_b)}{M_w C_w} \tag{6}$$

$$f(t) = \left[(\alpha\tau)_{eff} I(t) + (U_t + U_b)T_a\right] \tag{7}$$

$$(\alpha\tau)_{eff} = \left[(\alpha\tau)_w + (\alpha\tau)_b \frac{h_{Cn,bl-w}}{h_{Cn,bl-w} + h_{bl-a}}\right] Sec\beta \tag{8}$$

$$U_t = \left[\frac{1}{h_{t,w-gi}} + \frac{L_g}{K_g} + \frac{1}{h_{t,go-a}Sec\beta}\right]^{-1} \text{ And}$$

$$U_b = \left[\frac{1}{h_{Cn,bl-w}} + \frac{1}{h_{bl-a}}\right]^{-1} \tag{9}$$

where $L_g$ is the thickness of glass cover and $K_g$ is the thermal conductivity of the glass cover.

The solution of Eq. (5) with initial condition, namely $T_w = T_{w0}$ at t = 0, can be written as:

$$T_w = \frac{f(t)}{a}\left(1 - e^{-at}\right) + T_{w0}e^{-at} \tag{10}$$

With the following assumptions:

- For the time interval $\Delta t$ the function "$f$(t)" is constant, i.e., $f(t) = \overline{f(t)}$ "a" is constant during the time interval $\Delta t$.

After knowing $T_w$, $T_g$ can be evaluated from the Eqs. (1), (2) and (4). Then the hourly yield $\dot{m}_{ew}$ can be found by using the following equation:

$$\dot{m}_{ew} = h_{e,w-gi}\frac{\left(T_w - T_g\right)}{L} \times 3600 \tag{11}$$

Instantaneous efficiency of the experimental setup can be obtained through the following relation:

$$\eta_{it} = \frac{\dot{m}_{ew} \times L}{I(t) \times A_{st}} \times 100 \tag{12}$$

where $I(t)$ is the solar radiation falls on the experimental setup in a day, $A_{st}$ is the evaporation area, and $L$ is the latent heat of vapor (J/kg):

$$L = 3.1625 \times 10^6 + [1 - (7.616 \times 10^{-4} \times (T_v)); \text{ For } T_v > 70°\text{C} \tag{13}$$

And

$$L = 2.4935 \times 10^6 \left[1 - (9.4779 \times 10^{-4} \times (T_v) + 1.3132 \times 10^{-7} \times (T_v^2)) - 4.7974 \times 10^{-3} \times (T_v^3))\right]; \text{ For } T_v < 70°\text{C} \tag{14}$$

## 3.1 Solar Radiation on Tilt Surface

The value of solar intensity $I(t)$ falls on the different tilt angle of condensing cover is obtained by the following relation [10]:

$$I(t) = I_b R_b + I_d R_d + (I_b + I_d) R_r \tag{15}$$

where $R_b$ is the geometric factor for beam radiation $I_b$, $R_d$ in the geometric factor for diffuse radiation $I_d$ and $R_r$ in the factor of total reflected radiation from the ground, factors may be expressed as

$$R_b = \frac{\text{Cos }\theta}{\text{Cos }\theta_z} \tag{16}$$

$$R_d = \frac{1 + \text{Cos }\beta}{2} \tag{17}$$

$$R_d = \frac{\rho(1 + \text{Cos }\beta)}{2} \tag{18}$$

where β is the inclination angle of the glass cover and the value of "θ" can be obtained through the following relation [10]:

$$\begin{aligned}\text{Cos}\,\theta =&(\text{Sin}\,\theta\,\text{Cos}\,\beta - \text{Cos}\,\phi\,\text{Sin}\,\beta\,\text{Cos}\,\gamma)\text{Sin}\delta \\ &+ (\text{Cos}\,\phi\,\text{Cos}\,\beta + \text{Sin}\,\phi\,\text{Sin}\,\beta\,\text{Cos}\,\gamma)\text{Cos}\,\delta\,\text{Cos}\,\omega \\ &+ \text{Cos}\,\delta\,\text{Sin}\,\beta\,\text{Sin}\,\gamma\,\text{Sin}\,\omega\end{aligned} \tag{19}$$

The experimental setup is placed facing south direction, hence $\gamma = 0$. The above equation is expressed as

$$\text{Cos}\,\theta = \text{Cos}(\theta - \beta)\text{Cos}\,\delta\text{Cos}\,\omega + \text{Sin}(\theta - \beta)\text{Sin}\,\delta \tag{20}$$

By substituting the value of $\beta = 0$ in the above equation, the zenith angle ($\theta_z$) can be obtained

$$\text{Cos}\,\theta_z = \text{Cos}\,\delta\,\text{Cos}\,\phi\,\text{Cos}\,\omega + \text{Sin}\,\delta\,\text{Sin}\,\phi \tag{21}$$

# 4 The Internal Heat Transfer Coefficient

## 4.1 *Radiative Heat Transfer [9]*

$$h_{r,w-gi} = \varepsilon_{eff}\sigma\left[\frac{(T_w + 273)^4 - \left(T_{gi} + 273\right)^4}{T_w - T_{gi}}\right] \tag{22}$$

where $\varepsilon_{eff}$ is the effective emissivity and σ is Stefan–Boltzmann constant, which is ($5.6697 \times 10^{-8}$ W/m$^2$K$^4$).

$$\varepsilon_{eff} = \left[\frac{1}{\varepsilon_w} + \frac{1}{\varepsilon_g} - 1\right]^{-1} \tag{23}$$

where $\varepsilon_w$ is the emissivity of the water and $\varepsilon_g$ is the emissivity of the glass cover.

## 4.2 *Convective Heat Transfer Coefficient*

The following relation was given by Tiwari and Lawrence [11] for the different inclination angles of the condensing cover:

$$N_u = \frac{h_{Cn} L_s}{K_s}$$

$$= 1 + 1.44\left[1 - \frac{1708}{R_a \text{Cos}\,\beta}\right]^{+}\left(1 - \frac{\text{Sin}(1.8\beta)^{1.6} 1708}{R_a \text{Cos}\,\beta}\right) + \left[\left(\frac{R_a \text{Cos}\,\beta}{5830}\right)^{1/3} - 1\right]^{+} \tag{24}$$

$$h_{Cn} = \frac{K_s}{L_s}\left[1 + 1.44\left[1 - \frac{1708}{R_a \text{Cos}\,\beta}\right]^{+}\left(1 - \frac{\text{Sin}(1.8\beta)^{1.6} 1708}{R_a \text{Cos}\,\beta}\right) + \left[\left(\frac{R_a \text{Cos}\,\beta}{5830}\right)^{1/3} - 1\right]^{+}\right] \tag{25}$$

where β is the inclination angle of glass cover.

The meaning of the "+" exponent is that only the + values of the term in the square brackets are to be used, otherwise it is zero for negative values.

"Ra" is the Raleigh number:

$$R_a = \frac{g\beta_1 L_v^3 \Delta T}{v\delta} \tag{26}$$

$$P_w = \exp[25.317 - (5144/(T_w + 273.0)) \tag{27}$$

$$P_w = \exp[25.317 - \left(5144/\left(T_{gi} + 273.0\right)\right) \tag{28}$$

### *4.3 Evaporative Heat Transfer*

$$h_{e,w-gi} = 0.01623 \times h_{cn,w-gi}\left[\frac{P_w - P_{gi}}{T_w - T_{gi}}\right] \tag{29}$$

## 5 Results and Discussion

The experimental data are written in Tables 1, 2 and 3. The result found that productivity at high (41°) and low angles (11°) was lower than latitude angle (26°). There are two important reasons behind these results. The first one is on a setup with latitude angle (26°), solar radiation is normal to the glass cover, allowing the maximum amount of solar radiation falling on the basin surface. Solar radiation is not falling on other two solar still (41° and 11°) in high intensity because the sun is closer to the latitude angle. The second reason is that the cavity area is large at higher angle (41°), due to which a great amount of wall shadow covers the maximum area on the internal surface of the solar still, which reduces the productivity. Not enough tilt angle is found to slip the water droplets on the lower angle (11°), causing condensed water droplets

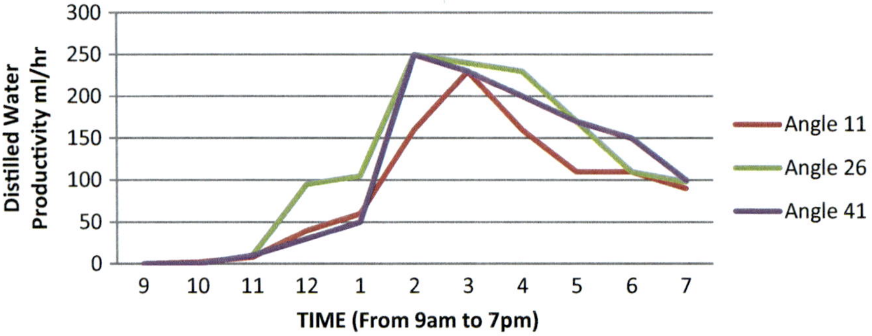


**Fig. 4** Hourly performance of solar stills, from 9 am to 7 pm

fall into the basin. During the experiment, it was found that the productivity was lowest at the morning and evening because of low solar intensity. Productivity was found to be highest in the afternoon. The maximum 290 ml distilled was achieved at noon on the lower angle, while the maximum productivity at latitude angle and higher angle was 250 ml and the maximum daily productivity of 1258 ml was achieved from solar still with latitude angle.

The different inclination of condensing cover (11°, 26° and 41°) has been chosen for winter, spring, rainy and summer season. The latitude of Gwalior city is 26.2183°N, in the summer season, the position of the sun is higher to latitude for which the lower angle of solar still will be received maximum amount of solar radiation and in cold weather the position of sun is below from the latitude, for which the higher angle will be appropriate. The maximum amount of radiation reaches the solar basin when the solar radiation falls normal to the glass cover, and the angle of inclination has been determined according to the position of the sun path angle for different season so that productivity can be obtained in all seasons.

A total of 4540 ml approximately 5 l of distilled water was collected from all the setups. Figure 4 shows the hourly production of the differential angle over time, implying that the productivity of all three solar stills is high in the afternoon. In the evening, the productivity of solar still decreased, due to the decreasing ambient temperature.

## 6 Conclusions

Based on experimental results, it is suggested that the angle of glass cover in solar still should be based on the latitude of the location because the solar radiation incident normal to the glass cover at latitude angle, allowing the maximum solar radiation reach to the basin surface and increase the evaporation rate. Productivity at the lower and higher angles was lower because the greater amount of radiations are reflected from the glass cover, as well as the large wall forming shadows on the basin surface and at the lower angle the condensate water droplets were not getting enough slope to slip down the water tray, causing the water droplets to fall into the basin.

## 7 Nomenclature

| | | | |
|---|---|---|---|
| A | Area (m) | Subscript | |
| $d_{np}$ | Size of nano-particles (nm) | np | Nano-particles |
| $C_{nf}$ | Specific heat capacity of nano-particles (J/kg-K) | nf | Nano-fluids |
| $\epsilon_{eff}$ | Effective emissivity | bs | Basin surface |
| $\epsilon_{bf}$ | Emissivity of base fluids | inl | Insulation |
| $F_n$ | Latent heat of vaporization of water | bf | Base fluids or basin fluids |
| h | Heat transfer coefficient (W/m$^2$°C) | go | Outer glass |
| $h_{T\ go\text{-}amb}$ | Total heat transfer coefficient from outer glass cover to ambient (W/m$^2$°C) | gi | Inner glass |
| K | Thermal conductivity (W/m°C) | amb | Ambient |
| $K_{inl}$ | Thermal conductivity of insulation (W/m°C) | sky | sky |
| $L_{inl}$ | Thickness of insulation (m) | Ep | Evaporation |
| $m_{bf}$ | Mass of base fluids | Cn | Convection |
| $m_{ss}$ | Mass of S | Rd | Radiation |
| T | Temperature (°C) | Greek words | |
| P | Partial pressure | $\rho$ | Density |
| Q | Heat transfer rate (W) | $\mu$ | Viscosity |
| SS | Solar still | $\beta$ | Thermal expansion co-efficient (K$^{-1}$) |
| | | $\sigma$ | Stefan–Boltzmann's constant (W/m$^2$K$^4$) |

## References

1. Thakur VK, Gaur MK, Sagar MK. Role of advance solar desalination technique for sustainable development. In: Pandit M, Srivastava L, Rao RV, Bansal JC, (eds) Intelligent Computing Applications for Sustainable Real-World Systems, Gwalior, India: Springer Nature Switzerland AG, ICSISCET 2019, PALO 13; 2020, https://doi.org/10.1007/978-3-030-44758-8_4, pp 28–38

2. Raj SV, Manokar AM (2017) Design and analysis of solar still. Mater Today Proc 4:9179–85. https://doi.org/10.1016/j.matpr.2017.07.275
3. Solar E, Performance S, Vacuum U (2019) Enhancing solar still performance using vacuum pump and geothermal energy 2019:1–13. https://doi.org/10.3390/en12030539
4. Dev R, Tiwari GN (2009) Characteristic equation of a passive solar still. DES 245:246–65. https://doi.org/10.1016/j.desal.2008.07.011
5. Cherraye R, Bouchekima B, Bechki D, Bouguettaia H (2020) The effect of tilt angle on solar still productivity at different seasons in arid conditions (south Algeria). Int J Ambient Energy 0:1–16. https://doi.org/10.1080/01430750.2020.1723689
6. Ahmad Z (2015) Performance analysis of single slope solar still performance analysis of single slope solar still 3:66–72
7. Azooz AA, Younis GG (2016) Effect of glass inclination angle on solar still performance 033702. https://doi.org/10.1063/1.4948625
8. Sharshir SW, Elsheikh AH, Peng G, Yang N, El-Samadony MOA, Kabeel AE (2017) Thermal performance and exergy analysis of solar stills—a review. Renew Sustain Energy Rev 73:521–44. https://doi.org/10.1016/j.rser.2017.01.156
9. Lawrence SA, Gupta SP, Tiwari GN (1990) Effect of heat capacity on the performance of solar still with water flow over the glass cover. Energy Convers Manag 30:277–85. https://doi.org/10.1016/0196-8904(90)90010-V
10. Ferna L, Rubio E, Porta-ga MA (2004) Modeling thermal asymmetries in double slope solar stills 29:895–906. https://doi.org/10.1016/j.renene.2003.11.001
11. Tiwari GN, Lawrence SA (1991) New heat and mass transfer relations for a solar still 31:201–203

# Chapter 18 Implementation and Performance Analysis of ECC-Based Text Encryption on Raspberry Pi 3

**Swati and Ramjee Prasad Gupta**

## 1 Introduction

Internet of things (IoT) has added internet connectivity to most of the systems and has made the crucial data available over a network. These vital private data using a public network need security, which could be achieved by encryption. Various encryption algorithms are used to serve this purpose of data confidentiality. ECC, RSA, and other primitive techniques are some of the encryption techniques for data encryption. Figure 1 presents the basic idea of cryptography. The sensor nodes used in the IoT are physically constrained devices with low computation power, leading to the need for a lightweight cryptographic scheme [1]. ECC is found to be a suitable candidate to cater this need of secure and lightweight solution of cryptography. This technique can be further optimized by using parallel processing [2].

ECC is a public-key infrastructure proposed by Neal Koblitz and Victor Miller in 1985 [4]. Accredited standards organizations such as ANSI [5, 6], IEEE [7], ISO [8, 9], and NIST have also included ECC in standards. The basic idea of any cryptographic system lies in transforming the message into a numerical value for performing mathematical operations. ECC operates on the points of an elliptic curve, so plaintexts are mapped onto a point on the curve, and the curve operations on the point yield a new point that serves as the corresponding ciphertext.

Swati (✉)
Indian Institute of Technology Patna, Bihta, Bihar, India

R. P. Gupta
B.I.T Sindri, Dhanbad, Jharkhand, India

© The Author(s), under exclusive license to Springer Nature Singapore Pte Ltd. 2022

H. M. Dubey et al. (eds.), *Artificial Intelligence and Sustainable Computing*, Algorithms for Intelligent Systems, https://doi.org/10.1007/978-981-16-1220-6_18

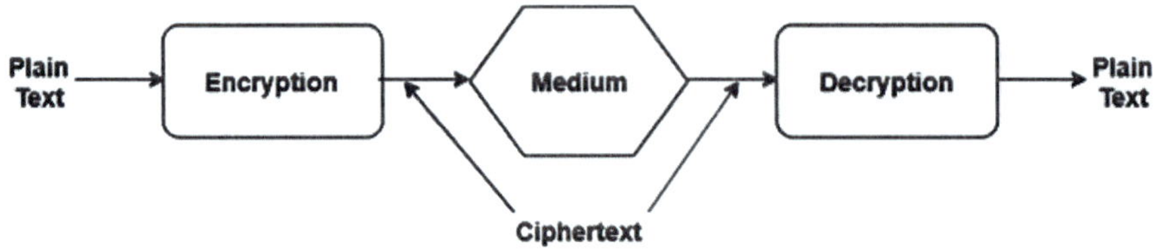


**Fig. 1** Simplified cryptographic model

## 2 Related Works

The merits of ECC have attracted a lot of research in this field. Unlike other crypto systems, the mapping of plaintext is not as simple as it seems to be. The difficulty lies in the fact that here the messages that are generally in the form of texts or numbers have to be plotted on the elliptic curve in the form of points [3].

In [10], Trappe et al. have discussed the fast probabilistic method proposed by Koblitz. In this scheme, the message M is transformed to x coordinate using the equation

$$x = (M.K + i) \bmod p,$$

where M = encoded value of alphabet in number, K = random public variable, and i varies from 1 to (k −1) to get an integral value of y.

Thus, m gets encoded as (x, y). The decoding of the message is also simple. The decrypted message can be obtained just by taking the floor of ((x−1)/K). The drawback of this scheme is its probabilistic nature. The probability of getting the corresponding y-coordinate for the message embedded in the x-coordinate of a point is only about ½. The failure rate could be minimized by selecting a large value of K, but still there exists about $1/2^k$ chance of failure. Secondly, there are huge chances of a collision attack if the size of the p is not sufficiently large.

Padma Bh et al. [11] had proposed another scheme for mapping. The message is first represented in the 8-bit ASCII form, and then each point on the curve is assigned to an ASCII value. These encode points are encrypted to two ciphertext points, which have to be decrypted and decoded at the receiver end.

The two ciphertext points sent are $\{kG, Pm + kP_\beta\}$

Where k = a random no., G = generator point, Pm = plaintext point on the curve, and $P_\beta$ = public key of the receiver.

The receiver recovers the plaintext by multiplying its private key with the first ciphertext point and subtracting the result with the second one.

$$\text{Decryption} : Pm + k\left(n_\beta\right) G - n_\beta.(kG) = Pm$$

The proposed method fails against chosen-plaintext attack as well as a known-plaintext attack. The frequency analysis will reveal the secret of mapping as for every repetition of plaintext, the same ciphertext will be generated.

Another better scheme was proposed by Hankerson et al. [12] in their book. Here the plaintext characters are mapped to the curve by multiplying the generator point to its ASCII value. This scheme is currently in use, as it is faster than other mapping techniques. The limitation of this technique lies in the fact that it is not immune to frequency analysis. The adversary can snoop the communication and go for the frequency analysis of the entire ciphertexts exposing it to vulnerabilities, viz., known-plaintext attack, chosen-plaintext attack, and chosen-ciphertext attack.

Another mapping method proposed by Amounas et al. [13] is based on matrices and the previous method of generator point-based mapping on the elliptic curve. This scheme used a matrix to permute the position of the points obtained by multiplying with the generator point. The plaintext is transformed to the points on the curve by using the generator point and placed in the form of a $3 \times r$ matrix after padding with space if needed. This matrix has to be multiplied with a $(3 \times 3)$ non-singular matrix to get the result set of points. The receiver end can get the decode of the ciphertext by using the elliptic curve decryption technique and the inverse of the matrix. The complexity and large computation time involved in this technique make it inefficient as time is the key factor in public-key cryptography.

The next scheme was proposed by Muthukuru et al. [14] in which they used the block mapping technique. The plaintext represented in ASCII has to be grouped in a fixed-length block, and the first block is XORed by an initial vector (IV), followed by multiplying with the generator point. Then the next blocks are XORed by the previous blocks, and the same process is continued. An array of points that are obtained is decrypted at the receiver end and divided by the generator point to obtain the blocks, which are then XORed with respective blocks to obtain the original plaintext. Clearly, for decryption, the IV is needed, and it has to be shared between the communicating parties, which can lead to man-in-the-middle attacks. Secondly, the XOR function used in this framework has eliminated the problem of frequency analysis, but still there exist the chances of a chosen-plaintext attack to extract the IV. Once it is revealed, an intruder can retrieve the whole message mapping.

The requirement of a mapping scheme satisfying all the criteria of a successful mapping, viz., secure and invertible points mapped on the elliptic curve, motivated us for the present framework. In the present framework, the plaintext is mapped to an affine point on the elliptic curve securely and encrypting the message by performing scalar multiplication using the sender's private key [15]. The original plaintext was recovered by using the decryption technique. The overhead of sharing common lookup tables between the communicating parties is avoided in this method and reverse order encryption provides more security for text encryption. An 192-bit sec curve was used, and the standard values of the message in ASCII and hexadecimal were transformed into affine points on the elliptic curve. Scalar multiplication was used to encrypt these points on the selected curve.

# 3 Preliminaries

The ECC technique is based on the field operations on the points of the elliptic curve involved. Efficient curve operations and computationally inexpensive steps of curve arithmetic are likewise crucial to performance. The general equation of an elliptic curve E over a field K can be defined as Eq. (1), and a graphical representation of two elliptic curves over finite field K is shown in Fig. 2:

$$E : y^2 + a_1xy + a_3y = x^3 + a_2x^2 + a_4x + a_6 \tag{1}$$

where $a_1, a_2, a_3, a_4, a_6 \in K$.

The definition of an elliptic curve also requires that the curve be non-singular and for that we need to find the discriminant of E.

Say, $\Delta$ is the discriminant of E and can be defined as Eq. (2)

$$\Delta = -d_2^2d_8 - 8d_4^3 - 27d_4^2 + 9d_2d_4d_6 \tag{2}$$

where

$$\begin{aligned} d_2 &= a_1^2 + 4a_2 \\ d_4 &= 2a_4 + a_1a_3 \\ d_6 &= a_3^2 + 4a_6 \\ d_8 &= a_1^2a_6 + 4a_2a_6 - a_1a_3a_4 + a_2a_3^2 - a_4^2 \end{aligned}$$

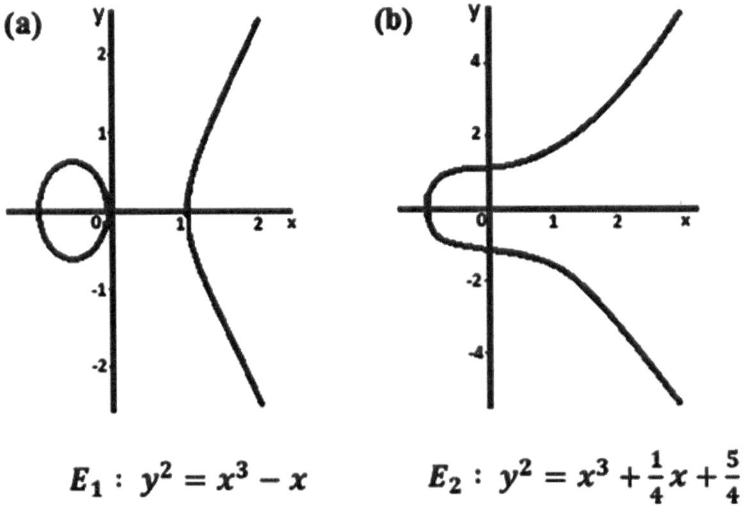


Fig. 2 Elliptic curves

Weierstrass curve is the most commonly used elliptic curve and its simplified equation is shown in Eq. (3).

$$y^2 = x^3 + ax + b, \text{ where a, b} \in \text{K} \tag{3}$$

The condition for the curve to be non-singular holds if and only if the discriminant

$$\Delta = -16\left(4a^3 + 27\,b^2\right)$$

## 3.1 *Point Representation*

The elliptic curve $y^2 = x^3 + ax + b$ is considered here over a field K and whose characteristic is neither 2 nor 3. In order to reduce the complexity, the points are represented using projective coordinates. Some of the most commonly used projective coordinates are discussed.

### 3.1.1 Standard Projective Coordinates

The projective points (X:Y:Z), Z $\neq$ 0, which corresponds to (X/Z, Y/Z) affine points, represent standard projective coordinates. The elliptic curve projective equation is

$$Y^2Z = X^3 + aXZ^2 + bZ^3 \tag{4}$$

(0:1:0) corresponded to the point of infinity $\infty$, while (X:Y:Z) is the negative of (X:Y:Z).

### 3.1.2 Jacobian Projective Coordinates

The projective points (X:Y:Z), Z $\neq$ 0, which corresponds to the (X/Z$^2$, Y/Z$^3$) affine points represent the Jacobian projective coordinates. The elliptic curve projective equation is

$$Y^2 = X^3 + aXZ^4 + bZ^6 \tag{5}$$

(1:1:0) corresponded to the point of infinity $\infty$, while (X:-Y:Z) is the negative of (X:Y:Z).

### 3.1.3 Chudnovsky Projective Coordinates

In this type, the Jacobian point (X:Y:Z) is represented as ($X$:$Y$:$Z$:$Z^2$:$Z^3$). In some point multiplication methods, where the projective coordinates are used for addition, the redundancy of this coordinate system proves to be beneficial.

## 3.2 Operations of ECC

The basic operations involved in ECC are point inverse, point addition, point doubling and scalar multiplication. Point inverse and scalar multiplication are the most expensive operations among all.

### 3.2.1 Point Inverse

For a point on an elliptic curve, be it J ($x_1$, $y_1$), its inverse can be represented as –J ($x_1$, $y_1$). Equation (6) can be used for calculating the inverse

$$-J(x_1, y_1) = J(x_1, p - y_1) \tag{6}$$

### 3.2.2 Point Addition

An abelian group under addition is formed from the elliptic curve over a finite field E(Fp). Two distinct points added on the elliptic curve is the point addition. If Q ($x_3$, $y_3$) be the point on the elliptic curve found after the addition of two points P ($x_1$, $y_1$) and N ($x_2$, $y_2$) on the same elliptic curve, then Q = P + N = Q ($x_3$, $y_3$). The following equation is used for the point addition

$$\begin{aligned} \lambda &= \frac{y_2 - y_1}{x_2 - x_1} \\ x_3 &= \lambda^2 - x_1 + x_2 \\ y_3 &= \lambda(x_1 - x_3) - y_1 \end{aligned} \tag{7}$$

### 3.2.3 Point Doubling

Much like point addition, a tangent is drawn to a single point on the curve. The interaction of the line with the curve is projected to find the double of the point on the same curve. Let Q ($x_3$, $y_3$) be the point on the point doubling result, which lies in the curve for a point P ($x_1, y_1$) on the elliptic curve. Point doubling can be calculated using Eq. (8):

$$\begin{aligned}\lambda &= \frac{3x_1^2 + a}{2y_1} \\ x_3 &= \lambda^2 - 2x_1 \\ y_3 &= \lambda(x_1 - x_3) - y_1\end{aligned} \tag{8}$$

#### 3.2.4 Scalar Multiplication

Let Q be a point on the curve obtained by the scalar multiplication of a scalar k with point p, then Q can be represented as Q = KP. If P ($x_1$, $y_1$) be a point on the curve, then the point k*P ($x_1$, $y_1$) can be calculated by computation of repeated point doublings and point additions where k is an integer in prime fields distinct additive, and there is existence of multiplication inverses.

### *3.3 Methodology*

#### 3.3.1 Mapping of Plaintext on the Curve

In the present framework, the plaintext is grouped into blocks of size M, followed by converting each character into 8-bit ASCII, resulting in an array of length 8 × M bits. The last block of the message is padded by "32" to maintain the size of the block as it is ASCII value of space. Append N number of 0 s at the end of each array where the values of M and N depend on the overlapping problem. More than one strings of message pointing to only one point on the elliptic curve lead to the problem of overlapping. To avoid such a problem, we have to find the maximum interval between two successive x-coordinates, which is equivalent to solving the elliptic curve discrete logarithmic problem (ECDLP). In literature, experimentally it has been established that the maximum gap cannot exceed a specific value, as the points are uniformly distributed throughout the curve [16, 17, 18]. So N = 8 bit is sufficient to avoid overlapping and M is selected by using Eq. (9)

$$\begin{aligned}\mathrm{M} &\le \left\lceil \frac{P-8}{8} \right\rceil, \text{ where P = size of the elliptic curve} \\ \mathrm{M} &= \text{no. of characters of the message considered at a time}\end{aligned} \tag{9}$$

Now the decimal value of each (M × 8 + N) array represents x-coordinate of the elliptic curve, and the corresponding y-coordinate is calculated using Eq. (10)

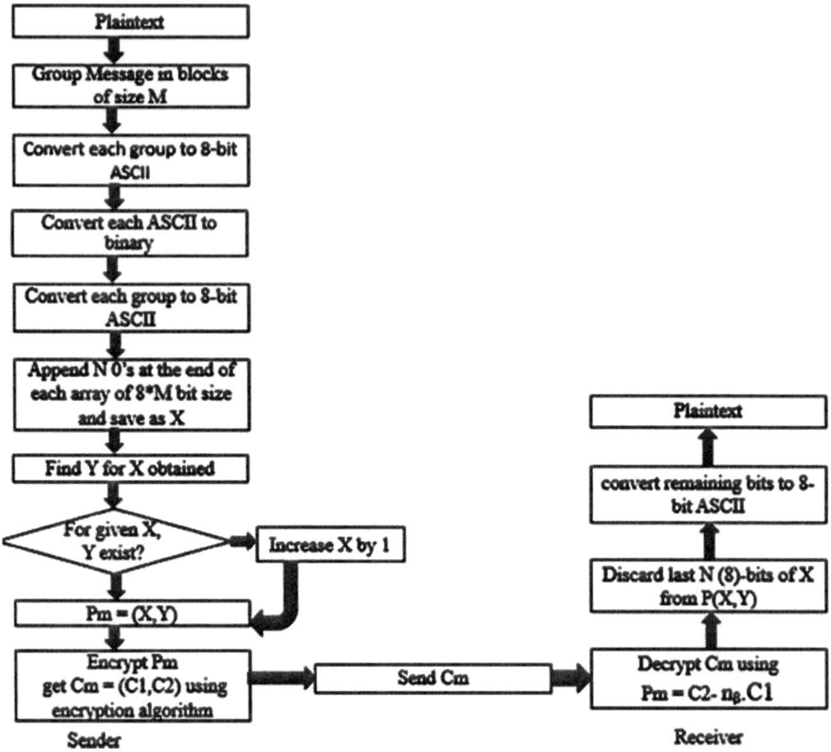


**Fig. 3** Block diagram of the mapping and reverse mapping of plaintext on elliptic curve

$$y^2 \equiv x^3 + ax + b \mod p \quad (10)$$

For the given x, if the equation does not have a solution, the value is increased by 1 and again checked for corresponding y. The process is repeated until we get a solution for the Eq. (10). For every block, the same process is repeated, and the respective points on elliptic curve are obtained.

The distinct points thus obtained are used for encryption using ECC. Figure 3 shows the block diagram of the implemented scheme that includes the mapping and reverse mapping of the plaintext on the elliptic curve.

## *3.4 Reverse Mapping*

To retrieve the message from the points mapped onto the elliptic curve, reverse mapping is applied at the receiver end. The y-coordinate is discarded as it does not have any participation in the encryption/decryption. The last 8 bits of the x-coordinate is dropped and the remaining bits are again converted to the 8-bit ASCII. The process is repeated for each ciphertext pair received.

### *3.5 Encryption*

In public-key cryptosystems like ECC, there are two keys involved in encryption, viz., public key and private key. A private key is generally used for encrypting/decrypting data, whereas public key is used to authenticate. The steps involved in encrypting the plaintext are:

Step 1 The message is projected on the curve using the technique discussed to obtain points (Pm).

Step 2 The sender chooses a random number "k" which serves as the session key.

Step 3 Public key of the receiver is $P_\beta$ which is equal to ($n_\beta$.G),
where $n_\beta$ = private key of receiver and G = generating point

Step4 A ciphertext (Cm) having two cipher points C1 and C2 is calculated and sent to the receiver.
Cm = {C1, C2}
where C1 = k.G
C2 = Pm + k.$P_\beta$
k = session key

### *3.6 Decryption*

Steps involved in decryption of the ciphertext at the receiver side are:

Step 1 At the receiver end, Pm is retrieved from the obtained cipher points by point multiplying private key with C1, and subtracting the result with C2.

Step 2 Now reverse map the points Pm to get the original plaintext

## 4 Implementation

This section includes the implementation of the encryption method discussed. The selected curve for encryption used is NIST recommended 192-bit, secp192r1 (P-192) curve [19]. The verifiably random elliptic curve domain parameters over $F_p$ secp192r1 are specified by the sextuple

T = (p, a, b, G, n, h) where the finite field Fp is defined by:
p = FFFFFFFF FFFFFFFF FFFFFFFF FFFFFFFE FFFFFFFF FFFFFFFF
= 2192 − 264 − 1
The curve E: $y^2 = x^3 + ax + b$ over Fp is defined by:

a = FFFFFFFF FFFFFFFF FFFFFFFF FFFFFFFE FFFFFFFF FFFFFFFC
b = 64210519 E59C80E7 0FA7E9AB 72243049 FEB8DEEC C146B9B1

E was chosen verifiably at random as specified in ANSI X9.62 [X9.62] from the seed:
S = 3045AE6F C8422F64 ED579528 D38120EA E12196D5

The base point G in compressed form is:
G = 03 188DA80E B03090F6 7CBF20EB 43A18800 F4FF0AFD 82FF1012
and in uncompressed form is:
G = 04 188DA80E B03090F6 7CBF20EB 43A18800 F4FF0AFD 82FF1012 07192B95 FFC8DA78 631011ED 6B24CDD5 73F977A1 1E794811
Finally, the order n of G and the cofactor are:
n = FFFFFFFF FFFFFFFF FFFFFFFF 99DEF836 146BC9B1 B4D22831
h = 01

## 4.1 Encryption

The message to be transmitted is:

How long do you want these messages to remain secret?

Plaintext mapping by the sender:

Step 1 Calculate the values of N and M. Here the curve used has a key size of 192 bits, so using Eq. (9), M = 23 and N = 8.

Step 2 Plaintext is grouped into blocks of 23 characters. The provided plaintext will be grouped as: {How long do you want th}, {ese messages to remain}, {secret?}. The last group is padded with "32" as it is ASCII value of "space" in order to maintain the length of the group.

Step 3 The ASCII values for these blocks of messages are:
[72, 111, 119, 32, 108, 111, 110, 103, 32, 100, 111, 32, 121, 111, 117, 32, 119, 97, 110, 116, 32, 116, 104],
[101, 115, 101, 32, 109, 101, 115, 115, 97, 103, 101, 115, 32, 116, 111, 32, 114, 101, 109, 97, 105, 110, 32],
[115, 101, 99, 114, 101, 116, 63, 32, 32, 32, 32, 32, 32, 32, 32, 32, 32, 32, 32, 32, 32, 32, 32]

Step 4 The ASCII values are converted to binary form.
['01001000', '01101111', '01110111', '00100000', '01101100', '01101111', '01101110', '01100111', '00100000', '01100100', '01101111', '00100000', '01111001', '01101111', '01110101', '00100000', '01110111', '01100001', '01101110', '01110100', '00100000', '01110100', '01101000']
['01100101', '01110011', '01100101', '00100000', '01101101', '01100101', '01110011', '01110011', '01100001', '01100111',

'01100101', '01110011', '00100000', '01110100', '01101111', '00100000', '01110010', '01100101', '01101101', '01100001', '01101001', '01101110', '00100000']
['01110011', '01100101', '01100011', '01110010', '01100101', '01110100', '00111111', '00100000', '00100000', '00100000', '00100000', '00100000', '00100000', '00100000', '00100000', '00100000', '00100000', '00100000', '00100000', '00100000', '00100000', '00100000', '00100000']

Step 5 Finally, N = 8 zeroes are appended at the end of strings of each block and decimal value of each block is assigned to $X_1$, $X_2$, $X_3$.
$\mathbf{X_1}$ = (0100100001101111011101110010000001101100011011110110
-1110011001110010000001100100011011110010000001111001011011
-1101110101001000000111011101100001011011100111010000100000
-01110100011010000000000000)$_2$
= (17761111214652491154086749810-916732607018729080
13732653056)$_{10}$
$\mathbf{X_2}$ = (0110010101110011011001010010000001101101011001010
-111001101110011011000010110011101100101011100110010000001
-110100011011110010000000111001001100101011011010110000101
10100101-1011100010-000000000000)$_2$
= (248756544171817962015096230l3-52706162749091
402485479645184)$_{10}$
$\mathbf{X_3}$ = (01110011011001010110001101110010011001010111
-10100001111100100000001000000010000000100000001000
0000100000001000000010-0000001000000010000000010000000
100000001000000010000000100000-0010000000000000000)$_2$
= (2829502880782835046226226112518403056l-
998630478328197365 76)$_{10}$

Step 6 The value of $Y_1$, $Y_2$ and $Y_3$ is computed using $X_1$, $X_2$ and $X_3$ in Eq. (10), respectively. If first iteration does not yield any value of Y, then X is incremented by 1.

Step 7 Two cipher points are calculated using the expression for C1 and C2. Finally, the ciphertext Cm = (C1, C2) is sent to the receiver end.

## 4.2 Decryption

The ciphertext is received at the other end of communicating parties. Encrypted message is decrypted to get the original plaintext using the following steps:

Step 1 All the received cipherpoints are decrypted, taking one at a time.

Step 2 Considering the first cipherpoint, Y coordinate is neglected, as it has no role in encryption or decryption calculation and X coordinate is converted to binary form

Step 3 The last N = 8 bits from the right end is discarded and 8-bits ASCII is regenerated from the remaining strings to recover the original message. We get the first block of message as "How long do you want th"

Step 4 Same steps are repeated for all the cipher points received.

## 5 Results and Discussions

All the implementation is done on Raspberry Pi 3 model B with Quad Core 1.2 GHz Broadcom BCM2837 64bit CPU, 1 GB RAM and BCM43438 wireless LAN and BLE on board, and using 192-bit Secp192r1 recommended elliptic curve parameter. The encryption/decryption technique is implemented in python and tested for different sizes of input messages. The time taken for encrypting and decrypting the example text is tabulated in Table 1. Encrypting the sample text takes 0.0128 s and decryption of the same text at the receiver end takes 0.0003 s. The framework is tested for 10, 20, 50, 100 and 500 Kb of text encryption and decryption.

The execution time varies with file size, which are enlisted in Table 2. During encryption, the text is plotted on the curve and is computationally more expensive compared to decryption. Encryption takes longer execution time compared to the decryption execution time.

The obtained data was plotted on a time versus size graph shown in Fig. 4. A linear non-weighted curve fitting graph was applied. Figure 4a shows the encryption time in seconds along y-axis versus text size in Kb along x-axis. Whereas, Fig. 4b shows the decryption time in seconds along y-axis versus text size in Kb along x-axis. Overall, both the encryption and decryption time versus size graph show a linear pattern.

**Table 1** Execution time (in sec) for plaintext "How long do you want these messages to remain secret?"

| | |
|---|---|
| Encryption time | 0.0128 |
| Decryption time | 0.0003 |

**Table 2** Performance w.r.t execution time (s) for various sizes of plaintext

| S. no. | Text size (Kb) | Encryption time (s) | Decryption time (s) |
|---|---|---|---|
| 1. | 10 | 0.7645 | 0.0157 |
| 2. | 20 | 1.6129 | 0.0339 |
| 3. | 50 | 4.0329 | 0.0858 |
| 4. | 100 | 8.2839 | 0.1730 |
| 5. | 200 | 17.2892 | 0.3473 |
| 6. | 500 | 46.0914 | 0.9034 |

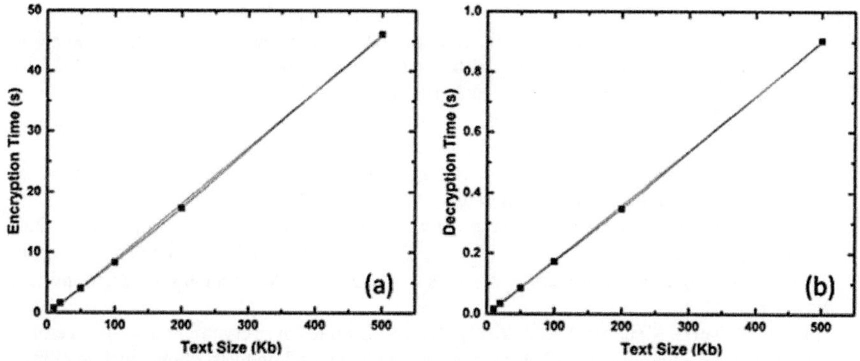


**Fig. 4** **a** Encryption time **b** Decryption time graph for varying text size

## 6 Conclusion

The implemented scheme is immune to many known attacks as known-plaintext attack, chosen-plaintext attack, chosen-ciphertext attack, collision attack and man-in-middle attack. The use of M characters at a time for mapping eliminates the scope of frequency analysis as there is only $\frac{1}{256^M}$ possiblilities of repeatition of string. Moreover, appending N number of 0 s at the end of the M*8 array of each block helps in getting every x-coordinate plotted on the curve by getting corresponding y-coordinate and at the time of decryption the original value of x-coordinate is recovered. This scheme is deterministic in nature rather than probabilistic which makes it immune to collision attack. It works well with the small text encryption. The relation between text size and encryption/decryption time is approximately linear. Abiding with the necessary conditions of a mapping scheme, it eliminates any requirement of sharing lookup table for encryption/decryption purpose. It is a fast, efficient and secure mapping scheme compared to the others.

## References


1. Washington, LC (2008) Elliptic curves: number theory and cryptography. Chapman and Hall/CRC
2. Jarvinen K, Skytta J (2008) On parallelization of high-speed processors for elliptic curve cryptography. IEEE Trans Very Large Scale Integr (VLSI) Syst 16(9):1162–1175
3. Keerthi K, Surendiran B (2017) Elliptic curve cryptography for secured text encryption. In: 2017 International conference on circuit, power and computing technologies (ICCPCT). IEEE
4. Hankerson, Darrel, Julio López Hernandez, and Alfred Menezes. "Software implementation of elliptic curve cryptography over binary fields." International Workshop on Cryptographic Hardware and Embedded Systems.Springer, Berlin, Heidelberg, 2000
5. ANSI X9.63, Public Key Cryptography for the Financial Services Industry: Elliptic Curve Key Agreement and Key Transport Protocols, working draft, August 1999

6. Brickell E, Gordon D, McCurley K, Wilson D (1993) Fast exponentiation with precomputation. In: Advances in Cryptology—Eurocrypt'92, LNCS 658, pp 200–207
7. IEEE P1363 (2000) Standard Specifications for Public-Key Cryptography
8. ISO/IEC 14888-3 (1998) Information technology—security techniques—digital signatures with appendix—part 3: certificate based-mechanisms
9. ISO/IEC 15946 (1999) Information technology—security techniques—cryptographic techniques based on elliptic curves, committee draft (CD)
10. Trappe W (2006) Introduction to cryptography with coding theory. Pearson Education India
11. Padma Bh, Chandravathi D, Prapoorna Roja P (2010) Encoding and decoding of a message in the implementation of Elliptic Curve cryptography using Koblitz's method. Int J Comput Sci Eng 2(5):1904–1907
12. Hankerson Darrel, Menezes Alfred (2011) Guide to Elliptic curve cryptography. Springer, US
13. Amounas F, El Kinani EH (2012) Fast mapping method based on matrix approach for elliptic curve cryptography. Int J Inf Netw Secur (IJINS) 1(2):54–59
14. Muthukuru J, Sathyanarayana B (2012) Fixed and variable size text based message mapping technique using ECC. Global J Comput Sci Technol
15. Silverman JH (2006) An introduction to the theory of elliptic curves. Brown University. June 19 2006
16. Burgess David A (1957) The distribution of quadratic residues and non-residues. Mathematika 4(2):106–112
17. Galbraith Steven D, McKee James (2000) The probability that the number of points on an elliptic curve over a finite field is prime. J London Math Soc 62(3):671–684
18. Hatley Jeffrey, Hittson Amanda (2009) Numerical evidence on the uniform distribution of power residues for elliptic curves. Invol J Math 2(3):305–321
19. Brown DRL (2010) Sec 2: Recommended elliptic curve domain parameters. Standars for Efficient Cryptography. https://secg.org/sec2-v2.pdf

# Chapter 19
# Approximate Partitional Clustering Through Systematic Sampling in Big Data Mining

**Kamlesh Kumar Pandey and Diwakar Shukla**

## 1 Introduction

The revolution of digital technology is changing the nature of the data concerning volume, variety, and velocity that defines the big data. Such trends in big data technologies are the internet of things, cloud computing, multiple connections systems, healthcare systems, retailing systems, government systems, and social networks [1]. The scalability, storage capacity, rigid data management, flexibility, performance, and securely granular access of massive data are the bases of data revolution [1]. Big data is the next frontier technique for productivity, innovation, and competition in the digital world [2]. The big data term was defined in 2001 with volume, variety, and velocity emerging characteristics of big data. After that, the Oracle identified the value, IBM introduced the veracity, and SAS coined the variability as fourth, fifth, and sixth characteristics of big data, respectively [3]. The next characteristic of big data is the visualization that visualizes big data-related problems and results as user exception [4]. These seven characteristics are known as Vs, characteristics, dimensional, and attributes of big data [4], where volume is the primary characteristic of big data. The volume, variety, and velocity are known as basic characteristics and value, veracity, variability, and visualization are known as supportable characteristics of big data [3, 5].

Traditional data analysis and mining techniques are not scalable for large-scale data because classical data mining techniques do not accommodate infinitely large datasets. Generally, the massive volume introduces scalability, computational cost, and uncertainty-related problems of data mining [2]. Big data mining uses classical machine learning methods with statistical inference for evolution, extended to utilizing traditional data mining techniques [6]. Big data mining not only mines the

K. K. Pandey (✉) · D. Shukla
Department of Computer Science and Applications, Dr. Harisingh Gour Vishwavidyalaya, Sagar, India

© The Author(s), under exclusive license to Springer Nature Singapore Pte Ltd. 2022

H. M. Dubey et al. (eds.), *Artificial Intelligence and Sustainable Computing*, Algorithms for Intelligent Systems, https://doi.org/10.1007/978-981-16-1220-6_19

hidden knowledge but also extracts the consistent patterns, systematic, and complicated relationships among the data through classification, clustering, association rule, regression, and other data mining techniques. It required more transparency than traditional data mining because massive volume data has held a lot of valuable information, data sources are traceable, and allow the data type diversity and velocity to make real-time analysis [7].

The literatures [8–12] addressed data reduction techniques for big data under a single and multiple machine execution environments. The most popular data reduction techniques are divide-and-conquer, condensation, feature selection, parallel processing, sampling, granular computing, incremental learning methods that reduced the data size for big data mining. Data sampling is one of the statistical techniques that extracts subsets of the entire dataset for analysis or mining [13]. The sampling strategies utilized the execution time, memory space, computation cost, scalability, and data size inside the large dataset mining [14, 15] and achieved high-quality mining results [16]. The selection of the sampling method depends on the cost efficiency issue that saves the time, resources, and effort of any algorithm [17, 18]. However, optimal sampling minimizes variance and maximizes accuracy of a specific sample size without bias. Inside the data mining, the sampling techniques provide an approximated result within the confidence limits [9]. The sampling-based data mining algorithm solves problems related to large volume or large-scale data [19]. The literatures [9, 20–22] described the sampling-based data mining techniques that achieved approximate results within a specific timeframe. Therefore, sampling-based data mining techniques are known as approximate data mining techniques. Some prevalent probability-based sampling methods are uniform random sampling, reservoir sampling, systematic sampling, progressive sampling, and stratified sampling [9, 23, 24].

In big data mining, various applications such as pattern recognition, artificial intelligence, image segmentation, wireless sensor networks, bioinformatics, social network analysis, customer segmentation, scientific data analysis, financial analysis, text analysis, vector quantization, target marketing, and so on. Clustering is one of the most widely used data mining techniques that identify the hidden patterns, knowledge, and class of the data through similarity and dissimilarity measures. The objective of clustering is to achieve high compaction and high separation between clusters. The minimum distance between the within-cluster of data objects defines the high compaction, and the maximum distance between-clusters of data objects defines the high separation [8]. Big data collects the structured, unstructured, and semi-unstructured data variety through heterogeneous sources. The structured data are organized in a well-structured format such as table, graph, and vector; the unstructured data used specific files format data such as text, images, audio, video format, while semi-structured data used the structured and unstructured data together such as XML. The clustering process has required a specific data format. Sometimes data is not organized in a proper data format for clustering, then at that time data conversion is required. Suppose the clustering process requires structured data, but clustering data is available in unstructured and semi-structured data formats, then for these reasons, unstructured and semi-structured data formats data are converted into pooled feature vectors through multi-way clustering, bimodal clustering, and co-clustering methods [11]. Clustering algorithms are classified basically into partitioning, hierarchical, grid, fuzzy, density, model, and graph class based on the working process, similarity, and dissimilarity measures [25, 26].

The traditional clustering algorithms are not suitable under large-scale data because they suffer from computation time, efficiency, performance, and quality-related issues [2, 27]. Consequently, computational efficiency is the biggest challenge of large-scale data, and it needs to improve accuracy and reliability. The motivation of big data mining is decreasing as the execution times of mining approaches [28]. The objective of this study is to improve the scalability, computational cost, efficiency, performance, and speed-up for big data mining. This paper achieves the paper objective through the combination of the systematic sampling and data mining approach using partitional clustering (K-means) inside the single machine environment.

The rest of the paper is organized as follows. Section 2 describes the partitional clustering and related works of the sampling-based data mining approach. Section 3 presents the systematic sampling process and its sampling formulation. Section 4 proposed a systematic sampling-based data mining model through the K-means algorithm known as SYK-means. Experimental results are shown and analyzed in Sect. 5 using some cluster internal measurements. Finally, conclusion and future work are drawn in Sect. 6.

## 2 Related Works

The partitional clustering groups the data into K clusters based on the similarity and dissimilarity. Suppose the dataset has N data points that consist of $\{n_1, n_2 \ldots\ldots n_n\}$ data points in the D-dimensional space, and partitioning clustering methods clustered the N into C = $\{c_1, c_2 \ldots\ldots c_k\}$ basis of predefined similarity function, cluster centroid, and statistical tools. The clustering algorithm is always satisfied with the $\{c_1 \cup c_2 \cup \ldots\ldots \cup c_k\} = C$ and $\{c_1 \cap c_2 \cap \ldots\ldots \cap c_k\} = \theta$ condition [10]. Partitional clustering algorithms optimize the objective function through the number of cluster K, initialization of cluster, center tendency, and statistical approach for center updating, and distance measures [12]. Based on the existing research, various K-based partitional clustering algorithms are available for data mining such as K-means, K-medians, K-modes, K-prototypes, and K-medoids [29]. The first section has introduced the advantages of sampling and clustering, and identified the volume-related problem of big data mining and their possible ways of solution.

David [30] resolved the vector quantization problems of KM and *K*-median clustering through sampling. David shows sampling-based clustering depends upon confidence and accuracy but is independent of the sample input size that is known as the approximation parameters. Aggarwal et al. [31] present the bi-criteria approximation approach for using adaptive sampling for KM clustering. The bi-criteria approximation approach provides the initial centroid of the KM clustering. Zhao et al. [16] proposed the cluster sampling arithmetic (CSA) algorithm that classified the datasets and extracted the quality sample data that analyzed the main body of data objects. The proposed CSA algorithm improved precision, recall, execution time, and F-measures over the SRS, DBS, and EASE algorithms. Aloise et al. [19] proposed the iterative sampling-based solution of the mini-max diameter clustering problem (MMDCP), which was identified as the NP-hard problem. The MMDCP algorithm is used for dimensionality reduction and noise-related data problems of large datasets. The MMDCP algorithm solved another partitioning problem of clustering that was described in the literature of [19]. Luchi et al. [32] proposed Rough-DBSCAN and I-DBSCAN samplings-based clustering algorithms for using DBSCAN clustering. The DBSCAN algorithm uses different shape clusters and isolates the noisy patterns on large datasets. Ben Hajkacem et al. [33] proposed the sampling-based clustering algorithm known as STiMR K-means. The proposed STiMR K-means algorithm has used reservoir random sampling, triangle inequality, and MapReduce acceleration techniques together. The performance of the STiMR K-means algorithm is achieved better than the K-means and MR K-means algorithms.

Some other sampling-based clustering algorithms are discussed inside the literature of [9, 16], and other sampling-based clustering algorithms are CURE [19], RSEFCM [19], CLARANS [19], eNERF [34], GOFCM [9], EM clustering [34], BIRCH [10], K-means sampler [35], MapReduce-based KM [33], single-pass FCM [9], MSERFCM [9], SDBSCAN [36], DBRS [37], DIDES [38], CLUSMASTER [39], PS-SC [40], TEP [41], INS [42], SSEFCM [9], etc.

## 3 Systematic Sampling for Big Data Mining

The existing research examination [9, 16, 33, 43, 44] and literature of the second section show the conceptual and practical views of the sampling plan that increased the speed-up, scale-up, performance, efficiency, and optimized the computational resources for any data mining techniques. The formal sampling is first dividing the entire dataset into the sample and un-sample data units through any sampling methods according to research requirement. The sample unit describes the selected data points for data mining that represent the entire dataset. The responsibility of sampling strategies is to produce a highly accurate sample for efficient data mining processes. The un-sample unit describes the unselected data points for data mining that used for sample extension, testing, and other purposes. After that, the data mining algorithm used the sample unit for extracting the knowledge. Thus, the data mining algorithm achieved the approximation result that is known as approximate data mining results. In this study, this paper used KM clustering with systematic sampling, which is known as the KM approximate clustering. This section presents the sampling views of the systematic sampling approach and discusses some advantages over conventional sampling such as simple random sampling (SRS), uniform random sampling, and so on.

In the literature of [45–47] shows the systematic sampling more precise, easier, and efficiencies estimators than simple random sampling (SRS). Systematic sampling is similar to the Hash Mod algorithm concerning the selection of samples for the reservoir [18]. The systematic sampling-related literature shows that systematic sampling is more efficient and operationally convenient in natural population-related problems. Systematic sampling is useful for real-time data because of systematic sampling dependent on the sampling interval and period length [45]. The systematic sampling selects the first data object randomly, and the rest of the data object selected according to a predetermined pattern. The use of the randomization inside the systematic sampling ensures the unbiased estimation and provides the equal opportunity of data selection in the real time or batch datasets [48].

Suppose the X dataset consists of N data objects with D-dimensional. Systematic sampling extracts n sample size from the dataset through two possibilities. The first possibility is similar to linear systematic sampling or $k$th systematic sampling, and the second possibility is similar to circular systematic sampling [45, 49–51]. Both of the possibilities extract the sample inside the basic characteristics of big data.

- The first possibility of systematic sampling is N = nk. In this case, n is the sample size and k is the predefined interval of data selection. This approach first generates a random number between 1 and k through uniform random distributions. Suppose the random distribution method generates the random number as i, then the first data item of the sample is ith. After that, select each $k$th data object for the sample. This $k$th systematic sampling approach suitable for a fixed-size sample, where the sample contained the $i, i+k, 1+2k, \ldots, i+(n-1)k$ index number data objects.
- The second possibility of systematic sampling is N $\neq$ nk. In this case, k is not defined in advance, then interval k has been found through the fraction of the length of the dataset and sample size such as k = N/n. In this approach first choose any data through uniform random distributions such as the random number and after that select every $k$th data for the sample. In this study, this paper adopts this approach for sample selection in the implementation because of this possibility used inside any sample size due to dynamic nature.

# 4 Proposed Systematic Sampling Model for Big Data Mining

This section presents a systematic sampling model for big data mining using the K-means clustering algorithm. The presented sampling model uses any data mining algorithm as the replacement of the K-means algorithm because the presented sampling model is the universal of any mining algorithm. The universally sampling-based big data mining algorithm is known as the approximate algorithm. For this reason, the K-means clustering algorithm is known as approximate clustering. The proposed sampling model is known as a systematic sampling-based data mining algorithm name such as SY (data mining algorithm name). This paper used the K-means known as SYK-means for using the presented model. The presented sampling model improves the computation efficiency of any traditional data mining algorithm for big data mining concerning scale-up, speed-up, computing time, and cost. It has achieved better computation effectiveness and efficiency than random sampling-based data mining that determined in the literature of [45–47]. The systematic sampling model of big data mining is shown in Algorithm 1.

**Algorithm 1 Systematic Sampling Model of Big Data Mining** (SYK-Means)

**Input:**
1. D dataset (N data units)
2. V'S validation setup (dataset mean, sample mean, sample variance, sample size, confidence interval, sampling error, sampling cost)

**Output:**
1. AC (approximate results) = $\{ac_1, ac_2 \ldots \ldots . ac_k\}$

**Methods:**
1. According to the research requirements define the sample size n
2. Extract the sample data S from the dataset
3. if $N = nk$ then used the first possibility of systematic sampling
4. if $N \neq nk$ then used the second possibility of systematic sampling
5. Validate the sample S data according to validation setup V'S.
6. Used the required data mining algorithms such as K-Means(S,K) and achieved the approximate results AC.
7. Validate the approximate data mining results as internal and external measurements.
8. Exit

# 5 Experimental Analysis

## 5.1 Experiment Environment (Tools) and Dataset

The experimental environment is implemented through the Jupyter notebook framework of python computing tool, Intel I3 processor, 320 GB hard disk, 4 GB DDR3 RAM, Windows 7 Operating System. For the experiment, this study used the computer-related Geo-Magnetic dataset that has 58,374 data points with ten attributes (https://archive.ics.uci.edu/ml/datasets.php).

## 5.2 Evaluation Criteria

Evaluation criteria measure the quality and efficiency of the proposed data mining methods that show the advantages of the proposed algorithm against the existing algorithm. Generally, big data clustering algorithms are measured by internal and external-related validation techniques. External validation techniques require some external information in advance, and internal validation techniques do not require any information. Internal measures validate the objective of clustering through intra- and inter-class similarity. The better clustering method obtains the highest intra-class similarities and lowest inter-class similarities. This paper validated the minimized SSE, intra- and inter-class similarity through R square (RS), root-mean-square standard deviation (RMSSTD), Davies Bouldin score (DB), Calinski Harabasz score (CH), Silhouette coefficient (SC) as internal validation [9, 47] and the CPU time (CT), number of iterations (IS) as efficiency (speed) validation [54–56]. The optimal results of the RS, SC, and CH indices are always maximized, and the RMSSTD, DB, CT, and IS indices are always minimized [52, 53, 56].

### *5.3 Results and Discussion*

The authors of [21, 35, 57] used the random sampling-based K-means inside of a single machine for cluster creation. This paper compares the performance of the proposed SYK-means algorithm against the random sampling-based K-means (RSK-means) and classical K-means algorithms under the single machine-based cluster creation. The implementation code of the RSK-means, SYK-means, and K-means algorithms are written inside the Jupyter notebook using the python language tool on single machine execution. The effectiveness (RS, CH, DB, SC) and efficiency (ET, IS) related reported results are shown in Tables 1 and 2, respectively. In this experiment, the number of clusters is fixed as three. Inside both tables, the effectiveness and efficiency-related reported results showed an average of ten trials, where the SYK-means algorithm achieved better computing performance than the RSK-means algorithm.

The RS criterion is used to validate the dissimilarity between-cluster, the CH criterion is used to measure the variance, compaction, and separation of the cluster, and the SC metric is to measure the similarity within-cluster. The DB criterion is to measure the cluster compaction without depending on the clusters. The RMSSTD criterion is to measure the cluster homogeneity with cluster variance through compactness. The CT criterion is to measure the CPU time of K-means algorithm execution without including the sampling time, and the IS criterion is to measure the convergence speed of the K-means algorithm.

Experimental observation of the Geo-Magnetic dataset and the reported results of RS, RMSSTD, DB, CH, SC, CT, and IS show that the SYK-means algorithm achieves better computing performance than the RSK-means algorithm. Table 1 summarized the proposed SYK-means algorithm, showed the cluster quality is never down than the RSK-means algorithm, and obtained the minimized SSE, highest intra-class similarity, and lowest inter-class similarity. Table 2 summarized the proposed SYK-means algorithm using the optimum iterations and CPU time that resolve the scalability, computation cost, and speed-up-related issues of the K-means algorithm.

**Table 1** Analysis of effectiveness measures ($means \pm std$)

| Measures | Sample size | RSK-Means | SYK-Means |
|---|---|---|---|
| RS | 5% | 0.92458 ± 0.062 | **0.96058 ± 0.049** |
| | 10% | 0.92509 ± 0.062 | **0.97227 ± 0.036** |
| | 15% | 0.9252 ± 0.062 | **0.96058 ± 0.049** |
| | 20% | 0.91406 ± 0.060 | **0.97227 ± 0.036** |
| RMSSTD | 5% | 14033451.341 ± 7194169.663 | **9916118.504 ± 5711455.031** |
| | 10% | 14008693.792 ± 7216243.975 | **8560487.908 ± 4280286.189** |
| | 15% | 13959054.121 ± 7153094.899 | **9913560.544 ± 5707755.732** |
| | 20% | 15289837.0847 ± 7002898.591 | **8558796.028 ± 4280863.214** |
| DB | 5% | 0.30599 ± 0.094 | **0.24984 ± 0.077** |
| | 10% | 0.30443 ± 0.096 | **0.23156 ± 0.058** |
| | 15% | 0.30463 ± 0.099 | **0.24986 ± 0.077** |
| | 20% | 0.3217 ± 0.094 | **0.23148 ± 0.058** |
| CH | 5% | 49388.029 ± 42352.630 | **73462.213 ± 33706.557** |
| | 10% | 100192.993 ± 85820.070 | **162951.672 ± 50568.591** |
| | 15% | 149299.826 ± 127497.413 | **209971.292 ± 96332.387** |
| | 20% | 168148.721 ± 167895.637 | **326025.857 ± 101179.704** |
| SC | 5% | 0.78839 ± 0.112 | **0.8529 ± 0.090** |
| | 10% | 0.78851 ± 0.113 | **0.87212 ± 0.069** |
| | 15% | 0.78745 ± 0.113 | **0.85142 ± 0.091** |
| | 20% | 0.7666 ± 0.1111 | **0.87413 ± 0.067** |

**Table 2** Analysis of efficiency measures ($means \pm std$)

| Measures | Sample size | RSK-means | SYK-means |
|---|---|---|---|
| CT | 5% | 1.14348 ± 0.973 | **0.85644 ± 0.582** |
| | 10% | 2.65824 ± 1.540 | **1.90632 ± 1.177** |
| | 15% | 4.17925 ± 2.511 | **2.67394 ± 1.329** |
| | 20% | 6.23689 ± 5.069 | **3.65821 ± 1.424** |
| IS | 5% | 5.7 ± 3.622 | **4.3 ± 2.057** |
| | 10% | 5.3 ± 3.128 | **3.8 ± 1.988** |
| | 15% | 5.4 ± 2.756 | **3.8 ± 1.398** |
| | 20% | 5.5 ± 4.222 | **3.2 ± 1.135** |

The comparative analysis of the K-means, RSK-means, and SYK-means algorithms is shown in Table 3 using the effectiveness (RS, RMSSTD, CH, DB, SC) and efficiency (ET, IS) measures that show the average results of ten trials using entire data. For this observation, the SYK-means algorithm achieves better computing performance than K-means and RSK-means algorithms. Table 3 shows that the proposed SYK-means algorithm achieves better variance, homogeny, compaction,

**Table 3** Analysis of effectiveness and efficiency measures ($means \pm std$)

| Measures | K-means | RSK-means | SYK-means |
|---|---|---|---|
| RS | 0.9372 ± 0.060 | 0.90213 ± 0.056 | **0.94889 ± 0.056** |
| RMSSTD | 12618862.702 ± 6989900.067 | 16679627.685 ± 6538452.804 | **11265274.375 ± 6538452.804** |
| DB | 0.28651 ± 0.094 | 0.34159 ± 0.088 | **0.26817 ± 0.08863** |
| CH | 1150346.404 ± 826221.586 | 670355.037 ± 772859.525 | **1310343.526 ± 772859.525** |
| SC | 0.80964 ± 0.111 | 0.74444 ± 0.104 | **0.82956 ± 0.104** |
| CT | 21.57192 ± 13.457 | 30.53555 ± 17.198 | **18.5785 ± 9.518** |
| IS | 4 ± 2.260 | 5.6 ± 2.836 | **3.5 ± 1.433** |

separation, convergence rate, CPU time, similarity, and dissimilarity than the K-means and RSK-means algorithms. The reported measure of the SYK-means algorithm shows the optimal results compared to the K-means and RSK-means algorithms. This observation is sufficient for the proposed method convenient for big data mining.

## 6 Conclusion

Data volume is the primary attribute of the big data mining that defines the other characteristics of big data mining. The sampling process is reduced to the data volume through randomized sampling strategies for big data mining process. The combination of sampling and data mining technique obtains the approximate results. The literature of sampling suggests that the performance of systematic sampling is better than the random sampling concerning sample variance, unbiased, and entire data covering. This study proposed the SYK-means algorithm for big data clustering with the help of systematic sampling and K-means clustering. The SYK-means algorithm achieved better sampling results than the RSK-means algorithm concerning selected sample sizes through effectiveness and efficiency measurement. Efficiency measures is used to validate scalability and speed-up and the effectiveness measures to validate quality-related issues of big data clustering. The SYK-means algorithm also achieved better efficiency and effectiveness than the RSK-means and K-means algorithms concerning the entire dataset. Further scope of this work is to use other data mining algorithms with different measurements under the single or multiple machines.

## References


1. Oussous A, Benjelloun F, Lahcen AA, Belfkih S (2017) Big Data technologies: a survey. J King Saud Univ Comput Inf Sci. https://doi.org/10.1016/j.jksuci.2017.06.001
2. Hariri RH, Fredericks EM, Bowers KM (2019) Uncertainty in big data analytics: survey, opportunities, and challenges. J Big Data 6:1–6. https://doi.org/10.1186/s40537-019-0206-3
3. Gandomi A, Haider M (2015) Beyond the hype: Big data concepts, methods, and analytics. Int J Inf Manage 35:137–144. https://doi.org/10.1016/j.ijinfomgt.2014.10.007
4. Sivarajah U, Kamal MM, Irani Z, Weerakkody V (2017) Critical analysis of Big Data challenges and analytical methods. J Bus Res 70:263–286. https://doi.org/10.1016/j.jbusres.2016.08.001
5. Lee I (2017) Big data: dimensions, evolution, impacts, and challenges. Bus Horiz 60:293–303. https://doi.org/10.1016/j.bushor.2017.01.004
6. Siddiqa A, Hashem IAT, Yaqoob I et al (2016) A survey of big data management: Taxonomy and state-of-the-art. J Netw Comput Appl 71:151–166. https://doi.org/10.1016/j.jnca.2016.04.008
7. Kacfah Emani C, Cullot N, Nicolle C (2015) Understandable Big Data: a survey. Comput Sci Rev 17:70–81. https://doi.org/10.1016/j.cosrev.2015.05.002
8. Khondoker MR (2018) Big data clustering. Wiley StatsRef Stat Ref Online 1–10. https://doi.org/10.1002/9781118445112.stat07978
9. Zhao X, Liang J, Dang C (2019) A stratified sampling based clustering algorithm for large-scale data. Knowl-Based Syst 163:416–428. https://doi.org/10.1016/j.knosys.2018.09.007
10. Pandove D, Goel S, Rani R (2018) Systematic review of clustering high-Dimensional and large datasets. ACM Trans Knowl Discov Data 12:1–68. https://doi.org/10.1145/3132088
11. Jain AK (2010) Data clustering: 50 years beyond K-means. Pattern Recognit Lett 31:651–666. https://doi.org/10.1016/j.patrec.2009.09.011
12. HajKacem MA Ben, N'Cir C-E Ben, Essoussi N (2019) Clustering methods for big data analytics. In: Unsupervised and semi-supervised learning, pp 1–23
13. Wu X, Zhu X, Wu G-Q, Ding W (2014) Data mining with big data. IeeexploreIeeeOrg, 1–26
14. Tsai C-W, Lai C-F, Chao H-C, Vasilakos AV (2016) Big Data analytics. In: Big Data technologies and applications, pp 1–400
15. Chen B, Haas P, Scheuermann P (2002) A new two-phase sampling based algorithm for discovering association rules. In: Proceedings of the eighth ACM SIGKDD international conference on Knowledge discovery and data mining. https://doi.org/10.1145/775107.775114, pp 462–468
16. Zhao J, Sun J, Zhai Y et al (2018) A novel clustering-based sampling approach for minimum sample set in big data environment. Int J Pattern Recognit Artif Intell 32:1–20. https://doi.org/10.1142/S0218001418500039
17. Ly T, Cockburn M, Langholz B (2018) Cost-efficient case-control cluster sampling designs for population-based epidemiological studies. Spat Spatiotemporal Epidemiol 26:95–105. https://doi.org/10.1016/j.sste.2018.05.002
18. Boicea A, Truică CO, Rădulescu F, Buşe EC (2018) Sampling strategies for extracting information from large data sets. Data Knowl Eng 115:1–15. https://doi.org/10.1016/j.datak.2018.01.002
19. Aloise D, Contardo C (2018) A sampling-based exact algorithm for the solution of the minimax diameter clustering problem. J Glob Optim 71:613–630. https://doi.org/10.1007/s10898-018-0634-1
20. Chen W, Oliverio J, Kim JH, Shen J (2019) The modeling and simulation of data clustering algorithms in data mining with big data. J Ind Integr Manag 04:1850017. https://doi.org/10.1142/s2424862218500173
21. Kollios G, Gunopulos D, Koudas N, Berchtold S (2003) Efficient biased sampling for approximate clustering and outlier detection in large data sets. IEEE Trans Knowl Data Eng 15:1170–1187. https://doi.org/10.1109/TKDE.2003.1232271
22. Xu Z, Wu Z, Cao J, Xuan H (2015) Scaling Information-Theoretic Text Clustering: A Sampling-based Approximate Method. In: Proceedings - 2014 2nd International Conference on Adv Cloud Big Data, CBD 2014. https://doi.org/10.1109/CBD.2014.56

23. Thompson SK (2012) Sampling, Third edn. Wiley Publication
24. Kim JK, Wang Z (2019) Sampling techniques for big data analysis. Int Stat Rev 87:S177–S191. https://doi.org/10.1111/insr.12290
25. Pandey KK, Shukla D (2019) A study of clustering taxonomy for big data mining with optimized clustering map reduce model. Int J Emerg Technol 10
26. Fahad A, Alshatri N, Tari Z et al (2014) A survey of clustering algorithms for big data: taxonomy and empirical analysis. IEEE Trans Emerg Top Comput 2:267–279. https://doi.org/10.1109/TETC.2014.2330519
27. Kumar S, Mohbey KK (2019) A review on big data based parallel and distributed approaches of pattern mining. J King Saud Univ Comput Inf Sci. https://doi.org/10.1016/j.jksuci.2019.09.006
28. Bakhthemmat A, Izadi M (2020) Decreasing the execution time of reducers by revising clustering based on the futuristic greedy approach. J Big Data 7:1–21. https://doi.org/10.1186/s40537-019-0279-z
29. Pandey KK, Shukla D, Milan R (2020) A Comprehensive Study of Clustering Algorithms for Big Data Mining with MapReduce Capability. In: Shukla RK, Agrawal J, Sharma S, et al (eds) Social networking and computational intelligence, Lecture Notes in Networks and Systems 100. Springer Nature Singapore Pte Ltd., pp 427–440
30. Ben-david S (2007) A framework for statistical clustering with constant time approximation algorithms for K-median and K-means 04:243–257. https://doi.org/10.1007/s10994-006-0587-3
31. Aggarwal A, Deshpande A, Kannan R (2009) Adaptive sampling for k-means clustering. Lecture Notes Computer Science (including Subser Lect Notes Artif Intell Lect Notes Bioinformatics) 5687 LNCS. https://doi.org/10.1007/978-3-642-03685-9_2, pp 15–28
32. Luchi D, Loureiros Rodrigues A, Miguel Varejão F (2019) Sampling approaches for applying DBSCAN to large datasets. Pattern Recognit Lett 117:90–96. https://doi.org/10.1016/j.patrec.2018.12.010
33. Ben Hajkacem MA, Ben Ncir CE, Essoussi N (2019) STiMR k-means: an efficient clustering method for big data. Int J Pattern Recognit Artif Intell 33:1950013. https://doi.org/10.1142/S0218001419500137
34. Wang L, Bezdek JC, Leckie C, Kotagiri R (2008) Selective sampling for approximate clustering of very large data sets. Int J Intell Syst 23:313–331. https://doi.org/10.1002/int.20268
35. Bejarano J, Bose K, Brannan T, Thomas A (2011) Sampling Within k-Means Algorithm to Cluster Large Datasets. Tech Rep HPCF-2011-12 1–11
36. Ji-hong G, Shui-geng Z, Fu-ling B, Yan-xiang H (2001) Scaling up the DBSCAN algorithm for clustering large spatial databases based on sampling technique. Wuhan Univ J Nat Sci 6:467–473
37. Wang X, Hamilton HJ (2003) DBRS : A Density-Based Spatial Clustering Method with Random Sampling. 563–575
38. Ros F, Guillaume S (2017) DIDES: a fast and effective sampling for clustering algorithm. Knowl Inf Syst 50:543–568. https://doi.org/10.1007/s10115-016-0946-8
39. Da Silva A, Chiky R, Hébrail G (2012) A clustering approach for sampling data streams in sensor networks. Knowl Inf Syst 32:1–23. https://doi.org/10.1007/s10115-011-0448-7
40. Jia H, Ding S, Du M (2017) A Nyström spectral clustering algorithm based on probability incremental sampling. Soft Comput 21:5815–5827. https://doi.org/10.1007/s00500-016-2160-8
41. Härtel P, Kristiansen M, Korpås M (2017) Assessing the impact of sampling and clustering techniques on offshore grid expansion planning. Energy Proc 137:152–161. https://doi.org/10.1016/j.egypro.2017.10.342
42. Zhan Q (2017) Improved spectral clustering based on Nyström method. https://doi.org/10.1007/s11042-017-4566-4, pp 20149–20165
43. Pandey KK, Shukla D (2020) Stratified Sampling-Based Data Reduction and Categorization Model for Big Data Mining. In: Gupta JC, Kumar BM, Sharma H, Agarwal B (eds) Communication and intelligent systems

44. Pandey KK, Shukla D (2019) Optimized sampling strategy for big data mining through stratified sampling. Int J Sci Technol Res 8:3696–3702
45. Mostafa SA, Ahmad IA, Ahmad IA (2017) Recent developments in systematic sampling: a review. J Stat Theory Pract ISSN. https://doi.org/10.1080/15598608.2017.1353456
46. Aune-lundberg L, Strand G (2014) Comparison of variance estimation methods for use with two-dimensional systematic sampling of land use/land cover data. Environ Model Softw 61:87–97. https://doi.org/10.1016/j.envsoft.2014.07.001
47. Kao F, Leu C, Ko C (2011) Remainder Markov systematic sampling. J Stat Plan Inference 141:3595–3604. https://doi.org/10.1016/j.jspi.2011.05.011
48. Larson L, Larson P, Johnson DE (2019) Differences in stubble height estimates resulting from systematic and random sample designs. Rangel Ecol Manag 72:586–589. https://doi.org/10.1016/j.rama.2019.03.007
49. Ziegel ER, Lohr SL (2000) Sampling: design and analysis. In: Technometrics, p 223
50. Shalabh (2019) Stratified sampling. In: Sampling theory, pp 1–27
51. Olufadi Y, Oshungade IO, Adewara AA (2012) On allocation procedures using systematic sampling. J Interdiscip Math 15:23–40. https://doi.org/10.1080/09720502.2012.10700783
52. Aggarwal CC, Reddy CK (2013) DATA custering algorithms and applications
53. Gan G, Ma C, Wu J (2007) Data clustering theory, algorithms, and applications
54. Peña JM, Lozano JA, Larrañaga P (1999) An empirical comparison of four initialization methods for the K-Means algorithm. Pattern Recognit Lett 20:1027–1040. https://doi.org/10.1016/S0167-8655(99)00069-0
55. Celebi ME, Kingravi HA, Vela PA (2013) A comparative study of efficient initialization methods for the k-means clustering algorithm. Expert Syst Appl 40:200–210. https://doi.org/10.1016/j.eswa.2012.07.021
56. Zahra S, Ghazanfar MA, Khalid A et al (2015) Novel centroid selection approaches for KMeans-clustering based recommender systems. Inf Sci (Ny) 320:156–189. https://doi.org/10.1016/j.ins.2015.03.062
57. Luchi D, Santos W, Rodrigues A, Varejao FM (2015) Genetic sampling k-means for clustering large data sets. In: CIARP 2015, LNCS 9423, pp 691–698

# Chapter 20
# Security of IoT-Based E-Healthcare System: A Blockchain Solution

**Ashish Singh, Punam Prabha, and Kakali Chatterjee**

## 1 Introduction

The Internet of things (IoT) is an intercommunication network of smart devices to serve meaningful objectives that make social work efficient. The general architecture of IoT comprises sensors/actuators, devices, gateways, protocols, Internet, and cloud server. In IoT-based e-healthcare system, different types of sensors are attached with the patient body. These sensors measure the physiological parameters such as body temperature, blood pressure, electrocardiogram, pulse oximeter oxygen saturation, etc. [1, 2]. The attached body sensor devices collect the information by sensing certain parameters of the patient's body and send it to the cloud server with the help of different types of gateways and protocols. IoT users can access the stored data from the cloud server by using smart devices like smartphones, computers, etc. IoT applications are cost-effective, as it needs low energy consumption. IoT offers an advanced level of services so that it can change the way people lead their day-to-day life.

Patient data security and privacy are a big challenging issue in the IoT-based healthcare system. In this type of healthcare system, the data are transferred from the patient body to the healthcare data center. During the traveling of the data, they have passed different security boundaries and used other protocols and standards.

A. Singh
School of Computer Engineering, KIIT Deemed to be University, Bhubaneswar, Odisha 751024, India
e-mail: ashish.singhfcs@kiit.ac.in

P. Prabha (✉) · K. Chatterjee
Department of Computer Science and Engineering, National Institute of Technology Patna, Patna, Bihar 800005, India
e-mail: punamp.phd18.cs@nitp.ac.in

K. Chatterjee
e-mail: kakali@nitp.ac.in

© The Author(s), under exclusive license to Springer Nature Singapore Pte Ltd. 2022 
H. M. Dubey et al. (eds.), *Artificial Intelligence and Sustainable Computing*,
Algorithms for Intelligent Systems, https://doi.org/10.1007/978-981-16-1220-6_20

Thus, it may be possible that the data may be changed during the transmission. The stored data at cloud storage is also not safe. An unauthorized user may modify it. Due to this, the treatment and diagnosis process of the patient may be affected. So, the e-healthcare system requires a high level of security mechanism. Several security standards and solutions are adopted by the e-healthcare system to provide the security of the patient data. Such standards include HIPAA, HITRUST CSF, COBIT, and DISHA. But, due to the constraint nature of IoT devices, increasing size and complexity of the healthcare data a new security solution needs to be developed that is capable of providing security and privacy to the e-healthcare system.

In recent years there have been tremendous efforts to deal with the security issues in IoT-based e-healthcare system using different technologies [3–9]. Among the various approaches proposed, some techniques purely focused on a specific layer. In contrast, some targeted at providing end-to-end security for IoT-based e-healthcare system. Security analysis presented in [10, 11] compares safety based on different cryptography algorithms and key management systems. Granjal et al. [12] analyzed security requirements of the IoT protocol and mechanisms which are required for secure communication between the devices. Sicari et al. [13] discussed security related to confidentiality, access control, and privacy for the IoT-based system. The identity-based authentication, network security, trust management, fault tolerance, and forensics implementation are surveyed in [14]. The survey area covered edge computing-based paradigms that include mobile cloud computing, edge computing, IoT, and fog computing. Alaba et al. [15] describe IoT security and privacy in terms of layered architecture where the threats are present in the network and hardware area. The security issues are classified based on perception, communication, and application aspect. In the IoT paradigm, preserving the privacy of IoT users is a significant task. Ray et al. [16] discussed blockchain solutions to mitigate the problem of decentralized computation and storage issues for IoT-based healthcare. They have also discussed most of the popular consensus algorithms used in IoT-based e-healthcare system. Conosenti et al. [17] identified blockchain solution that plays an essential role to perverse the privacy of decentralized and data-intensive applications that are running in distributed IoT devices. This research aim is to understand whether blockchain technology can be employed to maintain integrity, anonymity, and adaptability into IoT-based applications. Reyna et al. [18] present the challenges and potential benefits of the integration of blockchain with IoT applications. Tanwar et al. [19] proposed a model to maintain electronic healthcare records into healthcare applications. This model uses the blockchain technology for the creation of blocks of the patient records. We have used this concept in our proposed model for the design of blocks of the patient records.

From these studies, we have identified that the IoT-based e-healthcare security challenges can be solved through the integration of blockchain. The discussion also says data leaking and unauthorized modification of sensitive patient data are the most important challenges in the IoT security paradigm. The contribution of the work is summarized as follows:

1. This paper studies different security issues and challenges in IoT-based e-healthcare system.

2. We have proposed a blockchain solution for the IoT-based e-healthcare system to maintain the security and confidentiality of the patient data.
3. The security analysis and results demonstrate that model achieved the desired level of security and privacy.

The remaining parts of the paper are structured as follows: In Sect. 2, we have discussed back of the work in terms of security requirement and blockchain overview. Proposed blockchain solution for IoT-based e-healthcare system is discussed in Sect. 3. Implementation and results analysis of the proposed model is discussed in Sect. 4. The security analysis of the proposed model is illustrated in Sect. 5. The paper is closed with the conclusion and future work in Sect. 6.

## 2 Background

In this section, we have discussed the background of the work. First, we have discussed the security and privacy requirement, which is needed for IoT-based e-healthcare system [20–22]. Later, we have discussed the overview of blockchain [23, 24].

### *2.1 Security Issues and Requirements for IoT-Based E-Healthcare System*

In IoT-based e-healthcare system various smart objects and sensors are connected for transferring collected information from the physical environment and then take action on the physical environment. There are many security issues and challenges as a wide variety of devices and communication ranges are used. These devices most of the time remain unattended, so any type of physical attack can be possible. In the physical layer jamming attack, a lack of physical interface and sleep deprivation attack can be possible. Due to wireless communication, eavesdropping can easily happen. The other security issues are as follows:

1. Due to heterogeneous IoT devices, area, and data communication among the devices is a big challenge. Thus, many communication/network protocols exist by offering different network services.
2. Poor interoperability is another challenging problem due to the decentralized architecture and heterogeneity of the system.
3. There is a lack of privacy-preserving techniques to maintain the data privacy due to the complex decentralized architecture and constraint nature of devices.
4. IoT devices are resource-constrained devices that have low computation power. Hence, a highly secure encryption protocol cannot be implemented on the devices.
5. Key security is another issue for secure data transmission of these devices as the device memory is limited.

6. There is no access restriction in the devices as they most of the time remain unattended.
7. Node compromise attack is widespread due to the absence of a proper authentication mechanism before allowed to join them.

Therefore, it is essential to provide high-level security in this environment. The following security requirements have been identified from the above-discussed issues of IoT systems:

- The first and essential security requirement of IoT-based e-healthcare system is the protection of patient data from unauthorized access, data authentication, and data privacy.
- For device security, device authentication, software integrity, and tamper-proof are required.
- For key security, key management technique for issuance, storage, and revocation of the key is required.
- For user security, enrollment, identity management, authentication, and authorization are essential.
- For access control, some rules are necessary to control access into the network.

### 2.2 *Overview of Blockchain*

Blockchain is a decentralized peer-to-peer technology. The records of all the transactions are distributed among all the nodes of the blockchain network. There is no need for a central authority. Due to distributed technology, a database of records of all transaction is available with every node in the network, so it is not affected by the single-point failure. The blockchain is of three types: public, private, consortium, and hybrid blockchain. Blockchain is a chain of blocks in which each block is linked with its previous block using a crypto graphic hash. The first block of blockchain is called the genesis block. Blockchain is immutable technology, i.e., no one can alter or make changes in blocks of the blockchain. Each block of blockchain has two main parts: header and transaction counter. The header part contains the Block Version, Merkle-Tree Root, Timestamp, N-bits, Nonce, and Parent block hash. The transaction counter is the block body consisting of a list of all the transactions processed over data of respective block using asymmetric cryptography. The maximum number of transactions that a block can contain depends on the block size and size of each transaction. The basic structure of the blockchain is shown in Fig. 1.

## 3 Proposed IoT-Based E-Healthcare System

In this section, we have illustrated our proposed IoT-based e-healthcare system using blockchain technology. The model is divided into several parts: various medical

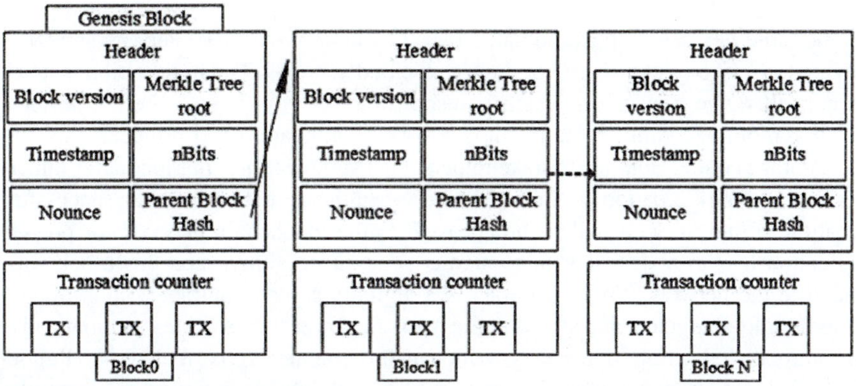


**Fig. 1** Basic structure of blockchain

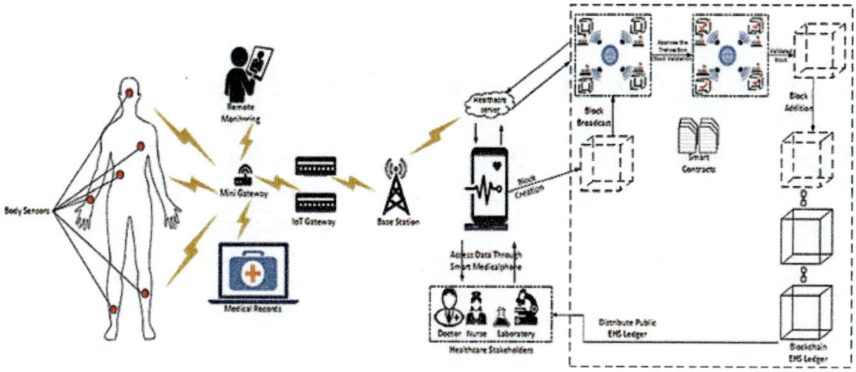

**Fig. 2** Proposed IoT-based e-healthcare system

sensor devices are attached with the human body which monitors the patient health condition, IoT gateway layer consists of various wireless protocols like Wi-Fi, ZigBee, and Bluetooth for communication. Collected data from medical sensor devices are stored into the medical server using blockchain technology. The proposed IoT-based e-healthcare system is shown in Fig. 2. The procedure for creation and addition of block into blockchain is discussed in Algorithm 1.

## *3.1 Network Model*

Human body sensors have been placed on the skin to sense the physiological parameters of the body. Mini Gateway can communicate to multiple body sensors that are situated in the range of 5 km and provide a link to a data server. Medical records are stored on the cloud server. The remote client can monitor from anywhere at any time via portable devices such as mobile, tablet, PC, laptop, etc. IoT Gateway supports device-to-device or end-to-end or device-to-cloud services. The base station is used as a small range transceiver which is responsible for connecting wireless devices (such as mobile, computer, etc.) to the network. Healthcare server provides a secure data storage center. Healthcare stakeholders must authorize the smartphone. A miner performed block creation in the IoT-based e-healthcare network. After successful creation of the block, it will be broadcasted to the blockchain network for proper validation by the nodes. If validation was done successfully, then this block was added to the network. Smart Contract is a digitally signed contract written in any programming language and stored in the network server that executes automatically to control illegal actions that are not written in the agreement. Distributed EHS Public ledger was introduced to provide access control and data management in electronic healthcare system so that information available in the ledger can be managed in a secure, fast and efficient way.

## *3.2 Working of the IoT-Based E-Healthcare System*

The whole system can be viewed in two parts. Part 1 was built with the help of different IoT devices. Part 2 was constructed by implementing blockchain technology. Communication between Part 1 and Part 2 was done by using Gateway, Base Station and Healthcare Server. After completing Part1, Part 2 was started which take created block as an input.

**Part 1**:

*Step 1*: Human body sensor continuously monitors the psychological behavior of human by placing a sensor on the skin of the body.
*Step 2*: Mini Gateway senses information gathered from human body sensors and passed it to the next hop (IoT Gateway).
*Step 3*: After collecting data from Mini Gateway, IoT Gateway transfers these data to base station. The base station accepts the request of IoT Gateway and allows connection to all wireless devices to the central hub.
*Step 4*: Base station stored the medical records to the healthcare server. Healthcare server can be used to store data and make accessing it easily at anytime from anywhere via the Internet through portable devices.
*Step 5*: Healthcare stakeholders can access data that are available on the healthcare server with the help of Smart Medical-phone.

**Part 2**:

*Step 1*: When healthcare actor initiates a transaction that time active miners are responsible for the creation of a block of the respective transaction.
*Step 2*: Now, the created block is broadcasted to the blockchain network.

*Step 3*: Miners available in the blockchain network accept this block and start to validate it by solving puzzle.
*Step 4*: To create a genuine block in the blockchain network consensus algorithm is applied. In the proposed model, we have applied a permissioned consensus algorithm for the validation of the block.
*Step 5*: After proper validation of the block by miner, the block is added to the blockchain network, and the request of healthcare actor is accepted.
*Step 6*: Each node in the blockchain network contains ledger, and their ledger is modified after introducing a new block in the blockchain network. This approach named as distributed EHS public ledger or distributed electronic healthcare system public ledger. It is responsible for providing fast, secure and efficient data services.
*Step 7*: All the blockchain information are distributed among multiple nodes. This information is easily accessible by all nodes and updated simultaneously if any things occurred in the network. A smart contract (digital contract) is also implemented to avoid illegal action taken by a dishonest node or miner.

**Algorithm 1** Algorithm for blockchain technology

**procedure Input:**(Created Block ($C_b$), Blockchain network ( $B_n$), Block version ($B_v$), Merkle Tree Root ($M_{rt}$) , Timestamp ($T_s$), N-bits ($N_b$), Nonce ($N$), and Parent block hash ($P_{hb}$), Transaction ($T_{xi}$), Distributed public EHS ledger ($D_{pehsl}$))
    **OUTPUT:** Updated Distributed public EHS ledger ($UD_{pehsl}$)
    **Initialization** Miners can create a block. After passing rigorous competition, a miner can be selected for adding their block in the blockchain network.
    **BEGIN**
    Miner ($M_{id}$, $C_b$)
    **while** (true) **do**
        Create block ($C_b$) $\leftarrow [B_v, M_{rt}, T_s, N, N_b, P_{hb}, T_{xi}]$
        **if** $M_{id}$ win completion **then**
            Add $C_b$ to $B_n$.
            Add Miner ($M_{id}$, $C_b$) to ledger $D_{pehsl}$
            update $D_{pehsl}$
        **else**
            Reject ($M_{id}$, $C_b$)
        **end if**
    **end while**
    Broadcast updated $UD_{pehsl}$
    **END**
**end procedure**

## 4 Implementation and Results

We have implemented blockchain technology using Hyperledger caliper blockchain benchmark tool [25]. It supports several Hyperledger family networks which allow users to test the performance results of different or predefined blockchain solutions.

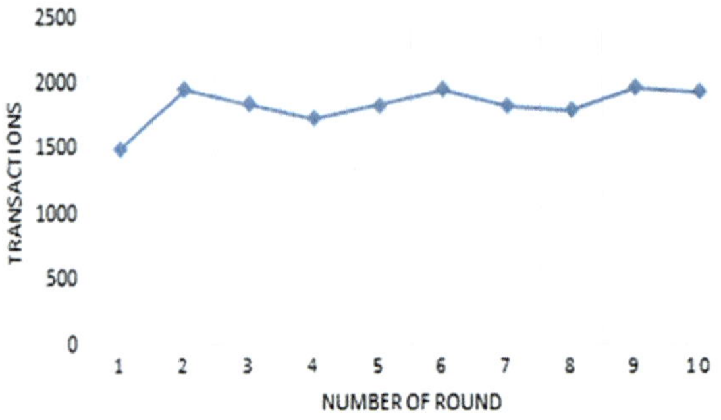


**Fig. 3** Transaction success rate

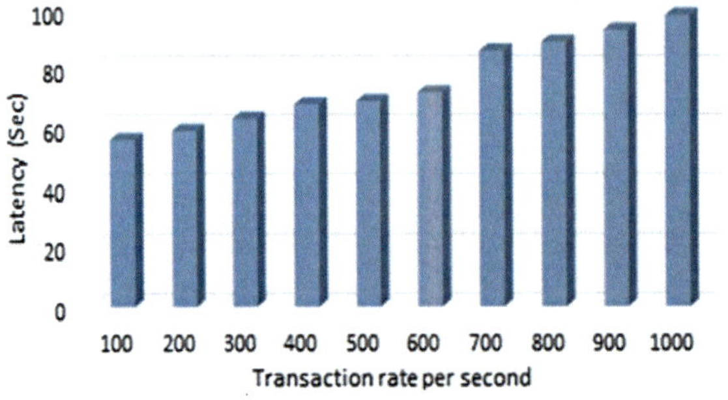


**Fig. 4** Latency of the transaction

For the implementation of the work, we have taken the help of GitHub Hyperledger Project source code that comes under the Creative Commons Attribution 4.0 International License (CC-BY-4.0) [26] which measure the success rate, transaction throughput, transaction latency, and resource consumption [27]. The simulation of the work is tested with the PC configuration of Intel(R) Core (TM) i5-8250U CPU @ 1.60 GHz 1.80 GHz; RAM is 8.00 GB (7.89 usable), 64-bit Operating System, x64-based processor, 520 GB SSD.

To implement the system, we have done 2000 transactions of ten times. The writing rate of the transaction is 100, 200, 300…1000 transactions per second. The first experiment is conducted to determine the success rate of the transactions, which defines the number of transactions successfully completed and added into the blockchain. This can be determined by the total number of transactions committed on the network divided by multiplication of invalid and failed transaction with total transaction time of the committed nodes. The success rate of the transaction of ten times is shown in Fig. 3. In some cases, the transaction rate is not good due to the unusual nature of the network. The second measurement metric is the latency which defines the time taken by the network to complete the transaction. It is the time difference between the submission of the transaction and completion of the transaction. The latency of the communication and transaction rate is shown in Fig. 4. This figure shows when

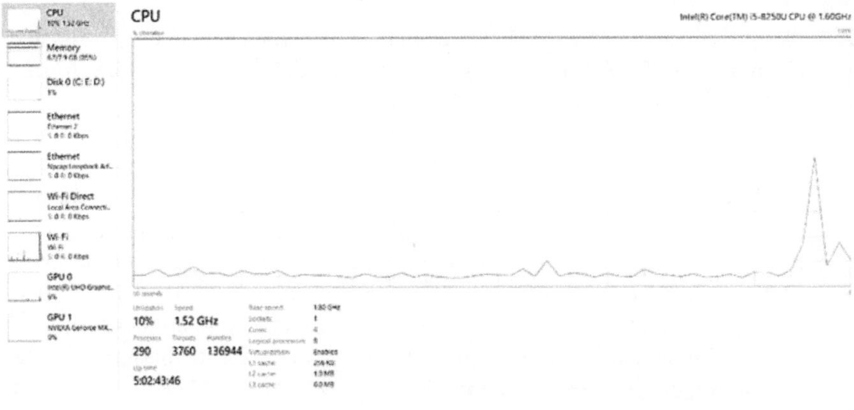


**Fig. 5** Resource consumption of the proposed model

the transaction rate increases parallelly the latency is also increased. To measure the resources consumption, we have executed Hyperledger caliper in our system.

CPU usage, GPU usage, memory, disk usage, Ethernet, and Wi-Fi usage are some parameters that are investigated in resources consumption category. The graphical results (Fig. 5) of the resource consumption show CPU usage is maximum up to 30%, memory usage is up to 85%, and disk usage is up to 1% at base speed 1.80 GHz.

## 5 Security Analysis

8. *Medical data privacy and secrecy*: All healthcare stakeholders have their access key using permission blockchain technology. The only user having authorized key could be allowed to access the system. Hence, the proposed system is safe from unauthorized access. Also, the proposed system is capable of handling data secrecy as encrypted data stored on each node in distributed public EHS ledger.
9. *Safety from unauthorized access*: In the proposed system, data are in encrypted form by using Hash functions and strong keys, so no one can obtain original data even if they access encrypted data.
10. *Safety from unauthorized data tampering*: The blockchain technology contains immutable property so that no one can modify any previous or present data because these data are chained with genesis block and replicated to all healthcare nodes. A smart contract is also implemented to avoid any illegal action taken by any malicious node.
11. *Safety from data loss*: All the blockchain information is distributed among multiple nodes via distributed public EHS ledger. In case loss of data occurred, at any node data could be retrieved easily from another node.

12. *Maintained a trusted data*: Data used in the blockchain network was validated by more than 51% miners using a consensus algorithm. Therefore, trusted medical data have maintained in the whole proposed system.
13. *Duplication and forgery*: The miner selection process will be passed through a highly complex process. It restricts that only one miner can produce only one block at one time by taking their unique identification details. Thus, duplication is not possible in the network. Also, all transactions are kept in distributed public EHS ledger, so the chance of forgery attack is impossible in the network.
14. *Secure transactions*: All the transactions in network use strongest asymmetric key cryptography-based algorithm (RSA, Zero-knowledge protocol, etc.) for the encryption and Hash function (such as SHA512) for the hashing. So, the transaction in the network is secured.

## 6 Conclusion and Future Work

In this paper, we have identified different security issues and challenges present in the IoT-based e-healthcare system. Blockchain technology can essentially offer a solution because it contains an immutable property to solve the identified issues. We have proposed a model for IoT-based e-healthcare system. The proposed model uses the blockchain technology for storing the data securely. All the collected data from medical sensors are stored in the form of blockchain at the medical server. So, it means the stored block of data is not modifiable by an attacker as well as the privacy of the data is maintained. It also improves the e-healthcare system and service in terms of security and privacy aspect.

In the future, we will try to make a more compact system so that the transmission and management of the data are handled more efficiently. This is important because the IoT devices have limited capability and low energy. Some more security vulnerability is needed to test in the proposed model. A more complex healthcare system with data privacy is investigated in the future.

## References

1. Gope P, Hwang T (2015) Bsn-care: a secure iot-based modern healthcare system using body sensor network. IEEE Sens J 16(5):1368–1376
2. Islam SR, Kwak D, Kabir MH, Hossain M, Kwak KS (2015) The internet of things for health care: a comprehensive survey. IEEE Access 3:678–708
3. Deng YY, Chen CL, Tsaur WJ, Tang YW, Chen JH (2017) Internet of things (IoT) based design of a secure and lightweight body area network (ban) healthcare system. Sensors 17(12):2919
4. Gervais A, Karame GO, Wüst K, Glykantzis V, Ritzdorf H, Cap-kun S (2016) On the security and performance of proof of work blockchains. In: Proceedings of the 2016 ACM SIGSAC conference on computer and communications security, pp. 3–16

5. Jiang S, Cao J, Wu H, Yang Y, Ma M, He J (2018) Blochie: a blockchain-based platform for healthcare information exchange. In: 2018 IEEE international conference on smart computing (smartcomp). IEEE, pp 49–56
6. Villalba MT, de Buenaga M, Gachet D, Aparicio F (2015) Security analysis of an IoT architecture for healthcare. In: International internet of things summit. Springer, pp. 454–460
7. Vora J, Italiya P, Tanwar S, Tyagi S, Kumar N, Obaidat MS, Hsiao KF (2018) Ensuring privacy and security in e-health records. In: 2018 International conference on computer, information and telecommunication systems (CITS). IEEE, pp 1–5
8. Wang EK, Sun R, Chen CM, Liang Z, Kumari S, Khan MK (2020) Proof of x-repute blockchain consensus protocol for iot systems. Comput Secur 101871
9. Wang J, Han K, Alexandridis A, Chen Z, Zilic Z, Pang Y, Jeon G, Piccialli F (2020) A blockchain-based e-healthcare system interoperating with wbans. Future Gener Comput Syst 110:675–685
10. Cirani S, Ferrari G, Veltri L (2013) Enforcing security mechanisms in the IP-based internet of things: an algorithmic overview. Algorithms 6(2):197–226
11. Roman R, Alcatraz C, Lopez J, Slaves N (2011) Key management systems for sensor networks in the context of the internet of things. Comput Electr Eng 37(2):147–159
12. Granjal J, Monteiro E, Silva JS (2015) Security for the internet of things: a survey of existing protocols and open research issues. IEEE Commun Surv Tutorials 17(3):1294–1312
13. Scare S, Rizzardi A, Grieco LA, Coen-Porisini A (2015) Security, privacy and trust in internet of things: the road ahead. Comput Netw 76:146–164
14. Roman R, Lopez J, Mambo M, Mobile edge computing, fog et al (2018) A survey and analysis of security threats and challenges. Future Gener Comput Syst 78:680–698
15. Alaba FA, Othman M, Hashem IAT, Alotaibi F (2017) Internet of things security: a survey. J Netw Comput Appl 88:10–28
16. Ray PP, Dash D, Salah K, Kumar N (2020) Blockchain for iot-based healthcare: background, consensus, platforms, and use cases. IEEE Syst J
17. Conoscenti M, Vetro A, De Martin JC (2016) Blockchain for the internet of things: a systematic literature review. In: 2016 IEEE/ACS 13th international conference of computer systems and applications (AICCSA). IEEE, pp 1–6
18. Reyna A, Martin C, Chen J, Solar E, Diaz M (2018) On blockchain and its integration with it. Challenges and opportunities. Future Gener Comput Syst 88:173–190
19. Tanwar S, Parekh K, Evans R (2020) Blockchain-based electronic healthcare record system for healthcare 4.0 applications. J Inform Secur Appl 50:102407
20. Maksimovi´c M, Vujovi´c V (2017) Internet of things based e-health systems: ideas, expectations and concerns. In: Handbook of large-scale distributed computing in smart healthcare. Springer, pp 241–280
21. Nasiri S, Sadoughi F, Tadayon MH, Dehnad A (2019) Security requirements of internet of things-based healthcare system: a survey study. Acta Inform Med 27(4):253
22. Tamizharas G, Sultanah HP, Balamurugan B (2017) IoT-based e-health system security: A vision archictecture elements and future directions. In: 2017 International conference of electronics, communication and aerospace technology (ICECA), vol 2. IEEE, pp 655–661
23. Priyadarshini I (2019) Introduction to blockchain technology. In: Cyber security in parallel and distributed computing: concepts, techniques, applications and case studies, pp 91–107
24. Zheng Z, Xie S, Dai H, Chen X, Wang H (2017) An overview of blockchain technology: Architecture, consensus, and future trends. In: 2017 IEEE international congress on big data (Big Data congress). IEEE, pp 557–564
25. Hyperledger Caliper, https://hyperledger.github.io/caliper/. Accessed 11 August 2020
26. Attribution 4.0 International (CC BY 4.0). https://creativecommons.org/licenses/by/4.0/. Accessed 10 August 2020
27. Lincoln N Hyperledger Caliper project. https://github.com/hyperledger/caliper. Accessed 10 Aug 2020

# Chapter 21
# Jaya Algorithm-Based Load Frequency Control in Nuclear Power Plant Incorporating Energy Storage Unit

**Nikhil Paliwal, Laxmi Srivastava, Manjaree Pandit, and Poonam Singh**

## 1 Introduction

Electric power system is a large, complex, and dynamic network of electrical components having a number of generators, transformers, transmissions line, and other equipments. Electric power system is deployed to generate, transmit, and use electric power. During the operation of power system, fluctuations may be caused due to change in load demand which affects the system frequency. Regulation of real power output in accordance with the changes in frequency within specified limits is done by load frequency controller [1].

The problem of frequency oscillation in the electrical power network has been increasing due to a number of reasons. Hence, load frequency control problem is becoming highly noteworthy nowadays in the field of electric power system and control [2, 3]. There are so many types of design methods which are reported in the literature for LFC problem, such as optimal method [4–7], classical method [8–10], and adaptive method [11, 12].

Like various real-time industrial problems, proportional-integral-derivative (PID) controller is preferred by the large community of researchers and engineers for LFC problem also. Tuning of PID controller parameters plays a vital role in LFC to achieve acceptable outcomes. Researchers and engineers have applied many classical and modern meta-heuristic optimization algorithms in the literature for tuning PID controller to meet specific objectives of LFC problem.

In this chapter, LFC of nuclear power plant is presented. PID controller is used in this paper for load frequency control in the under study power system. Various methods are implemented in the literature to address different power system problems [13–17]. In this study, optimal tuning problem of PID controller for LFC is addressed by Jaya optimization algorithm. Simulation results of Jaya algorithm-tuned PID controller for LFC are compared with existing methodology considering different

N. Paliwal (✉) · L. Srivastava · M. Pandit · P. Singh
Department of Electrical Engineering, MITS, Gwalior, MP, India

© The Author(s), under exclusive license to Springer Nature Singapore Pte Ltd. 2022

H. M. Dubey et al. (eds.), *Artificial Intelligence and Sustainable Computing*, Algorithms for Intelligent Systems, https://doi.org/10.1007/978-981-16-1220-6_21

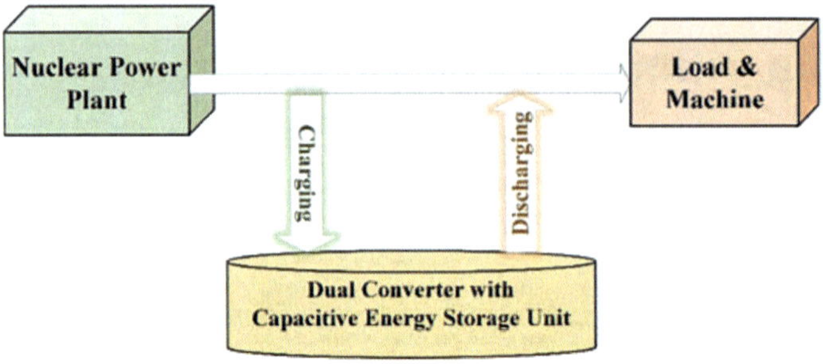


**Fig. 1** Integration of nuclear power generating station with CES unit

load disturbances in the under study system. Nuclear power generating station is also integrated with capacitive energy storage (CES) unit. The integration of nuclear power generating station with capacitive energy storage unit is as shown in Fig. 1. The effects of adding the CES unit in the under study system are also discussed in this article.

The highlights of this chapter are:

1. Optimization of PID controller using Jaya algorithm is done for LFC in single area nuclear power generating station for different load disturbances.
2. Analyze the performance of Jaya-PID controller against different load disturbances based on integral of time-weighted absolute error (ITAE).
3. Validated superiority and effectiveness of proposed control strategy with an existing methodology for LFC in single area nuclear power generating station.
4. Comparative statistical analysis of frequency deviation has been done for 1%, 2%, and 10% load disturbances.
5. Investigated the performance of under study system in the presence of capacitive energy storage unit.

## 2 Problem Formulation

### *2.1 System Under Study*

In this chapter, LFC of a single area nuclear power generating station has been carried out. Nuclear power generating plant consists of a governor, a high pressure (HP) turbine, and two low pressure (LP) turbines [18]. Single area nuclear power generating station is connected with mass and load [18]. The block diagram of nuclear power generating station with load and the PID controller is shown in Fig. 2.

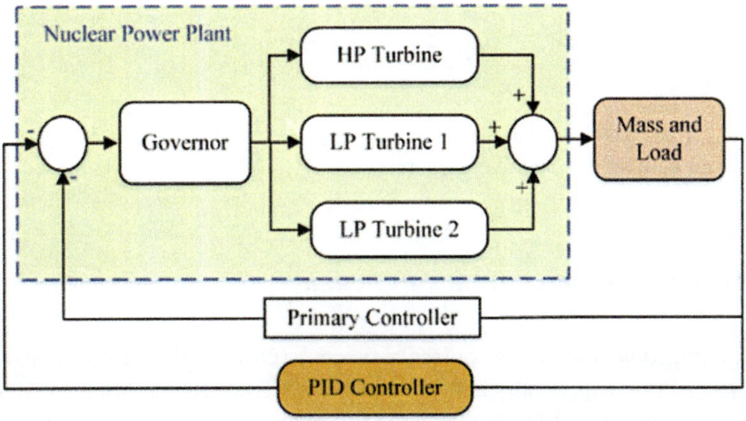


**Fig. 2** Single area nuclear power generating station with PID controller [18]

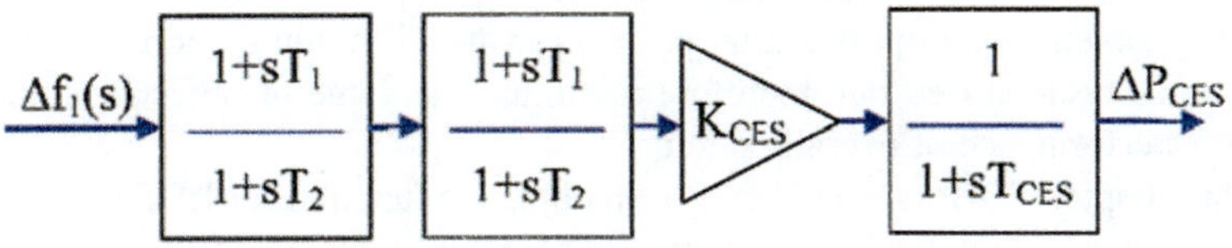


**Fig. 3** Block diagram of capacitive energy storage unit [19]

## 2.2 *Capacitive Energy Storage Unit*

CES unit consists of a supercapacitor and power conversion system [19]. A number of discrete capacitive units may be connected in parallel as per the need of requirement of energy storage capacity [19]. Block diagram of the capacitive energy storage unit is shown in Fig. 3. Block diagram of CES unit consists of time constant ($T_{CES}$) and gain block ($K_{CES}$). It also consists of two-phase compensation blocks which are having time constants $T_1$, $T_2$, $T_3$, and $T_4$, respectively [19].

## 2.3 *Overview of PID Controller*

PID controller is a closed control-loop mechanism. There are three gains in PID controller, namely integral gain, proportional gain, and derivative gain. Input control signal of PID controller in single area LFC problem is frequency deviation error. Output signal of PID controller is the sum of all the gains of controller.

Transfer function (TF) of PID controller may be represented as:

$$\mathrm{TF_{PID}} = \mathrm{K_P} + \frac{\mathrm{K_I}}{\mathrm{s}} + \mathrm{K_D s} \tag{1}$$

where $K_I$, $K_P$, and $K_D$ are the gains of the controller, namely integral gain, proportional gain, and derivative gain, respectively.

## 2.4 *Formulation of Objective Function*

In any optimization problem, the objective function (OF) has played a very vital role. Objective function is generally a mathematical expression which needs to be minimized or maximized. In LFC problem, the OF needs to be minimized for the better performance of the system. It is noteworthy to mention that optimized performance of the under study LFC system will be ensured at optimal PID controller gains. This may be guaranteed only at minimum value of employed objective function. Since the LFC problem is an optimization problem in which minimization of objective function has been carried out, therefore minimum the value of objective function, better the performance of overall system.

In this chapter, ITAE is considered as an objective function for LFC in the under study system. Mathematical expression of ITAE may be expressed as:

$$\text{Objective Function , J} = \text{ITAE} = \int_{0}^{T_{sim}} (\Delta \mathrm{f}).\mathrm{t}\, \mathrm{dt} \tag{2}$$

where $T_{sim}$ implies the time of simulation, and $\Delta f$ is the frequency deviation.

In load frequency control problem, bounds of parameters of controller are the constraints [20]. Hence, with the constraints taken into account, the optimization problem for LFC in single area nuclear power generating system may be designed as:

Minimize: Objective function, ITAE

Subjected to:

$$\begin{bmatrix} K_{Pmax} \geq K_P \geq K_{Pmin} \\ K_{Imax} \geq K_I \geq K_{Imin} \\ K_{Dmax} \geq K_D \geq K_{Dmin} \end{bmatrix} \tag{3}$$

where $K_{P\,min}$ and $K_{P\,max}$, $K_{I\,min}$ and $K_{I\,max}$, $K_{D\,min}$, and $K_{D\,max}$ refer to minimum and maximum values of PID controller gains.

# 3 Employed Optimization Algorithm

In this chapter, Jaya algorithm is employed for optimizing PID controller for LFC in under study system. Jaya optimization algorithm was developed in 2016 by Rao [21]. This algorithm was implemented to solve various problems like improvement of power quality using optimized solar photovoltaic-fed DSTATCOM [22], optimal designing of LQR controller for LFC [7], determining optimal power flow [23], and energy management controller designing [24]. Different studies have confirmed that Jaya algorithm performed better than conventional methods of optimization.

## *3.1 Jaya Algorithm and Its Mathematical Modeling*

Jaya algorithm works on the elementary principle of avoiding the worst solutions and deliberately search as well as update the best possible solutions as per the needs of the mentioned optimization problem.

In Jaya algorithm, total number of candidate solutions is represented by Q for a single solution agent. $N$ denotes the number of decisions in Jaya optimization algorithm. For an arbitrary $k$th iteration, the best among all the solutions of candidates is represented by $y_{best}^{k}$. On the other hand, the worst among all candidate solutions for $k$th iteration is represented by $y_{worst}^{k}$. Moreover, $j$th decision for an arbitrary $i$th candidate for $k$th iteration is denoted by $y_{i,j}^{k}$. Mathematical equation of Jaya algorithm for solution required in $k$th iteration [21] is represented as below:

$$Y_{i,j}^{k} = y_{i,j}^{k} + \mu_{i,j}^{k}\left(y_{j,best}^{k} - \left|y_{i,j}^{k}\right|\right) - \mu_{2,j}^{k}\left(y_{j,worst}^{k} - \left|y_{i,j}^{k}\right|\right) \quad (4)$$

where $\mu_{1,j}^{k}$ and $\mu_{2,j}^{k}$ denote random numbers in the range of 0 to 1 during $k$th iteration. Flowchart of Jaya optimization algorithm for optimal tuning of the PID controller is shown in Fig. 4.

# 4 Simulation Results

The considered single area nuclear power plant for LFC is modeled under MATLAB/Simulink environment. The data of nuclear power generating system and capacitive energy storage unit are taken into account for Simulink modeling from [18, 25]. Simulink block diagram for LFC of single area nuclear power station incorporating CES unit is shown in Fig. 5.

In this section, simulation results of Jaya algorithm-tuned PID controller are presented for LFC in single area nuclear power generating station. Optimal tuning of PID controller is done by using the intelligence of Jaya optimization algorithm. Maximum number of iteration and population size of Jaya algorithm are considered

**Fig. 4** Flowchart of Jaya algorithm

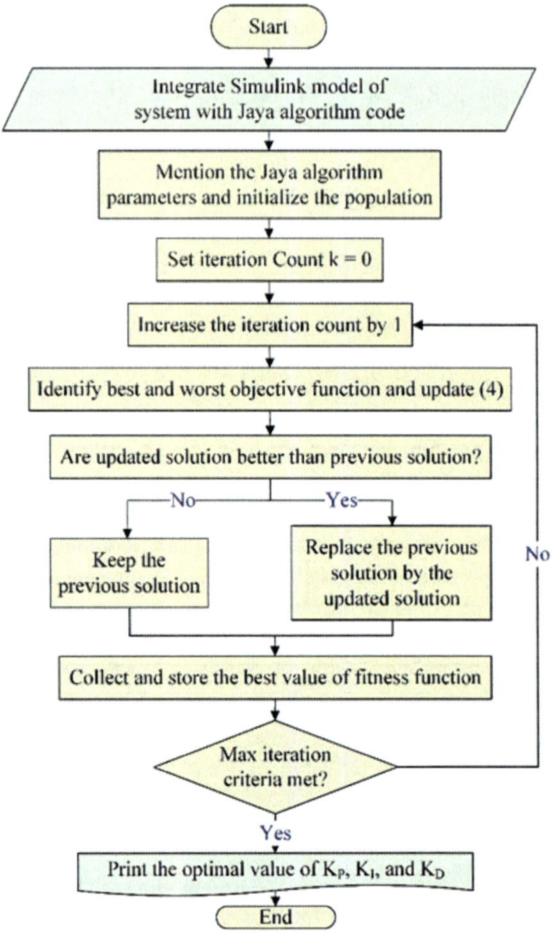


as 100 and 50, respectively. Simulation results of the proposed Jaya-PID controller are compared with ant colony optimization (ACO) tuned PID controller [18] for 1%, 2%, and 10% load disturbances in the under study system. The type of load considered in this paper is electrical load. The performances of ACO and Jaya-tuned PID controllers are compared based on the obtained value of ITAE in each case. In addition, the effects of population on objective function and related convergence curve have been presented for Jaya algorithm in case 1. The simulation time considered in this research paper is 10 s.

Three different cases are considered in this article. In case 1, optimization is done and simulation results are presented considering 1% load disturbance in under study system. In cases 2 and 3, load disturbances are considered as 2% and 10%, respectively, in single area nuclear power generating system. In all the three cases, ITAE is considered as an objective function for optimal tuning of PID controller.

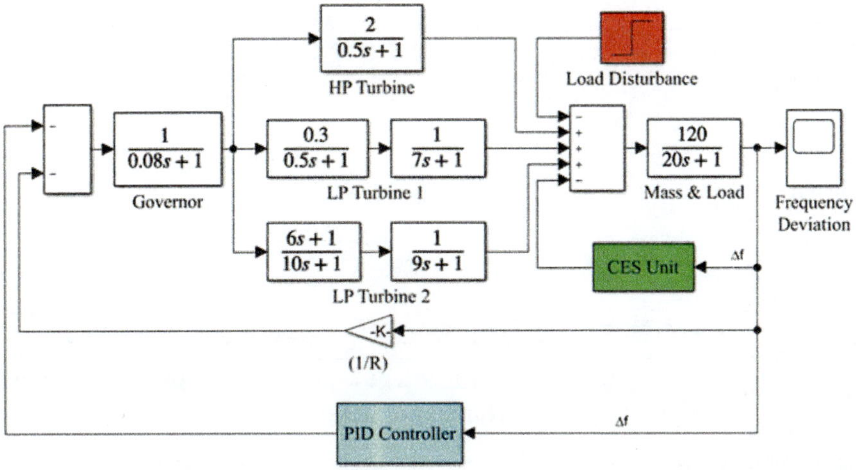


**Fig. 5** MATLAB/Simulink model of nuclear power plant with PID controller and CES unit

Comparative frequency response statistical analysis is performed in each case. The effects of adding CES unit in the under study are analyzed with proposed Jaya-PID controller for 1%, 2%, and 10% load disturbance. Comparison has been carried out between Jaya-tuned PI and PID controller for 10% load disturbance.

## *4.1 Case 1: Disturbance = 1%*

In this case, an experiment is conducted for load disturbance of 1% in the system. Optimization of PID controller is performed for 1% load disturbance using Jaya optimization algorithm. ITAE is minimized for obtaining optimized parameters of PID controller.

In this particular case, the effects of population on objective function and related convergence curve of Jaya algorithm have also been presented for optimal tuning of PID controller parameters. The optimal tuning of PID controller with different size of population has been done for LFC in the under study system. Different size of population considered for describing the effects of population on objective function are 20, 40, 50, and 60. The related convergence curve is shown in Fig. 6, which clearly shows the effects of population on objective function (i.e., ITAE). As observed from Fig. 6, with population size = 60, there is no significant improvement in the value of objective function, hence in this paper population size = 50 is considered for optimization process.

Nuclear power generating station is integrated with a capacitive energy storage unit for 1% load disturbance. The simulation results of LFC in single area nuclear power generating station with proposed technique for 1% load disturbance are compared with ACO-tuned PID controller [18]. The optimal tuned parameters of

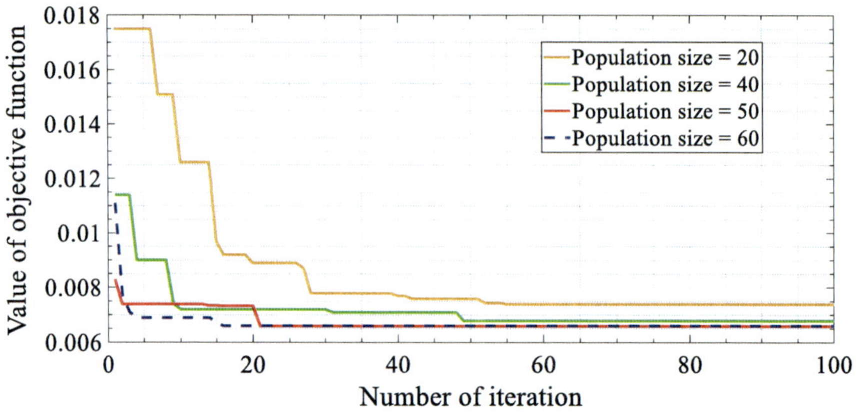


**Fig. 6** Convergence curve to show the effects of population on objective function

**Table 1** Optimized parameters and value of ITAE for 1% disturbance

| Controller | Optimized value | | | ITAE ($\times 10^{-3}$) |
|---|---|---|---|---|
| | $K_P$ | $K_I$ | $K_D$ | |
| ACO-PID [18] | 0.51 | 0.98 | 0.2 | 9.157 |
| Jaya-PID | 0.8507 | 0.9999 | 0.2197 | 6.670 |
| Jaya-PID with CES | 1.151 | 0.9999 | 0.2124 | **6.527** |

different controllers and value of ITAE are presented in Table 1. It has been observed from Table 1 that less value of ITAE is obtained with PID controller tuned by Jaya algorithm as compared to ACO-tuned PID controller [18]. ITAE is reduced further when CES unit is added in a single area nuclear power generating system.

Frequency response of under study system using ACO-PID [18], Jaya-PID, Jaya-PID with CES unit is as shown in Fig. 7. Comparative statistical frequency response analysis of ACO-PID [18], Jaya-PID, Jaya-PID with CES unit are presented in Table 4. From observing Table 4, it may be concluded that Jaya algorithm-tuned PID controller performs better as compared to ACO-tuned PID controller for load disturbance of 1%. With Jaya-PID controller, less values of peak overshoot, peak undershoot, peak to peak value, mean value, RMS value are obtained as compared to ACO-PID. Moreover, these values are reduced further by adding CES unit in the under study system. This shows the improved performance of system incorporating CES unit for 1% load disturbance.

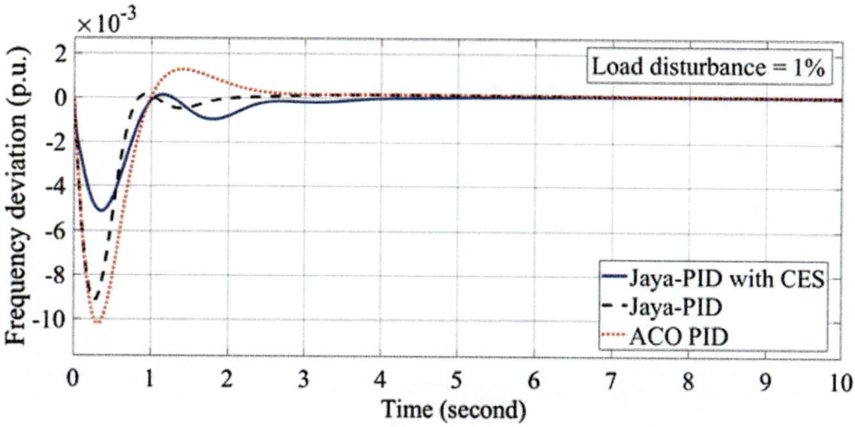


**Fig. 7** Frequency response of the system for 1% load disturbance

## *4.2 Case 2: Disturbance = 2%*

In this case, optimization is done for load disturbance of 2% in a single area nuclear power generating station. Optimization of PID controller is performed for 2% load disturbance using Jaya optimization algorithm. ITAE has been minimized for obtaining optimized parameters of PID controller. Nuclear power generating station also is integrated with capacitive energy storage unit for 2% load disturbance to the performance of load frequency controller. The simulation results of LFC in single area nuclear power generating station with the proposed technique for 2% load disturbance are compared with ACO-tuned PID controller [18]. The optimal tuned parameters of different controllers and value of ITAE are presented in Table 2. It has been observed from Table 2 that less value of ITAE is obtained with PID controller tuned by Jaya algorithm as compared to ACO-tuned PID controller [18]. The value of ITAE is reduced further when CES unit is added in a single area nuclear power generating system.

Frequency response of under study system using ACO-PID [18], Jaya-PID, Jaya-PID with CES unit is as shown in Fig. 8. Comparative statistical frequency response analysis of ACO-PID [18], Jaya-PID, Jaya-PID with CES unit are presented in Table 4. From Table 4, it has been concluded that Jaya algorithm-tuned PID controller

**Table 2** Optimized parameters and value of ITAE for 2% disturbance

| Controller | Optimized value | | | ITAE ($\times 10^{-2}$) |
|---|---|---|---|---|
| | $K_P$ | $K_I$ | $K_D$ | |
| ACO-PID [18] | 0.68 | 0.97 | 0.22 | 1.485 |
| Jaya-PID | 1.0442 | 0.998 | 0.2313 | 1.313 |
| Jaya-PID with CES | 1.500 | 1.000 | 0.2167 | **1.300** |

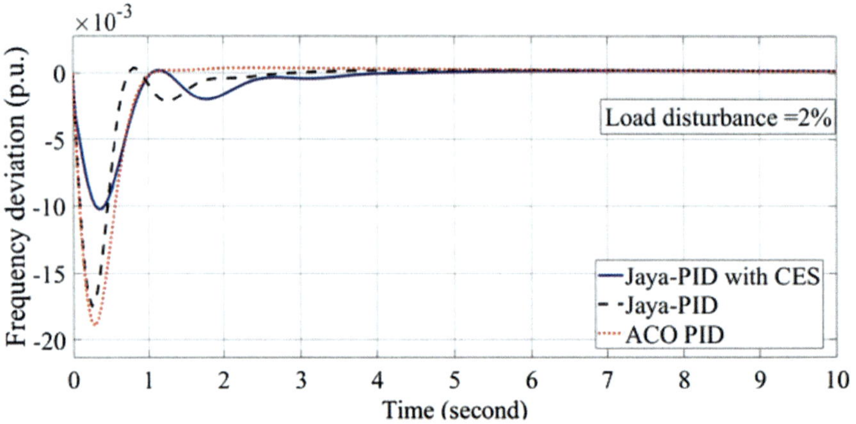


**Fig. 8** Frequency response of the system for 2% load disturbance

performs better as compared to ACO-tuned PID controller. With Jaya-PID controller, less values of peak overshoot, peak undershoot, peak to peak value, mean value, RMS value are obtained as compared to ACO-PID. Moreover, these values are reduced further by adding CES unit in the under study system. This indicates improved performance of the system with CES unit for 2% load disturbance.

### *4.3 Case 3: Disturbance = 10%*

In this case, simulation for LFC in a single area nuclear power generating system is done considering 10% change in load. PID controller is optimized for 10% load disturbance using Jaya optimization algorithm. Minimization of ITAE has been carried out for obtaining optimized parameters of PID controller. Nuclear power generating station is also integrated with capacitive energy storage unit for 10% load disturbance to enhance the performance of overall system. The simulation results of LFC in single area nuclear power generating station with proposed technique for 10% load disturbance are compared with ACO-tuned PID controller [18]. The optimal tuned parameters of different controllers and value of ITAE are presented in Table 3. From observing Table 3, it may be revealed that less value of ITAE is attained with Jaya-PID controller as compared to ACO-tuned PID controller [18] for 10% load disturbance. Value of objective function ITAE is reduced further when capacitive energy storage unit is added in single area nuclear power generating system.

The frequency response of under study system using ACO-PID [18], Jaya-PID, Jaya-PID with CES unit is as shown in Fig. 9. Comparative statistical frequency response analysis of ACO-PID [18], Jaya-PID, Jaya-PID with CES unit are presented in Table 4. From Table 4, it has been observed that Jaya algorithm-tuned PID

**Table 3** Optimized parameters and value of ITAE for 10% disturbance

| Controller | Optimized value | | | ITAE ($\times 10^{-2}$) |
|---|---|---|---|---|
| | $K_P$ | $K_I$ | $K_D$ | |
| ACO-PID [18] | 0.42 | 0.99 | 0.22 | 10.24 |
| Jaya-PID | 1.0445 | 1 | 0.2324 | 6.563 |
| Jaya-PID with CES | 1.4999 | 1 | 0.2257 | **6.507** |

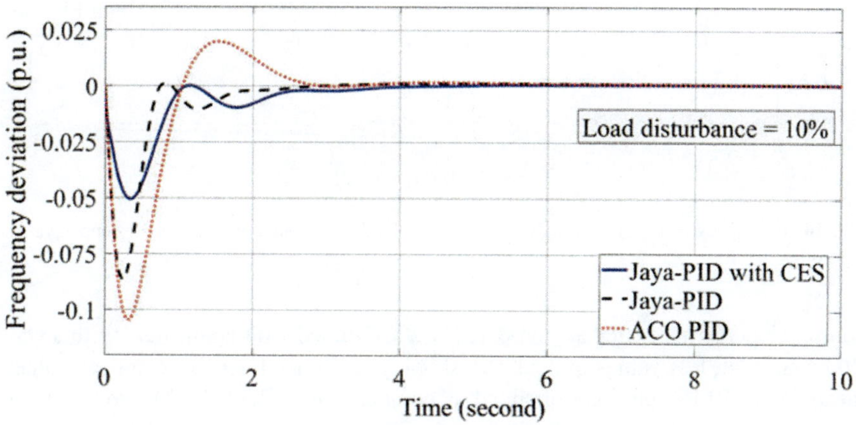


**Fig. 9** Frequency response of the system for 10% load disturbance

**Table 4** Statistical analysis of frequency response for 1, 2, and 10% load disturbance

| Load disturbance | Controller | Statistical analysis of frequency response curve | | | | |
|---|---|---|---|---|---|---|
| | | Peak overshoot (p.u.) ($\times 10^{-3}$) | Peak to peak (p.u.) ($\times 10^{-3}$) | Peak undershoot (p.u.) ($\times 10^{-3}$) | RMS (p.u.) ($\times 10^{-3}$) | Mean (p.u.) ($\times 10^{-4}$) |
| 1% | ACO-PID [18] | 1.266 | 11.503 | −10.243 | 2.851 | −8.563 |
| | Jaya-PID | 0.141 | 9.291 | −9.15 | 2.296 | −8.051 |
| | Jaya-PID with CES | 0.129 | 5.286 | −5.157 | 1.528 | −7.801 |
| 2% | ACO-PID [18] | 3.865 | 1.933 | −1.895 | 3.842 | −9.524 |
| | Jaya-PID | 3.761 | 1.786 | −1.749 | 3.303 | −9.398 |
| | Jaya-PID with CES | 1.978 | 1.0432 | −1.023 | 2.343 | −8.712 |
| 10% | ACO-PID [18] | 19.944 | 12.46 | −10.463 | 2.379 | −5.134 |
| | Jaya-PID | 1.593 | 8.885 | −8.726 | 1.652 | −4.702 |
| | Jaya-PID with CES | 0.9 | 5.151 | −5.061 | 1.165 | −4.433 |

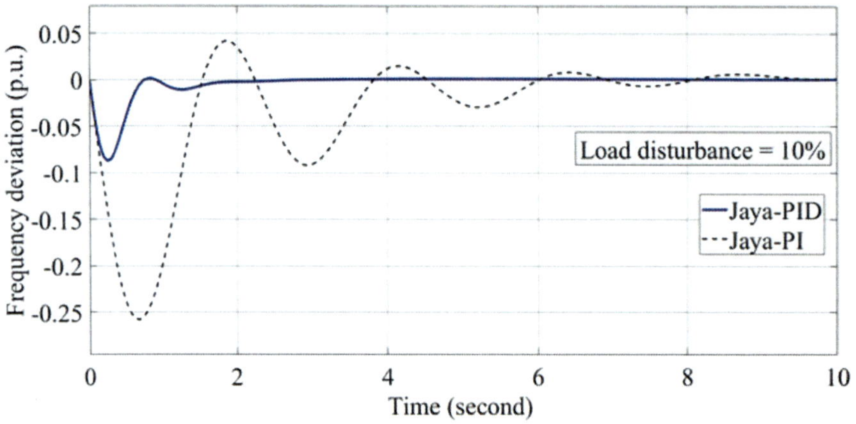


**Fig. 10** Frequency response of the system for 10% load disturbance with Jaya-PID and Jaya-PI controller

controller performs better as compared to ACO-tuned PID controller. With Jaya-PID controller, less values of peak overshoot, peak undershoot, peak to peak value, mean value, RMS value are obtained as compared to ACO-PID. Moreover, these values are reduced further by adding capacitive energy storage unit in the under study system. This indicates improved performance of the system when CES unit is incorporated in the under study system for 10% load disturbance.

**Comparison of Jaya-PI and Jaya-PID Controllers for 10% Load Disturbance**

Comparison of Jaya algorithm-tuned PI and PID controllers has been performed for LFC in single area nuclear power plant for 10% load disturbance to validate the superiority of Jaya-PID controller. The optimal values of PI controller tuned by Jaya algorithm are $K_P = 0.109$ and $K_I = 0.1128$ and the value of objective function (ITAE) is 0.526. From Table 3, it may be concluded that the value of ITAE obtained with Jaya-PID controller is much better as compared to that obtained with Jaya-PI controller. Frequency response of Jaya PI and Jaya-PID controller is shown in Fig. 10. From Fig. 10, it is evident that frequency response is better with Jaya-PID as compared to Jaya-PI controller for LFC in single area nuclear power plant for 10% load disturbance. Therefore Jaya-PID controller has been employed in this paper as compared to Jaya-PI controller.

## 5 Conclusion

In this chapter, Jaya optimization algorithm is employed in optimizing PID controller for LFC of nuclear power generating station. Integral of time-weighted absolute error is considered as objective function which has been minimized. Capacitive energy storage unit has been incorporated in the under study system for improved performance. Performance of Jaya algorithm-tuned PID controller is analyzed with 1%, 2%, and 10% load disturbances.

From the various simulation results obtained in this paper, it is obvious that Jaya algorithm-tuned PID controller performed better than the ACO algorithm-tuned PID controller. Less value of ITAE is obtained with Jaya-PID controller in comparison to ACO-PID controller for load disturbances 1%, 2%, and 10% in under study system. Also, the performance of the load frequency control is enhanced by adding capacitive energy storage unit. Simulation results also show that if CES unit is incorporated in the system, ITAE, peak overshoot, peak undershoot, peak to peak value, mean value, and RMS value of frequency response are reduced significantly. This proved the superiority of the proposed Jaya-PID controller for LFC in nuclear power plant.

**Acknowledgements** Authors are highly thankful and acknowledge for providing financial assistance to TEQIP-III. They are also thankful to the Director, MITS, Gwalior, India, for providing facilities for this research work.

## References

1. Mohanty B, Panda S, Hota PK (2014) Controller parameters tuning of differential evolution algorithm and its application to load frequency control of multi-source power system. Int J Electr Power Energy Syst 54:77–85. https://doi.org/10.1016/j.ijepes.2013.06.029
2. Kumar J, Ng KH, Sheble G (1997) AGC simulator for price-based operation, Part 1: a model. IEEE Power Eng Rev 17:51
3. Raineri R, Ríos S, Schiele D (2006) Technical and economic aspects of ancillary services markets in the electric power industry: an international comparison. Energy Policy 34:1540–1555. https://doi.org/10.1016/j.enpol.2004.11.015
4. Fosha E (1970) Megawatt-frequency control multiarea electric energy systems tif * DW. Power 556–563
5. Yamashita K, Taniguchi T (1986) Optimal observer design for load-frequency control. Int J Electr Power Energy Syst 8:93–100. https://doi.org/10.1016/0142-0615(86)90003-7
6. Das D, Nanda J, Kothari ML, Kothari DP (1990) Automatic generation control of a hydrothermal system with new area control error considering generation rate constraint. Electr Mach Power Syst 18:461–471. https://doi.org/10.1080/07313569008909490
7. Paliwal N, Srivastava L, Pandit M (2020) Jaya algorithm based optimal design of LQR controller for load frequency control of single area power system. In: Shukla R, Agrawal J, Sharma S, Chaudhari NSK (eds) Social networking and computational intelligence. Springer, Singapore, pp 595–603
8. Yazdizadeh A, Ramezani MH, Hamedrahmat E (2012) Decentralized load frequency control using a new robust optimal MISO PID controller. Int J Electr Power Energy Syst 35:57–65. https://doi.org/10.1016/j.ijepes.2011.09.007
9. Ali ES, Abd-Elazim SM (2013) BFOA based design of PID controller for two area load frequency control with nonlinearities. Int J Electr Power Energy Syst 51:224–231. https://doi.org/10.1016/j.ijepes.2013.02.030
10. Paliwal N, Srivastava L, Pandit M (2020) PSO-based PID controller designing for LFC of single area electrical power network. In: Pandit M, Dubey H, Bansal J (eds) Nature inspired optimization for electrical power system. Algorithms for Intelligent Systems, Springer, Singapore, pp 43–54

11. Toulabi MR, Shiroei M, Ranjbar AM (2014) Robust analysis and design of power system load frequency control using the Kharitonov's theorem. Int J Electr Power Energy Syst 55:51–58. https://doi.org/10.1016/j.ijepes.2013.08.014
12. Ali R, Mohamed TH, Qudaih YS, Mitani Y (2014) A new load frequency control approach in an isolated small power systems using coefficient diagram method. Int J Electr Power Energy Syst 56:110–116. https://doi.org/10.1016/j.ijepes.2013.11.002
13. Singh SN, Srivastava L, Sharma J (2000) Fast voltage contingency screening and ranking using cascade neural network. Electr Power Syst Res 53:197–205. https://doi.org/10.1016/S0378-7796(99)00059-0
14. Srivastava L, Singh SN, Sharma J (1998) Parallel self-organising hierarchical neural network-based fast voltage estimation. IEE Proc -Gener Transm Distrib 145:98. https://doi.org/10.1049/ip-gtd:19981741
15. Bisen D, Dubey H, Pandit M, Panigrahi BK (2012) Solution of large scale economic load dispatch problem using quadratic programming and GAMS: a comparative analysis. J Inf Comput Sci 7:200–211
16. Varshney S, Srivastava L, Pandit M (2012) ANN based integrated security assessment of power system using parallel computing. Int J Electr Power Energy Syst 42:49–59. https://doi.org/10.1016/j.ijepes.2012.03.016
17. Sharma M, Pandit M, Srivastava L (2010) Multi-area economic dispatch with tie-line constraints employing evolutionary approach. Int J Eng Sci Technol 2:132–149. https://doi.org/10.4314/ijest.v2i3.59182
18. Dhanasekaran B, Siddhan S, Kaliannan J (2020) Ant colony optimization technique tuned controller for frequency regulation of single area nuclear power generating system. Microprocess Microsyst 73: https://doi.org/10.1016/j.micpro.2019.102953
19. Ponnusamy M, Banakara B, Dash SS, Veerasamy M (2015) Design of integral controller for load frequency control of static synchronous series compensator and capacitive energy source based multi area system consisting of diverse sources of generation employing imperialistic competition algorithm. Int J Electr Power Energy Syst 73:863–871. https://doi.org/10.1016/j.ijepes.2015.06.019
20. Paliwal N, Srivastava L, Pandit M (2020) Application of grey wolf optimization algorithm for load frequency control in multi-source single area power system. Evol Intell. https://doi.org/10.1007/s12065-020-00530-5
21. Venkata Rao R (2016) Jaya: a simple and new optimization algorithm for solving constrained and unconstrained optimization problems. Int J Ind Eng Comput 7:19–34. https://doi.org/10.5267/j.ijiec.2015.8.004
22. Mishra S, Ray PK (2016) Power quality improvement using photovoltaic fed DSTATCOM based on JAYA optimization. IEEE Trans Sustain Energy 7:1672–1680. https://doi.org/10.1109/TSTE.2016.2570256
23. Warid W, Hizam H, Mariun N, Abdul-Wahab NI (2016) Optimal power flow using the jaya algorithm. Energies 9:. https://doi.org/10.3390/en9090678
24. Samuel O, Javaid N, Aslam S, Rahim MH (2018) JAYA optimization based energy management controller for smart grid: JAYA optimization based energy management controller. 2018 Int Conf Comput Math Eng Technol Inven Innov Integr Socioecon Dev iCoMET 2018 - Proc 2018-Janua:1–8. https://doi.org/10.1109/ICOMET.2018.8346337
25. Mohamed EA, Gouda E, Mitani Y (2019) Impact of SMES integration on the digital frequency relay operation considering high PV/wind penetration in micro-grid. Energy Procedia 157:1292–1304. https://doi.org/10.1016/j.egypro.2018.11.295

# Chapter 22
# DDoS Attacks Detection in Cloud Computing Using ANN and Imperialistic Competitive Algorithm

**Gopal Singh Kushwah and Virender Ranga**

## 1 Introduction

Nowadays cloud computing [1] has turned into a prominent technology as it offers various advantages like low cost, flexibility, and scalability over traditional infrastructure-based setups. This technology offers its resources in the form of services over the Internet. Therefore, the availability of these services becomes a paramount concern. Distributed denial of service attacks [2] can be used by the attacker for denial of these services to its users. In this attack, the attacker makes a network of compromised hosts that are used to attack a single victim. This attack is executed by exhausting the network resources like buffers in routers or server resources.

Several machine learning-based solutions to detect these attacks have been proposed in the literature. The challenges in attack detection include low values of detection accuracy, sensitivity, specificity, etc. In this paper, a technique based on ANN has been proposed. The ANN model is trained using an optimization algorithm inspired by imperialistic competition called the imperialistic competitive algorithm (ICA) [3]. It has been used in several areas for optimization problems including [4–7] and has shown good performance.

The organization of the remaining paper is as follows. In Sect. 2, some key contributions in recent years have been discussed in brief. The proposed attack detection technique is presented in Sect. 3. The results of various experiments are discussed in Sect. 4. In Sect. 5, conclusions are made.

G. S. Kushwah (✉) · V. Ranga
NIT Kurukshetra, Haryana 136119, India

V. Ranga
e-mail: virender.ranga@nitkkr.ac.in

© The Author(s), under exclusive license to Springer Nature Singapore Pte Ltd. 2022 
H. M. Dubey et al. (eds.), *Artificial Intelligence and Sustainable Computing*,
Algorithms for Intelligent Systems, https://doi.org/10.1007/978-981-16-1220-6_22

## 2 Related Works

In this section, some recently proposed works in this area are discussed. A detailed survey on proposed techniques in the last few years can be found in [8, 9]. In [10], the authors proposed an online sequential extreme learning machine (OS-ELM) based technique. For feature selection, it uses a combination of CFS subset eval, consistency subset eval, and filtered subset eval techniques. Alpha profiling is used to reduce time complexity, and beta profiling is used to reduce training dataset size. In [11], fuzzy c-means clustering is used to create training subsets to improve the learning capability of the classifier. These subsets are used to train ANNs. The results of all ANNs are combined using fuzzy aggregation. In [12], the authors used ANN for DDoS detection. For training ANN, blackhole optimization algorithm is used.

In [13], the authors used a deep learning-based approach that uses recurrent neural networks. In [14], the authors proposed a model for detecting rare types of attacks. Naïve Bayes and KNN classifiers are used for attack detection. It also used dimensionality reduction using linear discriminant analysis. In [15], the authors used Gaussian-Bernoulli-type restricted Boltzmann machine. It is trained with a dataset without labels. To improve the detection accuracy, the hyperparameters of the model are also optimized. In [16], the authors proposed a convolutional neural network-based model. It also resolves the problem of data imbalance. An ELM-based technique has been proposed in [17] and an ensemble of ELMs is used in [18] that uses majority voting for the final result calculation. In [19], the ANN model is used, which is trained using particle swarm optimization (PSO) and hybrid gravitational search. For feature selection, a combination of genetic algorithm (GA) and SVM is used. The GA uses multi-parent crossover and multi-parent mutation.

## 3 Proposed DDoS Attack Detection Technique (IC-ANN)

Figure 1 shows a hypothetical view of the cloud network with DDoS attack detector. The detector has two components preprocessor and classifier. The preprocessor is used to convert the captured network traffic into samples. It also performs operations like normalization and dealing with nominal features. Classification of network traffic into attack or normal is performed by the classifier. An ANN model is used as the classifier which is shown in Fig. 2. It is a single hidden layer ANN with *n-m-2* neuron architecture. First, it is trained and then used to detect attacks.

### 3.1 Training ANN Using ICA

The ANN model is trained with a dataset of labeled samples. For the $i^{th}$ training sample, features are represented as vector $X_i^t = [x_{i1}^t, x_{i2}^t, \ldots, x_{in}^t]$ and label as vector

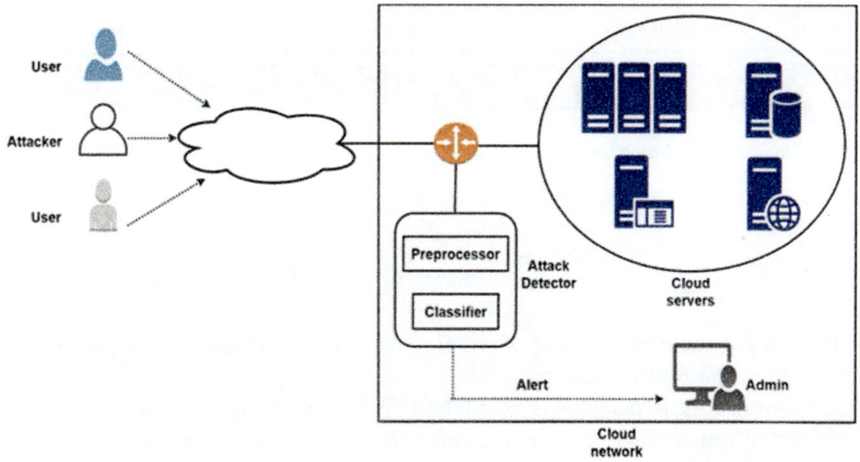


**Fig. 1** Cloud network with the proposed attack detector

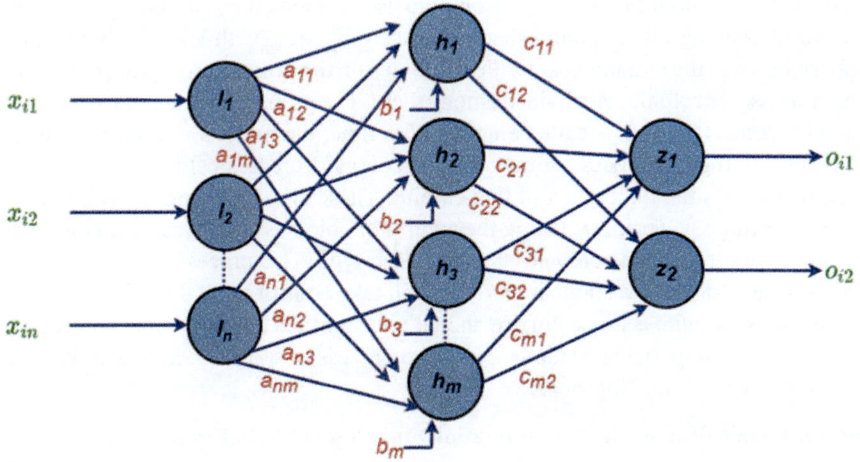

**Fig. 2** Artificial neural network model

$Y_i = [y_{i1}, y_{i2}]$. All features have continuous values in the range [0,1]. The label vector is a binary vector with one element being 1 and the other 0. The position of value 1 in the vector represents the class of that sample. The label vectors [0,1] and [1,0] represent the normal sample and attack sample, respectively.

In the process of training, the weights and biases of ANN are optimized in a way that the value of mean squared error (MSE) as given by the following equation should be very low.

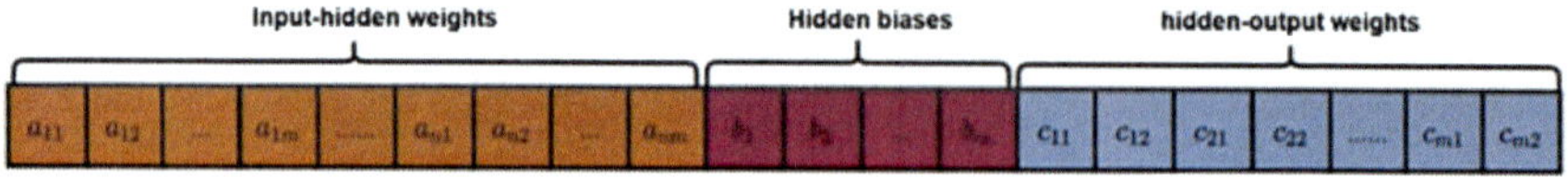


**Fig. 3** Encoding of solution for ANN training

$$\text{MSE} = \left(\frac{1}{N}\sum_{i=1}^{N}(Y_i - O_i)^2\right) \tag{1}$$

Here, $N$ represents the number of samples in the training dataset and $O_i$ represents the output vector of the $i^{\text{th}}$ sample.

To optimize these parameters of ANN, ICA has been used. It is a population-based optimization algorithm that uses an initial population of solutions and these solutions are evolved until the desired solution is found. In the case of ANN training, its weights and biases form a solution. These solutions are encoded in the form of vectors as represented by Fig. 3.

In ICA, the initial population of solutions is represented by all the countries in the world. Among these countries, some of the powerful (with less MSE) become imperialists and the remaining are called colonies. In the proportion of power of these imperialists, the colonies are divided among them. The group of an imperialist and the colonies controlled by it is called empire. With time, the imperialists improve their colonies. During this process, if the cost of a colony decreases below its imperialist cost, it becomes the imperialist of that empire. These empires compete with others to become more powerful by taking the control of colonies from weaker empires. If an empire has left with no colonies then it is destroyed. This process continues in the hope that at some time a single imperialist will take control of all the colonies. This is the ideal case with a single empire and all the countries have the same power and position as the imperialist of the empire. The process of training ANN with ICA is described by the following steps.

**Step 1: Generating the initial population and empires** In this step, $\text{IP}_{\text{size}}$ solutions of the form as shown in Fig. 3 are randomly initialized with values [−1,1], and these represent all countries in the world. Then, the power (inversely proportional to the cost, MSE) of each country is calculated. The top $\mu_{\text{imp}}$ powerful countries are called imperialists and the remaining $\mu_{\text{col}}$ are called colonies. Now the colonies are distributed among these imperialists and initial empires are formed. The number of colonies that each empire can get depends on the normalized power of its imperialist, which is defined as (for $i^{\text{th}}$ imperialist)

$$Pow_i^{norm} = \left|\frac{Cost_i^{norm}}{\sum_{j=1}^{\mu_{imp}} Cost_j^{norm}}\right| \tag{2}$$

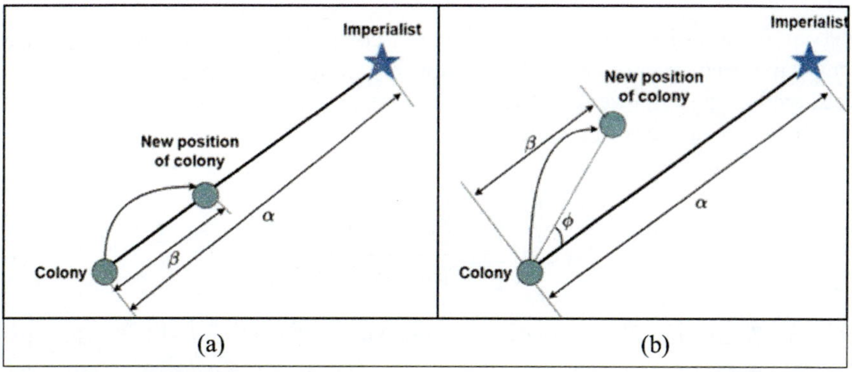


**Fig. 4** Colony movement

where $Cost_i^{norm}$ is the normalized cost (MSE) of $i^{\text{th}}$ imperialist defined as

$$Cost_i^{norm} = Cost_i - \max\{Cost_j\}, \text{ for } j = 1 \text{ to } \mu_{\text{imp}} \quad (3)$$

The $i^{\text{th}}$ empire gets $round\{Pow_i^{norm} * \mu_{\text{col}}\}$ colonies. These colonies are randomly chosen for each empire from available colonies.

**Step 2: Movement of colonies toward their imperialist in an empire** The fact of improving the colonies by their imperialists is modeled by the movement of colonies toward their respective imperialists. Figure 4a shows the movement of a colony toward its imperialist by $\beta$ units where $\beta \sim U(0, \lambda * \alpha)$. The value of $\lambda$ is taken as greater than 1 so that the colonies get closer to their imperialists. The colonies move on a path straight to the imperialist. The searching capability is improved by allowing some deviation from this direction. This deviation is defined as $\phi \sim U(-\zeta, \zeta)$ and shown in Fig. 4b.

The parameter $\zeta$ gives the range of deviation from the original direction of movement.

**Step 3: Position exchange** During the movement of colonies toward their imperialists, a colony can reach a position where it gets an improvement in cost over the imperialist. In this case, the positions of that colony and imperialist are exchanged. Now the colonies of that empire start moving toward the new imperialist.

**Step 4: Calculation of power** The power of an empire is modeled by calculating the total cost of the empire as follows:

$$Total\ Cost_i = Cost_{imp} + \eta * mean\{Cost(colonies\ of\ empire)\} \quad (4)$$

Here, $\eta$ is a positive number less than 1, which defines the participation of colonies in the calculation of the power for an empire.

**Step 5: Imperialistic competition** There is competition among empires to take control of the colonies of others. In this process, strong empires become stronger and weak empires become weaker. To model this competition, one colony (the weakest) at a time from the weakest empire is chosen. Now, a probability of getting control of this colony is calculated for each empire.

$$Prob_i = \left| \frac{Total\ Cost_i^{norm}}{\sum_{j=1}^{\mu_{imp}} Total\ Cost_j^{norm}} \right| \tag{5}$$

$$Here,\ Total\ Cost_i^{norm} = Total\ Cost_i - \max\{Total\ Cost_j\},\ \text{for } j = 1 \text{ to } \mu_{\text{imp}} \tag{6}$$

Suppose, the probabilities of all the empires are represented by vector A = [$\text{prob}_1$, $\text{prob}_2$, …, $\text{prob}_{\mu\text{imp}}$]. Then another vector B = [$r_1$, $r_2$, …, $r_{\mu\text{imp}}$] of same length with elements randomly chosen from (0,1) is created. Now a vector C = A−B is calculated, which shows the probability of getting the colony by the respective empire. The colony is given to the empire with the highest value in the vector C.

**Step 6: Elimination of empires and convergence** During the competition, the power of empires change. The powerless empires are eliminated from the competition when they do not have any colony left. In the end, only a single empire is left that has possession of all colonies. This is the ideal world where all the countries have the same power and position which means there is no discrimination between imperialists and colonies. All colonies are also imperialists and they are controlled by themselves.

At this point, ICA gives the optimal solution (optimal values of weights and biases) for the problem of ANN training. The pseudocode of the training processes is shown in Algorithm 1.

Algorithm 1: Training ANN using ICA

```
1.  Initialize IPsize number of countries (encoded solutions)
2.  Create initial imperialists and empires
3.  Move colonies toward their imperialists
4.  If any colony achieves lower cost than imperialist
5.          Exchange position of colony and imperialist
6.  End if
7.  Compute power of all empires
8.  Give the weakest colony from weakest empire to the empire having the
    highest possession probability
9.  Eliminate empires with no colony
10. If there is only one empire left
11.         Stop and return the solution (weight & biases of ANN)
12. Else
13.         Goto step 3
14. End if
```

### 3.2 Attack Detection Using ANN

For detecting attacks, the samples are prepared from the traffic data. These samples have the same features as used for training. The prepared samples of the form $X_i = [x_{i1}, x_{i2},\ldots, x_{in}]$ are applied to the ANN, and output $O_i$ is calculated. If the output $O_i = [0,1]$ then the sample belongs to normal traffic, and if $O_i = [1,0]$ then the sample belongs to attack traffic. In the case of attack detection, the cloud admin is informed by sending an alert.

## 4 Experimental Results and Discussions

We used Windows 10 operating system running on Intel Core i5 CPU with 16 GB RAM for experiments. The implementation is done in MATLAB. The NSL KDD dataset [20] is used in our experiments which is a widely used dataset. It has a training subset KDDTrain+ with 125973 samples and two test subsets KDDTest+ and KDDTest-21 with 22544 and 11850 samples, respectively. The dataset contains samples with five classes: normal, probe, DoS, U2R, and R2L. All types of attack samples are combined as we evaluate the proposed technique for binary classification. The samples have 41 features and we used all these features. The following performance metrics have been used.

$$Accuracy = \frac{N_{TP} + N_{TN}}{N_{TP} + N_{FP} + N_{TN} + N_{FN}} * 100 \tag{7}$$

$$Sensitivity = \frac{N_{TP}}{N_{TP} + N_{FN}} * 100 \tag{8}$$

$$Specificity = \frac{N_{TN}}{N_{TN} + N_{FP}} * 100 \tag{9}$$

$$Precision = \frac{N_{TP}}{N_{TP} + N_{FP}} * 100 \tag{10}$$

$$F1 - Score = 2*\frac{Precision * Sensitivity}{Precision + Sensitivity} * 100 \tag{11}$$

Here $N_{FP}$, $N_{TP}$, $N_{FN}$, and $N_{TN}$ represent the number of false positives, true positives, false negatives, and true negatives, respectively.

First, the ANN is trained with ICA. The number of initial countries and imperialists is taken as 50 and 20, respectively. The number of decades (iterations) is 100. With these parameter values, the number of hidden neurons is varied from 10 to 100 and the best performance in this range is recorded.

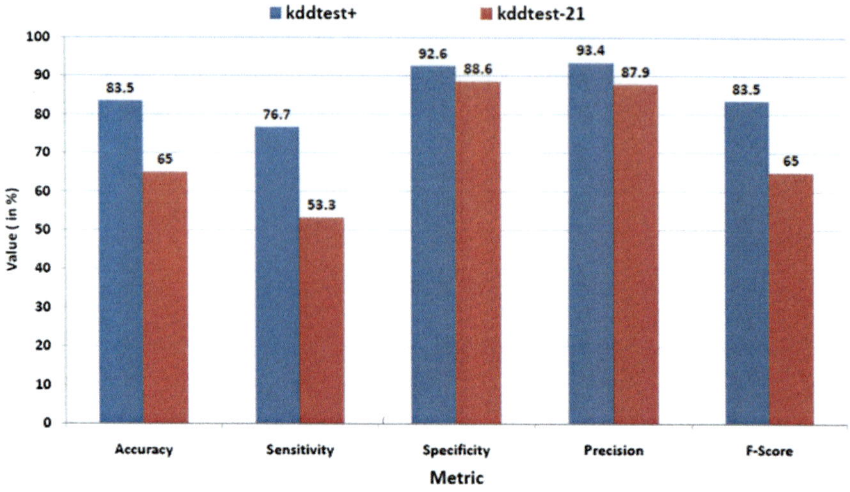


**Fig. 5** Performance of proposed technique

The values of accuracy, sensitivity, specificity, precision, and F-score achieved by our technique with KDDTest+ dataset are 83.5%, 76.7%, 92.6%, 93.4%, and 83.5%, respectively. With KDDTest-21 dataset, our technique shows values of 65%, 53.3%, 88.6%, 87.9%, and 65%, respectively, for these metrics. The performance of the proposed technique (IC-ANN) is shown in Fig. 5. Next, the performance of the proposed technique with some other machine learning-based techniques namely backpropagation ANN (BP-ANN), decision tree, and support vector machine is compared. Figures 6 and 7 show this comparison using KDDTest+ and KDDTest-21 datasets, respectively. Our proposed technique gives better performance than all of these, as observed from the figures. At last, performance comparison with some state-of-the-art works is also presented in Table 1.

## 5 Conclusions

Developing solutions for the problem of DDoS attack detection in the cloud is a challenging task. Machine learning-based techniques play a significant role in the detection of these attacks. In this work, we propose an ANN-based detection technique. For training the ANN, imperialistic competitive algorithm has been used. Our proposed technique achieves accuracy, sensitivity, specificity, precision, f-score of 83.5%, 76.7%, 92.6%, 93.4%, 83.5% with KDDTest+ and 65%, 53.3%, 88.6%, 87.9%, 65% with KDDTest-21 datasets, respectively.

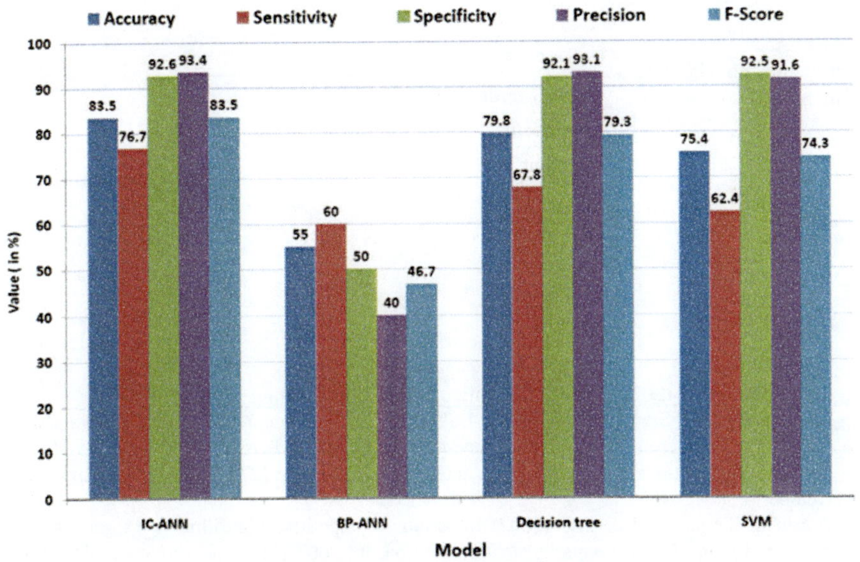


**Fig. 6** Performance comparison with other techniques using KDDTest+ dataset

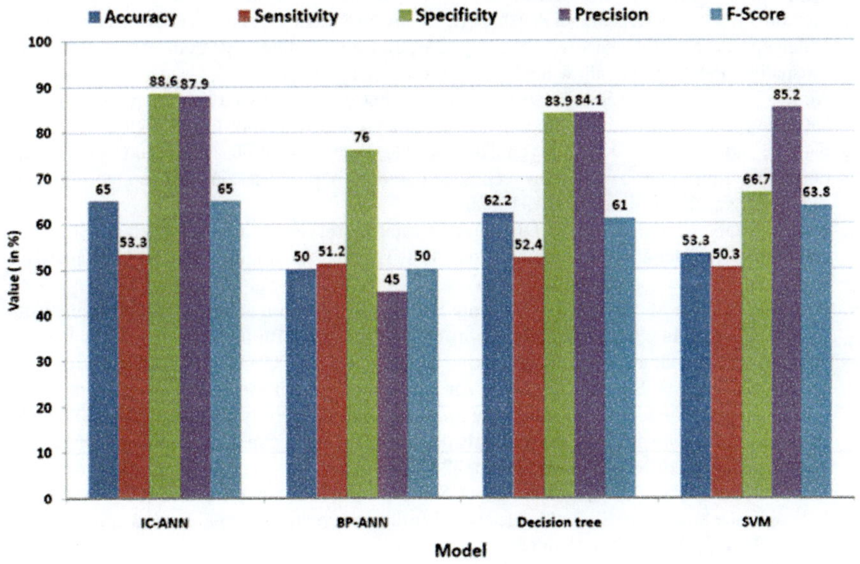


**Fig. 7** Performance comparison with other techniques using KDDTest-21 dataset

**Table 1** Accuracy comparison with other works (using KDDTest+ dataset)

| Research work | Accuracy |
|---|---|
| Imamverdiyev et al. [15] | 73.23 |
| Wu et al. [16] | 79.48 |
| Pajouh et al. [14] | 83.24 |
| Yin et al. [13] | 83.28 |
| Proposed work(IC-ANN) | 83.50 |

## References


1. Mell P, Grance, T (2011) The NIST definition of cloud computing
2. Lau F, Rubin SH, Smith MH, Trajkovic L (2000) Distributed denial of service attacks. IEEE international conference on systems, man and cybernetics. 'cybernetics evolving to systems, humans, organizations, and their complex interactions', vol 3, pp 2275–2280. (cat. no. 0) (2000, October)
3. Atashpaz-Gargari E, Lucas C (2007) Imperialist competitive algorithm: an algorithm for optimization inspired by imperialistic competition. In: 2007 IEEE congress on evolutionary computation, pp 4661–4667. (2007, September)
4. Fathy A, Rezk H (2017) Parameter estimation of photovoltaic system using imperialist competitive algorithm. Renew Energy 111:307–320
5. Tien Bui D, Shahabi H, Shirzadi A, Chapi K, Hoang ND, Pham BT, Saro L (2018) A novel integrated approach of relevance vector machine optimized by imperialist competitive algorithm for spatial modeling of shallow landslides. Remote Sen 10(10), 1538
6. Aliniya Z, Mirroshandel SA (2019) A novel combinatorial merge-split approach for automatic clustering using imperialist competitive algorithm. Expert Syst Appl 117:243–266
7. Fang Q, Nguyen H, Bui XN, Nguyen-Thoi T (2020) Prediction of blast-induced ground vibration in open-pit mines using a new technique based on imperialist competitive algorithm and M5Rules. Nat Resour Res 29(2):791–806
8. Kushwah GS, Ranga V (2019) Distributed denial of service attacks and defense in cloud computing. In: Handbook of research on the IoT, cloud computing, and wireless network optimization. IGI Global, pp 41–59
9. Potluri S, Mangla M, Satpathy S, Mohanty SN (2020) Detection and prevention mechanisms for DDoS attack in cloud computing environment. In: 2020 11th international conference on computing, communication and networking technologies (ICCCNT), pp 1–6. (2020, July)
10. Singh R, Kumar H, Singla RK (2015) An intrusion detection system using network traffic profiling and online sequential extreme learning machine. Expert Syst Appl 42(22):8609–8624
11. Pandeeswari N, Kumar G (2016) Anomaly detection system in cloud environment using fuzzy clustering based ANN. Mobile Netw Appl 21(3):494–505
12. Kushwah GS, Ali ST (2017) Detecting DDoS attacks in cloud computing using ANN and black hole optimization. In: IEEE 2nd international conference on telecommunication and networks (TEL-NET), pp 1–5. (2017, August)
13. Yin C, Zhu Y, Fei J, He X (2017) A deep learning approach for intrusion detection using recurrent neural networks. Ieee Access 5:21954–21961
14. Pajouh HH, Dastghaibyfard G, Hashemi S (2017) Two-tier network anomaly detection model: a machine learning approach. J Intell Inf Syst 48(1):61–74
15. Imamverdiyev Y, Abdullayeva F (2018) Deep learning method for denial of service attack detection based on restricted boltzmann machine. Big Data 6(2):159–169
16. Wu K, Chen Z, Li W (2018) A novel intrusion detection model for a massive network using convolutional neural networks. IEEE Access 6:50850–50859
17. Kushwah GS, Ali ST (2019) Distributed denial of service attacks detection in cloud computing using extreme learning machine. Intern J Commun Netw Distrib Syst 23(3):328–351

18. Kushwah GS, Ranga V (2020) Voting extreme learning machine based distributed denial of service attack detection in cloud computing. J Inf Secur Appl 53
19. Hosseini S, Zade BMH (2020) New hybrid method for attack detection using combination of evolutionary algorithms, SVM, and ANN. Comput Netw 107168
20. Tavallaee M, Bagheri E, Lu W, Ghorbani AA (2009) A detailed analysis of the KDD CUP 99 data set. In 2009 IEEE symposium on computational intelligence for security and defense applications, pp 1–6. (2009, July)

# Chapter 23
# Analysis and Review of the Kinetic Façades in Kolding Campus, South Denmark University

**Rishika Sood and Anjali S. Patil**

## 1 Kinetic Architecture

Buildings have always been influenced and shaped by natural and dynamic forces such as time, weather and the ever-changing human needs. Consequently, there has been a need to develop buildings that can adjust and deal with these dynamic circumstances and surroundings. Ever since the onset of industrialization and innovation of new technologies, architects and engineers have been altering the built environment to introduce the dimension of time in it. To achieve this, movement is created in buildings either mechanically or by using natural means such as air and water. In the recent decades, a new typology of architecture, "Kinetic Architecture" that incorporates such mechanisms has gained much popularity.

Kinetic architecture relies on the design of buildings in which transformative and mechanized structures aim to change the shape of buildings to match the needs of people on the inside and adapt to the elements on the outside [17]. Thus, kinetic architecture leads to the creation of spaces that can physically reconfigure themselves to ensure maximum comfort for the occupants. While individual parts maybe allowed to move, the overall structural integrity of the building is maintained.

Over the years, through experimentation, kinetic mechanisms have been incorporated in a variety of components of a building that range from roofs, façades to the interiors.

R. Sood (✉) · A. S. Patil
MITS Gwalior, Gwalior, India

A. S. Patil
e-mail: anjalipatil@mitsgwalior.in

© The Author(s), under exclusive license to Springer Nature Singapore Pte Ltd. 2022 
H. M. Dubey et al. (eds.), *Artificial Intelligence and Sustainable Computing*,
Algorithms for Intelligent Systems, https://doi.org/10.1007/978-981-16-1220-6_23

### 1.1 Kinetic Façades

Different from a conventional façade that is static, a kinetic façade allows movement to occur on the building's surface. This helps to create what the architect Buckminster Fuller called a "skin-like articulation" effect, and is an extension of the idea that a building's envelope is an active system rather than just a container [22].

If equipped with computerized mechanisms, a kinetic façade can measure the weather conditions of the external environment and thereby control the intake of daylight, wind and solar energy into the building. To ensure a comfortable interior environment, these façades use movable and transformable elements that respond to the climatic and environmental factors. While the initial cost of setting up such façades maybe high, this cost can usually be recovered by the excellent energy savings offered by these systems in their operation.

This paper shall analyze and review the aspects of kinetic façade with reference to the Kolding Campus, South Denmark University. The qualitative and quantitative data for this analysis have been taken from various secondary sources. The qualitative analysis method has been used to analyze the data about the architectural, sustainable and kinetic façade features of the Kolding Campus from literature reviews and case studies. In addition to this, in order to analyze the climatic data and the energy consumption, quantitative analysis methods have been used.

## 2 Kolding Campus, South Denmark University

The Kolding Campus is a part of the South Denmark University which offers courses for communications, design, culture and languages. It is located at the heart of the city Kolding and is in close proximity to the Kolding river. Built by the Henning Larsen Architects, based at Copenhagen; the construction of this campus was completed in 2014.

The five-storied University building has an iconic triangular plan. The plan of the university campus aims to create dynamic study spaces that are different from the conventional classrooms. To achieve this, the position of the staircases has been displaced on all floors to create a sense of community and cohesion between the students, teachers and researchers in the building. Thus, when the triangular form repeats itself in varied patterns on all floors, a variety of spaces that range from individual study spaces, formal and informal group spaces to circular semi open rooms for group studying are created. The triangles in the plan also cut into the floor slabs creating a five-storied atrium. This gives the impression that the interior space twists as it rises toward a triangular skylight.

The circulation inside the Kolding Campus not only encourages frequent social interactions but also promotes interdisciplinary learning among students. The orientation of the building on site ensures that the activities in the university open up toward the town. This way the exterior and the interior spaces of the campus become

**Fig. 1** The exterior view (left) and interior view (right) of the Kolding Campus. *Source* [1]

one interconnected urban space. The engaging design of the building is well appreciated by both students and academicians who believe that the Kolding Campus may pave the way for future of the educational institutions across the globe.

# 3 Climatic Data: Kolding, Denmark

## *3.1 Temperature and Rainfall*

Kolding has a warm and temperature climate with frequent rainfall throughout the year. The average temperature in Kolding is 7.8 °C or 46.0°F. It receives about 750 mm of precipitation annually.

**Table 1** Kolding weather by month// weather averages. *Source* [7]

| | Jan | Feb | March | April | May | June | July | Aug | Sept | Oct | Nov | Dec |
|---|---|---|---|---|---|---|---|---|---|---|---|---|
| Avg. Temp (°C) | −0.3 | −0.1 | 2.3 | 6.2 | 10.9 | 14.3 | 15.8 | 15.6 | 12.8 | 9 | 4.3 | 2.5 |
| Min Temp (°C) | −3.2 | −3.3 | −0.8 | 2.7 | 7 | 10.4 | 12 | 11.5 | 9.3 | 6.1 | 1.2 | 0.9 |
| Max Temp (°C) | 2.7 | 3.1 | 5.4 | 9.8 | 14.9 | 18.3 | 19.6 | 19.8 | 16.4 | 12 | 7.4 | 4.2 |
| Min Temp (°C) | 31.5 | 31.8 | 36.1 | 43.2 | 51.6 | 57.7 | 60.4 | 60.1 | 55.0 | 48.2 | 39.7 | 36.5 |
| Min Temp (°F) | 26.2 | 26.1 | 30.6 | 36.9 | 44.6 | 50.7 | 53.6 | 52.7 | 48.7 | 43.0 | 34.2 | 33.6 |
| Max Temp (°F) | 36.9 | 37.6 | 41.7 | 49.6 | 58.8 | 64.9 | 67.3 | 67.6 | 61.5 | 53.6 | 45.3 | 39.6 |
| Precipitation (mm) | 59 | 40 | 48 | 44 | 50 | 57 | 70 | 70 | 75 | 83 | 85 | 69 |

From Table 1 it can be inferred that the months of July and August are the warmest months when the maximum temperature may go up to 19.8 °C or 67.6°F. Likewise, the months of January and February are the coolest when the minimum temperature may fall to −3.3 °C or 26.1°F. The months of October and November receive the highest precipitation which may range from 83 mm to 85 mm, respectively.

### 3.2 Sun Shading Chart

From the sun shading chart of Kolding, it is evident that the sun is low in the sky for most part of the year. Only in the months of July and August is the sun at its highest position in the sky in the Kolding region. Due to the low position of the sun, the sunlight will strike the ground at a slanted angle and will cause less warmth.

It can also be inferred that the days are shorter during the winter months and have only 7 h of the sun. During the summer months, the days are comparatively longer with almost 15 h of the sun.

### 3.3 Requirements for Buildings Based on Climatic Data

- The buildings in this region should be designed to let in maximum sunlight during the winter months.
- At the same time, the entry of excessive sunlight in the building during the summer months should be avoided.
- Artificial heating mechanisms may have to be additionally provided in the building during the winter months. The need for these artificial mechanisms can also be avoided by using passive solar building design.
- The choice of a vertical shading element would prove to be more effective than a horizontal shading element because of the low position of the sun in the sky for most part of the year.

## 4 Sustainable Features and Their Compatibility with Local Climate

### 4.1 Passive Design Features

The building's orientation has been done to let in maximum daylight into the building. The front (main) façade that faces the road has been oriented in the northwest direction. The rest of the two façades face the south and the northeast direction, respectively. The façades on all sides of the building have been equipped with vertical kinetic shading device to regulate the amount of daylight and heat entering the building. The shape and structure of the building have also been designed to minimize energy-intensive technology. The triangular geometry of the building ensures the best utilization of every square meter. An atrium has been given in the building immediately adjacent to the front façade to let in daylight and minimize the need of artificial lighting systems during the daytime hours.

The landscaping in the building has also been done keeping in mind the climatic needs of the region. Trees have been planted away from the façades to avoid any obstruction for channeling in the daylight. The green areas on front are low heighted and have been tied together in an ecological infrastructure.

The building has been part of a three-year development project which examines how thermal properties of concrete can be increased—and the energy consumption for heating and cooling thus reduced. In order to make optimal use of the thermal properties of concrete, the slabs are exposed where possible. This prevents large fluctuations in temperature and improves the indoor quality [18]–2).

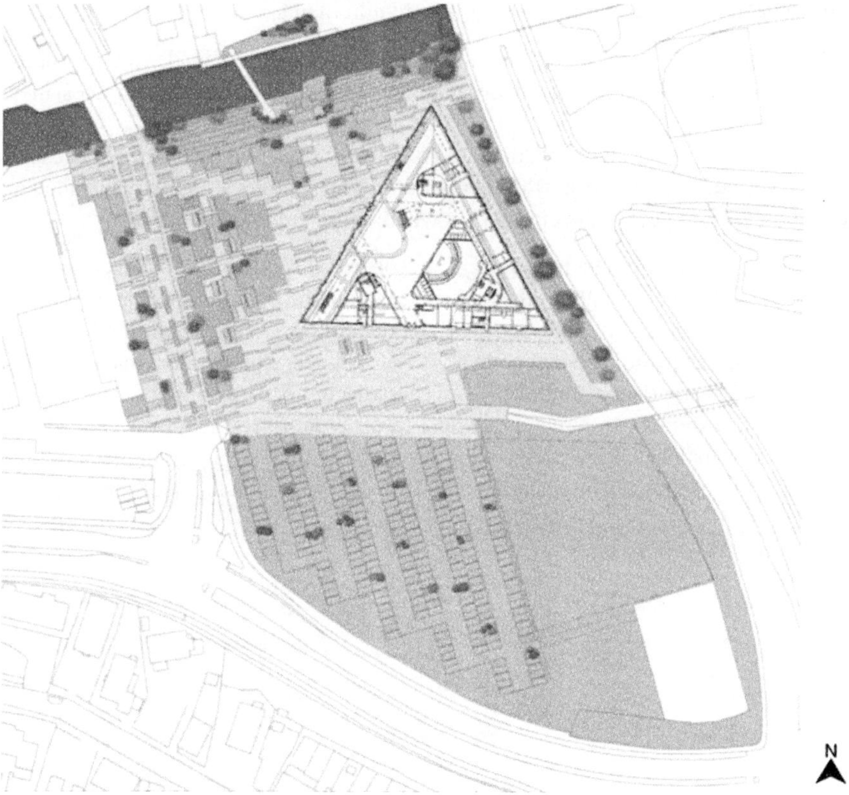


**Fig. 2** The site plan of the Kolding University SDU. *Source* [1]

## *4.2 Kinetic Façade Features*

The façades in the Kolding Campus are covered with kinetic vertical shading devices made of 1600 perforated aluminum panels that also act as shutters. When not in use, these shutters lie flat along the façade. When opened, they protrude from the façade at certain angles depending upon the need of daylight and heat in the building. These panels are mounted on a galvanized steel frame that projects outwards to a distance of 600 mm from the building. This 600 mm gap is also used as a maintenance service corridor for the kinetic façade. The double-skin composition comprising the steel frame and panels also provides flexibility for the future addition of pipes and ducts without affecting the landmark appearance of the building [5].

**Orientation of shutters based on direction of façades**. Shutters have been arranged on the façades keeping in mind the climatic requirements.

*South and Northeast Façade.* The shutters on most part of the façades are movable. At places where these shutters provide shading to toilets, staircases and terraces, they are fixed at 30°–60°.

*Northwest Façade.* There are comparatively fewer shutters on this façade because of the lower heat load in the northwest direction. The shutters are mostly installed at an angle of 90°, expect at the classrooms where these are movable in nature.

**Operation**. The kinetic shading devices are operated by sensors that measure the light and heat levels throughout the day and regulate the opening mechanism of the shutters. These sensors monitor and send the data to the Central Building Management System. The Building Management System triggers small motors that are either hidden at the base of each shutter or between adjacent pair of shutters. The motors then cause the shutters to move and adjust to different daylight scenarios, thereby optimizing the balance between the natural and the artificial lighting in the building. These shutters can also be operated manually depending upon the needs of the users

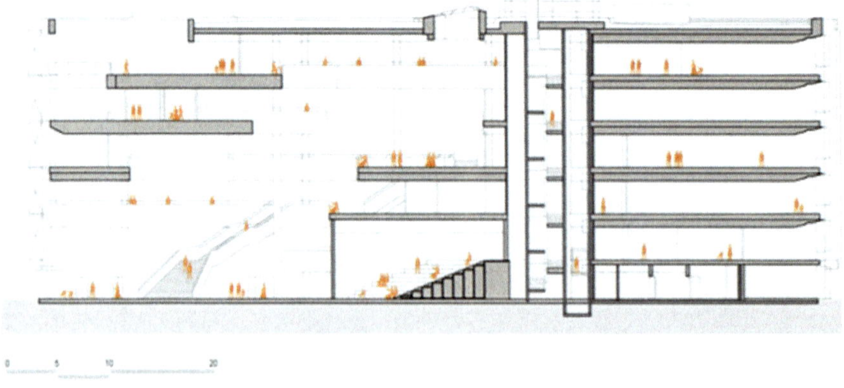

**Fig. 3** Section of the building. *Source* [1]

in the building. Though their operation is controlled by electric motors, their contribution to the overall energy consumption of the building is significantly less as they move gradually, remaining at one fixed position for several hours [8].

**Design of Shutters**. The aluminum shutters are made of approximately 4,500 $m^2$ of perforated aluminum sheets. The perforations in the façade are additionally designed and adapted to an opening angle of approximately 30%. Engineers and architects have conducted analysis and calculations to establish this as the optimal opening angle in relation to the amount of light and energy let in and out of the building, while at the same time providing users with optimal views to the outside urban space (McManus, Kolding Campus University of Southern Denmark, 2020).

Even when the shutters are fully closed, the perforations allow a controlled amount of light to enter the building. Additionally, the shutters also contribute visually to emphasis on the kinetic nature of the building's façades. The shutters are additionally a part of an artwork created by German Artist Tobias Rehberger. The artwork revolves around the dynamic nature of the building and is themed on time. Shutters have been painted in different colors as a part of the artwork. These shutters further close or open to indicate 1 min, 10 min or 60-minute intervals and help visitors to determine the time of the day by just looking at the façade. The varied adjustments of the shutters based upon the different weather conditions daily additionally render a unique look to the building every day.

The overall kinetic system is less complex when compared to other shading systems [10]. The system manages to balance the inflow of daylight into the building without any negative consequences such as heat loss or increased requirement for cooling. Additionally, these shutters also act as windbreakers in the winter months [15, p. 469].

**Fig. 4** The colored shutters on the façade. *Source* [1]

# 5 Energy Consumption

## *5.1 Energy Consumption of Kolding Campus*

Almost 70–75% of the energy consumption of the building was locked in the early design phases through the means of passive design strategies and the kinetic façade. The building with a floor area of 13,700 $m^2$ uses a total energy of 38 $KWh/m^2$/year and meets the requirements of Energy Class 1 Buildings, making the Kolding Campus one of the world's first low energy universities.

Campus Kolding is the first university building in Denmark to meet stringent energy consumption demands described in 2015 building [8]. The University Campus was also awarded the prestigious LEAF Award in the category of Best Sustainable Development of the year 2015.

## *5.2 Comparison of Energy Consumption with Other Buildings*

The energy consumption of Kolding Campus will be compared to other buildings in the European region with similar climatic conditions. For this comparison, the total floor area and the primary energy consumption has been taken into account that measures the total domestic energy demand for the building.

**Cologne Oval Office, Germany**. The office consists of two sibling office buildings that stand in stark contrast to their surroundings in the business district of Cologne. Each building has a double-skin ventilated façade that supports automated vertical louvers. With a total area of 43,000 $m^2$, the offices have a total energy consumption of 105 $KWh/m^2$/year.

**Palazzo Lombardia, Milan, Italy**. Palazzo Lombardia's distinctive form is composed of sinuous interweaving strands of linear office space, seven to nine stories in height. Each office building has green roofs and active climate walls with vertical blades that rotate to provide shade. With a total floor area of 72,000 $m^2$, the building has a total energy consumption of 91 $KWh/m^2$/year.

**KfW Westerkade, Frankfurt, Germany**. The 56 m high office building located in the west end of Frankfurt contains conference facilities and office spaces. The building has a double-layered wind-pressurized façade that has high insulation values and offers natural ventilation. With a total floor area of 39,000 $m^2$, the building has a total energy consumption of 100 $KWh/m^2$/year.

**Conclusion**. The comparison of the energy consumption of the Kolding Campus with other buildings with kinetic façades suggests that its energy consumption is roughly equal of the energy consumption of other buildings in Europe such as Cologne Oval Office Germany and KfW WesterKade Frankfurt. However, its energy savings can be optimized by introducing more energy-efficient mechanisms as done in Palazzo Lombardia in Milan.

## 6 Effect of Sustainable Features on the University Students

Since a lot of students willing to take admission in the Kolding Campus will only know the university by images available online, the design has to be visually appealing in addition to being sustainable [10]. Therefore, although the façade of the Kolding Campus is kinetic in nature, attention to the visual characteristics of the building has also been given. This has in turn created a unique image of the Kolding Campus in front of the students.

In his article on "Perforated Triangular screens in Kolding Campus", published on the website of the RIJAB Journal, Stephen Cousins compares the setting of the University building with a science fiction novel. He says that watching a time-lapse video of the façades gives an impression of watching an obscure time peace of some advanced alien civilization. He further describes the movement of the three façades as a kaleidoscope of butterflies that are about to take flight from the building.

Thus, incorporating dynamic mechanisms like these in the design of educational institutes can greatly help students understand the importance of sustainable lifestyles in a playful manner.

## 7 Conclusion

The paper sheds light on the aspect of kinetic architecture and its subsequent incorporation in the Kolding Campus University in the form of kinetic façades. The harmonious blend of the kinetic mechanisms of the façade along with the overall design renders an exceptional look to the building. Further, the compatibility of the sustainable design elements with the local climate drastically reduces the energy consumption. This excellent integration of innovative technologies in architecture subtly exhibits the true potential of kinetic architecture in the buildings of the present and the future. It can also be concluded that the scope of these transformable mechanisms is not merely limited to the tangible components of the built environment. If used wisely in the designs, these mechanisms can play a major role in creating awareness and changing the mindsets of the general public, as seen in the case of the University students at Kolding Campus.

## References

1. Arcdog (2017) SDU University of Southern Denmark Campus Kolding. Arcdog. http://arcdog.com/portfolio/sdu-university-of-southern-denmark-campus-kolding/
2. Archdaily (2013) Kfw Westarkade/ Sauerbruch Hutton. Archdaily. https://www.archdaily.com/316143/kfw-westarkade-sauerbruch-hutton
3. Archello (n.d.) Kolding Campus- Henning Larsen as Architects. Archello. https://archello.com/project/kolding-campus
4. Architects HL (2015) BEST SUSTAINABLE DEVELOPMENT OF THE YEAR: SDU CAMPUS KOLDING. State of the Green. https://stateofgreen.com/en/partners/henning-larsen-architects/news/best-sustainable-development-of-the-year-sdu-campus-kolding/
5. Asefi FF(2019) Environmentally responsive kinetic faÇade for educational buildings. J Green Build, 167–185
6. BUILDING CASE STUDIES- SDU Building (2018) Open Scholarship. https://openscholarship.wustl.edu/bcs/380/
7. Climate-Data.Org (2015) KOLDING CLIMATE (DENMARK). Climate-Data.Org. https://en.climate-data.org/europe/denmark/region-of-southern-denmark/kolding-300/#:~:text=Kolding%20climate%20summary,climate%20is%20classified%20as%20Cfb
8. Cousins S (2015). Perforated triangular sun screens move throughout the day to control light and heat at Henning Larsen Architects' highly energy efficient university building. RIJAB J. https://www.ribaj.com/products/campus-kolding-university-of-southern-denmark
9. Gavelle P (2017) *SDU* UNIVERSITY – Facade. More More Exless. http://moremorexless.blogspot.com/2017/01/sdu-university-of-southern-denmark.html
10. Kotryna Zvironaite AK (2014) A playful way through the world of moving facades. In: KINETICA, pp 1–31)
11. Larsen H (Director) (2014) Visiting Campus Kolding [Motion Picture]
12. Library B (n.d.) Oval Offices. Buildings Library. https://www.ajbuildingslibrary.co.uk/projects/display/id/3475
13. Mara F (2011) Oval Offices, Cologne, by Sauerbruch Hutton. Architect's J. https://www.architectsjournal.co.uk/archive/oval-offices-cologne-by-sauerbruch-hutton
14. McManus D (2020) Kolding Campus University of Southern Denmark. E-Architect. https://www.e-architect.co.uk/denmark/kolding-campus-university-southern-denmark
15. Nitisha Vedula a) LA (2018) Designing adaptive facades with a new holistic eco design approach. Facade 2018-Adaptive, pp 465–471
16. Palazzo Lombardia (2010) Wikipedia. https://en.wikipedia.org/wiki/Palazzo_Lombardia
17. Razaz ZE (2018) Sustainable vision of kinetic architecture. J Build Apprais, pp 341–356
18. SDU - University of Southern Denmark, Campus (2018) Skyfish.com. https://www.skyfish.com/p/henninglarsenarchitects/574600/16805170/preview#:~:text=SDU%20Campus%20Kolding%20is%20the,integrated%20in%20the%20architectural%20design
19. SDU Campus Kolding / Henning Larsen (2015). Archdaily. https://www.archdaily.com/590576/sdu-campus-kolding-henning-larsen-architects
20. Sejr K (2020) UNIVERSITY OF SOUTHERN DENMARK–CAMPUS KOLDING. Danish Association of Architectural Firms. https://www.danskeark.com/content/university-southern-denmark-campus-kolding
21. University of Southern Denmark–Campus Kolding (n.d.). Henning Larson. https://henninglarsen.com/en/projects/featured/0942-sdu-campus-kolding
22. Wiki DB (2020) Kinetic Facade. Designing Buildings Wiki. https://www.designingbuildings.co.uk/wiki/Kinetic_facade#:~:text=A%20kinetic%20fa%C3%A7ade%20is%20one,occur%20on%20a%20building's%20surface

# Chapter 24
# Performance Analysis and ANN Modelling of Apple Drying in ETSC-Assisted Hybrid Active Dryer

**Gaurav Saxena, M. K. Gaur, and Anand Kushwah**

## 1 Introduction

Using the power of the Sun for the preservation of fruits and other foodstuff from agricultural has been in practice for centuries. Among fruits, apple is one of the most commercially sold fruit and is consumed as raw, fresh/preserved fruit juice, thin dried slices etc. The consumption of apple is done in various ways as fresh fruit, preserved juice, slices etc. Doymaz developed a dryer based on electrical energy for studying the drying behaviour of green apples. Heated air for drying from an electric resistance and a temperature controller was supplied by means of a centrifugal fan. Pre-treatment of slices was done at room temperature in 4% citric acid and later blanched in hot water for 3 min duration. The results obtained from thin-layer mathematical modelling showed that the Wang and Singh, logarithmic, and Verma et al. models provided most suitable fit with the experimental data. Pre-treated slices in the citric acid solution showed greater effective moisture diffusivity in comparison to the other samples [1].

Polat et al. determined the drying characteristics and other quality parameters by using eight drying treatments based on microwave power, vacuum treatments and pulse ratios. Page model showed best curve fitting for drying kinetics [2]. Yaun et al. investigated the drying shrinkage-deformation mechanism of apple slices during process of heat and mass transfer. Image analysis was used to obtain the shrinkage deformation values of apple slices. The software solver COMSOL using mathematical models of stress–strain as well as heat-mass transfer was used to simulate the

G. Saxena (✉) · M. K. Gaur · A. Kushwah
Department of Mechanical Engineering, MITS, Gwalior, Madhya Pradesh, India 475005
e-mail: gauravsaxena@rjit.org

M. K. Gaur
e-mail: gmanojkumar@rediffmail.com

A. Kushwah
e-mail: anand.kushwah1989@gmail.com

© The Author(s), under exclusive license to Springer Nature Singapore Pte Ltd. 2022

H. M. Dubey et al. (eds.), *Artificial Intelligence and Sustainable Computing*, Algorithms for Intelligent Systems, https://doi.org/10.1007/978-981-16-1220-6_24

drying process by hot air. The result showed moisture stress as the key parameter causing the shrinkage deformation of the apple slices during drying process [3].

The first solar-based dryer was developed by Everitt and Stanley in 1976 [4]. In the subsequent decades, several researchers have worked on the performance and designs of solar drying systems. From the advent the dependency on solar energy as the single input energy is favourable only for low capacity dryers where the possibility of damage to the crops due to unfavourable weather is less [5–7]. The review showed the following major limitations of conventionally used hot air solar dryer with no auxiliary power source to compensate non-constant supply of solar heating:

a. Low outputs due to lesser moisture removal rate,
b. Influence of climatic conditions on the rate of drying and the quality of final dried products,
c. Dependency on available sunshine hours of the day,
d. No measure to control the temperature of the dryer as per the drying requirements,
e. Slower drying rates and lesser efficiencies.

The researchers mainly used two approaches to overcome the above limitations which are derived from non-continuous accessibility of solar energy. These are enhancing the efficiency of existing dryers and use of auxiliary energy source. The later approach of the usage of auxiliary system such as solar collectors, geothermal system, biomass, waste heat recovery or a system of two or more, i.e. hybrid drying as presented in Fig. 1, assist in achieving the sustainable solar drying.

Considerable research had already been carried out on the solar dryers based on various types of collector and designs. To ensure continuous drying until moisture content get reduced to desirable level and to mitigate the intermittent nature of the solar energy received by the solar dryers, integration of auxiliary heating devices or heat storage devices is considered favourable.

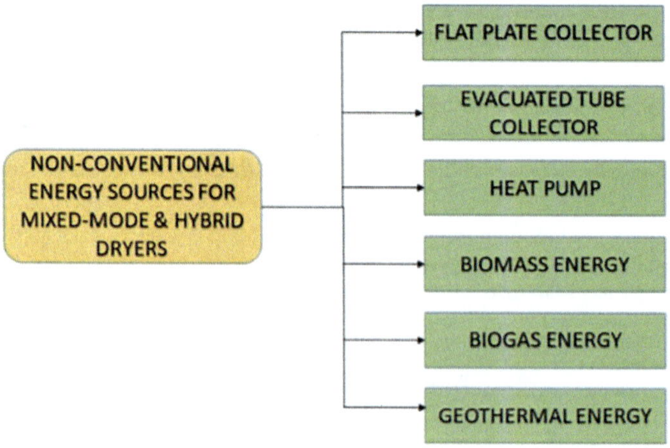


**Fig. 1** Strategies for non-conventional-assisted solar drying system

### 1.1 Evacuated Tube Hybrid Drying Systems

Flat plate type solar collector is widely used in solar dryers for generating hot air to dry the products as compared to solar evacuated collector. Recently some researchers have used evacuated tube heating as an assistance for solar dryers and achieved favourable results as.

Solar vacuum tubes were used by Mahesh et al. to heat, stream flow of atmospheric air to higher temperature. The setup was tested for drying of various agricultural samples of fruit and vegetable. The result showed that the direct sun drying takes almost 1.5 times as compared to the designed vacuum tubes type dryer [8].

An experimental investigation for the thermal performance of evacuated tube-based solar dryer having shell and tube heat exchanger and drying chamber as main components was performed by Singh, Vyas, and Yadav (2018). The result showed 35.4 °C as the maximum difference in temperature between the ambient air and hot air and 55% as the maximum evaluated efficiency of the setup [9]. Iranmanesh et al. used evacuated tube collector and thermal storage system in a solar cabinet dryer for apple drying. The performance of the dryer was determined at three flow rates: 0.025, 0.05 and 0.09 kg/s of air for system with and without thermal storage. The result showed that the integration of thermal storage material with evacuated tube collector increases the input thermal energy by about 1.72% for the air flow rates of 0.025 and provided an efficiency of 39.9% [10]. Daghigh et al. determined the best model for drying of Tarkhineh (an Iranian food) using evacuated tube (ET) collector with photovoltaic thermal (PV/T) in the designed setup of indirect solar dryer. The developed new models provided best results for both open sun drying and drying in the dryer. The payback period of ET integrated system was 2.5 years as compared to the PV/T solar dryer [11].

The review showed that the desired drying air temperature is achieved by using various conventional as well as modern hybrid systems in mixed-mode drying. Both flat plate solar collectors followed by the solar evacuated tube heating system are generally used for the development of solar-assisted hybrid drying systems. Aim of the present work is to determine the performance of greenhouse dryer assisted with evacuated tube solar heaters and also to determine drying kinetics for drying fruits (apple). Besides this, the most suitable model for thin layer drying is also determined through curve fitting and neural network analysis.

## 2 Experimentation

### 2.1 Equipment

The complete assembly of hybrid greenhouse dryer (HSD) with evacuated water tube collector is installed on the roof of MITS, Gwalior, India (26.2183° N, 78.1828° E). Referring to Fig. 2a and b shows actual view of main components comprising

(a) (b)

**Fig. 2** Actual view of (**a**) installed hybrid solar drying system (**b**) drying tray loaded with the apple slices

arrangement for series flow of heated water inside the wire and tubes type drying platform placed inside the greenhouse dryer. The complete setup has two sections: Section-I has solar heating systems for maintaining supply of solar heated water inside the coils/tubes of greenhouse dryers and Section-II is the solar greenhouse dryer assembly equipped with solar heated drying platforms.

The schematic view of complete setup is presented in Fig. 3. The greenhouse setup consists of two-layer drying platform placed inside the greenhouse with floor area 200 × 185 $cm^2$ each. Each layer of drying platform has 17 arrays of U-type copper tubes, each having length of 196 cm for series flow of heated water. Steel wire mesh of size 1.2 × 1.2 mm and wire diameter 0.3 mm is placed over the copper tubes. This wire mesh gains heat by direct contact with copper tubes and transfer it to the drying product placed over it. Besides this, the greenhouse environment air also receive this additional secondary heat (primary source is greenhouse heating). The fresh air enters from a 15 mm height wire mesh passage at the bottom of the greenhouse. The desired greenhouse temperature is maintained by controlling the flow rate of heated water inside the copper tubes through a flow regulating valve concept, and the design is under patent [12]. The pumped circulation of solar heated water inside the copper tubes and active circulation of air inside the greenhouse are maintained

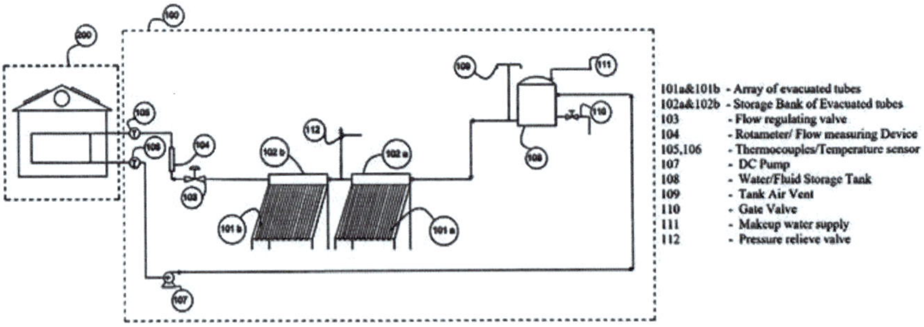


**Fig. 3** Schematic flow diagram of the design setup

by a 12 V/17 W DC pump and 4 numbers of 12 V/0.35 A DC fans, respectively. Both the devices are powered by 2 numbers of multi-crystalline 12 V/75 W solar PV panels.

## 2.2 Instrumentation and Measurements

A computerized data logger, Universal dataTaker DT85 equipped with dEX software was used to record the readings of fully functioning weather station equipped with hygrometer, anemometer, pyranometer and 09-channel thermocouples. Besides this, portable measuring devices like temperature gun, portable weight machine and portable anemometer is also for measurement of observations. The specifications of instruments used for data observations are listed in Table 1. The experimental observations of influencing surrounding conditions and greenhouse environment are recorded by the data logger of weather monitoring station after each regular interval of 30 min. The drying parameters for apple drying are presented in terms of observations of mass variation, moisture content M (kg of water per kg of dry matter), dimensionless moisture ratio (Mt/Mo) and drying rate in g/s for both greenhouse and open sun drying in Table 2.

**Table 1** Specifications of measuring instruments

| Instrument | General specification | Range | Accuracy | Uncertainty |
|---|---|---|---|---|
| 1-Weather monitoring station | | | | |
| a-Anemometer | 3 cup rotor attached to a chopper and IR emitter/detector circuit | 0–60 m/s | ±2% of full scale reading | —— |
| b-Pyranometer | 72 element copper constantan thermopile | 0–1500 W/m$^2$ | …………… | ……….. |
| c-Thermocouples | Standard Platinum RTD element (PT-1000, or PT-100 covered by a weather shield) | −40 °C to + 80 °C | 0.1 °C | ±0.25 °C |
| 2-Temperature gun thermometer | Infrared non-contact digital temperature Gun | −50 °C to 380 °C/− 58 °F to 716 °F | ±1.5 °C | ±0.1 °C |
| 3-Weight machine | Digital shop weighing machine | 0.001–30 kg | ——— | ±0.0001 kg (0.1 gm) |
| 4-Data logger | Make:-DataTaker; dEx software interface | —— | ——— | ——— |

**Table 2** Drying performance parameter for apple drying

| Time of the day (hrs) | Open sun drying | | | | Solar hybrid greenhouse drying | | | |
|---|---|---|---|---|---|---|---|---|
| | Mean sample Temp. (°C) | Moisture content, M (gram of water per gram of dry matter) | Dimensionless moisture ratio (Mt/Mo) | Mean drying rate[a] gm/min | Mean sample Temp. (°C) | Moisture content, M (gram of water per gram of dry matter) | Dimensionless moisture ratio (Mt/Mo) | Mean drying rate[a] gm/min |
| *Day-1* | | | | | | | | |
| 11:00 | 26 | 6.14 | 1.00 | 0.767 | 31.5 | 6.14 | 1.00 | 0.933 |
| 11:30 | 28.2 | 5.32 | 0.87 | 0.833 | 29.9 | 5.14 | 0.84 | 1.033 |
| 12:00 | 28.9 | 4.43 | 0.72 | 0.267 | 31.6 | 4.04 | 0.66 | 0.533 |
| 12:30 | 28.2 | 4.14 | 0.67 | 0.467 | 35.1 | 3.46 | 0.56 | 0.433 |
| 01:00 | 24.7 | 3.64 | 0.59 | 0.333 | 35.8 | 3.00 | 0.49 | 0.400 |
| 01:30 | 30.2 | 3.29 | 0.54 | 0.367 | 36.2 | 2.57 | 0.42 | 0.467 |
| 02:00 | 27.2 | 2.89 | 0.47 | 0.300 | 33.5 | 2.07 | 0.34 | 0.333 |
| 02:30 | 27.9 | 2.57 | 0.42 | 0.200 | 31.7 | 1.71 | 0.28 | 0.200 |
| 03:00 | 25.3 | 2.36 | 0.38 | 0.167 | 27.8 | 1.50 | 0.24 | 0.167 |
| 03:30 | 25.2 | 2.18 | 0.35 | 0.067 | 30.8 | 1.32 | 0.22 | 0.067 |
| 04:00 | 24.6 | 2.11 | 0.34 | ……… | 30.7 | 1.25 | 0.20 | ……… |
| *Day-2* | | | | | | | | |
| 11:00 | 27.6 | 2.11 | 0.34 | 0.233 | 40.5 | 1.250 | 0.20 | 0.100 |
| 11:30 | 31.5 | 1.86 | 0.30 | 0.200 | 36.5 | 1.143 | 0.19 | 0.167 |
| 12:00 | 34.2 | 1.64 | 0.27 | 0.233 | 41.5 | 0.964 | 0.16 | 0.133 |
| 12:30 | 31.7 | 1.39 | 0.23 | 0.200 | 43.2 | 0.821 | 0.13 | 0.133 |
| 01:00 | 33.5 | 1.18 | 0.19 | 0.200 | 41.9 | 0.679 | 0.11 | 0.167 |
| 01:30 | 33.1 | 0.96 | 0.16 | 0.233 | 39.2 | 0.500 | 0.08 | 0.033 |
| 02:00 | 35.5 | 0.71 | 0.12 | 0.033 | 42.5 | 0.464 | 0.08 | 0.033 |
| 02:30 | 32.6 | 0.68 | 0.11 | 0.033 | 40.6 | 0.429 | 0.07 | 0.033 |
| 03:00 | 34.3 | 0.64 | 0.10 | 0.033 | 40 | 0.393 | 0.06 | 0.000 |
| 03:30 | 30 | 0.61 | 0.10 | 0.000 | 38.2 | 0.393 | 0.06 | 0.000 |
| 04:00 | 30.3 | 0.61 | 0.10 | ……… | 36.4 | 0.393 | 0.06 | ……… |

[a]The value of drying rates corresponding to each 30 min interval is constant

## 3 Dryer Performance Parameters

The performance of the mentioned concept is evaluated by two indexes: surface temperature of crop and drying performance.

### 3.1 Surface Temperature of Crop

The thermal performance of the dryer is expressed in terms of temperature rise of crop surface with respect to time. This parameter is selected for representing the thermal performance of the system as it directly affects the drying rate. The relative rise in greenhouse environment temperature is used for comparative analysis of present setup with the designs developed by the other researchers.

### 3.2 Drying Performance

The performance of the present dryer assisted by auxiliary solar system is represented in terms of mass reduction and its influence on moisture content and drying rate. For research purpose under engineering, the dry basis moisture content is preferred, since the weight change associated with each percentage point of moisture reduction on dry basis becomes constant [13]. It is defined as the weight of moisture present in a produce or sample per unit weight of the bone dry material in the produce and is given by Eq. 1 as:

$$M_{db} = \frac{W_t - W_d}{W_d} = \frac{W_m}{W_d} \tag{1}$$

where $W_t$, $W_d$ and $W_m$ are the weight of the sample at time t, bone dry weight and weight of moisture at time t, respectively.

The drying rate is evaluated using Eq. 2:

$$\dot{m} = {}^{dm}\!/_{dt} = \frac{M_t - M_{t+\Delta t}}{\Delta t} \tag{2}$$

where $M_t$ and $M_{t+\Delta t}$ are the moisture content at time $t$ and $t + \Delta t$, respectively. Here, the values are determined for each consecutive time intervals of 30 min.

## 4 Mathematical Modelling of Thin-Layer Drying Equation

For the investigation of drying characteristics of apple crops, dimensionless moisture ratios of samples during drying are calculated by the following Eq. 3 as:

$$M_R = \frac{M_t}{M_o} = \frac{m - m_d}{m_o - m_d} \tag{3}$$

where $M_t$ and $M_o$ are the moisture content of the product (dry basis) at time t and t = 0, respectively; here $m_o$ and $m$ represent the measured starting mass of the product at time t, respectively. Whereas $m_d$ represents the mass of the dry matter of the product.

Suitable mathematical model is derived, under direct conductive heat transfer from drying bed as well as auxiliary heat supply to the greenhouse environment. Table 3 shows models derived from two theories, viz., Newton's law of cooling and Fick's second law of diffusion as well as other empirical relations which are used for regression through MATLAB.

The best results to fit the model could be obtained with the biggest values of coefficient of determination $R^2$ and minimum values of the root mean square error. This can be estimated as:

$$R^2 = 1 - \frac{\sum_{i=1}^{N}\left(MR_{pre,i} - MR_{exp,i}\right)^2}{\sum_{i=1}^{N}\left(\overline{MR_{pre}} - MR_{exp,i}\right)^2} \tag{4}$$

$$RMSE = \left[\frac{1}{N}\sum_{i=1}^{N}\left(MR_{pre,i} - MR_{exp,i}\right)^2\right]^{1/2} \tag{5}$$

Here, $MR_{exp,i}$ is experimental moisture ratio, $MR_{pre,i}$ the predicted moisture ratio and $N$ is the number of observations.

## 5 ANN Model for Dimensionless Moisture Ratio

In order to obtain the best prediction from the expected values obtained from most suitable MATLAB simulations, several networks were trained and evaluated using the input and target values through ANN tool of MATLAB R 2018a. Various numbers of multilayer feed-forward back propagation (FFBP) were designed and tested using combination of different hidden layers (1 or 2) and neurons. Levenberg-Marquardt (LM) learning algorithms were used for training the network in ANN. The activation/transfer-sigmoid function for the first hidden layers was selected to be logarithmic (Log) while tangent (Tan) functions suited best for the second hidden layer and output layer. This arrangement of transfer functions in neural network approximation problems is common and provides better results [25]. The performance of predicted ANN models is measured using mean square error (MSE) and correlation coefficient (R) between the predicted values of the network and the target or experimental values, as given by the Eqs. 6 and 7:

$$MSE = \frac{\sum_{i=1}^{N}(MR_{pre.o,i} - MR_{exp.o,i})^2}{N} \tag{6}$$

**Table 3** Empirical models to derive drying kinetics

| Model no. | Model name | Model | References |
|---|---|---|---|
| *I—Models Resulting from Newton's Law of Cooling* | | | |
| 1 | Lewis | $M_R = e^{-kt}$ | Lewis [14] |
| 2 | Page | $M_R = e^{-kt^n}$ | Page [15] |
| *II—Models Resulting from Fick's Second Law of Diffusion* | | | |
| 3 | Henderson and Pabis | $M_R = ae^{-kt}$ | Henderson and Pabis [16] |
| 4 | Modified Henderson and Pabis | $M_R = ae^{-kt} + be^{-gt} + ce^{-ht}$ | Karathanos [17] |
| 5 | Logarithmic (Asymptotic) | $M_R = ae^{-kt} + c$ | Chandra and Singh [18], Yagcioglu [19] |
| 6 | Two-term | $M_R = ae^{-k_1 t} + be^{-k_2 t}$ | Henderson [20] |
| 7 | Midilli and Kucuk | $M_R = ae^{-kt^n} + bt$ | Midilli et al. [21] |
| 8 | Modified Midilli model | $M_R = ae^{-kt^n} + b$ | Ghazanfari et al. [22] |
| 9 | Approximate Diffusion model | $M_R = ae^{-kt} + (1 - a)e^{-kbt}$ | Sharaf-Eldeen et al. [23] |
| *III—Other Empirical Model* | | | |
| 10 | Wang and Singh | $M_R = 1 + at + bt^2$ | Wang and Singh [24] |

$a, b, c, h, g, k, k_1,$ $and$ $k_2$ are the empirical constant in drying models; t is the drying time in hours

$$R = \sqrt{1 - \frac{\sum_{i=1}^{N}\left(MR_{pre.o,i} - MR_{exp.o,i}\right)^2}{\sum_{i=1}^{N}\left(MR_{pre.o,i} - \overline{MR_{exp.o}}\right)^2}} \quad (7)$$

where $MR_{pre.o,i}$ is the predicted output from observation i, $MR_{exp.o,i}$ is the experimental (target) output from observation i, $\overline{MR_{exp.o}}$ is the average value of experimental output, and total number of data observations is N.

# 6 Uncertainty Analysis

In the present experimentation work, investigation of observations like sample weight, temperature, solar isolation, etc. is carried out. Each measuring instrument utilized for the measurement of observations had its own accuracy (mentioned in Table 1), and this results in an uncertainty in the calculated values of experimental outcomes which are dependent. Different parameters were measured in the present experimental work for the evaluation of moisture content and other drying performance parameter. The result ($R$) is the specified function of the independent variables $x_1, x_2, x_3 \ldots x_n$. If the given parameter has uncertainty in the same order, then the resultant/final uncertainty will be evaluated as Eq. 8 [26, 27].

$$\delta_R = \sqrt{\left[\left(\frac{\partial R}{\partial x_1}\delta_{x1}\right)^2 + \left(\frac{\partial R}{\partial x_2}\delta_{x2}\right)^2 + \left(\frac{\partial R}{\partial x_3}\delta_{x3}\right)^2 + \ldots + \left(\frac{\partial R}{\partial x_i}\delta_{xi}\right)^2 + \ldots + \left(\frac{\partial R}{\partial x_n}\delta_{xn}\right)^2\right]} \quad (8)$$

Here, "$\delta_{xi}$" is the uncertainty in the $i$th variable and "$\frac{\partial R}{\partial x_i}\delta_{xi}$" is the contribution to the total uncertainty from each of the $i$th variable [28, 29].

The total/resultant uncertainty in the temperature measurement can result from digital thermometer, thermocouples, connection points of measuring instruments/sensors and readings. The total uncertainty values in temperature measurement will be expressed as Eq. 9:

$$\delta_{Temp,total} = \delta_{Tc} = \delta_{Tgh} =$$
$$\delta_{Ta} = \sqrt{\left[(\delta_{thermocouples})^2 + \left(\delta_{digitalthermometer}\right)^2 + \left(\delta_{connectionpoints}\right)^2 + \left(\delta_{reading}\right)^2\right]}$$
$$= \sqrt{\left[(0.25)^2 + (0.1)^2 + (0.1)^2 + (0.25)^2\right]} = \pm 0.381 \quad (9)$$

The total/resultant uncertainty in the time measurement can result from oscillation of time meter and reading. The total uncertainty in the time measurement may be calculated as Eq. 10:

$$\delta_{Time,total} = \delta_{tT} = \delta_{tml} = \sqrt{\left[(\delta_{oscillation})^2 + \left(\delta_{reading}\right)^2\right]}$$
$$= \sqrt{\left[(0.0003)^2 + (0.1)^2\right]} = \pm 0.1 \quad (10)$$

The total/resultant uncertainty in the measurement of loss in mass ($\delta_{ml}$) can result from digital machine, friction of tray and reading. The total uncertainty in the measurement of loss in mass will be expressed as Eq. 11:

$$\delta_{ml} = \sqrt{\left[\left(\delta_{digital\,machine}\right)^2 + \left(\delta_{reading}\right)^2 + \left(\delta_{friction}\right)^2\right]}$$
$$= \sqrt{\left[(0.1)^2 + (0.1)^2 + (0.5)^2\right]}$$
$$= \pm 0.52 \quad (11)$$

The total/resultant uncertainty in the measurement of the moisture content ($\delta_{mc}$) can be caused from moisture analyser and reading. This uncertainty may be expressed as Eq. 12:

$$\delta_{mc} = \sqrt{\left[\left(\delta_{moisture\,analyser}\right)^2 + \left(\delta_{reading}\right)^2\right]} = \sqrt{\left[(0.001)^2 + (0.001)^2\right]} = \pm 0.0014 \quad (12)$$

The propagation effect of above uncertainties on calculated values of moisture ratio ($\delta_{MR}$) and drying rate ($\delta_{DR}$) of hygroscopic crops may be evaluated from Eq. 13:

$$\delta_{MR} = \delta_{DR} = \sqrt{\left[(\delta_{tml})^2 + (\delta_{ml})^2 + (\delta_{mc})^2\right]}$$
$$= \sqrt{\left[(0.1)^2 + (0.52)^2 + (0.0014)^2\right]} = \pm 0.53 \quad (13)$$

The above results of final uncertainties in the measured parameters and calculated experimental parameters during the experimentation work showed that all uncertainties are in the permissible limit.

# 7 Results and Discussion

## *7.1 Dryer Performance Parameters*

The results of the observation recorded through experimentation and calculations are used to present the performance of the hybrid system in terms of two parameters: surface temperature of crop and drying performance. The maximum air temperature of 59.34 °C and a maximum relative rise in temperature of 33.4 °C are achieved at the upper zone of the greenhouse

**Surface temperature of crop:** As compared to the open sun drying the increase in mean crop surface temperature in terms of relative temperature differences are presented at every half hourly intervals for both the crops in Fig. 4. The maximum relative increase in crop surface temperature achieved was 12.9 °C at the solar intensity of 605.32 $W/m^2$ and dimensionless moisture ratio is 0.2. The result shows that as the dried product gets converted into partial solid state due to loss of bound and unbound moisture a thermal equilibrium state is achieved for both the crops, and temperature states for complete solid become nearly the same. This is due to cease in two processes, viz., cease in transfer of moisture to the solid surface and its subsequent evaporation due to the energy applied and further cease in the transfer of energy (mostly thermal) from the surrounding environment required to remove the moisture from the surface.

### Drying Performance

*Mass Reduction Rate:* The drying performance of the designed drying system is evaluated in terms of reduction in mass and its derived effect on moisture content and rate of drying. The reduction in mean sample mass or rate of reduction in bound and unbound moisture from the crops under consideration is shown in Fig. 5. The drying curve for reduction in mass shows complete drying time for apple drying reduced by 1.5 h for present sample size.

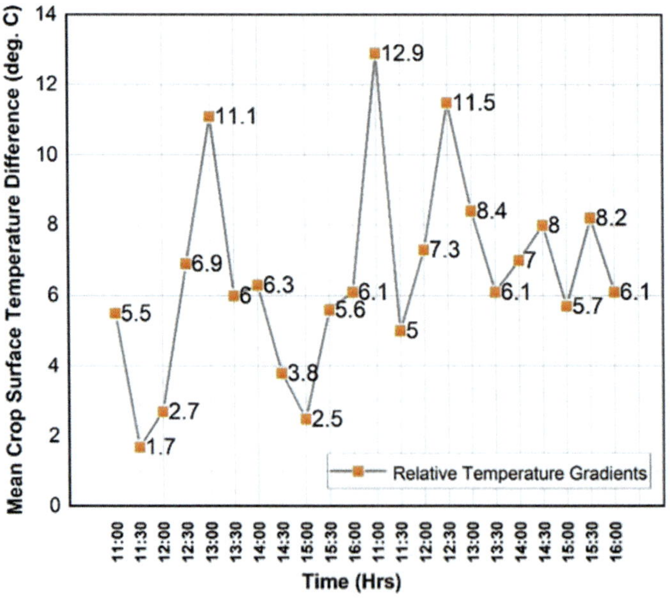


**Fig. 4** Increase in mean crop surface temperature as compared to the open sun drying

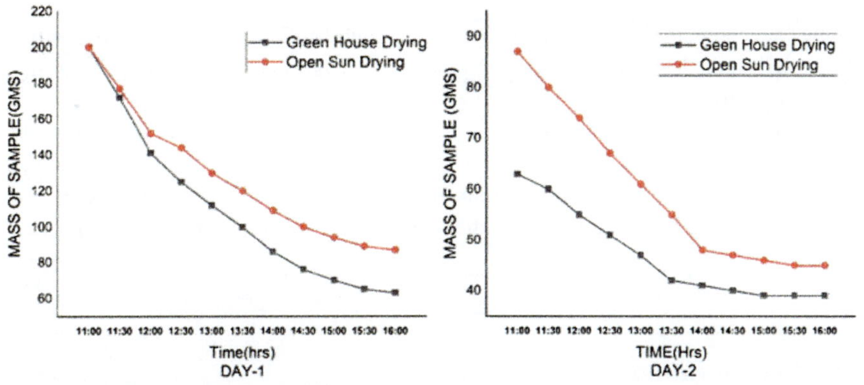


**Fig. 5** Variation in weight with time

*Drying Rate Periods:* The drying rate periods are presented by the variation curves of moisture content and drying rate. The drying rate for each 30-min interval is assumed constant during calculation as well as presentation. For fruit under consideration, the drying rate periods show change in slope of the curve in Fig. 6. The moisture phenomenon takes place in two stages: firstly from inner portion of the material to its surface through diffusion and later surface evaporation of moisture takes place. Apple being the high moisture content crop follows initially the constant drying rate period, thereafter the drying process takes place in the falling rate period. Similar observation was reported by Ahmed, Shivhare, and Singh [30]; Alibas [31].

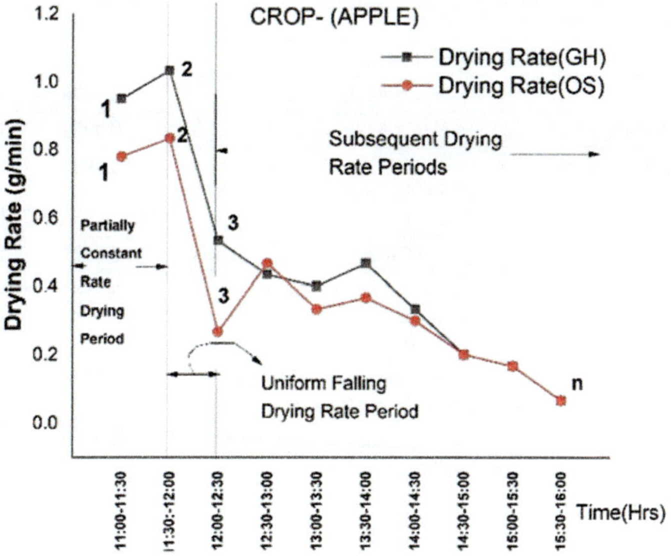


**Fig. 6** Drying rate periods for apple drying

Drying rate characteristics curve of day-1 were found sufficient to provide the initial phase of constant drying rate followed by subsequent variable-I and variable-II falling drying rate periods.

The intermittent transient variations of slope are considered as single drying rate period for presentation. The results showed that the moisture release rate for the hygroscopic fruits is dependent on solar intensity, but for present hybrid system the change in slope is comparatively steadier due to backup supply from auxiliary heat source.

## 7.2 Drying Kinetics

Curve fitting analysis is used for the determination of the constants of the mathematical models of Table 3 for corresponding practical values of dimensionless moisture ratio $\left({}^{M_t}/{}_{M_o}\right)$ and are presented in Table 4. The empirical relations are developed using these constant to determine the corresponding expected values, and followed by this, the experimental dimensionless moisture ratio is plotted against cumulative increment in time for each individual case of crop drying under hybrid greenhouse dryer. The curve are obtained by nonlinear least square method in the form

**Table 4** Results of curve fitting analysis

| Mathematical model | Constants and coefficients | Goodness of fit | |
|---|---|---|---|
| | | R-square | RMSE |
| *Crop-I (Apple)* | | | |
| Lewis | k = 0.3426 | 0.9924 | 0.02365 |
| Page | k = 0.3964; n = 0.8846 | 0.9976 | 0.01368 |
| Henderson and Pabis | a = 0.9664; k = 0.3304 | 0.9939 | 0.02174 |
| Modified Henderson and Pabis | a = 0.6841; b = −0.02737; c = 0.3438;g = 40.93; h = 0.8267; k = 0.2587 | **0.9981** | **0.01352** |
| Logarithmic | a = 0.9462; c = 0.04301; k = 0.3815 | 0.9972 | 0.015 |
| Two-term | a = 0.3798; b = 0.6245; k1 = 0.6848; k2 = 0.2492 | **0.998** | **0.01297** |
| Midilli and Kucuk | a = 1.006; b = 0.001603; k = 0.3983; n = 0.9061 | **0.9979** | **0.0136** |
| Modified Midilli | a = 0.9846; b = 0.02115; k = 0.4052; n = 0.9146 | **0.9979** | **0.01352** |
| Approx. diffusion | a = 0.7737; b = 0.9957; k = 0.343 | 0.9924 | 0.02493 |
| Wang and Singh | a = −0.2491; b = 0.0163 | 0.9546 | 0.05914 |

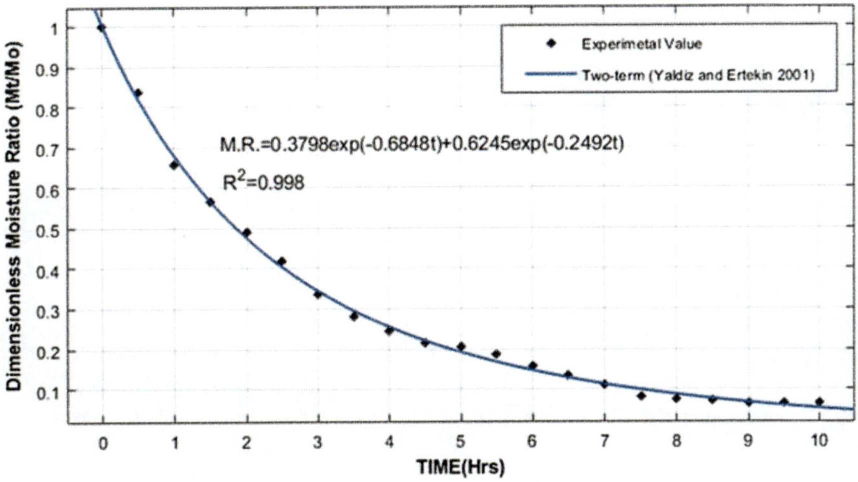


**Fig. 7** Plot of MR versus time for most suitable correlation

of mathematical drying models using MATLAB. The largest value of coefficient of determination $\left(R^2\right)$ shows close agreement to the experimental results.

The mathematical models providing closest curve fitting and having $R^2$ value nearest to 1 are shown in bold in Table 4. The most suitable regressed fitted curve obtained for drying of crop under hybrid greenhouse environment is shown in Fig. 7. The result shows that under present solar-assisted drying conditions models resulting from Fick's second law of diffusion, i.e., Two-term model is the most suitable for estimation of expected value of dimensionless moisture ratios.

## 8 ANN Model

The predicted values of most suitable mathematical correlation as input along with the included time (hour) and respective mean sample temperature (°C) are used in ANN to develop a static prediction model for determination of moisture ratio (MR). In this way configuration with 03 inputs and 01 target parameters are used in this study. The input, hidden and output layers are expressed in the index of neurons i, j and k, respectively. The input neurons receive the data and pass the output to the next hidden layer as shown in Fig. 8. The training results of the static model to predict the moisture content are shown in Table 5. This summarised table shows that the neural network with 8 and 8 neurons in the first and second hidden layers, respectively, provided best results for kinetic modelling of apple crop.

This model had highest values of RMSE and lowest values of R for train, validation and test. Similar research for development of ANN model was done by Motevali et al. using drying conditions and time as input for modelling and prediction of the moisture

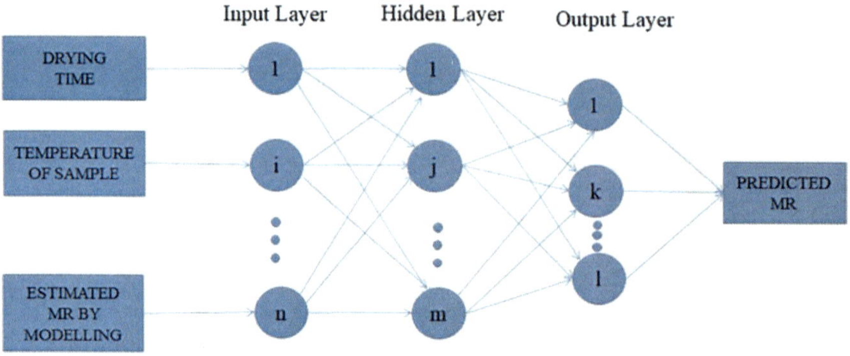


**Fig. 8** Configuration of neural network for predicting moisture ratio (MR) with hidden layer

ratio and drying rate [32]. The correlation obtained from optimum topologies based on the highest values of R and the lowest values of mean square errors is shown in Fig. 9.

## 9 Conclusion

The present solar-assisted PVT hybrid greenhouse dryer is tested for apple drying, the dryer utilizes the heat received from the evacuated tubes to maintain and control the temperature of the greenhouse environment according to the requirements of the crop to be dried. The flow rate of heated water regulated manually in the drying trays maintains the desirable backup supply of solar heat from the integrated system. The result of present experimental work showed maximum achieved air temperature of 59.34 °C in moderate winter season of November month and a maximum relative rise in temperature of 33.4 °C at the upper zone of the greenhouse, whereas nearly same rise in maximum relative temperature rise with respect to the ambient air of 35.4 °C was achieved by Singh et al. [9] using vacuum tube as air heating device. The drying curve for reduction in mass shows complete drying time reduced by 1.5 h for the present sample size of an apple. The drying rate curve shows the complete drying process takes place initially with a constant rate drying period followed by a variable falling rate period for both the cases. Models resulting from 04 models, viz., Two-term, Modified Henderson and Pabis, Midilli and Kucuk, and Modified Midilli showed good correlation for prediction of the expected value of dimensionless moisture ratio. The values obtained from the most suitable mathematical model derived from Two-term provided the best results for the determination of moisture ratio (MR) for a static neural network with 8 and 8 neurons, respectively, in the two hidden layers. The present hybrid dryer having provision for the flow of solar-heated water to achieve a wide range of the desired temperature at different drying platforms makes it suitable for drying quality and quantity-based drying of the apple crop.

**Table 5** Summary of ANN models

| Activation/transfer function | Neurons (First hidden layer) | Neurons (Second hidden layer) | Coefficient of correlation (R) | | | Mean square E | | |
|---|---|---|---|---|---|---|---|---|
| | | | Training | Validation | Test | Training | Validation | Test |
| Log/Tan-Sigmoid | 2 | – | 0.99045 | 0.99961 | 0.99784 | 0.003264 | 0.000265 | 0.0001641 |
| Log/Tan-Sigmoid | 4 | – | 0.99958 | 0.99996 | 0.99791 | 0.00008339 | 0.00000274 | 0.0001331 |
| Log/Tan-Sigmoid | 6 | – | 0.99964 | 0.99844 | 0.99986 | 0.0000477 | 0.0001317 | 0.001785 |
| Log/Tan-Sigmoid | 8 | – | 0.99998 | 0.99996 | 0.99999 | 0.000003223 | 0.000002251 | 0.00152 |
| Log/Tan-Sigmoid | 10 | – | 0.99991 | 0.99999 | 0.9994 | 0.000009939 | 0.0001929 | 0.001373 |
| Log/Tan/Tan-Sigmoid | 5 | 5 | 0.9987 | 0.99998 | 0.9998 | 0.0002968 | 0.000009773 | 0.00007247 |
| Log/Tan/Tan-Sigmoid | 5 | 6 | 0.9936 | 0.99972 | 0.98372 | 0.0006694 | 0.00004731 | 0.02708 |
| Log/Tan/Tan-Sigmoid | 6 | 5 | 0.99085 | 0.99837 | 0.99278 | 0.00179 | 0.00119 | 0.003138 |
| Log/Tan/Tan-Sigmoid | 6 | 7 | 0.93811 | 0.9999 | 0.99974 | 0.01408 | 0.00007968 | 0.00004632 |
| Log/Tan/Tan-Sigmoid | 7 | 8 | 0.91805 | 0.93067 | 0.99731 | 0.01455 | 0.000395 | 0.0006878 |
| Log/Tan/Tan-Sigmoid | **8** | **8** | **0.99993** | **1** | **0.99976** | **0.00001523** | **0.000002** | **0.000009801** |
| Log/Tan/Tan-Sigmoid | 8 | 10 | 0.9974 | 0.99986 | 0.99567 | 0.0006838 | 0.00004066 | 0.0003673 |

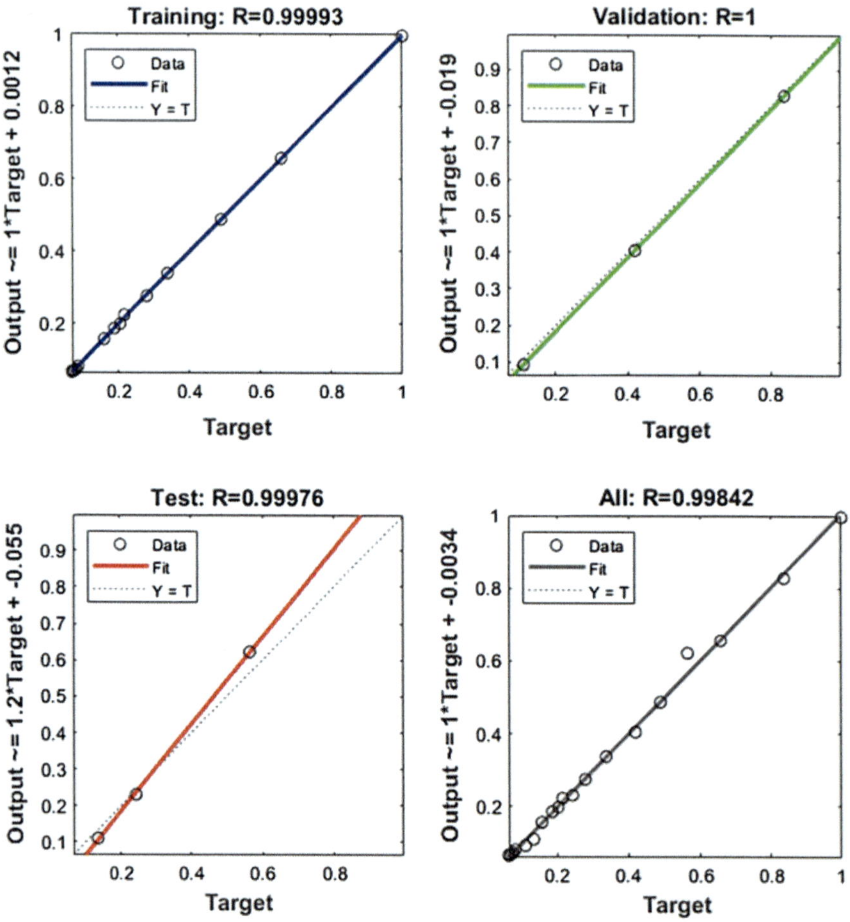


**Fig. 9** Correlation between experimental and predicted values of moisture ratios for optimum topologies

**Acknowledgements** The authors acknowledge the financial assistance provided by the All India Council for Technical Education, New Delhi to the second author for carrying out the present research work.

## References

1. Doymaz İbrahim (2009) An experimental study on drying of green apples. Dry Technol 27(3):478–485. https://doi.org/10.1080/07373930802686065
2. Polat A, Taskin O, Izli N, Asik BB (2019) Continuous and Intermittent Microwave-vacuum drying of apple: drying kinetics, protein, mineral content, and color. J Food Process Eng

42(3):e13012. https://doi.org/10.1111/jfpe.13012

3. Yuan Yuejin, Tan Libin, Yingying Xu, Yuan Yueding, Dong Jixian (2019) Numerical and experimental study on drying shrinkage-deformation of apple slices during process of heat-mass transfer. Int J Therm Sci 136:539–548. https://doi.org/10.1016/j.ijthermalsci.2018.10.042
4. Daghigh R, Ruslan MH, Sulaiman MY, Sopian K (2010) Review of solar assisted heat pump drying systems for agricultural and marine products. Renew Sustain Energy Rev 14(9):2564–2579. https://doi.org/10.1016/j.rser.2010.04.004
5. Ibrahim M, Sopian K, Daud WRW, Alghoul MA (2009) An experimental analysis of solar-assisted chemical heat pump dryer. Intern J Low-Carbon Technol 4(2):78–83. https://doi.org/10.1093/ijlct/ctp016
6. Amer BMA, Gottschalk K, Hossain MA (2018) Integrated hybrid solar drying system and its drying kinetics of chamomile. Renew Energy 121: 539–547. https://doi.org/10.1016/j.renene.2018.01.055
7. Fadhel MI, Sopian K, Daud WRW (2010) Performance analysis of solar-assisted chemical heat-pump dryer. Sol Energy 84(11):1920–1928. https://doi.org/10.1016/j.solener.2010.07.001
8. Mahesh A, Sooriamoorthi CE, Kumaraguru AK (2012) Performance study of solar vacuum tubes type dryer. J Renew Sustain Energy 4(6): https://doi.org/10.1063/1.4767934
9. Singh Prashant, Vyas Shubham, Yadav Avadhesh (2018) Experimental comparison of open sun drying and solar drying based on evacuated tube collector. Int J Sustain Energ 38(4):348–367. https://doi.org/10.1080/14786451.2018.1505726
10. Iranmanesh M, Akhijahani HS, Jahromi MSB (2020) Corrigendum to 'CFD modeling and evaluation the performance of a solar cabinet dryer equipped with evacuated tube solar collector and thermal storage system' [Renew Energy 145 (2020) 1192–1213]. Renew Energy 147:2525. https://doi.org/10.1016/j.renene.2019.10.012
11. Daghigh Roonak, Shahidian Roonak, Oramipoor Hooman (2020) A multistate investigation of a solar dryer coupled with photovoltaic thermal collector and evacuated tube collector. Sol Energy 199:694–703. https://doi.org/10.1016/j.solener.2020.02.069
12. Gaur MK, Pandit RK, Saxena G, Kushwah A, Saxena P (2019) Method and apparatus for controlling the temperature of solar dryer, Application No. 201921001878 A. Off J Indian Pat Of 05/2019: 4191)
13. Ekechukwu OV (1999) Review of solar-energy drying systems I: an overview of drying principles and theory. Energy Convers Manag 40(6):593–613. https://doi.org/10.1016/s0196-8904(98)00092-2
14. Lewis WK (1921) The rate of drying of solid materials. J Ind Eng Chem 13(5):427–432. https://doi.org/10.1021/ie50137a021
15. Page G (1949) Factors influencing the maximum rates of air-drying shelled corn in thin layers. M.S. Dissertation, Lafayette, IN Purdue University
16. Henderson SM, Pabis S (1961) Grain drying theory I: temperature effect on drying coefficient. J Agric Eng Res 6:169–174
17. Karathanos VT (1999) Determination of water content of dried fruits by drying kinetics. J Food Eng 39(4): 337–344. https://doi.org/10.1016/s0260-8774(98)00132-0
18. Chandra PK, Paul Singh R (2017) Applied numerical methods for food and agricultural engineers. https://doi.org/10.1201/9781315137650
19. Yagcioglu ADACF (1999) Drying characteristic of laurel leaves under different conditions. In: Proceedings of the 7th International Congress on Agricultural Mechanization and Energy, pp 565–569
20. Henderson SM (1974) Progress in Developing the thin layer drying equation. Trans. ASAE 17(6):1167–1168. https://doi.org/10.13031/2013.37052
21. Midilli A, Kucuk H, Yapar Z (2002) A new model for single-layer drying. Drying Technol 20(7):1503–1513. https://doi.org/10.1081/drt-120005864
22. Ghazanfari A, Emami S, Tabil LG, Panigrahi S (2006) Thin-layer drying of flax fiber: II. Modeling drying process using semi-theoretical and empirical models. Drying Technol 24 (12):1637–1642 (2006). https://doi.org/10.1080/07373930601031463

23. Sharaf-Eldeen YIs, Blaisdell JL, Hamdy MY (1980) A model for ear corn drying. Trans ASAE 23(5):1261–1265. https://doi.org/10.13031/2013.34757
24. Wang GY, Paul Singh R (1978) Single layer drying equation for rough rice. Paper-American Society of Agricultural Engineers
25. Motevali A, Minaei S, Khoshtaghaza MH, Kazemi M, Nikbakht AM (2010) Drying of pomegranate arils: comparison of predictions from mathematical models and neural networks. Int J Food Eng 6(3). https://doi.org/10.2202/1556-3758.1889
26. Moffat RJ (1985) Using uncertainty analysis in the planning of an experiment. J Fluids Eng 107(2):173–178. https://doi.org/10.1115/1.3242452
27. Moffat RJ (1982) Closure to 'discussions of "contributions to the theory of single-sample uncertainty analysis. J Fluids Eng 104(2):259–260. https://doi.org/10.1115/1.3241821
28. Akpinar EK (2010) Drying of mint leaves in a solar dryer and under open sun: modelling, performance analyses. Energy Convers Manag 51(12):2407–2418. https://doi.org/10.1016/j.enconman.2010.05.00
29. Ahmad A, Prakash O (2020) Performance evaluation of a solar greenhouse dryer at different bed conditions under passive mode. J Solar Energy Eng 142:1. https://doi.org/10.1115/1.4044194
30. Ahmed J, Shivhare US, Singh G (2001) Drying characteristics and product quality of coriander leaves. Food Bioprod Process 79(2):103–106. https://doi.org/10.1205/096030801750286258
31. Alibas Ilknur (2006) Characteristics of chard leaves during microwave, convective, and combined microwave-convective drying. Drying Technol 24(11):1425–1435. https://doi.org/10.1080/07373930600952776
32. Motevali A, Abbaszadeh A, Najafi GH, Minaei S, Ghobadian B (2012) Drying of jujube ('zizyphus jujube mill') fruit: comparison of prediction from mathematical models and artificial neural networks. Aust J Crop Sci 6(2):210

# Chapter 25
# Equilibrium Optimizer-Based Optimal Allocation of SVCs for Voltage Regulation and Loss Minimization

**Prachi Mafidar Joshi and H. K. Verma**

## 1 Introduction

Present-day, increasing electrical energy demand introduces the challenge to the electrical operator to satisfy the load demand in economic, efficient, stable and reliable manner. As the power transfer takes place through long transmission lines, it causes many problems related to power transfer, such as reactive power flow issues, frequency collapse and voltage collapse, which reduce the actual power transfer capacity of the lines [1, 2]. For voltage profile improvement, the flow of reactive power has to be regulated. Because of advancement in the technologies of power electronics, the Flexible AC Transmission System (FACTS) is the new solid state technology that is sufficiently able to regulate transmission system parameters [2, 3]. FACTS is also capable of minimizing the voltage variation and real power losses in the transmission lines.

Mostly used FACTS devices are Thyristor Controlled Series Compensators (TCSC), Static Var Compensators (SVC), Unified Power Flow Controllers (UPFC), Static Compensators (STATCOM), etc. Due to its superior performance in the system with low cost, it is extensively used for compensation in the power system. The SVCs are connected in shunt to regulate the flow of reactive power as per requirement so as to maintain the bus voltages within safe limits. Numerous researches have been done to solve optimal SVC placement problem [4–7].

In the present paper, an equilibrium optimizer is proposed for multiple SVCs placement in the system economically to reduce the voltage deviations along with reduced power losses. For this purpose, IEEE 30 bus test system with different loadings is considered. To prove the strength of algorithm, the comparative analysis of the results has been performed with well-established optimization techniques, PSO and GWO. This paper is arranged as: Sect. 2 explains the optimal allocation of

P. M. Joshi (✉) · H. K. Verma
Department of Electrical Engineering, Shri G. S. Institute of Technology and Science, Indore, India

© The Author(s), under exclusive license to Springer Nature Singapore Pte Ltd. 2022

H. M. Dubey et al. (eds.), *Artificial Intelligence and Sustainable Computing*, Algorithms for Intelligent Systems, https://doi.org/10.1007/978-981-16-1220-6_26

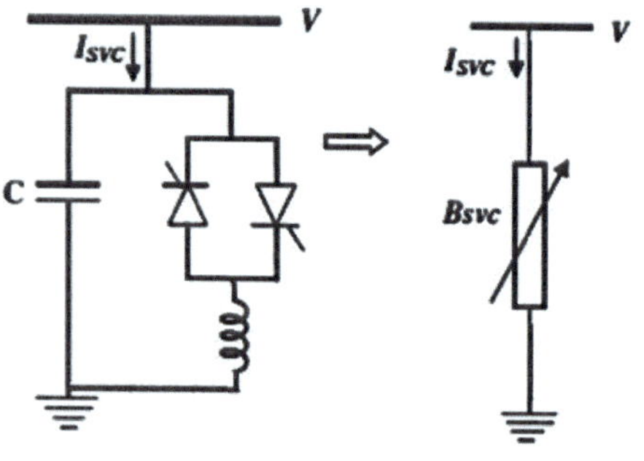


**Fig. 1** Model of SVC

SVCs, particle swarm optimization (PSO) is expressed in Sects. 3, and 4 illustrates the Grey Wolf Optimizer (GWO), Sect. 5 explains the Equilibrium Optimizer (EO), Sect. 6 describes the results and discussion and Sect. 7 gives the conclusion.

## 2 Optimal Allocation of Static VAR Compensators

### 2.1 *Static VAR Compensator (SVC) Model*

The SVC is a shunt-connected FACTS device so as to control the flow of reactive power $Q_{SVC}$ at a selected bus. The SVC is modeled as shown in Fig. 1 [8].

Where $B_{SVC} = -\frac{1}{X_{SVC}}$ and current through the SVC is $I_{SVC} = jB_{SVC}V$.

The reactive power drawn by the SVC is given as

$$Q_{SVC} = -B_{SVC}V^2 \tag{1}$$

where,

$$Q_{\min} \leq Q_{SVC} \leq Q_{\max} \tag{2}$$

### 2.2 *Problem Formulation*

The main objective functions for optimal SVC placement problem are given as follow:

**Voltage deviation minimization. For maintaining the bus voltages in the stable range, the voltage deviation minimization can be computed using** Eq. (3) [8]:

$$\mathbf{VD} = \sum_{i=1}^{N_b} |\boldsymbol{V}_b - \boldsymbol{V}_{\mathbf{ref}}| \tag{3}$$

where, $V_b$ is the bus voltage, $V_{\text{ref}}$ is the reference voltage, which is generally taken as unity and $N_b$ is the number of buses in the considered system.

**Minimization of active power losses. For efficient power transfer, minimization of active power losses can be computed using the following relation** [8]:

$$\boldsymbol{P}_{\mathbf{Loss}} = \sum_{k=1}^{N_L} \left[ \boldsymbol{G}_k \left( \boldsymbol{V}_i^2 + \boldsymbol{V}_j^2 - 2\boldsymbol{V}_i \boldsymbol{V}_j \cos \boldsymbol{\delta}_{ij} \right) \right] \tag{4}$$

where, $N_L$ is the number of branches, $G_k$ is conductance of $k$th branch, $V_i$ and $V_j$ are the voltages at $i$th and $j$th bus and $\delta_{\text{ij}}$ is the phase angle between the voltages of $i$th and $j$th bus.

**Minimizing the total installation cost of SVCs. Installation cost of SVCs can be mathematically formulated as** [9]:

$$\mathbf{IC} = \boldsymbol{C}_{\mathbf{SVC}} \times \boldsymbol{S} \times 1000 \tag{5}$$

where, IC represents the installation cost of SVCs in \$ and $C_{\text{SVC}}$ is the installation cost of SVCs in \$/kVAR

$$\boldsymbol{C}_{\mathbf{SVC}} = 0.0003\boldsymbol{S}^2 - 0.3051\boldsymbol{S} + 127.38 \tag{6}$$

$$\boldsymbol{S} = |\boldsymbol{Q}_2 - \boldsymbol{Q}_1| \tag{7}$$

where, S is the range of operation for SVCs in MVAR, $Q_1$ represents the flow of reactive power in the branch before the installation of SVCs and $Q_2$ is the flow of reactive power reactive through the line after the installation of SVCs.

The aforementioned three functions are to be minimized together, for which a combined multi-objective function can be described as fitness function given by Eq. (8), that is solved using different optimization algorithms under consideration, where $\lambda_1$ and $\lambda_2$ are the coefficients to normalize the function and have the values 4 $\times$ $10^6$ and $10^7$ respectively.

$$\mathbf{minFT} = \mathbf{IC} + \boldsymbol{\lambda}_1(\mathbf{VD}) + \boldsymbol{\lambda}_2(\boldsymbol{P}_{\mathbf{Loss}}) \tag{8}$$

# 3 Particle Swarm Optimization (PSO)

PSO is the bioinspired optimization technique introduced by James Kennedy and Russel Eberhert in 1995 [10]. It is motivated by the social behavior of birds and fish. [10] has adopted this property to solve optimization problems. Every particle of the population acts as a bird that has some fitness value determined by the fitness function and the velocity with which particles move. If maximum and minimum ranges are known (dependent on design variables and variety of problem), the set of particle positions known as population and their velocities can be determined using the following relations:

$$X_i^t = X_{\min} + (\mathbf{rand}(X_{\max} - X_{\min})/V_{\max}) \quad (9)$$

$$V_i^t = 0.4\mathbf{rand}V_{\max} \quad (10)$$

Swarm maintains the trace of the best position obtained in the previous iteration, which is known as personal or local best $p_{\text{best}}$ value and also the best position obtained so far among the population is known as global best $g_{\text{best}}$ value. The particles update their positions by means of the present velocities, current positions and the $p_{\text{best}}$ and $g_{\text{best}}$ values as given below:

$$V_i^{t+1} = \mathbf{W}\mathbf{V}_i^t + C_1 r_1 \left(p_{\text{best},i}^t - X_i^t\right) + C_2 r_2 \left(g_{\text{best}}^t - X_i^t\right) \quad (11)$$

$$X_i^{t+1} = X_i^t + V_i^{t+1} \quad (12)$$

In Eq. (11), inertia weight is represented by W that controls the effect of past velocities and can have a value around 0.8–1.2 calculated using Eq. (13). $C_1$ is the cognitive coefficient and $C_2$ is the social coefficient. These can have the values between 0 and 2. For the problem under consideration, the values of $C_1$ and $C_2$ have been considered equal to 2.

$$W = W_{\max} - \frac{(W_{\max} - W_{\min})}{\mathbf{iter}_{\max}} \times \mathbf{iter}_{\text{current}} \quad (13)$$

## 3.1 Step by Step Algorithm

(i) Generate randomly the positions of particles as population within their prescribed limits using uniform distribution, as in Eq. (9).
(ii) Generate the velocities that determine the particle movement, within their range, as in Eq. (10).

(iii) Calculate the fitness function for each particle position.
(iv) Find $p_{best}$ and $g_{best}$ values.
(v) Update the velocities and particle positions using Eq. (11).
(vi) If solution is converged, then stop, else go to step (iii).

# 4 Grey Wolf Optimizer (GWO)

GWO is the algorithm based on swarm intelligence, developed by Seyedali Mirjalili, Seyed Mohammad Mirjalili and Andrew Lewis in 2014 [11]. It is inspired by grey wolves, which pursue an exact social hierarchy. The top-most level of hierarchy includes the alpha wolves, which act as a leader. The next level of hierarchy contains the beta wolves, which are lower than alpha and these help the alpha in decision-making. Betas are the second-best candidates, which are able to take place of alpha, if required. Deltas are the third-best wolves in the group, which obey the instructions of alphas and betas and accordingly instruct the remaining lower raked wolves named as omegas of the group. The social hierarchy contains the process of tracking, encircling and attacking the prey.

To mathematically model this hierarchy, alpha ($\alpha$) represents the best solution, beta ($\beta$) and delta ($\delta$), respectively, are the second and third best solutions. Rest of the solutions are taken as omega ($\omega$).

*Encircling the prey.* Before hunting, first grey wolves encircle the prey. This can be represented mathematically as follows:

$$\vec{D} = \left|\vec{C}.\overrightarrow{X_p}(t) - \vec{X}(t)\right| \tag{14}$$

$$\vec{X}(t+1) = \overrightarrow{X_p}(t) - \vec{A}.\vec{D} \tag{15}$$

The vectors $\vec{A}$ and $\vec{C}$ can be given as:

$$\vec{A} = 2\vec{a}.\overrightarrow{r_1} - \vec{a} \tag{16}$$

$$\vec{C} = 2.\overrightarrow{r_2} \tag{17}$$

where, $\overrightarrow{r_1}$ and $\overrightarrow{r_2}$ represent the random vectors in [0, 1] elements of $\vec{a}$ are linearly decreased from 2 to 0.

*Hunting.* After encircling, alpha candidates direct the approximate position of the prey to the other wolves, as the accurate position of the prey in the entire search space is unknown. To mathematically model the process of hunting, it is supposed that the actual location of the prey is known to the alpha, beta and delta. While updating positions, the positions of alpha beta and delta are used as given below:

$$\overrightarrow{D_\alpha} = \left|\overrightarrow{C_1}.\overrightarrow{X_\alpha} - \vec{X}\right|, \overrightarrow{D_\beta} = \left|\overrightarrow{C_2}.\overrightarrow{X_\beta} - \vec{X}\right|, \overrightarrow{D_\delta} = \left|\overrightarrow{C_3}.\overrightarrow{X_\delta} - \vec{X}\right| \tag{18}$$

$$\overrightarrow{X_1} = \overrightarrow{X_\alpha} - \overrightarrow{A_1}.\left(\overrightarrow{D_\alpha}\right), \overrightarrow{X_2} = \overrightarrow{X_\beta} - \overrightarrow{A_2}.\left(\overrightarrow{D_\beta}\right), \overrightarrow{X_3} = \overrightarrow{X_\delta} - \overrightarrow{A_3}.\left(\overrightarrow{D_\delta}\right) \tag{19}$$

$$\vec{X}(t+1) = \frac{\overrightarrow{X_1} + \overrightarrow{X_2} + \overrightarrow{X_3}}{3} \tag{20}$$

*Attacking the prey (exploitation).* The process of hunting ends, as the prey stops moving, then grey wolves attack the prey. In mathematical modeling, the movement of wolves in direction of prey takes place, as $\vec{a}$ is decreased linearly from 2 to 0 with an increase in iterations. The value of A varies [−a, a].

*Searching the prey (exploration).* In the searching process, first, alpha, beta and delta explore in the search space to find a better solution and then exploit to attack the prey. The divergence of wolves takes place when $|\vec{A}| > 1$. $\vec{C}$ is weight that varies randomly in [0, 2]. The impact of prey is emphasized when C > 1 and it is deemphasized as C < 1 in Eq. (14).

### 4.1 Step by Step Algorithm

(i) Generate the grey wolves' population within the predefined range using uniform distribution.
(ii) Initialize the values of a, A and C according to the Eq. (16–17).
(iii) Obtain the fitness values for each candidate.
(iv) Obtain the first three best agents as $X_\alpha$, $X_\beta$ and $X_\delta$ respectively according to the fitness values.
(v) Update the position of current candidate using Eq. (20).
(vi) Update a, A and C (16–17) and for these updated values, evaluate the fitness values.
(vii) Update the values of $X_\alpha$, $X_\beta$ and $X_\delta$.
(viii) Check if convergence achieved, if not, repeat steps (v) to (vii).

## 5 Equilibrium Optimizer (EO)

One of the metaheuristic optimization techniques inspired by the laws of physics, named as equilibrium optimizer (EO), developed by [12] for solving optimization problems. The mathematical modeling of EO can be performed as follows.

*Initialization.* In EO, concentration vector is represented by the group of particles, which are having the solution for optimization problem. The concentration vector is initially generated randomly using uniform distribution with the following expression.

$$\boldsymbol{C}_i^{\textbf{initial}} = \boldsymbol{C}_{\textbf{min}} + \textbf{rand}_i(\boldsymbol{C}_{\textbf{max}} - \boldsymbol{C}_{\textbf{min}}) \tag{21}$$

where, $C_i^{initial}$ is the initial concentration vector of the particles. $C_{min}$ and $C_{max}$ are the minimum and maximum values for each design variable of the problem respectively. $rand_i$ is the random number in [0,1].

*Equilibrium pool and the candidates.* Equilibrium state is considered as condition for convergence of the algorithm, at which global optimum solution is obtained. This algorithm completes the process of exploration and exploitation by means of a vector named as equilibrium pool, which contains the best four particles that are at equilibrium state responsible for exploration and the fifth candidate responsible for exploitation, is the average of the best four particles. The equilibrium pool can be represented as below:

$$\overrightarrow{\boldsymbol{C}_{\textbf{eq,pool}}} = \left\{\overrightarrow{\boldsymbol{C}_{\textbf{eq(1)}}}, \overrightarrow{\boldsymbol{C}_{\textbf{eq(2)}}}, \overrightarrow{\boldsymbol{C}_{\textbf{eq(3)}}}, \overrightarrow{\boldsymbol{C}_{\textbf{eq(4)}}}, \overrightarrow{\boldsymbol{C}_{\textbf{eq(ave)}}}\right\} \tag{22}$$

*Concentration update.* The term $\vec{F}$ maintains the proper balance of expedition and exploitation. The turnover rate varies with respect to time.

$$\vec{F} = e^{-\vec{\lambda}(t-t_0)} \tag{23}$$

where, $\vec{\lambda}$ represents a random vector in [0,1] and t decreases over the iterations and can be represented by the following relation:

$$t = \left(1 - \frac{\textbf{iter}}{\textbf{Max\_iter}}\right)^{\left(a_2 \frac{\textbf{iter}}{\textbf{Max\_iter}}\right)} \tag{24}$$

where, iter and Max_iter are, respectively, represent the present and total number of iterations. $a_2$ is a constant, such that, higher value of $a_2$ gives reduced exploration and better exploitation. To ensure the convergence, search speed must be reduced. To obtain this, $\vec{t}_0$ can be evaluated as follows:

$$\vec{t}_0 = \frac{1}{\vec{\lambda}} \ln\left(-a_1 \textbf{sign}(\vec{r} - 0.5)\left[1 - e^{-\vec{\lambda}t}\right]\right) + t \tag{25}$$

where, $a_1$ is a constant, such that, higher value of $a_1$ helps in better exploration. [12] suggested $a_1 = 2$ and $a_2 = 1$, which have been considered for solving the problem under consideration.

$$\vec{F} = a_1 \mathbf{sign}(\vec{r} - 0.5)\left[e^{-\vec{\lambda}t} - 1\right] \tag{26}$$

The generation rate $\vec{G}$ is an another important parameter that helps in attaining optimum solution and can be evaluated as:

$$\vec{G} = \overrightarrow{G_0} e^{-\vec{\lambda}(t - t_0)} = \overrightarrow{G_0}\vec{F} \tag{27}$$

where,

$$\overrightarrow{G_0} = \overrightarrow{\mathbf{GCP}}\left(\overrightarrow{C_{\mathbf{eq}}} - \vec{\lambda}\vec{C}\right) \tag{28}$$

and

$$\overrightarrow{\mathbf{GCP}} = \begin{cases} 0.5 r_1 & r_2 \geq \mathrm{GP} \\ 0 & r_2 < \mathrm{GP} \end{cases} \tag{29}$$

where, $\overrightarrow{\mathrm{GCP}}$ is generation rate control parameter that shows the contribution probability of the generating terms to update the concentration process. GP is the generation probability that informs the particles to update their state and is set to 0.5 to obtain appropriate balance between exploration and exploitation. Hence, the final concentration updating takes place using following equation.

$$\vec{C} = \overrightarrow{C_{\mathbf{eq}}} + \left(\vec{C} - \overrightarrow{C_{\mathbf{eq}}}\right).\vec{F} + \frac{\vec{G}}{\vec{\lambda} V}\left(1 - \vec{F}\right) \tag{30}$$

### 5.1 Step by Step Algorithm

(i) Generate the candidates within the predefined limits using uniform distribution.
(ii) Initialize the values of $a_1$, $a_2$ and $GP$.
(iii) Obtain the fitness values for each candidate.

(iv) Set the equilibrium pool Eq. (22) by first four best agents as $\overrightarrow{C_{eq(1)}}, \overrightarrow{C_{eq(2)}}, \overrightarrow{C_{eq(3)}}$ and $\overrightarrow{C_{eq(4)}}$ respectively and fifth one is the average of these four agents.
(v) Choose one candidate from equilibrium pool randomly.
(vi) Obtain $\vec{F}$ and $\vec{G}$ and then update the concentration using Eqs. (26–30).
(vii) Check whether the convergence is achieved, if not, repeat steps (iii) to (vi).

## 6 Results and Discussion

In order to demonstrate the strength of EO compared with PSO [13] and GWO for solving optimal SVC placement problem, it is implemented on IEEE 30 bus system. Also to closely analyze the performance of the algorithms, these are tested with different loading conditions.

Tables 1, 2 and 3 demonstrate the performance of PSO, GWO and EO on optimal SVC placement problem with loading factors of 1.35, 1.45 and 1.55 respectively. The population size and number of iterations for each algorithm are set to 30 and 500 respectively. The results obtained are the best values out of 20 runs. These algorithms are also compared with one of the traditional methods that is modal analysis for the identification of weak bus [14].

Table 1 illustrates that the installation cost obtained with PSO is 2645000 $, which is much larger as compared to that obtained from GWO is 2137000 $, whereas, EO shows the promising capability of convergence by converging it up to 1813000 $. In Table 2, installation cost obtained from PSO and GWO is, respectively, 2913600 $ and 2760000 $, while EO gives better convergence by converging it up to 1837000 $. Table 3 shows that cost obtained by EO with higher loading conditions is also superior to that obtained from the other two algorithms under consideration.

**Table 1** Results for optimal SVCs placement on IEEE 30 bus system with loading factor 1.35

| | Traditional approach (modal analysis) | PSO | GWO | EO |
|---|---|---|---|---|
| SVC1(location1) (p.u.) | 0.002(6) | 0.0037(17) | 0.0191(4) | 0.0422(25) |
| SVC2(location2) (p.u.) | 0.10(28) | 0.0576(12) | 0.0481(28) | 0.0124(26) |
| SVC3(location3) (p.u.) | 0.07(26) | 0.1133(6) | 0.0948(23) | 0.0125(12) |
| SVC4(location4) (p.u.) | 0.11(30) | 0.0373(4) | 0.0085(12) | 0.0772(14) |
| Total installation cost ($) | 3510500 | 2645000 | 2137000 | 1813000 |
| Deviation of voltage (p.u.) | 0.6940 | 0.693 | 0.5959 | 0.5469 |
| Total real power losses (p.u.) | 0.0390 | 0.0365 | 0.0352 | 0.0331 |
| Fitness value | 6325500 | 5782000 | 4872600 | 4381600 |

**Table 2** Results for optimal SVCs placement on IEEE 30 bus system with loading factor 1.45

| | Traditional method (modal analysis) | PSO | GWO | EO |
|---|---|---|---|---|
| SVC1(location1) (p.u.) | 0.004(6) | 0.079(21) | 0.0921(22) | 0.0058(21) |
| SVC2(location2) (p.u.) | 0.07(26) | 0.1005(12) | 0.0109(12) | 0.0171(12) |
| SVC3(location3) (p.u.) | 0.09(28) | 0.0533(26) | 0.0901(26) | 0.0674(26) |
| SVC4(location4) (p.u.) | 0.18(30) | 0.0005(28) | 0.0280(29) | 0.0559(23) |
| Total installation cost $ | 4245400 | 2913000 | 2763000 | 1837000 |
| Deviation of Voltage (p.u.) | 0.780 | 0.7748 | 0.7685 | 0.6978 |
| Total real power losses(p.u.) | 0.0391 | 0.0388 | 0.0383 | 0.0361 |
| Fitness value | 7756400 | 6400900 | 6220000 | 4989900 |

**Table 3** Results for optimal SVCs placement on IEEE 30 bus system with loading factor 1.55

| | Traditional method (modal analysis) | PSO | GWO | EO |
|---|---|---|---|---|
| SVC1(location1) (p.u.) | 0.14(26) | 0.2412(12) | 0.0304(26) | 0.1034(25) |
| SVC2(location2) (p.u.) | 0.16(19) | 0.0084(20) | 0.0958(12) | 0.2324(26) |
| SVC3(location3) (p.u.) | 0.08(30) | 0.1721(27) | 0.0604(15) | 0.0071(12) |
| SVC4(location4) (p.u.) | 0.1(18) | 0.0168(19) | 0.2114(29) | 0.0008(29) |
| Total installation cost ($) | 5928804 | 5322600 | 4895200 | 4185600 |
| Deviation of Voltage (p.u.) | 0.8006 | 0.7701 | 0.6767 | 0.666 |
| Total real power losses (p.u.) | 0.0476 | 0.0451 | 0.0431 | 0.0427 |
| Fitness value | 9607204 | 8854000 | 8033000 | 7276600 |

For better understanding of the efficiency of EO, Figs. 2, 3 and 4 demonstrate the convergence graph of optimal SVC placement using PSO, GWO and EO with various loading cases.

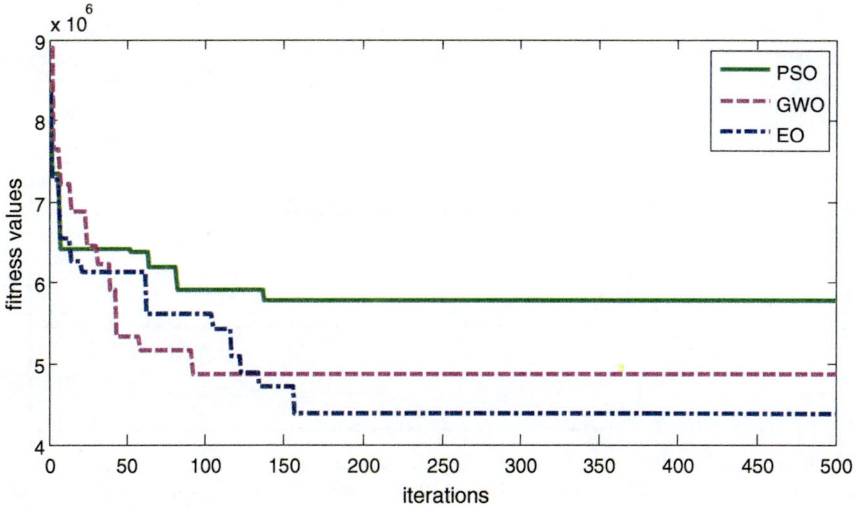


**Fig. 2** Convergence graph for loading factor of 1.35

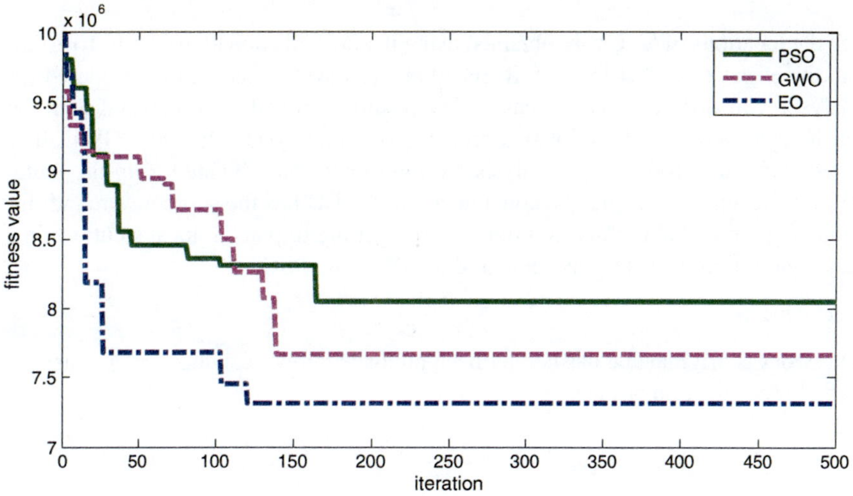


**Fig. 3** Convergence graph for loading factor of 1.45

# 7 Conclusion

The recently introduced EO has been successfully executed on multi-objective optimal SVC placement problem. In this paper, the total installation cost of SVCs has been minimized along with minimizing the real power losses in the system while maintaining the system voltage in the stable range. For this problem, both the sizes

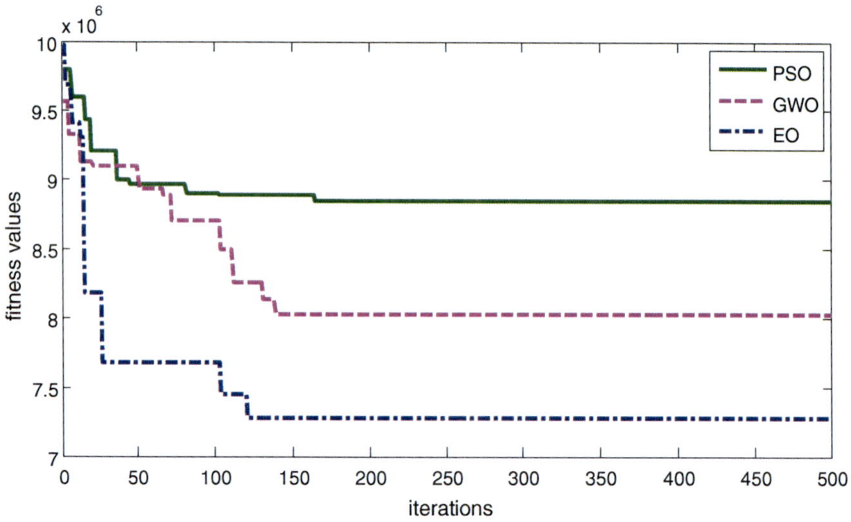


**Fig. 4** Convergence graph for loading factor of 1.55

and the locations of SVCs are obtained using the optimization algorithms. To establish the effectiveness of the EO, IEEE 30 bus test system has been considered for analysis with different loading cases. The results obtained are compared with the results achieved from traditional approach (modal analysis), PSO and GWO. The results obtained from modal analysis are inferior to that obtained from the meta-heuristic techniques. It may be concluded that the EO has the supreme strength for achieving the global optimal solution without getting trapped in local solutions in a lesser number of iterations as compared to GWO and PSO.

**Declaration**

This work is original and has neither been published previously nor under consideration for publication elsewhere.

## Appendix

See Fig. 5.

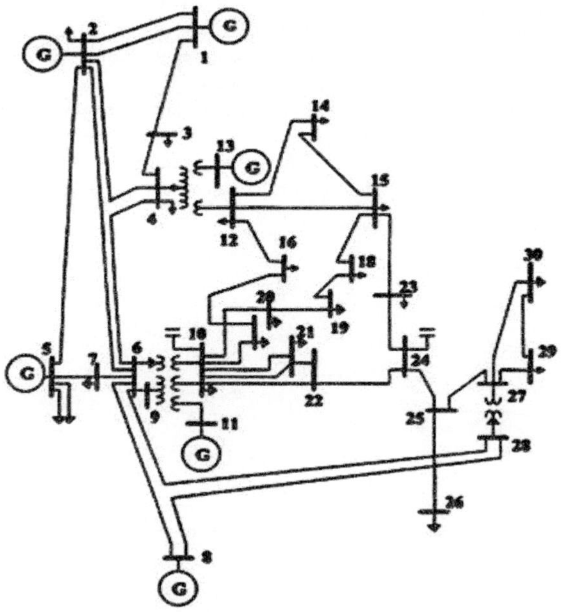

**Fig. 5** IEEE 30 bus test system—single line diagram [15]

## References


1. Yoshida et al (2000)A particle swarm optimization for reactive power and voltage control considering voltage security assessment. IEEE Trans Power Syst 151232–1239
2. Hingorani G (1988) Power electronics in electrical utilities: role of power electronics in future power systems. In: Proceedings of the 1988 IEEE, vol 76, No 4 Apr 1988, pp 481–482
3. Singh et al (2018) Enhancement of voltage profile by incorporation of SVC in power system networks by using optimal load flow method in MATLAB/Simulink environments. Energy Rep 4:418–434
4. Verma et al (2013) TLBO based voltage stable environment friendly economic dispatch considering real and reactive power constraints. J Inst Eng India Ser B 94(3):193–206, (Sept–Nov 2013)
5. Rashed et al (2012) Optimal location and parameter setting of TCSC for loss minimization based on differential evolution and genetic algorithm. ICMPBE Phys Procedia 33:1864–1878
6. Ravi et al (2013) Optimal location of FACTS devices using Improved Particle Swarm Optimization. Electr Power Energy Syst 49:333–338
7. Rao Nartu et al (2019) A fuzzified Pareto multiobjective cuckoo search algorithm for power lossesminimization incorporating SVC, vol 23, pp 10811–10820
8. Gerbex et al (2001) Optimal location of multi-type FACTS devices in a power system by means of genetic algorithms. IEEE Trans Power Syst 16(3):537–544
9. Rashed et al (2008) Evolutionary optimization techniques for optimal location and parameter setting of TCSC under single line contingency. Appl Math Comput 205(1):133–147
10. Kennedy et al (1995) Particle swarm optimization. In: Proceedings of IEEE international conference on neural networks (ICNN'95), vol 4, pp 1942–1948
11. Mirjalili et al (2014) Grey wolf optimizer. Adv Eng Softw 69, 46–61
12. Faramarzi et al (2020) Equilibrium optimizer: a novel optimization algorithm. Knowl-Based Syst 191:

13. Joshi et al (2019) An improved TLBO based economic dispatch of power generation through distributed energy resources considering environmental constraints. Sustain Energ Grids Netw 18(2019):1–19
14. Zakaria et al (2015) Identification of voltage collapse vulnerable buses using modal analysis. In: 2015 international conference on computing, control, networking, electronics and embedded systems engineering (ICCNEEE), Khartoum, pp 392–396
15. https://al-roomi.org/multimedia/Power_Flow/30BusSystem/IEEE30BusSystemDATA2.pdf

# Chapter 26
# LGCBPNN: Determine the Multilayer Neural Network Using Local Gravitational Clustering

**Tanya Shrivastava and Punit Kumar Johari**

## 1 Introduction

AI is a popular term for the area of science and technology [1], which is used to include computers with the capacity to execute functions like logic, reasoning, planning, learning or perception, also, define AI simply as an ensemble of difficult problems that people and other living things may solve but does not have the algorithm to solve [2].

NN is a major field of technology, where one can use error as well as an estimation to incorporate "human brain decision-making power" in computer programs. Usually, NN comprises of three-layer data forms: input Layer (IL), HL and output layer (OL). There is another question: How many HLs should be used to overcome a complex problem [3]. The selection of hidden neurons utilizing NN is one of the key problems faced by researchers. The problem occurs when the networks match the data so precisely that they lack their capacity to generalize the test results [4].

NN reliability is determined by error. The lowest error is more stable and the higher error is worse [5]. In that context, it is important to determine the correct number of hidden neurons to prevent fitting [6]. Until learning starts, training or generalization errors usually are high. The accuracy of training depends on the study of parameters. Parameters comprise NN design, number of HL neurons, activation function, input or weight change. Figure 1 shows the general architecture of neural network that includes input, hidden and output layers. It is also vital to find the neuron number in HL of artificial NNs [7]. NNs use a supervised approach [8].

The residuals of the paper are structured, as is the similar analysis described in Sect. 2 of an existing area. Section 3 presented a proposed approach. The experimental studies are included in Sect. 4. Section 5 concludes the paper.

T. Shrivastava (✉) · P. K. Johari
Madhav Institute of Technology and Science, Gwalior 474009, Madhya Pradesh, India

P. K. Johari
e-mail: pkjohari@mitsgwalior.in

© The Author(s), under exclusive license to Springer Nature Singapore Pte Ltd. 2022

H. M. Dubey et al. (eds.), *Artificial Intelligence and Sustainable Computing*, Algorithms for Intelligent Systems, https://doi.org/10.1007/978-981-16-1220-6_27

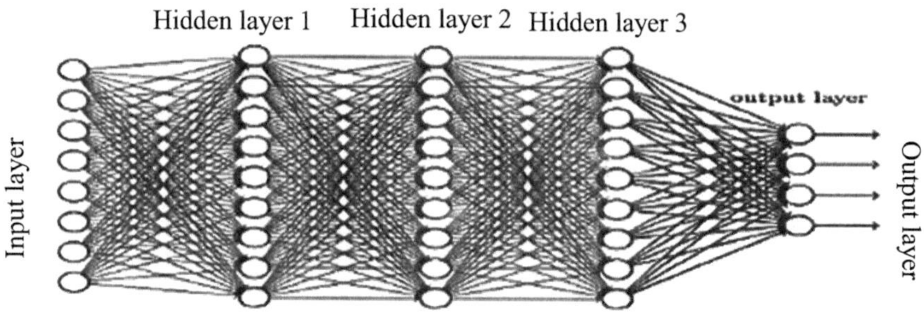


**Fig. 1** Neural network

## 2 Literature Review

Gheche et al. [9] proposed that node clustering issues, which usually consider only network information, be expanded to an issue where the network information, as well as node characteristics, are considered together to allow the clustering of function [9].

Zhou et al. [10] Throughout this document, the Particle Swarm Optimization (PSO) Big Data Mining (DM) method Backpropagation (BP) NN suggests a non-linear parallel optimizer model with the Apache Spark or the Hadoop HDFS methods on the data set for on balance Sheet element or off Balance sheet products for financial risk control throughout commercial IoT banks [10].

Kimura [11] explored a new form of clustering, rendered possible by a neural feed network. The method proposed is compatible with deep neural networks, as opposed to self-organization maps and growing neural gas networks [11].

Tej and Holban [12] discussed methods for evaluating the optimum design of NN by utilizing pattern recognition or DM [12].

Tej and Holban [13] offered regression techniques for evaluating the optimal layout of NN. To define the regression function, use results derived from clustering input data as independent variables [13].

Wu and Peng [14] in this report, a DM technique consisting of K-means clustering or NN bagging for short-term WPF is proposed to resolve the complexities of the training samples and also to increase the prediction accuracy [14].

Chamatkar and Butey [15] The NN is not directly suitable for DM, as it does not explicitly state how the classifications are made as symbolic rules for verifying or analyzing human nature [15].

Wang [16] NNs have become more and more prevalent in DM. NNs have strong noisy data tolerance, moderate exactness and are ideally adapted to data mining. Concentrations on the NN-based DM method [16].

# 3 Proposed Methodology

## *3.1 Problem Statement*

Hidden layers play a vital role in the performance of Backpropagation Neural Network. If the number of neurons is less as compared to the complexity of the problem data then "Underfitting" may occur. If unnecessary more neurons are present in the network then "Overfitting" may occur.

NNs have challenging structures, as optimal number of HLs for an MLNN does not have an analytical method available. To address this problem, we proposed an original method for determining NNs as well as optional HL neurons with LGC clustering for optimal architecture.

In this work, we have used six different types of datasets from UCI Machine Learning Repository namely, Landsat [20], Sonar [21], ECG, P-wave [22], T-wave [22] and QRS complex [22]. First, we have to cluster the data into groups using the LGC clustering method. Each data grouped into three or four clusters and these clusters are responsible for deciding the number of HL. Then calculate the number of neurons for the HL by different–different formulas. After that train the 70% of data and finally test the 15% data and perform validation on 15% data as well. Therefore, we have achieved a new model (i.e. LGCBPNN) to determine the NN.

## *3.2 Local Gravitation in Clustering*

The physical lawed LGC algorithm measures LRF from any data point. In this manner, the data points in the central area are explicitly separated from data points in the border region. For each data point, during the first step, LGC calculates the LRF area. In the second step, LGC carries out the clustering process according to data given by LRF, first by dividing it into interiors, boundary, and unlabeled points the data points then by gently linking the points identified as interiors. The data points are then allocated to the groups, named as boundary points [17].

Border data points are generally relatively little CE (typically CE $< 0$) values. Thus, LGC uses data encapsulation in the magnitude as well as in the directions of LRFs to carry out cluster tasks. More precisely, interior points may be linked to their neighbors and inserted into their clusters while the borders are called obstacles and the connecting efficiency is reduced. There might be mislabeling of certain data points in the central area of clusters as boundary points, but given that occurrence of these points is generally smaller than in borders, there may be gentle relation of internal points. Data points at border areas, on the other hand, frequently had very large magnitudes of LRF, and these data points normally end the soft-connection process.

### 3.3 Number of Neurons in Hidden Layers

In NN's HLs, neurons number can be calculated using each of the following equations:

$$(N_h) = (1/2)((N_i) + (N_o)) \tag{1}$$

$$(N_h) = \sqrt{((1/2)((\mathrm{N_i}) + (\mathrm{N_o})))}, \tag{2}$$

here,

$N_h$ is the number of neurons in HLs, $N_i$ is the number of neurons in ILs, No is the number of neurons in OLs. This method for estimating the amount of HLs is often proposed in other studies:

$$(N_h) = ((N_i) + (N_o)) \times 2/3. \tag{3}$$

Only by the number of inputs or outputs, the optimal structural design of a NN cannot be resolute, as it depends very much on the size of the training set and function complexity that is necessary to learning [18]. The mathematical method is applied to evaluate the first HL number of neurons:

$$(N_{fh}) = (N_t)/(10((N_i) + (N_o))), \tag{4}$$

where, $N_{fh}$ is number of neurons in first HLs, $N_t$ is number of training items.

If multiple (two or three) layers are hidden, number of neurons in previous HL frequently ranges from 1/3 to 2/3. It is recommended that each layer should have the same number of neurons in one HL or 1/2 in the previous layer. This method can be used to measure the number of neurons in the HL:

$$(N_h) = ((N_t) - (N_o))/(N_i) + (N_o) + 1. \tag{5}$$

Analyze the number of neurons in HLs with this formula for simulating and training NLs:

$$(N_h) = 1/2((N_i) + (N_o)) + \sqrt{(N_t)}. \tag{6}$$

If multiple HL is present, number of neurons determined to (6) shall be equally separated by number of HLs.

### 3.4 Backpropagation Algorithms for Multilayer Feed-Forward NN (MLFFNN)

The training of FNN has done mainly through backpropagation (BP). BPNN algorithm is Rumelhart, Hinton and Williams' most popular and older supervised multilayer learning feed-forward neural network algorithm. To adjust the connection weight, the BPLAs use the gradient descent search method. The output layer contains an MLFFNN with I input neurons, the m neurons of the cached layer and the n output neurons l-m-n. With the estimation of the activation function, the input values are translated to output signals [19]. The proposed algorithm for this work is described below and the flowchart for a model is depicted in Fig. 2.

**Proposed Algorithm**

| | |
|---|---|
| Step 1. | Start |
| Step 2. | Collect dataset |
| Step 3. | Load the data |
| Step 4. | Preprocess the data to remove the missing value and replace them by a mean value |
| Step 5. | Use Min-Max normalization to normalize the data |
| Step 6. | Apply center of Gravitational clustering algorithm; On the training dataset a cluster analysis is performed to categorize data into groups |
| Step 7. | Determine no. of HL of NN on basis of get no. of clusters. |
| Step 8. | Determination no. of neurons in HLis performed using one of the equations (1)–(6). |
| Step 9. | Generate a neural network |
| Step 10. | Train the 70% data via neural network |
| Step 11. | Perform testing on 15% data as well as validation on 15% data |
| Step 12. | Measure Accuracy and MSE |
| Step 13. | End |

## 4 Experimental Analysis

Experimental studies were performed to analyze neural network data sets were performed until MLNN architecture can have been distinct. The simulation has done by using MATLAB 2018.

### 4.1 Dataset Description

This paper presents six datasets. The following data sets are included:

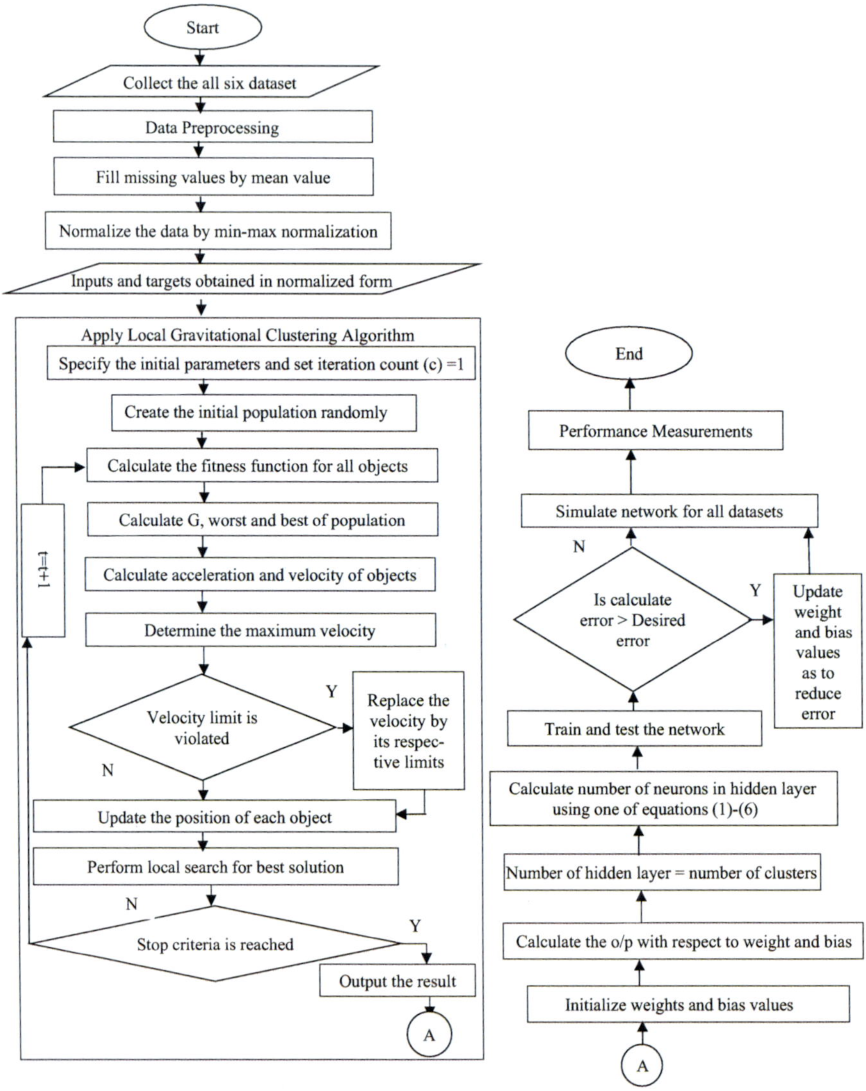


**Fig. 2** Proposed model

### (1) **Landsat Satellite**

A dataset of Landsat Satellite images [20] was generated from image data from Landsat Multispectral Scanner. Every pixel class is coded as number in the sample dataset. The Landsat satellite data set comprises 4,435 training set records representing 36 attributes.

(2) **Sonar data**

The sonar signal [21]; a vector form comprising 60 features has been sampled to extract the signal. The name of every record is the letter "R," if an object is a rock (metal cylinder) or "M" if the object is mine.

(3) **ECG**

ECG signal analysis [22] also includes a data set. Certain parameters to be measured, together with heart rate, calculation of rate time, the accuracy rate (constant or oscillation) number of P waves in cardiac cycles, as well as estimation of intervals or segments.

(4) **P wave, QRS complex and T wave**

ECG signal is composed of three major elements: P, QRS and T. The study findings and other clinical data of these components are used in the diagnosis of different cardiovascular disease disorders.

### *4.2 Screenshots of Simulated Result*

Here, we represent the screenshot for the sonar dataset result after applying the proposed approach.

Menu for the sonar dataset is shown in Fig. 3 to the proposed method and in Fig. 4 represented the graph of generated number of clusters over the data points.

Number of clusters is 3 for the sonar dataset in Fig. 4. The number of HLs is therefore equal to 3. The results are representing in the form of a confusion matrix shown in Fig. 5. The network has trained using backpropagation neural network shown in Fig. 6.

Similarly, the screenshot of the result for all remaining datasets has been achieved. Table 1 indicated the number of hidden layers that are equal to the generated number of clusters.

### *4.3 Experiment for Optimal Number of HL Neurons*

The second analysis is dependent on the MATLAB experiment previously specified in Table 2.

As shown in Table 2, number of neurons with each HL equivalent to 15 or 2, all provided an accuracy of approximately 88%, was the best accuracy for the Sonar satellite dataset. The best number of the hidden neurons is two, meaning that formula (6) is the best choice.

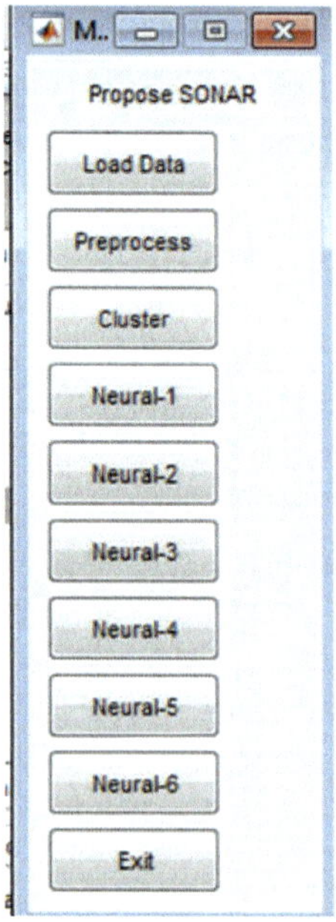


**Fig. 3** Menu for sonar data

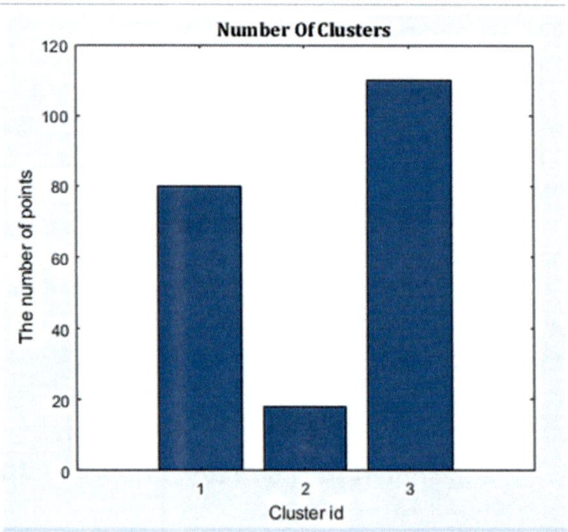


**Fig. 4** Number of clusters

We utilized to calculate the best accuracy values for each data set among the six formulas, with the outcome given in Table 3. The table's outcome is depicted as bar graph in Fig. 7. From Fig. 7, we can see that the proposed method (indicated by light orange color) has high accuracy values in comparison of the existing neural network method (indicated by dark orange color).

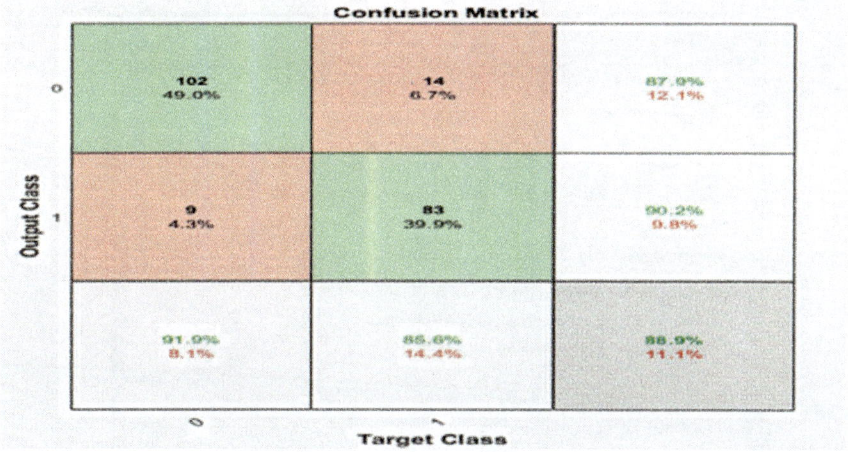


**Fig. 5** Confusion matrix for sonar data

Calculated least error/epoch values for each data set among the six formulas, with the outcome given in Table 4. The table's outcome depicted a line graph in Fig. 8. From Fig. 8, we can see that the proposed method (indicated by orange color) has the least MSE values in comparison to the existing neural network method (indicated by blue color).

## 5 Conclusion

The main aim of the clustering analysis is to divide the set of data points into different clusters on the behalf of the distance parameter. The main issue was the selection of an appropriate number of hidden layers and the number of hidden neurons via hidden layers. So, local gravitation clustering is used to resolve this issue. This clustering is useful to determine the number of clusters and the hidden number of layers is very similar to such a number of clusters. At last, the hidden neurons have been calculated through various formulas for determining hidden layers. A multilayer feed-forward back propagation neural network is used to train the model. The experiments have been performed on different six types of datasets using MATLAB 2018. A multi-layer neural network makes it possible for input data to be linked more complex and non-linear with output. From the experimental analysis, we have achieved better accuracy and lower mean square value for each dataset in comparison to the existing model.

This work will further enhance this model to identify an optimum number of hidden layers and hidden neurons. The approach to improve network performance changes in connectivity weights, learn rates and number of hidden layers required can be extended to Nemours hidden layers.

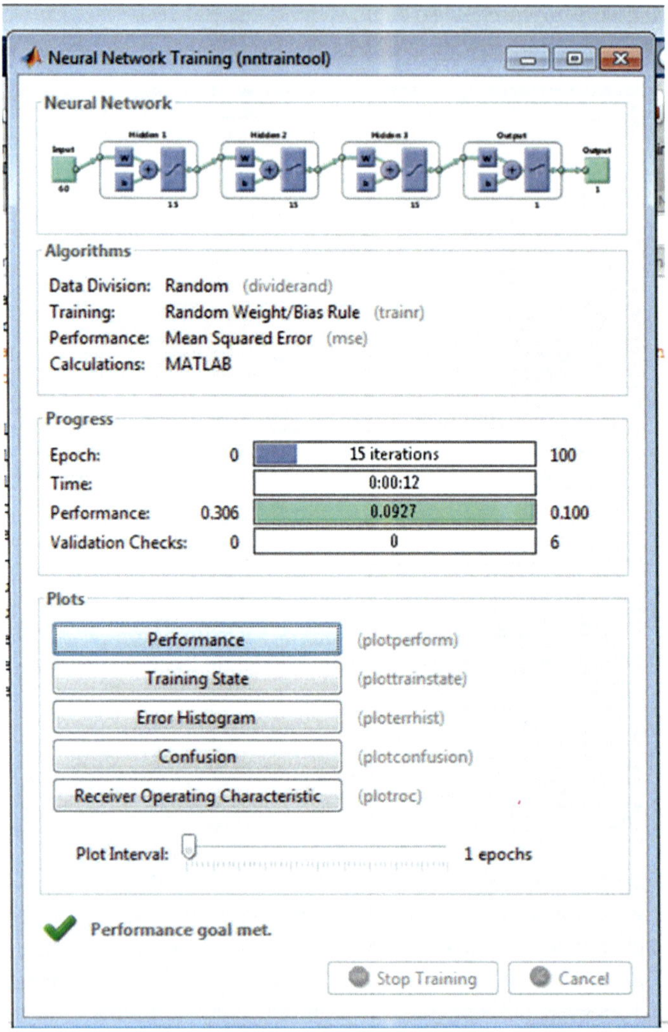


**Fig. 6** Train the network for sonar data

**Table 1** Comparison of the number of clusters gets by clustering method and best number of hidden layers as per various neural network architectures' comparison

| Data set | Number of clusters | Best number of HLs as per various NN architectures' comparison |
|---|---|---|
| Landsat satellite | 3 | 3 |
| Sonar | 3 | 3 |
| ECG | 4 | 4 |
| P-Wave | 4 | 4 |
| QRS | 4 | 4 |
| T-Wave | 4 | 4 |

**Table 2** Accuracy and Error/Epoch values for the numbers of neurons as calculated on the basis of Eqs. (1–6)

| Data set | Formula | Number of HL | Number of neurons/HL | Accuracy (in %) | Error/Epoch |
|---|---|---|---|---|---|
| Landsat Satellite | 5 | 3 | 6 | *98.492* | *0.014737* |
| | 6 | 3 | 2 | 98.321 | 0.025033 |
| | 7 | 3 | 8 | 98.135 | 0.018502 |
| | 8 | 3 | 1 | 98.368 | 0.017670 |
| | 9 | 3 | 1 | 98.477 | 0.016124 |
| | 10 | 3 | 13 | 98.445 | 0.016345 |
| Sonar | 5 | 3 | 15 | *88.942* | *0.099038* |
| | 6 | 3 | 2 | 88.461 | 0.097544 |
| | 7 | 3 | 20 | 86.538 | 0.100749 |
| | 8 | 3 | 0 | – | – |
| | 9 | 3 | 3 | 85.096 | 0.119905 |
| | 10 | 3 | 22 | 87.092 | 0.108811 |
| ECG | 5 | 4 | 6 | 65.486 | 0.215590 |
| | 6 | 4 | 2 | 64.601 | 0.224029 |
| | 7 | 4 | 8 | 67.921 | 0.205285 |
| | 8 | 4 | 91 | 62.611 | 0.267063 |
| | 9 | 4 | 11 | *71.902* | *0.186182* |
| | 10 | 4 | 16 | 70.575 | 0.187276 |
| P-wave | 5 | 4 | 4 | 64.601 | 0.225083 |
| | 6 | 4 | 2 | 54.867 | 0.248084 |
| | 7 | 4 | 5 | *64.823* | *0.220276* |
| | 8 | 4 | 162 | 57.079 | 0.309229 |
| | 9 | 4 | 19 | 63.274 | 0.224898 |
| | 10 | 4 | 14 | 63.938 | 0.233167 |
| QRS | 5 | 4 | 3 | 61.725 | 0.234685 |
| | 6 | 4 | 1 | 69.469 | 0.193179 |
| | 7 | 4 | 4 | 69.690 | 0.204383 |
| | 8 | 4 | 189 | 68.362 | 0.228661 |
| | 9 | 4 | 22 | *72.787* | *0.189015* |
| | 10 | 4 | 13 | 69.247 | 0.202683 |
| T-Wave | 5 | 4 | 4 | 65.044 | 0.221790 |
| | 6 | 4 | 2 | 66.150 | 0.228947 |
| | 7 | 4 | 5 | 63.716 | 0.239887 |
| | 8 | 4 | 162 | *70.132* | *0.206423* |
| | 9 | 4 | 19 | 69.690 | 0.210890 |
| | 10 | 4 | 14 | 68.362 | 0.207322 |

**Table 3** Comparison of accuracy value between the existing and proposed model

| Dataset | Existing | Proposed |
|---|---|---|
| Landsat satellite | 87.925 | 98.492 |
| Sonar | 84.134 | 88.942 |
| ECG | 62.168 | 71.902 |
| P-Wave | 62.611 | 64.823 |
| QRS | 64.159 | 72.787 |
| T-Wave | 60.841 | 70.132 |

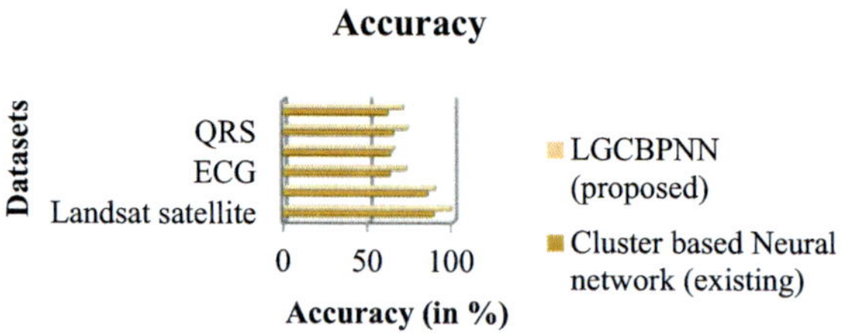


**Fig. 7** Accuracy bar graph

**Table 4** Comparison of mean squared error value between the existing and proposed model

| Dataset | Existing (Clustering based Neural Network) | Proposed (LGCBPNN) |
|---|---|---|
| Landsat satellite | 0.102164 | 0.014737 |
| Sonar | 0.133009 | 0.097544 |
| ECG | 0.226183 | 0.186182 |
| P-Wave | 0.232422 | 0.220276 |
| QRS | 0.228613 | 0.189015 |
| T-Wave | 0.238919 | 0.206423 |

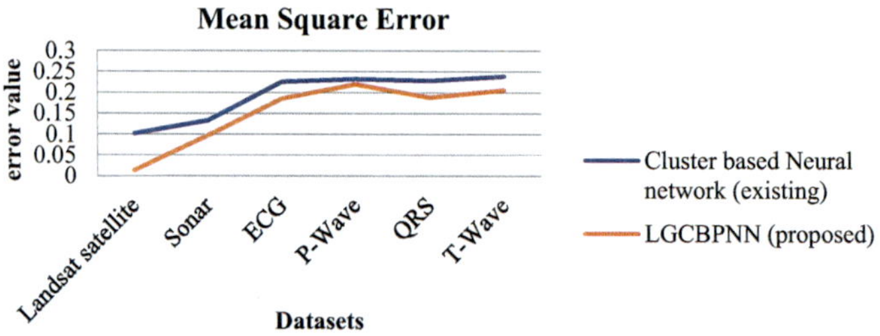


**Fig. 8** MSE line graph

## References


1. Perez JA, Deligianni F, Ravi D, Yang GZ (2017) Artificial intelligence and robotics. arxiv
2. Nwadiugwu M (2015) Neural network, artificial intelligence, and the computational brain. Hindawi Publishing Corporation
3. Karsoliya S (2012) Approximating number of hidden layer neurons in multiple hidden layer BPNN architecture. Int J Eng Trends Technol 3(6):714–717
4. Sheela KG, Deepa SN (2013) Review on methods to fix number of hidden neurons in neural networks. Hindawi Pub Corp Math Prob Eng 1–11 (2013)
5. Jinchuan K, Xinzhe L (2008) Empirical analysis of optimal hidden neurons in neural network modeling for stock prediction. Proc Pacif-Asia Workshop Comput Intell Ind Appl 2:828–832
6. Curry B, Morgan PH (2006) Model selection in neural networks: some difficulties. Eur J Oper Res 170(2):567–577
7. Pitambare DP, Kamde PM (2014) Efficiently exploring clusters using genetic algorithm and fuzzy rules. In: International conference on computer communication and informatics, Coimbatore, pp 1–6
8. Pitambare DP (2016) Survey on optimization of number of hidden layers in neural networks. Int J Adv Res Comput Commun Eng 5(11):537–540
9. Gheche ME, Chierchia G, Frossard P (2020) OrthoNet: multilayer network data clustering. IEEE Trans Signal Inf Proc over Netw 6:152–162
10. Zhou H, Sun G, Fu S, Liu J, Zhou X, Zhou J (2019) A big data mining approach of PSO-Based BP neural network for financial risk management with IoT. IEEE Access 7:154035–154043
11. Kimura M (2018) Auto clustering: a feed-forward neural network-based clustering algorithm. In: IEEE international conference on data mining workshops (ICDMW), Singapore, pp 659–666
12. Tej ML, Holban S (2018) Determining neural network architecture using data mining techniques. In: International conference on development and application systems (DAS), pp 156–163. Suceava
13. Tej ML, Holban S (2018) Determining optimal neural network architecture using regression methods. In: International conference on development and application systems (DAS), Suceava, pp 180–189
14. Wu W, Peng M (2017) A data mining approach combining $K$-Means clustering with bagging neural network for short-term wind power forecasting. IEEE Int Things J 4(4):979–986
15. Chamatkar AJ, Butey PK (2015) Implementation of different data mining algorithms with neural network. In: International conference on computing communication control and automation, Pune, pp 374–378
16. Wang S (2008) Research on a new effective data mining method based on neural networks. In: International symposium on electronic commerce and security, Guangzhou City, pp 195–198
17. Wang Z et al (2018) Clustering by local gravitation. IEEE Trans Cybern 48(5):1383–1396
18. Kaastra I, Boyd M (1996) Designing a neural network for forecasting financial and economic time series. Neurocomputing 10(3):215–236
19. Vora K, Yagnik S (2014) A survey on back propagation algorithms for feed forward neural networks. Int J Eng Dev Res (IJEDR) 193–197
20. Srinivasan A (1993) Statlog (Landsat Satellite) data set. In: Dua D, Karra Taniskidou E (eds) UCI machine learning repository. University of California, School of Information and Computer Science, Irvine, CA. https://archive.ics.uci.edu/ml/datasets/Statlog+%28Landsat+Satellite%29
21. Sejnowski T (1988) Connectionist bench (Sonar, Mines vs. Rocks) Data set. In: Dua D, Karra Taniskidou E (eds) UCI machine learning repository. University of California, School of Information and Computer Science, Irvine, CA. https://archive.ics.uci.edu/ml/datasets/Connectionist+Bench+%28Sonar%2C+Mines+vs.+Rocks%29

22. Guvenir HA, Acar B, Muderrisoglu H (1998) Arrhythmia data set. In: Dua D, Karra Taniskidou E (eds) UCI machine learning repository. University of California, School of Information and Computer Science, Irvine, CA. https://archive.ics.uci.edu/ml/datasets/Arrhythmia

# Chapter 27
# Combined Heat–Power Economic Emission Dispatch for a Microgrid Using Coyote Optimization

**Vishal Chaudhary, Hari Mohan Dubey, and Manjaree Pandit**

## Nomenclature

| | | | |
|---|---|---|---|
| DER | Distributed energy resources | BCS | Best compromise solution |
| RES | Renewable energy resources | ACSA | Ant colony search algorithm |
| CSA | Cuckoo search algorithm | IWO | Invasive weed optimization |
| IGSO | Improved group search optimization | SSA | Salp Swarm algorithm |
| IHS | Improved Harmony Search | WOA | Whale optimization algorithm |
| BLPSO | Biogeography-based learning PSO | KKO | Kho-Kho optimization |
| GWO | Grey wolf optimization | MVO | Multi-verse optimization |
| $N_P$ | Number of group | $N_C$ | Number of coyote in each group |
| IBEA | Indicator & crowding distance-based evolutionary algorithm | OTLBO | Oppositional teaching learning based optimization |
| DE | Differential evolution | $Lb_j, Ub_j$ | Lower and upper limits of variable $j$ |
| $\mathcal{R}^{p,t}$ | Ranked Social condition of coyote of group $p$ at time $t$ | $C_{r1}$ | Cultural difference between random coyote and alpha ($\alpha$) |
| $r_1, r_2$ | Random coyote within $N_P$ group | $\delta_a, \delta_g$ | Alpha factor, group factor |
| $R_j$ | Random number within design variable limits | $C_{r2}$ | Difference between random coyote and cultural tendency of group |

(continued)

V. Chaudhary (✉) · H. M. Dubey · M. Pandit
Department of Electrical Engineering, Madhav Institute of Technology and Science, Gwalior, MP, India

© The Author(s), under exclusive license to Springer Nature Singapore Pte Ltd. 2022

H. M. Dubey et al. (eds.), *Artificial Intelligence and Sustainable Computing*, Algorithms for Intelligent Systems, https://doi.org/10.1007/978-981-16-1220-6_28

(continued)

| | | | |
|---|---|---|---|
| $\rho_s, \rho_a$ | Scattered and association probability | $r_a, r_g$ | Random number ∈ (0, 1) |
| $\text{rand}_j$ | Random number between (0,1) | $D$ | Dimension of problem |

# 1 Introduction

To meet the ever-increasing power demand, creating a new thermal power plant or expanding an existing one requires a long time and large capital investment. From the economic and sustainable operation of the electric power system, the new concept for it needs to be investigated. Considering global warming and climate change, collective effort is needed from researchers and power producer utilities to tackle issues related to the prices of fossil fuel resources and their environmental impact. It can be possible by applying the concept of cogeneration, where multiple outputs can be possible through a single system and by integrating RES in the existing system. Cogeneration often called combined heat and power (CHP), which helps to reduce the cost of power generation and also increases the overall efficiency of the system. CHP helps to minimize emissions due to particular matter, which would lead to the greenhouse effect. Unfortunately, integration of RES in the existing power grid creates several operation issues due to stochastic nature of wind speed and solar irradiation. CHP economic dispatch (CHPED) is an optimization problem; its objective is to find the most cost-effective scheduling of available power generating unit to meet both customer power and heat demand [1]. Due to complicated operational constraints associated with thermal generating units as ramp rate limits, prohibited operating zones, and valve point loading effect, makes problem formulation highly nonlinear, discontinuous, and non-convex in nature. Moreover, an integration of RES, the probabilistic constraints are needed to be embedded in the objective function, which makes problem formulation more complex to solve. Finding a solution for such a complex problem is a difficult task, and it needs a robust optimization tool. Literature suggests the nature-inspired metaheuristic has the ability to find the global/near-global solution for a complex constrained optimization problem by exploring and exploiting the search space in a decent amount of time irrespective of the nature of the problem [2].

The solution of CHPED has been reported in various literatures either using a conventional optimization approach or by using metaheuristic. Among the conventional optimization, Lagrange relaxation embedded with surrogate sub-gradient multiplier (LRSS) [3] as updating mechanism, Lagrange relaxation with an alternating iterative method (LRAI) [4] and benders decomposition (BD) [5] were used for the solution of these problem having different dimensions and complexity level. Among metaheuristic ACSA [6], self-adaptive PSO [7], CSA [8], IWO [9], IGSO [10], OTLBO [11], IHS [12], WOA [13], BLPSO [14] and KKO [15]. In recent years, more attention has been paid to both energy conservation as well as issues related to environmental protection. Therefore, the minimization of emission as another objective is also considered with CHPED. As a result, the objective becomes a combined heat power economic emission dispatch (CHPEED) problem. The multi-objective heat and power economic emission dispatch problem was investigated using DE [16], PSO [16], GWO [17], MVO [18], and IBEA [19]. In [20], PSO has been used for optimum CHPEED scheduling of a microgrid considering demand response. Multi-region CHPEED problem with solar and wind integration was investigated in [21], using SSA and by OQNLP [22] of GAMS. Details of the various optimization methods, objectives, nature of objective function and dimension of the problem considered for analyses are summarized in Table 1.

In this paper, an attempt is made to investigate a CHPEED problem of a microgrid comprises of five DERs. The impact of fuel cell plant integration on generation cost and reduction in emission is also analyzed here. For optimization of the problem, a new population-based method, COA is implemented with aim to analyze how the thermal and electric demands could be shared among the different DER under the different scenario. The contents of the paper are organized in the following five sections. Following the introduction of Sect. 1, the problem formulation presented

**Table 1** The method, dimension, and complexity of the cogeneration problem

| Method | Year | Objective | Dimension | Complexity |
|---|---|---|---|---|
| LR [1] | 1996 | CPED | 4 | Convex |
| LRSS [3] | 2013 | CHPED | 4 | Convex |
| LR with AI [4] | 2019 | CHPED | 4 and 48 | Convex, VPL |
| BD [5] | 2013 | CHPED | 4 and 5 | Convex |
| ACO [6] | 1999 | CPED | 4 | Convex |
| SPSO [7] | 2014 | CHPED | 4 | Convex |
| COA [8] | 2015 | CHP | 4 and 5 | Convex, Cubic |
| IWO [9] | 2014 | CHP | 4 and 5 | Convex, Cubic |
| IGSO [10] | 2014 | CHPED | 24,48 | Convex, VPL & POZ |
| OTLBO [11] | 2014 | CHPD | 7,24,48 | Convex, VPL & POZ |
| IHS [12] | 2019 | CHPD | 4,7 | Convex |
| WOA [13] | 2019 | CHPED | 24,84,96 | VPL & POZ |
| BLPSO [14] | 2020 | CHPED | 5,7,24,48 | VPL & POZ |
| KKO [15] | 2020 | CHPED | 4,5,7,24 | VPL |
| DE, PSO [16] | 2012 | MO_CHPEED | 5 | Convex |
| GWO [17] | 2015 | MO_CHPEED | 4, 7, 24, 11 | Convex, VPL |
| MVO [18] | 2020 | MO_CHPEED | 7140 | VPL |
| IBEA [19] | 2020 | MO_CHPEED | 5, 7 | VPL |
| PSO [20] | 2020 | CHPEED with DR | 5 | linear |
| SSA [21] | 2019 | MA-CHPED | 104 | VPL, POZ, probabilistic |
| OQNLP [22] | 2020 | MA-CHPEED | 48 | VPL, POZ |

in Sect. 2, the concept behind the COA is in Sect. 3. The detailed simulation results under various test conditions and discussions on it are presented in Sect. 4. Finally, the concluding remarks are drawn in Sect. 5.

## 2 Formulation of CHPEED Problem

The fuel cost of DER is evaluated using

$$f_{\text{DER}} = \sum_{i=1}^{N} \left(a_i P_i^2 + b_i P_i + c_i\right) \tag{1}$$

where $f_{\text{DER}}$, is fuel cost of power thermal generation in ($/h), $a_i$, $b_i$, and $c_i$ are the fossil fuel cost coefficients of DER units.

The amount of emission released in g/kWh can be expressed as [16]:

$$f_E = \sum_{i=1}^{N} \left(\alpha_i P_i^2 + \beta_i P_i + \gamma_i\right) \tag{2}$$

The operational cost of fuel cell plant ($f_{FC}$) can be written as [23]:

$$f_{FC} = \lambda \times \left(\sum_{i=1}^{N_{FC}} P_{FC,i} / \eta_{FC,i}\right) \tag{3}$$

where $\lambda$ is the cost of natural gas ($/kg), $P_{FC,i}$ is the fuel cell generation from unit $i$, $\eta_{FC,i}$ is the efficiency of $i$ th unit and $N_{FC}$ represents the total number of fuel cell plants.

Here the objective is to minimize ($f$);

$$f = w \times f_C + (1 - w) \times h \times f_E \tag{4}$$

where

$$f_C = f_{\text{DER}} + f_{FC} \tag{5}$$

$$h_i = \frac{\left(a_i P_i^2 + b_i P_i + c_i\right)}{\left(\alpha_i P_i^2 + \beta_i P_i + \gamma_i\right)} \tag{6}$$

Subjected to associated constraints:

$$\sum_{i=1}^{N} P_i - P_d - P_l = 0 \tag{7}$$

where,

$$P_l = \sum_{i=1}^{N}\sum_{j=1}^{N} P_i B_{ij} P_j + \sum_{i=1}^{N} B_{0i} P_i + B_{00} \tag{8}$$

$$P_i^{\min} \le P_i \le P_i^{\max} \tag{9}$$

$$H_0 = \sum_{i=1}^{N} \theta_i P_i \tag{10}$$

where,

$$H_0 \ge H_d \tag{11}$$

and

$$\theta_i = \frac{\text{heat rate}\left(\frac{\text{kJ}}{\text{kWh}}\right)}{3600} \times \eta_i \times \eta_{ex} \tag{12}$$

## 2.1 *Topsis*

TOPSIS stands for technique of order preference by simulation to ideal solution. It is a multi-criteria decision scheme, here the solution is closest to the ideal solution and the farthest from the worst solution selected as the BCS. Here it is assumed that there is $m$ alternative option with $n$ criteria for the decision-making. Let $s_{ij}$ is the score of option $i$ with respect to criteria $j$. Where $S = s_{ij}$ is a $m \times n$ matrix. The step associated to find out the BCS is described below.

***Step 1***: Realize normalized score/decision matrix as follows:

$$S_{ij}^{n} = s_{ij} \Big/ \sum_{i=1}^{m} \sqrt{s_{ij}^{2}} \quad i = 1, 2, \ldots m \text{ and } j = 1, 2, 3, \ldots n \tag{13}$$

Here $m$ is the number of alternative solution and $n$ is the number of alternatives.

***Step 2***: Compute the weighted normalized matrix and computed as below:

$$S_{ij}^{w} = \Psi_j \times S_{ij}^{n}. \tag{14}$$

Here,$\Psi_j$ is the weight assigned to a particular objective, $\sum_{j=1}^{n} \Psi_j = 1$.

***Step 3***: Positive ideal and negative ideal solutions are computed using (15) and (17) as:

$$X^{+} = \left\{S_1^{+}, S_2^{+}, \ldots S_n^{+}\right\} \tag{15}$$

where,

$$S_j^{+} = \left\{\max\left(s_{ij}\right) \text{ if } j \in J \text{ and } \min\left(s_{ij}\right) \text{ if } j \in J'\right\} \tag{16}$$

$$X^{-} = \left\{S_1^{-}, S_2^{-}, \ldots S_n^{-}\right\} \tag{17}$$

where,

$$S_j^{-} = \left\{\min(s_{ij}) \text{ if } j \in J \text{ and } \max(s_{ij}) \text{ if } j \in J'\right\}. \tag{18}$$

***Step 4***: The distance of separation from positive ideal and negative ideal solution is computed as:

$$D_i^{+} = \sqrt{\sum_{j=1}^{n}\left(s_{ij} - s_j^{+}\right)^2} \tag{19}$$

$$D_i^{-} = \sqrt{\sum_{j=1}^{n}\left(s_{ij} - s_j^{-}\right)^2} \tag{20}$$

***Step 5***: The relative closeness of each solution is verified by (21) as:

$$RC_i^{*} = \left(\frac{D_i^{-}}{D_i^{+} - D_i^{-}}\right); 0 < RC_i^{*} < 1. \tag{21}$$

The best result is selected with $RC_i^{*}$ closest to 1.

## 3 Coyote Optimization Algorithm

Coyote optimization algorithm (COA) is a nature-inspired population-based meta-heuristic inspired by the canine species' social behavior. In COA, the population size $N$ is split in $N_P$ group with $N_C$ coyote in each group. Therefore,

$$N = N_P \times N_C \tag{22}$$

During optimization, each coyote represents a possible solution, and its social behavior represents the cost of the objective function. The social condition soc (set of decision variable) of coyote $c$ in a group $p$ at any instant of time $t$ can be written as [23]:

$$\text{soc}_c^{p,t} = \chi = \chi_1 + \chi_2 + \cdots + \chi_D \tag{23}$$

Adoption of the coyote in its environment is considered as the cost of the objective function.

Initially, the search agents (coyotes) are generated randomly within upper and lower boundaries of the design variable $j$:

$$\text{soc}_{c,j}^{p,t} = Lb_j + r_j \times \left(Ub_j - Lb_j\right) \tag{24}$$

where random number $r_j \in (0, 1)$, the fitness function of each coyote can be written as [23]:

$$\text{fit}_c^{p,t} = f\left(\text{soc}_c^{p,t}\right) \tag{25}$$

At starting, coyote randomly involves in the group and go out of their pack with probability ($\rho$):

$$\rho = 0.005N_c^2 \tag{26}$$

The leader of coyote ($\alpha$) is considered as best coyote adapted to the environment and represented as:

$$\alpha^{p,t} = \text{soc}_c^{p,t}, \quad \text{for} \quad \min \text{fit}_c^{p,t} \tag{27}$$

The COA algorithm retrieval of information is carried out using cultural transformation among coyotes within the group. It can be modeled as [23]:

$$\text{cul}_j^{p,t} = \begin{cases} \mathcal{R}_{\frac{N_c+1}{2},j}^{p,t} & , N_c \text{ is odd} \\ \cdot & \cdot \quad \cdot \\ \dfrac{\mathcal{R}_{\frac{N_c}{2},j}^{p,t} + \mathcal{R}_{\left(\frac{N_c+1}{2}\right),j}^{p,t}}{2} & , \text{otherwise} \end{cases} \tag{28}$$

The birth of coyotes influenced by the social behavior of two parents that are selected on a random basis within the search space and the environmental factor. On the basis of the above biological fact, the birth of new coyotes is written as:

$$\text{pup}_j^{p,t} = \begin{cases} \text{soc}_{r1,j}^{p,t}, & \text{rand}_j < \rho_s \text{ or } j = j_1 \\ \text{soc}_{r2,j}^{p,t}, & \text{rand}_j \geq \rho_s + \rho_a \text{ or } j = j_2 \\ R_j, & \text{otherwise} \end{cases}. \tag{29}$$

where

$$\rho_s = 1/D \tag{30}$$

$$\rho_a = (1 - \rho_s)/2 \tag{31}$$

For the representation of cultural effect inside the group, two factors $\delta_a$ and $\delta_g$ are considered in COA.

$$\delta_a = \alpha^{p,t} - \text{soc}_{c_{r1}}^{p,t} \tag{32}$$

$$\delta_g = \text{cul}^{p,t} - \text{soc}_{c_{r2}}^{p,t} \tag{33}$$

The social behavior of coyote is updated by the $\alpha$ coyote and the group influence as:

$$\text{new\_soc}_c^{p,t} = r_a\delta_a + r_g\delta_g \tag{34}$$

It can be evaluated as:

$$\text{new\_fit}_c^{p,c} = f\left(\text{new\_soc}_c^{p,t}\right) \tag{35}$$

Above fitness values are analyzed for better fitness values than before as:

$$\text{new\_soc}_c^{p,t+1} = \begin{cases} \text{new\_soc}_c^{p,t}, & \text{new\_fit}_c^{p,t+1} < \text{fit}_c^{p,t} \\ \text{soc}_c^{p,t}, & \text{otherwise} \end{cases} \tag{36}$$

The social condition of coyote, which adapts best itself to environmental condition, is considered as the best solution to the problem. The steps associated with COA for optimization of an objective function are described using the flowchart in Fig. 1.

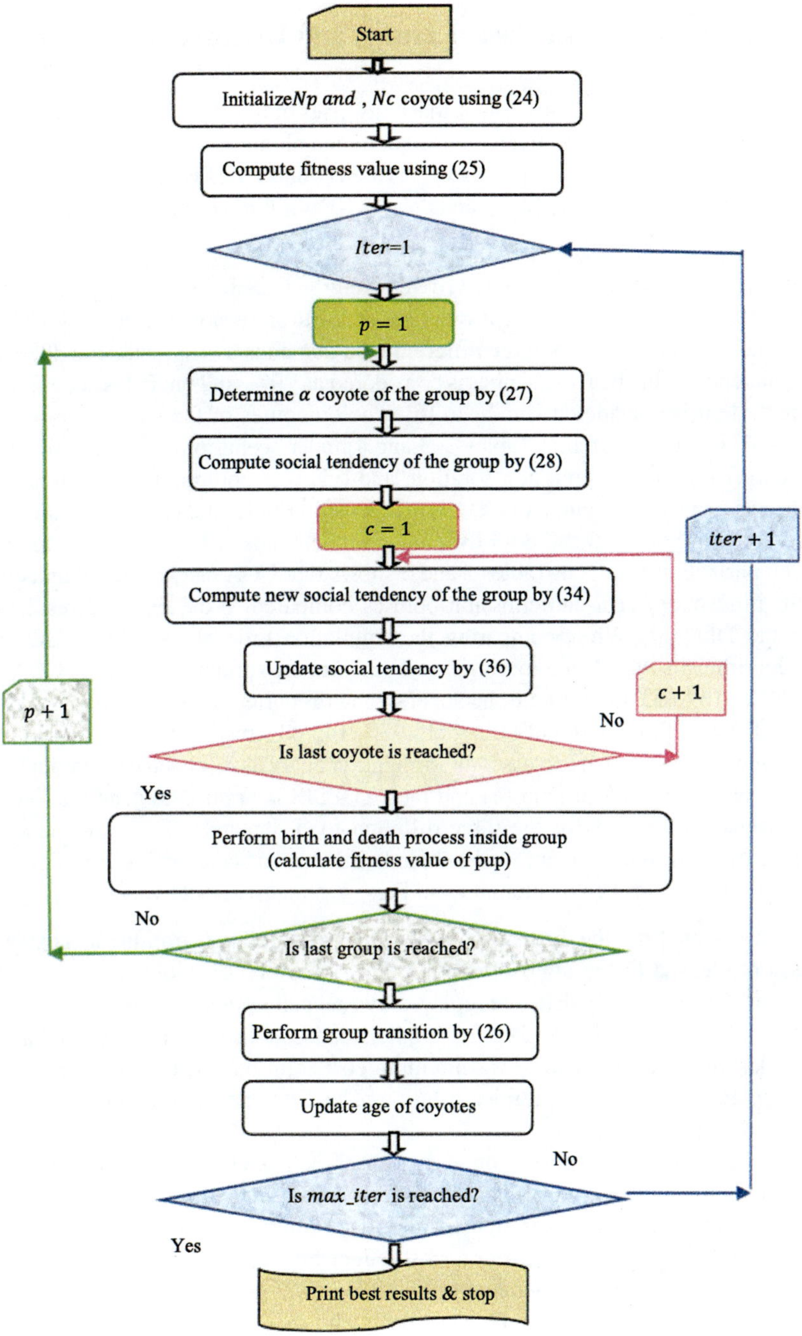


**Fig. 1** Flow chart of COA

## 4 Description of Test Case, Results, and Discussion

The CHPEED problem of a microgrid that comprises of five different DERs is solved using COA in MATLAB R2013a environment and simulations were performed on CPU with configuration as i7 processor, 3.10 GHz and 2 GB RAM. The parameter of COA is considered as $N_P = 5$ and $N_c = 10$, with maximum iteration of 1500 as stopping criteria.

**Case 1**: This test case is comprised of five conventional DERs of a microgrid. Its fuel cost coefficient, emission and heat rate data of DER are adapted from [16]. COA's performance is evaluated for three different load demands as 169, 248, and 338 kW. The efficiency of the heat exchanger is considered as 90%, and the B-loss coefficient of the system is considered similar to [6]. The outcomes of the simulation results for case 1 are summarized in Tables 2, 3 and 4 under scenario of (i) minimum cost solution, (ii) minimum emission solution, and (iii) best compromise solution. The stable cost convergence curve of COA is presented in Fig. 2, and pareto optimal front is presented in Figs. 4, 5 and 6 for different load demands.

Comparison of results in Tables 2 and 3 shows that COA can able to attain better results in terms of cost and emission both as compared to the reported results of PSO and DE both. While comparing the simulation time of the COA, PSO and DE algorithm in Fig. 3, the minimum and maximum simulation time of COA as 110.201 s, 107.333 s is found to be superior than reported simulation time of PSO as 134.078 s, 320.625 s and of DE as 115.97 s, 143.63 s respectively.

For the multiobjective test case, the simulation analysis has been carried out with step increment of 0.05 of $w$ in (4) and then TOPSIS is applied to get BCS, which is presented in Table 4 for the three different load demands. The corresponding optimal front for the different load demands in terms of cost versus emission and then emission versus heat is plotted from Figs. 4, 5 and 6 respectively.

**Case 2**: To analyze the impact of renewable resource integration on cost and emission released to the atmosphere, the third units of DERs of case 1 and then third and fourth units of DERs are replaced by fuel cell system. The data of fuel cell system are listed in Table 8 [24]. For similar load condition, simulation is carried out under the same scenario of (i) minimum cost solution, (ii) minimum emission solution, and (iii) best compromise solution. The outcome of simulation results is presented in Tables 5 and 6.

By comparison of best cost solution, it is observed that cost is reduced by 6.2794$ (26.22%) with single fuel cell plant integration and reduced by 8.3984$ (35.05%) with two fuel cell plant integration for the load demand of 169 kW. For the load demand of 248 kW power generation, cost reduced by 79,275$ (26.81%), 13.6226$ (46,622%) and for the load demand of 338 kW power generation, cost reduced by 8.0928$ (22.38%), 13.9707 (38.64%) respectively.

**Table 2** Results of optimal fuel cost

| $P_d$ (kW) | Method | DER output (kW) | | | | | Cost ($/h) | $NO_X$ (g/kWh) | Heat (kWh) | $P_l$ (kW) |
|---|---|---|---|---|---|---|---|---|---|---|
| | | $P_1$ | $P_2$ | $P_3$ | $P_4$ | $P_5$ | | | | |
| 169 | COA | 0.00 | 60.5427 | 80 | 23.3447 | 6.0 | 23.9714 | 54.4254 | 191.5292 | 0.8874 |
| | DE [16] | 0.00 | 63.2 | 80 | 20.3 | 6.3 | 23.9689 | 54.7162 | 191.716 | – |
| | PSO [16] | 0.00 | 63.86 | 80 | 20 | 6.3 | 23.96 | 54.74 | 191.487 | – |
| 248 | COA | 0.00 | 96.6218 | 80 | 43.2998 | 30 | 29.4246 | 49.7827 | 273.7992 | 1.9216 |
| | DE [16] | 0.00 | 110.10 | 80 | 29.10 | 30 | 29.18 | 50.95 | 273.510 | – |
| | PSO [16] | 0.00 | 108.70 | 80 | 30.48 | 30 | 29.39 | 50.82 | 273.490 | – |
| 338 | COA | 0.00 | 153.5382 | 80 | 81.0725 | 30 | 36.1567 | 45.1739 | 350.9057 | 6.6107 |
| | DE [16] | 0.00 | 166.30 | 80 | 64.3 | 30 | 35.8974 | 45.8467 | 348.048 | – |
| | PSO [16] | 0.00 | 166.68 | 80 | 63.89 | 30 | 35.897 | 45.87 | 348.00 | – |

**Table 3** Results of optimal emission

| $P_d$ (kW) | Method | DER output (kW) | | | | | Cost ($/h) | $NO_X$ (g/kWh) | Heat (kWh) | $P_l$ (kW) |
|---|---|---|---|---|---|---|---|---|---|---|
| | | $P_1$ | $P_2$ | $P_3$ | $P_4$ | $P_5$ | | | | |
| 169 | COA | 0.00 | 68.7279 | 16.00 | 81.5386 | 7.4186 | 25.8180 | 50.4585 | 155.6652 | 4.6851 |
| | DE [16] | 0.00 | 63.90 | 19.40 | 81.60 | 8.20 | 25.75 | 50.490 | 157.740 | – |
| | PSO [16] | 0.00 | 67.85 | 17.08 | 82.11 | 6.00 | 25.74 | 50.490 | 154.74 | – |
| 248 | COA | 0.00 | 118.633 | 33.787 | 88.467 | 13.811 | 31.1477 | 47.2179 | 237.5130 | 6.6986 |
| | DE [16] | 0.00 | 113.30 | 35.10 | 88.30 | 15.40 | 30.9433 | 47.303 | 237.243 | – |
| | PSO [16] | 0.00 | 112.72 | 36.17 | 88.48 | 14.78 | 30.915 | 47.300 | 237.430 | – |
| 338 | COA | 0.00 | 172.9046 | 57.5042 | 96.3632 | 21.1694 | 37.2928 | 44.7166 | 333.6450 | 9.9413 |
| | DE [16] | 0.00 | 166.50 | 58.30 | 96.10 | 21.50 | 36.851 | 44.82 | 329.79 | – |
| | PSO [16] | 0.00 | 166.20 | 58.64 | 96.07 | 21.66 | 36.84 | 44.82 | 330.07 | – |

**Table 4** Results of best compromise solution

| $P_d$ (kW) | Method | DER output (kW) | | | | | Cost ($/h) | $NO_X$ (g/kWh) | Heat (kWh) | $P_l$ (kW) |
|---|---|---|---|---|---|---|---|---|---|---|
| | | $P_1$ | $P_2$ | $P_3$ | $P_4$ | $P_5$ | | | | |
| 169 | COA | 0.00 | 40.0001 | 71.4064 | 52.5646 | 6.0000 | 24.3257 | 52.2208 | 185.4896 | 0.9711 |
| | DE [16] | 0.00 | 40.55 | 75.44 | 48.28 | 6.23 | 24.24 | 52.53 | 188.55 | – |
| | PSO [16] | 0.00 | 40.00 | 73.57 | 51.11 | 6.00 | 24.3028 | 52.33 | 187.2958 | – |
| 248 | COA | 0.00 | 88.3017 | 80.00 | 76.8422 | 6.0000 | 29.5579 | 48.1435 | 256.9700 | 3.1439 |
| | DE [16] | 0.00 | 87.84 | 80.00 | 76.48 | 6.36 | 29.525 | 48.155 | 256.705 | – |
| | PSO [16] | 0.00 | 89.25 | 80.00 | 75.42 | 6.00 | 29.516 | 48.17 | 256.490 | – |
| 338 | COA | 0 | 154.5005 | 77.1329 | 92.9876 | 21.0681 | 36.4368 | 44.8861 | 343.6694 | 7.6621 |
| | DE [16] | 0 | 150.54 | 80 | 90.92 | 20.55 | 36.072 | 45.02 | 341.7225 | – |
| | PSO [16] | 0 | 150.20 | 80 | 89.86 | 21.95 | 36.06 | 45.03 | 342.72 | – |

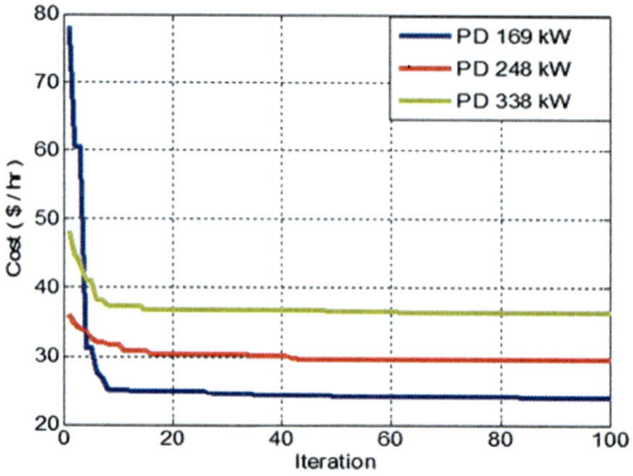


**Fig. 2** Cost convergence curve of COA

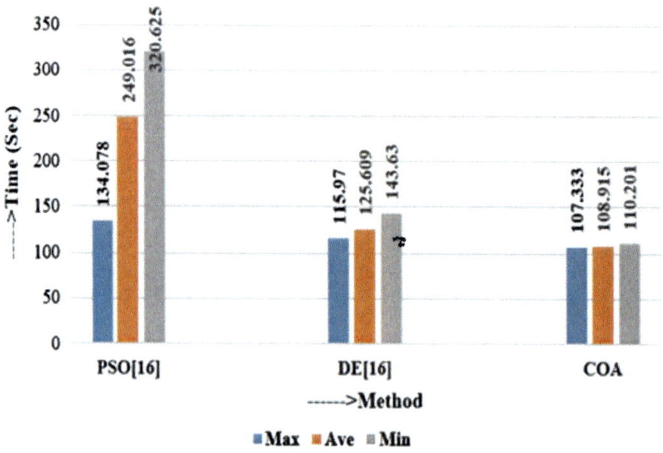


**Fig. 3** Comparison of CPU time

On the other hand while comparing impact of fuel cell plant integration on emission release, it is observed that emission is reduced by 13.611 g/kWh (27%), 16.0589 g/kWh (31.825%) for single, two fuel cell plant integration and power demand of 169 kW. Similarly, reduction in emission of 12.53 g/kWh (26.53%), 14.2166 g/kWh (30.108%) for load demand of 248 kW and reduced by 11.4844 g/kWh (25.68%), by 13.066 g/kWh (29.219%) for load demand of 338 kW respectively.

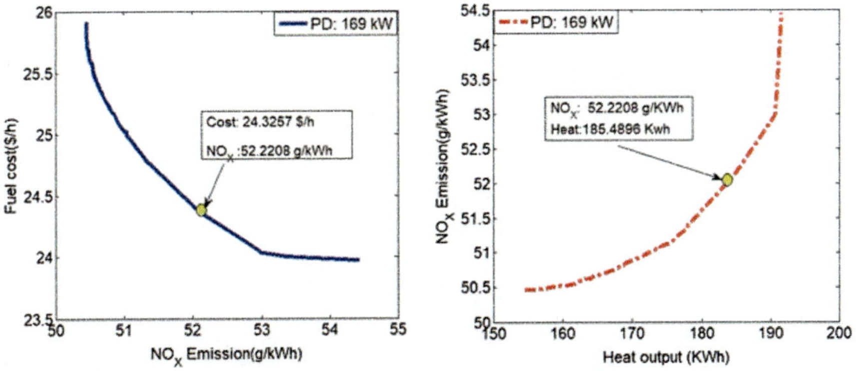


**Fig. 4** Pareto optimal front for PD: 169 kW

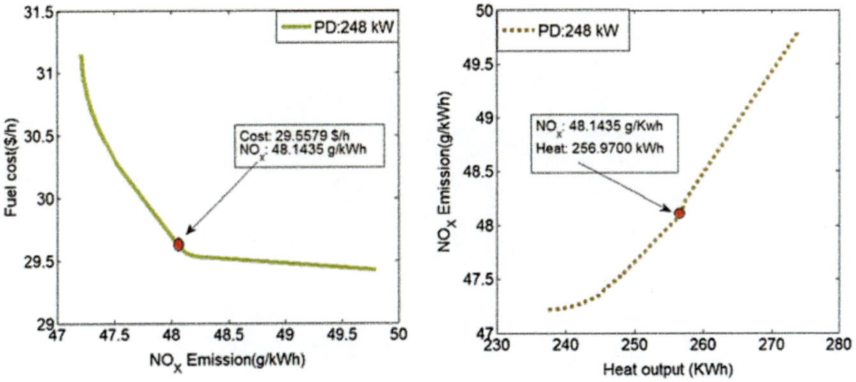


**Fig. 5** Pareto optimal front for PD: 248 kW

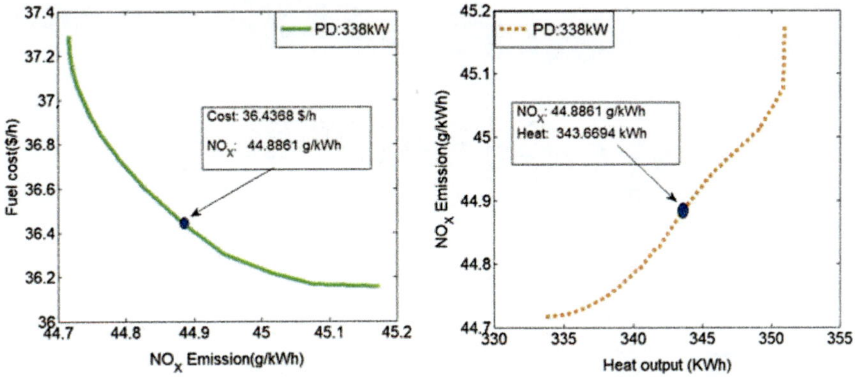


**Fig. 6** Pareto optimal front for PD: 338 kW

**Table 5** Summary of simulation results of COA with single fuel cell integration

| $P_d$ (kW) | Case | DER output (kW) | | | | | Cost ($/h) | $NO_X$ (g/kWh) | Heat (kWh) | $P_l$ (kW) |
|---|---|---|---|---|---|---|---|---|---|---|
| | | $P_1$ | $P_2$ | $P_3$ | $FC_1$ | $P_3$ | | | | |
| 169 | $w = 1$ | 0.00 | 40.00 | 28.6093 | 100.00 | 6.00 | 17.6920 | 40.9787 | 162.2917 | 5.6093 |
| | $w = 0$ | 0.00 | 121.1489 | 37.1738 | 0.0002 | 13.5058 | 24.1597 | 36.8474 | 173.7404 | 2.8286 |
| | BCS | 0.00 | 79.7603 | 43.5863 | 40.2184 | 6.0000 | 21.0609 | 38.4972 | 170.4884 | 1.5650 |
| 248 | $w = 1$ | 0.00 | 66.3504 | 80.000 | 100.00 | 6.00 | 21.6304 | 38.4460 | 257.1701 | 4.3504 |
| | $w = 0$ | 0.00 | 173.5675 | 59.5481 | 0.00 | 20.9036 | 29.5241 | 34.6839 | 260.2234 | 6.0192 |
| | BCS | 0.00 | 128.1407 | 63.9542 | 43.7561 | 15.6072 | 26.0023 | 35.8652 | 255.5264 | 3.4582 |
| 338 | $w = 1$ | 0.00 | 135.2869 | 80.00 | 100.00 | 30.00 | 28.0639 | 35.2019 | 350.8127 | 7.2869 |
| | $w = 0$ | 61.5202 | 190.6545 | 68.0589 | 0.00 | 23.5337 | 38.0531 | 33.2322 | 449.1080 | 5.7673 |
| | BCS | 0 | 184.1704 | 79.6658 | 56.2664 | 25.1234 | 31.3943 | 34.1458 | 348.6827 | 7.2260 |

**Table 6** Summary of simulation results of COA with two fuel cell integration

| $P_d$ (kW) | Case | DER output (kW) | | | | | Cost ($/h) | $NO_X$ (g/kWh) | Heat (kWh) | $P_l$ (kW) |
|---|---|---|---|---|---|---|---|---|---|---|
| | | $P_1$ | $P_2$ | $FC_1$ | $FC_2$ | $P_3$ | | | | |
| 169 | $w = 1$ | 0.00 | 40.00 | 67.3978 | 56.7648 | 6.00 | 15.5730 | 39.3284 | 183.1240 | 1.1626 |
| | $w = 0$ | 0.00 | 155.5088 | 0.00 | 0.00 | 18.4040 | 24.4593 | 34.3996 | 156.8069 | 4.9128 |
| | BCS | 0.00 | 105.1846 | 29.9588 | 26.1755 | 9.9318 | 20.2293 | 36.2024 | 165.5917 | 2.2507 |
| 248 | $w = 1$ | 0.00 | 45.5638 | 100.00 | 100.00 | 6.00 | 15.9356 | 39.0335 | 268.3897 | 3.5638 |
| | $w = 0$ | 37.2834 | 192.4074 | 0.0000 | 0.0000 | 23.8403 | 31.6157 | 33.0013 | 291.7778 | 5.5311 |
| | BCS | 0 | 147.2872 | 69.5618 | 15.9246 | 19.3547 | 23.9263 | 34.5789 | 262.9751 | 4.1283 |
| 338 | $w = 1$ | 0.00 | 113.6896 | 100.00 | 100.00 | 30.00 | 22.1860 | 35.4435 | 361.3610 | 5.6896 |
| | $w = 0$ | 128.7562 | 190.9099 | 0.00 | 0.00 | 23.7575 | 42.0662 | 31.6506 | 526.2740 | 5.4235 |
| | BCS | 0 | 194.9866 | 66.4322 | 58.4409 | 26.5436 | 27.9834 | 33.5250 | 342.7150 | 8.4033 |

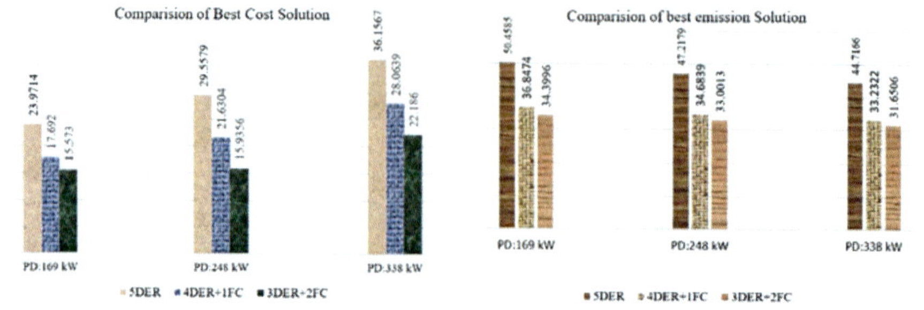


**Fig. 7** Impact of single and multiple fuel plant integration on cost and emission

**Table 7** Comparison of BCS under scenario with/without fuel cell integration

| Case | Cost | NOx | Mu_cost | Mu_emission | TOPSIS |
|---|---|---|---|---|---|
| *PD: 169 kW* | | | | | |
| Case I | 24.3257 | 52.2208 | 0.8179 | 0.6371 | 0.7275 |
| Case II | 21.0609 | 38.4972 | 0.4791 | 0.6007 | 0.5399 |
| Case III | 20.2293 | 36.2024 | 0.4760 | 0.6342 | 0.5551 |
| *PD: 248 kW* | | | | | |
| Case I | 29.5579 | 48.1345 | 0.9227 | 0.6453 | 0.7840 |
| Case II | 26.0023 | 35.8652 | 0.4462 | 0.6860 | 0.5661 |
| Case III | 23.9263 | 34.5789 | 0.4904 | 0.7385 | 0.6144 |
| *PD: 338 kW* | | | | | |
| Case I | 36.4368 | 44.8861 | 0.7523 | 0.6302 | 0.6913 |
| Case II | 31.3943 | 34.1458 | 0.6666 | 0.5362 | 0.6014 |
| Case III | 27.9834 | 33.5250 | 0.7084 | 0.5058 | 0.6071 |

The above comparison in terms of chart for different test conditions is presented in Fig. 7.

The comparison of BCS under the scenario of with/without fuel cell integration for different load demands and its corresponding TOPSIS rank is shown in Table 7. Here also it is observed that there is a significant saving in generation cost as well as reduction in green house emission can be possible if fuel cell plant is integrated with the thermal DER.

## 5 Conclusion

In this paper, COA is successfully implemented for solving a Microgrid scheduling problem to provide particular power demand and its corresponding range of heat demand. Here TOPSIS is used to locate BCS among the conflicting objectives. For sake of confirmation, obtained results are also compared with the reported results by PSO and DE. The impact of fuel cell plant integration on cost and emission reduction is also analyzed with three different power demands. It has been observed that COA performs better under all conditions. This work can be extended with wind and solar integration under the dynamic environment.

**Acknowledgements** This reserch work is financially supported by AICTE-RPS project File No. 8-228/RIFD/RPS/POLICY-1/2018-19 dated 20 March 2020 and IRS/2020/R&D/CE/MKT/07/615 dated 09 March 2020 of TEQIP III. Authors are thankful to the Director MITS, Gwalior for providing essential facilities to carry out this work.

## Appendix

See Table 8.

**Table 8** Generation capacity, cost coefficients and efficiency of fuel cell plant

| Type | $P^{min}$(kW) | $P^{max}$ (kW) | $a_i$ (\$/ kW$^2$ h) | $b_i$ (\$/kWh) | $c_i$ (\$/h) | $\eta$(%) |
|---|---|---|---|---|---|---|
| $f_{FC}$ | 0 | 100 | 0 | 0.07 | 0 | 85 |

## References

1. Guo T, Henwood MI, Van Ooijen M (1996) An algorithm for combined heat and power economic dispatch. IEEE Trans Power Syst 11:1778–1784
2. Dubey HM, Pandit M, Panigrahi BK (2018) An overview and comparative analysis of recent bio-inspired optimization techniques for wind integrated multi-objective power dispatch. Swarm and Evol. Comp 38:12–34
3. Sashirekha A, Pasupuleti J, Moin N (2013) Tan C: Combined heat and power (CHP) economic dispatch solved using Lagrangian relaxation with surrogate sub gradient multiplier updates. Int J Electr Power Energy Syst 44:421–430
4. Chena J, Zhang Y (2019) A lagrange relaxation-based alternating iterative algorithm for non-convex combined heat and power dispatch problem. Electr Power Syst Res 177:
5. Abdolmohammadi HR, Kazemi A (2013) A benders decomposition approach for a combined heat and power economic dispatch. Energy Convers Manag 71:21–31
6. Song Y, Chou C, Stonham T (1999) Combined heat and power economic dispatch by improved ant colony search algorithm. Electr Power Syst Res 52:115–121
7. Ramesh V, Jayabarathi T, Shrivastava N, Baska A (2009) A novel selective particle swarm optimization approach for combined heat and power economic dispatch. Electr Power Components Sys. 37:1231–1240
8. Mellal MA, Williams EJ (2015) Cuckoo optimization algorithm with penalty function for combined heat and power economic dispatch problem. Energy. 93:1711–1718

9. Jayabarathi T, Yazdani A, Ramesh V, Raghunathan T (2014) Combined heat and power economic dispatch problem using the invasive weed optimization algorithm. Front Energy 8:25–30
10. Hagh MT, Teimourzadeh S, Alipour M, Aliasghary P (2014) Improved group search optimization method for solving CHPED in large scale power systems. Energy Convers Manag 80:446–456
11. Roy PK, Paul C, Sultana S (2014) Oppositional teaching learning based optimization approach for combined heat and power dispatch. Int J Electr Power Energy Syst 57:392–403
12. Zohra BF, Lahouari AK, Mostefa R (2019) An improved harmony search algorithm for solved the combined heat and power economic dispatch. Int J Electr Eng Informatics 11:440–450
13. Nazari-Heris M, Mehdinejad M, Mohammadi-Ivatloo B, Babamalek-Gharehpetian G (2019) Combined heat and power economic dispatch problem solution by implementation of whale optimization method. Neural Comput Appl 31:421–436
14. Chen X, Li K, Xu B, Yang Z (2020) Biogeography-based learning particle swarm optimization for combined heat and power economic dispatch problem. Knowl-Based Syst 208:
15. Srivastava A, Das DK (2020) A new Kho-Kho optimization algorithm: an application to solve combined emission economic dispatch and combined heat and power economic dispatch problem. Engg Appl Artif Intell 94:
16. Basu AK, Bhattacharya A, Chowdhury S, Chowdhury SP (2012) Planned Scheduling for economic power sharing in a CHP-based micro-grid. IEEE Trans Power Systems 27(1):30–38
17. Jayakumar N, Subramanian S, Ganesan S, Elanchezhian E (2015) Combined heat and power dispatch by grey wolf optimization. Int J Energy Sect Manag 9:523–546
18. Sundaram A (2020) Multiobjective multi-verse optimization algorithm to solve combined economic, heat and power emission dispatch problems. Appl Soft Comp 91:106195
19. Sun J, Deng J, Li Y (2020) Indicator & crowding distance-based evolutionary algorithm for combined heat and power economic emission dispatch. Appl Soft Comp 90:106158 (2020)
20. Zeng X, Berti S (2019) New optimization method based on energy management in microgrids based on energy storage systems and combined heat and power. Comp Intell 1–26. https://doi.org/10.1111/coin.12238
21. Basu M (2019) Squirrel search algorithm for multi-region combined heat and power economic dispatch incorporating renewable energy sources. Energy 182:296–305
22. Nazari-Heris M, Mohammadi-Ivatloo B, Zare K, Siano, P (2020) Optimal generation scheduling of large-scale multi-zone combined heat and power systems. Energy 210:118497
23. Pierezan J, Coelho LDS (2018) Coyote optimization algorithm: a new metaheuristic for global optimization problems. IEEE Cong Evol Comp (CEC 2018). https://doi.org/10.1109/cec.2018.8477769
24. Basu M, Chowdhury A (2013) Cuckoo search algorithm for economic dispatch. Energy 60(1):99–108

# Chapter 28
# Sparse Solutions in the Identification of Output Error Models

**Vikram Saini and Lillie Dewan**

## 1 Introduction

From the identification perspective, the estimation of linear dynamic models requires the input–output data and prior knowledge about the model structure. The prediction accuracy of model with a large number of parameters introduces the high variance in the estimates and increases the model complexity [1]. The model complexity can be reduced by selecting the model using as few parameters as possible, which correspondingly increases the bias. Standard estimation methods are not regularized and therefore tend to produce statistical errors in the estimation. It is well known that the search for sparse parameters and model orders makes the problem computationally intensive as it requires the search for all combinations of model orders. Alternatively, there are methods which by minimizing the squared error along with penalty on parameter vector, leads to sparse solutions, that is, forcing the insignificant parameters to zero [2–4]. Sparse estimation techniques using the regularization of the parameters can be utilized to optimize the bias-variance trade-off. The motivation for this approach is that a low-dimensional model is, sometimes, sufficient to represent the system efficiently.

The Non-Negative Garrote (NNG) [5] and least absolute shrinkage and selection operator (LASSO) [6] are some early approaches of the penalized least squares. A detailed review of the estimation methods for the sparse solutions of linear systems can be found here [7]. Sparse estimation methods have numerous applications in the literature to construct sparse solutions. The sparse optimization has been used for the identification of impact force acting on airfoil and the wind turbine blade [8]. Sparse estimation of truncation errors arising in the solutions of partial differential equations

V. Saini (✉)
MITS, Gwalior 474005, Madhya Pradesh, India

L. Dewan
NIT Kurukshetra, Kurukshetra 136119, Haryana, India
e-mail: l_dewan@nitkkr.ac.in

© The Author(s), under exclusive license to Springer Nature Singapore Pte Ltd. 2022

H. M. Dubey et al. (eds.), *Artificial Intelligence and Sustainable Computing*,
Algorithms for Intelligent Systems, https://doi.org/10.1007/978-981-16-1220-6_29

has been studied [9]. Also, the sparse solutions for the sparse dynamical systems from a single sample trajectory have been considered [10]. It is well known that rational function models of discrete-time system provide a non-convex optimization problem [1]. Therefore, the convex optimization algorithms such as penalized least square with $l$-1 norm cannot be explicitly implemented for sparse estimation of such models.

In this paper, the application of regularized estimation of sparse linear dynamic models with noise has been studied [11–13]. First, the least square (LS) method penalized with $l_1$ norm is presented for the sparse estimation. This approach has been studied in literature in the compressive sensing setting [11, 14, 15]. The $l_1$ norm is the closest possible convex function to the function $l_0$ and provides a convex optimization problem provided that the cost function without $l_1$ norm of parameter vector is convex [16]. In this paper, the low data assumption has been relaxed to study the other sparsity measure. Sparsity measure in terms of $l_q$ norm $(0 < q < 1)$ of the parameter vector is utilized to recover the sparse estimates.

This paper is organized as follows. Section 2 describes the OE model and associated estimation problem formulation. The two-stage algorithm using the data filtering approach is presented in Sect. 3 along with the regularized least square with $l_1$ and $l_q$ approach. Section 4 presents numerical simulation through Monte-Carlo study. Finally, the last section describes some conclusions.

## 2 System Model and Background

The behavior of the discrete linear time-invariant system can be written in rational function representation [1]. The OE model is represented by the following equation

$$y(k) = \frac{B\left(q^{-1}\right)}{A\left(q^{-1}\right)}u(k) + \varepsilon(k) \tag{1}$$

where $u(k) \in \mathbb{R}$ represents the system input; $y(k) \in \mathbb{R}$ represents the output of the linear system. The signal $\varepsilon(k) \in \mathbb{R}$ is white noise sequence with zero mean and variance $\sigma_\varepsilon^2$. In addition, $A(q^{-1})$ and $B\left(q^{-1}\right)$ are polynomial functions

$$A\left(q^{-1}\right) = 1 + a_1 q^{-1} + a_2 q^{-2} + \cdots + a_{n_a} q^{-n_a} \tag{2a}$$

$$B\left(q^{-1}\right) = b_1 q^{-1} + b_2 q^{-2} + \cdots + b_{n_b} q^{-n_b} \tag{2b}$$

The OE model in (1) may be expressed in the following linear regression form

$$y(k) = \boldsymbol{\psi}(k)\boldsymbol{\theta} + v(k) \tag{3}$$

The parameter vector $\boldsymbol{\theta} := \left[\boldsymbol{a}^T\ \boldsymbol{b}^T\right]^T \in \mathbb{R}^{n_a+n_b}$ and the information vector $\boldsymbol{\psi}(k) := \left[\phi_y(k)\ \phi_u(k)\right] \in \mathbb{R}^{n_a+n_b}$, where,$\phi_y(k) := [-y(k-1) \ldots - y(k-1)], \phi_u(k) := [u(k-1) \ldots u(k-n_a)]$,$\boldsymbol{a} = \left[a_1 \ldots a_{n_a}\right]^T \in \mathbb{R}^{n_a}$ and $\boldsymbol{b} = \left[b_1 \ldots b_{n_b}\right]^T \in \mathbb{R}^{n_b}$. The noise term in (3) is given as:

$$v(k) = A\left(q^{-1}\right)\varepsilon(k) \tag{4}$$

The noise sequence $v(k)$ is a moving average of $n_a + 1$ samples of the white noise $\varepsilon(k)$, which introduces correlation in the output $y(k)$ and noise sequence $v(k)$.

System identification aims to estimate the values of parameters $a_1, \ldots, a_{n_a}, b_1, \ldots, b_{n_b}$ and the model orders ($n_a$ and $n_b$) such that the estimated model uniquely represents the behavior of system. Furthermore, the search for true model orders makes the problem computationally intensive as it requires the search among all combinations of model orders [1]. The parameter estimation solves the following optimization problem

$$\widehat{\boldsymbol{\theta}} = arg\ \underbrace{min}_{\boldsymbol{\theta}\in\mathbb{R}^n}\ \boldsymbol{j}(\boldsymbol{\theta}) \tag{5a}$$

$$\boldsymbol{j}(\boldsymbol{\theta}) := (\boldsymbol{y}(k) - \boldsymbol{\varphi}(k)\boldsymbol{\theta})^2, \boldsymbol{Y} \in \mathbb{R}^L, k = 1, \ldots, L \tag{5b}$$

This squared error is a measure of energy and thus solves the minimization of energy problem. In this paper, it is assumed that Supp($\boldsymbol{\theta}$) $<$ dim($\boldsymbol{\theta}$), where Supp($\boldsymbol{\theta}$) $= \boldsymbol{\theta}_0 = \left\{i \in \mathbb{Z}^+ : \theta_i \neq 0\right\}$ i.e. the number of non-zero components of vector $\boldsymbol{\theta}$. This situation naturally arises when the model orders are unknown as well as linear system exhibits sparse model representation. Sparse estimation techniques using the regularization of the parameters can be utilized. It is convenient to introduce the sparsity measure in the optimization problem. The simple and intuitive approach is to consider the following minimization problem

$$\text{arg}\ \underbrace{\text{min}}_{\boldsymbol{\theta}\in\mathbb{R}^n}\ \boldsymbol{\theta}_0 \tag{6a}$$

$$\text{Subject to}(\boldsymbol{y}(k) - \boldsymbol{\varphi}(k)\boldsymbol{\theta})^2 < \zeta, \boldsymbol{Y} \in \mathbb{R}^L, k = 1, \ldots, L \tag{6b}$$

for some $\zeta > 0$. This is an NP hard problem, and several approximations are given in the literature to solve this problem. The $l_1$ and $l_q$, $0 < q < 1$ are such approximations that solve the above problem as a convex optimization problem at least using $l_1$ approach.

## 3 Sparse Identification

This section introduces the $l_1$ and $l_q$ sparsity measures. Further, the two-stage algorithm is derived using the $l_1$ and $l_q$ norm of parameter vector.

### 3.1 $l_1$ Approach

The best possible approximation is to consider the minimization of $l_1$ norm of parameter vector. Consider the following convex optimization problem [6].

$$\widehat{\boldsymbol{\theta}} = arg \underbrace{min}_{\boldsymbol{\theta} \in \mathbb{R}^n} \boldsymbol{j}(\boldsymbol{\theta}) \tag{7a}$$

$$\text{Subject to } \boldsymbol{\theta}_1 < \zeta \tag{7b}$$

for some $\zeta > 0$. Instead of Eq. (7a, 7b), the following optimization problem has been solved to compute the sparseness in estimates.

$$\hat{\boldsymbol{\theta}} = \arg \underbrace{\min}_{\boldsymbol{\theta} \in \mathbb{R}^n} \frac{1}{2} \|y - \psi\theta\|^2 + \lambda \|\theta\|_1 \tag{8}$$

where $\boldsymbol{\psi} = \left[\varphi^T(1) \ldots \varphi^T(L)\right]^T$ and $\boldsymbol{Y} = [y(1) \ldots y(L)]^T$. It is found that the solution to problem given in Eq. (7a, 7b) can be found using the optimization problem defined in (8). This problem can be interpreted as a penalized least-squares or regularized least square ($l_1$-RLS) where the parameter vector has the minimum norm among all solutions. It is reasonable to expect that the solution is sparse. However due to numerical imp-precision, the solution is not exact and therefore to recover the support, the parameters that are close to zero are taken as zero. The complexity of the model is related to the number of variables involved in the final model.

### 3.2 $l_q$ Approach

In literature, it was pointed that the $l_1$ penalty tends to produce biased estimates for large coefficients. It is suggested to replace the $l_1$ penalty with other penalized functions that lead to sparse models. Among all other penalties to recover the sparse estimates, one such penalty is $l_q$ penalty $0 < q < 1$ [17, 18]. The estimation thus solves the following optimization problem

$$\hat{\theta} = \arg \underbrace{\min}_{\theta \in \mathbb{R}^n} \frac{1}{2} Y - \psi\theta^2 + \lambda\theta_q \quad (9)$$

for some $\lambda > 0$. Now there are two tuning parameters for the optimization problem that is $\lambda$ and $q$. The parameter $q$ controls the penalty function and therefore it is important to select a suitable value. Both parameters $\lambda$ and $q$ are usually computed by grid search over all possible values.

### *3.3 Consistency of the Estimates*

To identify the OE model structure, a two-stage algorithm is adopted. The motivation behind the two-stage algorithm is that a reasonable estimate of noise is sufficient to form the required convex cost. In the case of OE model, a reasonable estimate of the denominator polynomial of rational function representation is sufficient to provide the information of the noise prior to the sparse estimation. In instrumental variable-assisted sparse estimation, initial estimates are obtained using the IV method (see [19, 20]) consisting of observed variables $u(k)$ and $y(k)$ in the first stage. The IV method provides unbiased estimates in the presence of correlated noise. The instrumental variables are given by

$$\zeta^T(k) = \left[\hat{y}(k-1) \ldots - \hat{y}(k-n_{a'})u(k-1) \ldots u(k-n_b)\right] \quad (10)$$

where $n_{a'} > n_a$ and $y(k)$ is the estimate of the model. The parameters estimate using the IV method is obtained according to the following steps.

1. Obtain the LS estimates of the model defined in (1).
2. Obtain the estimates of model $y(k)$ using parameters estimated in the first step.
3. Obtain the IV estimates using the instrumental variables defined in Eq. (10).

The IV method does not provide the minimal possible parameter representation and therefore in the second stage, the penalized optimization problem is solved using the filtered data set available from the IV method. Let the estimate of the polynomial obtained using the IV method be $\hat{A}(q^{-1})$ Further let $u_{\text{fil}}$ and $y_{\text{fil}}$ represent the input and output after filtering operation. The regularized least square with $l_1$ and $l_q$ penalty has been solved using the filtered equations. The matrix $\psi$ is replaced with $\psi_{fil}$ and vector $Y$ is replaced with $Y_{\text{fil}}$ wherein $\psi_{\text{fil}} = \left[\varphi^T_{\text{fil}}(1) \ldots \varphi^T_{\text{fil}}(L)\right]^T$ and $Y_{\text{fil}} = \left[y_{\text{fil}}(1) \ldots y_{\text{fil}}(L)\right]^T$. The detailed steps of the $l$-1 IVRLS are as follows:

Step 1: Collect the data set ($u(k)$,$y(k)$) for $k \in T, \text{Card}(T) = N$, and $T \subset \mathbb{Z}^+$. Initialize the parameter vector $\hat{\theta}$ and assumes an initial order of the model $((\hat{n}_a = \hat{n}_b = c : c > 0))$.

Step 2: Fit the data set ($u(k)$,$y(k)$) using the IV method such that the following property holds

$$E(\boldsymbol{\zeta}(k)v(k)) = 0 \tag{11}$$

Select the estimate with minimum $l$-1 norm. Let the estimates obtained are $\hat{A}(q^{-1}$ and $\hat{B}(q^{-1}$,

Step 3: Obtain the filtered data set

$$y_{\text{fil}}(k) = \frac{1}{\hat{A}_{IV}(q^{-1})}y(k),\ y_{\text{fil}}(k) = \frac{1}{\hat{A}_{IV}(q^{-1})}u(k)$$

Step 4: Construct the matrix $\boldsymbol{\psi}$ and vector $\boldsymbol{Y}$.

Step 5: Compute the initial solution by setting $\hat{\theta}(p) = \hat{\theta}$ for $p = 0$. Iterate $p \rightarrow p+1$ and compute the sparse estimates using penalized least-square solution.

## 4 Simulation Study

This section presents a numerical example to show the effectiveness of the proposed approach compared with other methods using two noise levels with different amplitudes. The analysis of stochastic properties is shown using the Monte-Carlo simulations with 300 runs. The convergence of estimates toward the true values is shown using the criterion of statistically estimation error (SEE)

$$\text{SEE} = \sqrt{\frac{\left\| m(\hat{\theta}) - \theta_s^\circ \right\|_2}{\left\| \theta_s^\circ \right\|_2}} \tag{12}$$

where $m(\hat{\theta})$ represents the sample mean of 300 simulation runs, and $\boldsymbol{\theta}_s^\circ$ denotes the true parameter vector of the model.

**Example** Consider the following discrete OE system

$$A\left(q^{-1}\right) = 1 - 0.50q^{-1} - 0.30q^{-4} \tag{13a}$$

$$B\left(q^{-1}\right) = 0.18q^{-1} - 0.25q^{-3} \tag{13b}$$

It is evident that the true parameter vector say $\boldsymbol{\theta}_s^\circ$ is a sparse vector and the model with $n_a = n_b = 5$ ($\boldsymbol{\theta} \in \mathbb{R}^{10}$) is considered for the sparse estimation. Two different noise levels have been considered the corresponding Signal to Noise Ratio (SNR) $=$ 11.03 dB, and 04.52db. The stochastic properties are investigated using 300 Monte-Carlo simulation study for the considered example. For the algorithm described, the optimization is carried out in two stages. In the first stage, the extended IV method is applied to obtain the initial estimate which is the unbiased estimate for the considered model. The obtained unbiased estimates are used to estimate the noise-free data variables. Further, the information obtained in the previous step is utilized in $l_1$-RLS and $l_q$-RLS optimization to obtain the sparse estimates using the interior point method [12, 21]. The SEE, MC and parameter estimates, for example, are shown in Tables 1, 2, and 3.

In each example, two cases have been investigated with different noise levels. A data set with L = 1600 is gathered from the linear discrete-time OE system defined in the example. The following methods are considered for the comparative study, the LS method, the IV [19] method (with $n_{a'} = 10$), prediction error method available in MATLAB as function "pem" the $l_1$-RLS method without using the IV method in the first stage and the $l_q$-RLS method without using the IV method in the first stage. The regularization parameter $\lambda$ is searched over the grid from 3 to 8 in steps of 0.5. The parameter $\lambda$ is found as 4.5 and 3.2 for the $l_1$-RLS and $l_q$-RLS method, respectively.

It is observed that $l_1$-RLS with IV provides high estimation accuracy with lower SEE and MC in low as well as high SNR case. Also, the estimates have low variance than obtained using the PEM and IV method (for instance see the standard deviation of parameters a1, a2, a5, b2 and b4) in case of 11.03 dB SNR. For SNR = 04.52 dB, the estimates of a2, a4, b1, b2 and b5 contribute large variance using the PEM method. However, the performance of the IV method is similar to $l_1$ -RLS (employed with IV in the first stage) in terms of the obtained standard deviation of estimates. The performance of the regularized method using the IV method provides lower SEE than the $l_1$ -RLS method when employed without using the IV method.

It is seen that from the results that the $l_q$-RLS with IV provides lower SEE and MC as compared to the $l_1$-RLS with IV in both cases of SNR. For low SNR, the performance of $l_q$-RLS and $l_q$-RLS with IV is identical in terms of SEE and MC in high SNR case. However, in high SNR case, the estimates obtained using the $l_q$-RLS with IV are more consistent in the context of sparseness.

## 5 Conclusions

In this paper, a two-stage algorithm to estimate the sparse and unbiased parameters for linear time-invariant discrete system has been presented. This method employs the instrumental variables in the estimation process combined with the penalized estimation using sparsity measure. The IV method provides an estimate of noise, which is further used to filter the input and output data. The filtered data are used for the sparse estimation of linear models using the $l_1$ and $l_q$ sparsity to recover the true support of the model structure. The results are validated using a simulation example with two different cases of noise level through Monte-Carlo simulations. It is shown that the $l_q$ sparsity provides promising results in finding the sparse model. This is evident from the obtained estimates having lower SEE and better bias-variance trade-off using the presented approach.

**Table 1** Monte-Carlo simulation results for example with SNR = 11.03 dB

| Parameters | | $a_1$ | $a_2$ | $a_3$ | $a_4$ | $a_5$ | $b_1$ | $b_2$ | $b_3$ | $b_4$ | $b_5$ |
|---|---|---|---|---|---|---|---|---|---|---|---|
| **LS** | Mean | −0.0777 | −0.0728 | −0.0640 | −0.2412 | −0.1363 | 0.1798 | 0.0762 | −0.2250 | −0.1044 | −0.0235 |
| | Std | 0.0223 | 0.0165 | 0.0137 | 0.0088 | 0.0114 | 0.0021 | 0.0045 | 0.0039 | 0.0063 | 0.0046 |
| **PEM** | Mean | −0.3158 | −0.0897 | −0.0025 | −0.2994 | −0.0560 | 0.1799 | 0.0334 | −0.2496 | −0.0460 | −0.0008 |
| | Std | 0.6187 | 0.3151 | 0.0270 | 0.0160 | 0.1846 | 0.0020 | 0.1111 | 0.0064 | 0.1545 | 0.0079 |
| **IV** | Mean | −0.4835 | 0.0201 | −0.0156 | −0.2966 | −0.0066 | 0.1799 | 0.0032 | −0.2482 | −0.0040 | −0.0202 |
| | Std | 0.1691 | 0.0934 | 0.0291 | 0.0170 | 0.0523 | 0.0024 | 0.0306 | 0.0067 | 0.0420 | 0.0078 |
| **$l_1$-RLS** | Mean | −0.0819 | −0.0674 | −0.0551 | −0.2322 | −0.1316 | 0.1720 | 0.0690 | −0.2166 | −0.0968 | −0.0162 |
| | Std | 0.0225 | 0.0164 | 0.0136 | 0.0088 | 0.0114 | 0.0021 | 0.0045 | 0.0039 | 0.0063 | 0.0045 |
| **$l_1$-RLS IV** | Mean | −0.4809 | 0.0852 | −0.0040 | −0.2530 | −0.0016 | 0.1579 | 0.0000 | −0.2133 | −0.0051 | −0.0051 |
| | Std | 0.0530 | 0.0505 | 0.0294 | 0.0216 | 0.0179 | 0.0091 | 0.0035 | 0.0099 | 0.0078 | 0.0056 |
| **$l_q$-RLS** | Mean | −0.0787 | −0.0719 | −0.0624 | −0.2394 | −0.1356 | 0.1784 | 0.0748 | −0.2235 | −0.1030 | −0.0221 |
| | Std | 0.0225 | 0.0164 | 0.0139 | 0.0087 | 0.0115 | 0.0022 | 0.0047 | 0.0037 | 0.0065 | 0.0049 |
| **$l_q$-RLS IV** | Mean | −0.4888 | 0.0016 | −0.0045 | −0.2941 | −0.0072 | 0.1570 | 0 | −0.2246 | −0.0011 | 0 |
| | Std | 0.1332 | 0.0764 | 0.0252 | 0.0217 | 0.0405 | 0.0111 | 0.0175 | 0.0130 | 0.0256 | 0.0022 |
| **True value** | | **−0.50** | **0** | **0** | **−0.30** | **0** | **0.18** | **0** | **−0.25** | **0** | **0** |

**Table 2** Monte-Carlo simulation results for example with SNR = 04.52 dB

| Parameters | | $a_1$ | $a_2$ | $a_3$ | $a_4$ | $a_5$ | $b_1$ | $b_2$ | $b_3$ | $b_4$ | $b_5$ |
|---|---|---|---|---|---|---|---|---|---|---|---|
| **LS** | Mean | −0.0454 | −0.0274 | −0.0499 | −0.1789 | −0.1134 | 0.1798 | 0.0818 | −0.2135 | −0.1047 | −0.0237 |
| | Std | 0.0231 | 0.0182 | 0.0177 | 0.0136 | 0.0145 | 0.0047 | 0.0058 | 0.0058 | 0.0081 | 0.0070 |
| **PEM** | Mean | −0.3768 | −0.0626 | 0.0027 | −0.3023 | −0.0350 | 0.1798 | 0.0220 | −0.2496 | −0.0305 | −0.0001 |
| | Std | 0.0046 | 0.1171 | 0.0147 | 0.1626 | 0.0176 | 0.6480 | 0.3189 | 0.0608 | 0.0355 | 0.2004 |
| **IV** | Mean | −0.4860 | 0.1056 | −0.0405 | −0.2672 | −0.0162 | 0.1797 | 0.0026 | −0.2336 | −0.0030 | −0.0154 |
| | Std | 0.1305 | 0.1073 | 0.0653 | 0.0393 | 0.0453 | 0.0053 | 0.0244 | 0.0156 | 0.0317 | 0.0165 |
| **$l_1$-RLS** | Mean | −0.0485 | −0.0168 | −0.0320 | −0.1614 | −0.1020 | 0.1651 | 0.0685 | −0.1972 | −0.0902 | −0.0104 |
| | Std | 0.0227 | 0.0162 | 0.0176 | 0.0133 | 0.0146 | 0.0049 | 0.0060 | 0.0056 | 0.0083 | 0.0063 |
| **$l_1$-RLS IV** | Mean | −0.2908 | 0.0996 | −0.0155 | −0.1368 | −0.0021 | 0.1407 | 0.0055 | −0.1701 | −0.0068 | −0.0024 |
| | Std | 0.1111 | 0.1095 | 0.0603 | 0.0576 | 0.0229 | 0.0185 | 0.0084 | 0.0225 | 0.0167 | 0.0098 |
| **$l_q$-RLS** | Mean | −0.0463 | −0.0264 | −0.0478 | −0.1767 | −0.1123 | 0.1779 | 0.0799 | −0.2115 | −0.1027 | −0.0218 |
| | Std | 0.0234 | 0.0180 | 0.0178 | 0.0137 | 0.0146 | 0.0047 | 0.0060 | 0.0059 | 0.0084 | 0.0073 |
| **$l_q$-RLS IV** | Mean | −0.4853 | 0.0083 | −0.0024 | −0.2665 | −0.0096 | 0.1480 | −0.0008 | −0.2088 | −0.0006 | −0.0021 |
| | Std | 0.1014 | 0.0837 | 0.0512 | 0.0364 | 0.0319 | 0.0138 | 0.0083 | 0.0141 | 0.0125 | 0.0066 |
| **True value** | | **−0.50** | **0** | **0** | **−0.30** | **0** | **0.18** | **0** | **−0.25** | **0** | **0** |

**Table 3** SEE and MC for example with SNR = 11.032 dB and SNR = 04.52 dB

| | SNR = 11.03 dB | | SNR = 04.52 dB | |
|---|---|---|---|---|
| | SEE | MC | SEE | MC |
| **LS** | 0.8506 | 10 | 0.8767 | 10 |
| **PEM** | 0.5774 | 8 | 0.8714 | 8 |
| **IV** | 0.2389 | 7 | 0.4298 | 9 |
| $l_1$ **-RLS** | 0.8433 | 10 | 0.8714 | 10 |
| $l_1$ **-RLS IV** | 0.4053 | 5 | 0.6718 | 6 |
| $l_q$ **-RLS** | 0.8491 | 10 | 0.8755 | 10 |
| $l_q$ **-RLS IV** | 0.2386 | 4 | 0.3142 | 4 |

However, when the noise level increases, the performance of the $l_1$ regularized estimation suffers. It is also noticed that the sparse estimation depends on the initialization. This suggests that the solution to the penalized estimation problem depends on the penalty function and the number of parameters.

**Acknowledgements** The authors gratefully acknowledge to Ministry of Human Resource Development (MHRD), New Delhi for providing the financial support under TEQIP (TEQIP-II & TEQIP-III).

## References

1. Ljung L (1999) System Identification: theory for the User, 2nd edn. Prentice-Hall, Upper Saddle River, NJ
2. Akaike H (1974) A new look at the statistical model identification. IEEE Trans Autom Control 19(6):716–723
3. Schwarz G (1978) Estimating the dimension of a model. Ann Statist 6(2):461–464
4. Efon B, Hastie T, Johnstone I, Tibshirani R (2004) Least angle regression. Ann Statist 32(2):407–451
5. Breiman L (1995) Better subset regression using the nonnegative garrote. Technometrics 37(4):373–384
6. Tibshirani R (1996) Regression shrinkage and selection via the LASSO. J Roy Statist Soc ser B 58(1):267–288
7. Bruckstein AM, Donoho DL, Elad M (2009) From sparse solutions of systems of equations to sparse modeling of signals and images. SIAM Rev 51:34–81
8. Qiao B, Liu J, Liu J, Yang Z, Chen X (2019) An enhanced sparse regularization method for impact force identification. Mech Syst Signal Proc 126:341–367
9. Thaler S, Paehler L, Adams NA (2019) Sparse identification of truncation errors. J Comput Phys 397:
10. Fattahi S, Matni N, Sojoudi S (2019) Learning sparse dynamical systems from a single sample trajectory. In: 2019 IEEE 58th conference on decision and control (CDC), Nice, France, pp 2682–2689
11. Rojas CR, Tóth R, Hjalmarsson H (2014) Sparse estimation of polynomial and rational dynamical models. IEEE Trans Autom Control 59(11):2962–2977

12. Vikram, Dewan L (2016) Sparse identification of output error models using l-1 regularized least square. In: 2016 IEEE first international conference on control, measurement and instrumentation (CMI), Kolkata
13. Saini V, Dewan L (2018) Sparse parameter estimation of LTI models with lp sparsity using genetic algorithm. Int J Model Identif Control 29(1)
14. Cotter SF, Rao BD, Engan K, Kreutz-Delgado K (2005) Sparse solutions to linear inverse problems with multiple measurement vectors. In: IEEE Transactions on Signal Processing, vol 53, no 7, pp 2477–2488, July 2005
15. Gasso G, Rakotomamonjy A, Canu S (2009) Recovering sparse signals with a certain family of nonconvex penalties and DC programming. IEEE Trans Signal Proc 57(12):4686–4698
16. Rockafellar RT (1970) Convex analysis. Princeton University Press, Princeton, NJ
17. Frank I, Friedman J (1993) A statistical view of som chemometrics regression tools (with discussion). Technometrics 35:109–148
18. Fu WJ (1998) Penalized regression: the bridge versus the lasso. J Comput Graph Stat 7:397–416
19. Soderstrom T, Stoica P (1983) Instrumental variable methods for system identification. Springer, Berlin
20. Stoica P, Nehorai A (1988) Non-iterative optimal min-max instrumental variable method for system identification. Int J Control 47(6):1759–1769
21. Wright SJ (1997) Primal-Dual interior-point methods. SIAM, PA

# Chapter 29
# Techno-Economic Assessment of a Hybrid Renewable Energy System Using Dynamic Search Space

**Poonam Singh, Manjaree Pandit, Laxmi Srivastava, and Nikhil Paliwal**

## Nomenclature

| | |
|---|---|
| DSS | Dynamic search space |
| CRF | Capacity Recovery factor |
| $I_{cc}$ | Initial capital cost |
| ANC | Annualized cost |
| $M_{NC}$ | Operating and maintains cost |
| $cf_{nt}$ | Net capacity factor |
| n | Number of years |
| i | Interest rate |
| $N_{PV}$ | Number of solar PV |
| $N_{WT}$ | Number of wind turbine |
| $N_{DG}$ | Number of DG |
| $N_{ST}$ | Number of storages |
| $N_S$ | Number of sources |
| $P_{gen}$ | Total power generation |
| $P_d$ | Load demand |
| $P_{losses}$ | Power losses |
| $P_{PV}$ | Power of solar |
| $P_{WT}$ | Power of wind |
| $P_{DG}$ | Power of DG |
| $P_{st}$ | Power of storage |

P. Singh (✉) · M. Pandit · L. Srivastava · N. Paliwal
Department of Electrical Engineering, Madhav Institute of Technology and Science, Gwalior, Madhya Pradesh, India

M. Pandit
e-mail: manjaree_p@hotmail.com

L. Srivastava
e-mail: srivastaval@hotmail.com

© The Author(s), under exclusive license to Springer Nature Singapore Pte Ltd. 2022

H. M. Dubey et al. (eds.), *Artificial Intelligence and Sustainable Computing*, Algorithms for Intelligent Systems, https://doi.org/10.1007/978-981-16-1220-6_30

## 1 Introduction

With social and lifestyle changes, the demand for electrical power has significantly increased over the years, particularly in developing countries. This has led to more attention toward reducing the environmental pollution. Clean energy from renewable sources is slowly replacing pollution-causing and fast depleting fossil fuels. However, maintaining power balance under constantly changing loads with uncertain renewable power availability poses a great challenge for utilities [1–4]. Therefore, hybrid energy systems having mixed generation and adequate storage are increasingly being proposed worldwide. The ease of integrating distributed generating units in the smart microgrids has opened many new dimensions and a plethora of challenges for researchers. Energy management aims to meet the continuously varying demand for energy in the most efficient, economical, environment-friendly and reliable manner [5].

The higher maintenance and operating costs, maintenance, transmission losses and difficulties in integrating renewable power generation units pose problems in grid expansion to remote areas [6]. Economic feasibility and geographical issues also play an important role in decision-making and policy planning in such cases. Power generation through diesel comes out with various real-life difficulties like complex maintenance of machines, fuel carrying issues and emissions of harmful pollutants. The fundamental requirement of electricity in remote areas is for domestic purposes rather than for industrial purposes. Therefore, locally available small renewable energy (RE) sources are usually found to be economically feasible solutions for meeting the power requirements of isolated areas. To ensure minimum loss of power supply probability (LPSP), a mix of renewable power units along with adequate storage can be employed [7]. This can be realized by isolated or utility-connected hybrid microgrids.

The energy produced from RE sources like solar, wind, tidal etc. has non-linearity in nature and the same is the case with load demand, so it is important to bridge the gap between load demand and energy available from generating sources using suitable ES [8]. The optimal allocation of ES in HRES is very important from techno-economic point of view [9]. The RE sources alone may not be sufficient to meet the load demand and hence the generation capacity needs to be augmented by using some conventional sources of energy such as diesel- or gas-powered generators etc. along with RE sources [10]. Then maintaining emission within permissible limits is a major factor while planning and designing the HRES. The transportation of fuel to remote areas is a big problem and the optimal HRES sizing model must consider this factor [11].

HRES is a combination of various renewable energy sources (RES) like solar photovoltaic (PV) panels, wind power (WT), diesel generator (DG) and energy storage (ES) system. The optimization of HRES includes the optimal sizing of these constituent systems for techno-economics and reliability. This may be achieved by employing various methods and techniques. Paliwal et al. [12] develop the relationship for the size of ES, cycle count and replacement over lifetime, they also optimize HRES of PV-Wind-DG and fulfill techno-eco-social criterion. Askarzadeh [13] develops an algorithm discrete harmony search simulated annealing (DHSSA) for the optimization of PV-Wind-ES system and compares the result obtained of annual cost with discrete harmony search (HS) and harmony search simulated annealing (HSSA) algorithms. Marie et al. [14] optimize HRES of PV-Wind-DG-ES (various battery configurations like lead acid, vanadium redox flow, lithium ion battery). Kumar et al. [15] proposed the optimization solution for HRES of PV-Wind using BBO algorithm and compare the optimized result from GA, PSO, HOMER.

This paper discusses the optimization of hybrid renewable system of wind, solar, DG and battery using interior point algorithm (IPA) and particle swarm optimization (PSO). The battery is used to meet supply and demand gap arise from the uncertainty of load and variation in power generation from renewable energy sources. In this paper, the life cycle cost and load profile of all four seasons are taken into consideration for analysis. The techno-economic analysis has also been carried out to find the effect of dynamic search space (DSS) on LCE for HRES. Section 2 presents the economic model, Sect. 3 presents the methodology using PSO, Sect. 4 presents objective function and constraints, Sect. 5 presents results and discussion and Sect. 6 concludes the findings.

## 2 Economic Modeling

Here as shown in Fig. 1, the configuration proposed of HRES comprising of solar, wind and diesel. The LCE is based on the ratio of life cycle cost (LCC) to lifetime energy production for a given HRES [16–18]. The life cycle cost (LCC) comprising of capital investment, annual maintenance and operation cost [19, 20]. The LCOE is shown in (1).

$$\text{LCE} = \frac{\text{Life cycle cost}}{\text{Life time energy production}} \tag{1}$$

$$\text{LCE} = \frac{(\text{CRF} * I_{\text{cc}} + (\text{ANC} + M_{\text{NC}} * n))}{8760 * \text{cf}_{\text{nt}}} \tag{2}$$

Capacity recovery factor (CRF) can be used to calculate the annualized cost of the system component. It is used to get current worth and can be calculated as.

$$\text{CRF} = \frac{i(1+i)^n}{(1+i)^n - 1} \tag{3}$$

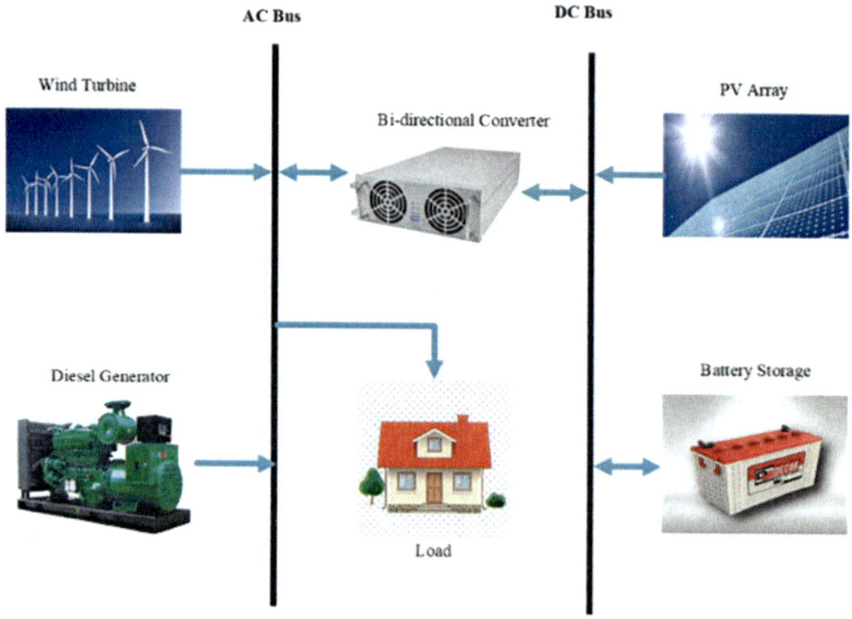


**Fig. 1** Energy flow diagram hybrid renewable energy system

## 3 Methodology

Initially, Eberhart and Kennedy have proposed the technique PSO in 1995 and later many different versions were presented [21]. The PSO is a modern metaheuristic and nature-inspired population-based algorithm inspired by the movement of birds when searching for food. It is a very powerful but simple optimization tool and provides solutions for problems by scattering random particles in the search domain. Each particle represents one solution of the problem and can be represented in the n-dimensional search space by a vector. The position of each particle is updated using velocity, which consists of three terms: (i) inertia term representing previous velocity, (ii) cognitive term that shares knowledge from the best position of particle achieved so far (pbest) and (iii) social term that takes update information from the global best solution (gbest) [22]. The optimization flow chart is shown in Fig. 2.

## 4 Objective Function and Constraints

The objective of this study is the minimization of the levelized cost of energy for the hybrid renewable energy system, which consists of solar, wind, DG and storage. Decision variables of this problem are the number of units of each power supply

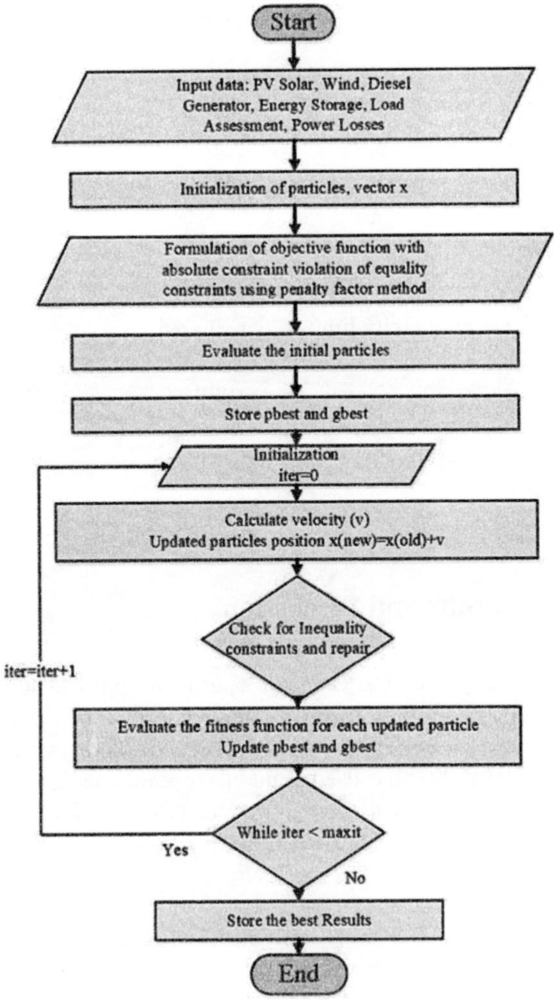


**Fig. 2** Optimization process with PSO

system such that all binding constraints along with power balance are satisfied at each time [23, 24].

$$\text{Min LCE}(\text{N}_{\text{PV}}, \text{N}_{\text{WT}}, \text{N}_{\text{DG}}, \text{N}_{\text{ST}}) \tag{4}$$

Generation constraint for the HRES is given by:

$$\sum_{S=1}^{N_s} P_{\text{gen}} = P_d + P_{\text{losses}} \mp P_{\text{st}} \tag{5}$$

Upper and lower limits of power supply units are given by:

$$N_{\mathrm{pv}}^{\min} \leq N_{\mathrm{pv}} \leq N_{\mathrm{pv}}^{\max} \tag{6}$$

$$N_{\mathrm{wt}}^{\min} \leq N_{\mathrm{wt}} \leq N_{\mathrm{wt}}^{\max} \tag{7}$$

$$N_{\mathrm{dg}}^{\min} \leq N_{\mathrm{dg}} \leq N_{\mathrm{dg}}^{\max} \tag{8}$$

$$N_{\mathrm{st}}^{\min} \leq N_{\mathrm{pv}} \leq N_{\mathrm{st}}^{\max} \tag{9}$$

Storage capacity limits are given by:

$$P_{\mathrm{st}}^{\min} \leq P_{\mathrm{st}} \leq P_{\mathrm{st}}^{\max} \tag{10}$$

For achieving the best results, the search space given by (6)–(9) is dynamically varied.

## 5 Results and Discussion

This study attempts to find an optimal design and sizing of various components of the HRES. Table 1 gives the complete economic information of the different power sources [24, 25]. Here, converter cost and efficiency are not considered for simplicity. MATLAB 2018b is used to find the optimal results. Figure 3 shows the load profile for all four seasons [18, 24]. Here PSO and IPA are used for optimization. The parameters of the PSO and IPA are as follow:

PSO: Population size = 20; trials = 10; c1, c2 = 2; w1 = 0.9; w2 = 0.4; iterations = 100;

IPA: x0 = 0.5; trials = 10; maxfun_eval = 20000; maxiter = 200.

Two test cases have been considered for study, analysis and comparison with or without diesel generator.

Test Case I: The HRES configuration for this case is solar–wind–diesel–storage

Test Case II: The HRES configuration for this case is solar–wind–storage.

**Table 1** Economic data of power sources in the HRES [18]

| | Solar PV | WT | DG | Storage |
|---|---|---|---|---|
| IC(€/kW) | 2832 | 5832 | 148 | 148 |
| AC(€/kW) | 333.00 | 685.02 | 70.01 | 39.04 |
| O&M (€/kW/year) | 56.70 | 116.64 | 6.40 | 2.96 |
| Life (year) | 20.0 | 20.0 | 20.0 | 5.0 |
| Rating (kW/kWh) | 0.53 | 1.5 | 2.5 | 1.0 |
| CF | 0.3 | 0.6 | 0.8 | 0.9 |

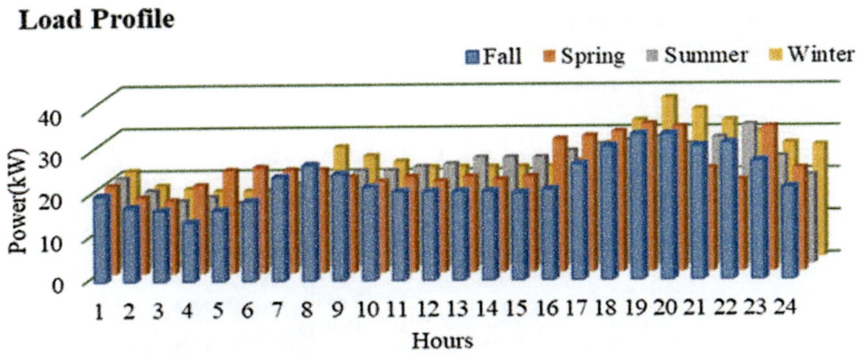


**Fig. 3** Hourly load profile for the four seasons [18, 24]

## *5.1 Dynamic Search Space (DSS)*

To find the sizes of various HRES units such that the life cost over 20 years is minimized while the seasonal load demands are met, the developed model includes power balance constraints on an hourly basis. The problem constraints given by (6)–(9) were found to influence the results quite significantly as the selected upper bound puts a limit on the search capacity of the algorithm. As a result, the optimizer converges to a higher LCE. Therefore, the search space was dynamically varied (within 0–30) in a random manner to facilitate the optimizer to select the best combination/number of power units in the HRES. The best hourly upper limits for the fall season (which produce the least value of LCE) are shown in Figs. 4 and 5 for IPA and PSO respectively. The convergence characteristics of IPA and PSO are shown in Fig. 6.

The optimal results of IPA and PSO for Test Case I are compared in Table 2. The PSO produces better results for LCE and it is found to be about 11 times faster.

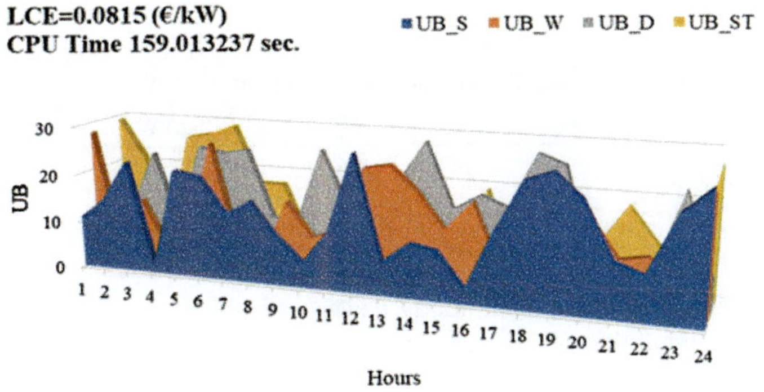


**Fig. 4** Selection of optimal search space using DSS for IPA

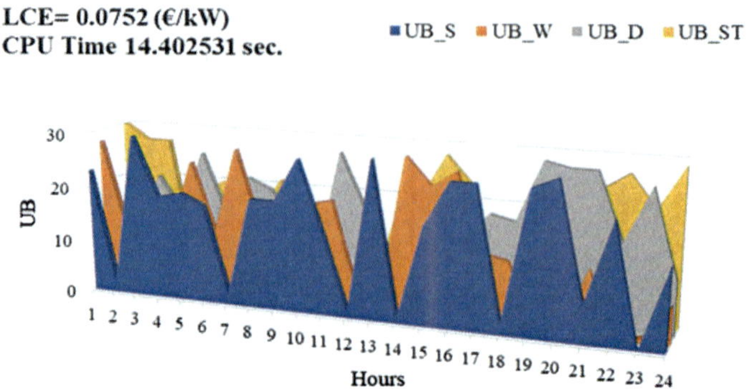


**Fig. 5** Selection of optimal search space using DSS for PSO

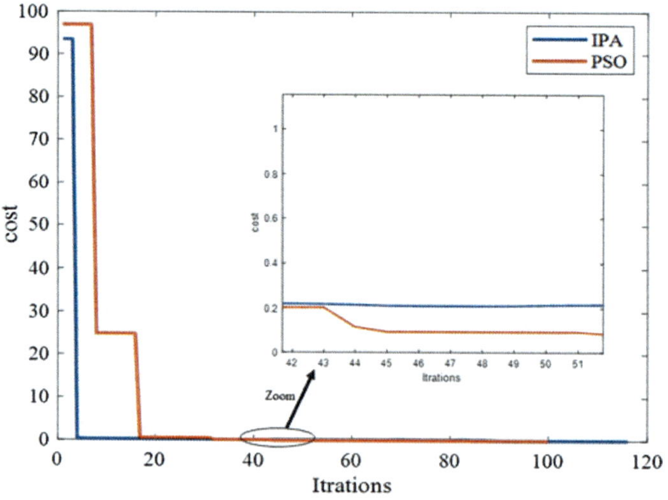


**Fig. 6** Convergence curve of PSO and IPA for the test case I

**Table 2** Comparison of results of IPA and PSO using DSS for test case I (Fall)

| Algorithm | $N_{PV}$ | $N_{WT}$ | $N_{DG}$ | $N_{ST}$ | LCE(€/kW) | CPU time |
|---|---|---|---|---|---|---|
| IPA | 11 | 12 | 14 | 26 | 0.0815 | 159.0132 |
| PSO | 26 | 18 | 12 | 30 | 0.0752 | 14.4025 |

Figures 7 and 8 show the optimal hourly power schedule computed using IPA and PSO for 1 day in fall season, respectively.

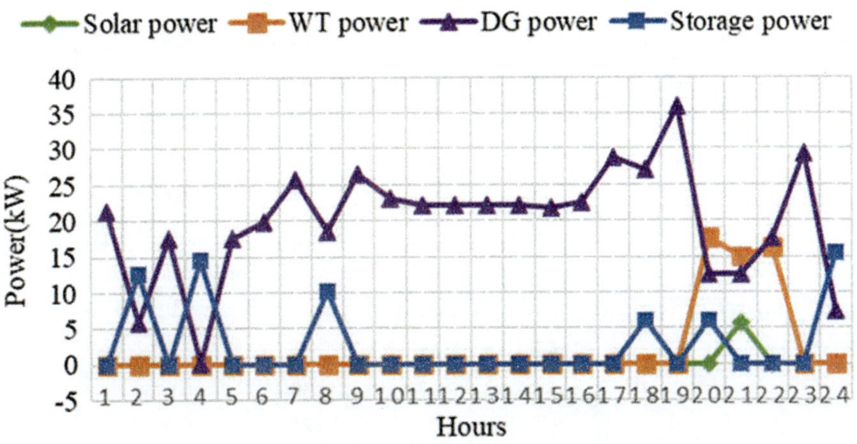


**Fig. 7** Optimal hourly power schedule computed by IPA for a sample day

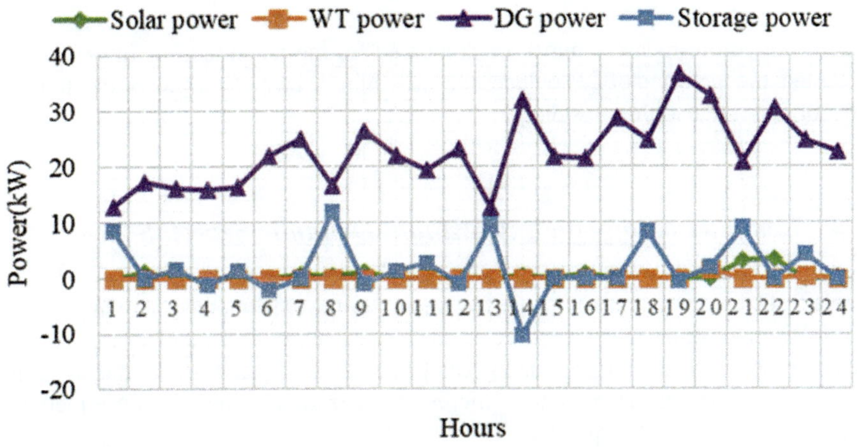


**Fig. 8** Optimal hourly power schedule computed by PSO for a sample day

## *5.2 Optimal Sizing Considering Seasonal Loads*

The effect of seasonal loads on optimal sizing of HRES computed using the PSO is shown in Table 3 for Test Case I.

**Table 3** Optimal sizing considering the seasonal variation of demand using PSO (Case-I)

| Seasons | $N_{PV}$ | $N_{WT}$ | $N_{DG}$ | $N_{ST}$ | LCE(€/kW) |
|---|---|---|---|---|---|
| Fall | 26 | 18 | 12 | 30 | 0.0752 |
| Spring | 30 | 17 | 14 | 23 | 0.0755 |
| Summer | 23 | 24 | 15 | 26 | 0.0769 |
| Winter | 28 | 19 | 10 | 29 | 0.0791 |

**Table 4** Effect of DG on optimal sizing and LCE using PSO

| Test Case | $N_{PV}$ | $N_{WT}$ | $N_{DG}$ | $N_{ST}$ | LCE(€/kW) |
|---|---|---|---|---|---|
| I | 26 | 18 | 12 | 30 | 0.0752 |
| II | 30 | 20 | – | 21 | 0.3607 |

## 5.3 *Effect of Diesel Generator on LCE*

The results of Test Case I and II are compared for the fall season in Table 4. It is clear that the inclusion of DG reduces the LCE significantly. This may be due to less cost of diesel due to the subsidies.

## 5.4 *Comparison of with and Without Dynamic Search Space*

Table 5 shows the superiority and effectiveness of the DSS method in reducing the LCE. By allowing the optimizer to select the best value from a dynamically changing search space, a significant cost saving can be achieved. Figure 9 shows the value of LCE using different methods and different optimization techniques. DSS method minimizes LCE 67% compared to without DSS.

**Table 5** Comparison of with and without DSS method

| Technique | Methods | $P_{PV}$ | $P_{WT}$ | $P_{DG}$ | LCE(€/kW) |
|---|---|---|---|---|---|
| IPA | With_DSS | 5.576 | 48.832 | 479.493 | 0.0815 |
| | Without_DSS | 2.0310 | 290.6310 | 300.000 | 0.2462 |
| PSO | With_DSS | 11.5574 | 1.9008 | 540.6249 | 0.0752 |
| | Without_DSS | 6.9323 | 270.8842 | 298.0139 | 0.2417 |

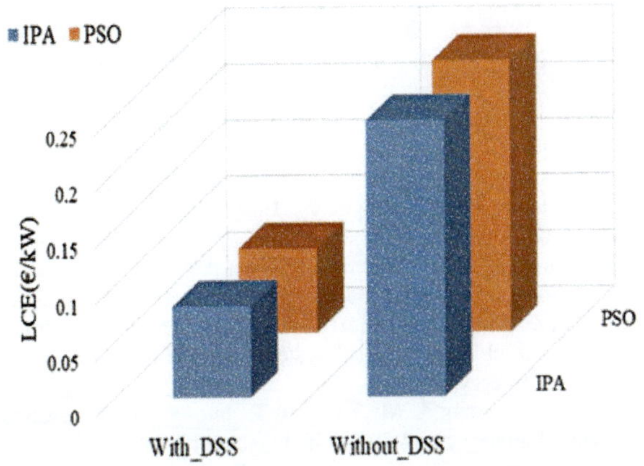


**Fig. 9** Comparison of efficacy of the DSS method using IPA and PSO

# 6 Conclusion

The aim of this paper is the sizing and allocation of a hypothetical HRES. This paper investigates the optimization of HRES by taking LCE as optimization function. The LCE concept establishes techno-economic viability of the hybrid power system over its lifetime.

The optimal LCE is computed using dynamic search space (DSS) method employing an interior point algorithm (IPA) and the results are compared and validated using the PSO model. The effect of different seasons on the optimal sizing and composition of the HRES is also analyzed. The effect of DG on LCE is studied and it is observed that the DSS method is very effective in optimal sizing with about 67% reduction in life cycle cost. The study is expected to be useful for optimizing a large number HRES, which are emerging worldwide.

**Acknowledgements** The authors acknowledge and thankful to the Director MITS Gwalior for facilitating to carry out this research work. The authors are thankful to MHRD New Delhi for TEQIP-III financial assistance program.

# References

1. Ramesh M, Saini R (2020) Effect of different batteries and diesel generator on the performance of a stand-alone hybrid renewable energy system. In: Energy Sources, Part A: Recovery, Utilization, and Environmental Effects, pp 1–23
2. Curto D, Favuzza S, Franzitta V, Musca R, Navarro Navia MA, Zizzo G (2020) Evaluation of the optimal renewable electricity mix for Lampedusa island: the adoption of a technical and economical methodology. J Clean Prod 263:121404
3. Curto D, Franzitta V, Viola A, Cirrincione M, Mohammadi A, Kumar A (2019) A renewable energy mix to supply small islands. A comparative study applied to Balearic Islands and Fiji. J Clean Prod 241:118356

4. Ma T, Javed MS (2019) Integrated sizing of hybrid PV-wind-battery system for remote island considering the saturation of each renewable energy resource. J Energy Conver Manag 182:178–190
5. Fathy A, Kaaniche K, Alanazi T (2020) Recent approach based social spider optimizer for optimal sizing of hybrid PV/Wind/Battery/Diesel Integrated Microgrid in Aljouf region. IEEE Access 8:57630–57645
6. Al-falahi MD, Jayasinghe S, Enshaei H (2017) A review on recent size optimization methodologies for standalone solar. Energy Conv Manag 143:252–274
7. Razmjoo A, Shirmohammadi R, Davarpanah A, Pourfayaz F (2019) Stand-alone hybrid energy systems for remote area power generation. Energy Rep 5:231–241
8. Li C, Zhou D, Wang H, Lu Y, Li D (2019) Techno-economic performance study of stand-alone wind/diesel/battery hybrid system with different battery technologies in the cold region of China. Energy 192:
9. Javed MS, Song A, Ma T (2019) Techno-economic assessment of a stand-alone hybrid solar-wind-battery system for a remote island using genetic algorithm. J Energy 176:704–717
10. Kaabeche A, Ibtiouen R (2014) Techno-economic optimization of hybrid photovoltaic/wind/diesel/battery generation in a stand-alone power system. Sol Energy 103:171–182
11. Lau KY, Yousof M, Arshad S, Anwari M, Yatim A (2010) Performance analysis of hybrid photovoltaic/diesel energy system under Malaysian conditions. Energy 35:3245–3255
12. Paliwal P, Patidar NP, Nema RK (2014) Determination of reliability constrained optimal resource mix for an autonomous hybrid power system using particle swarm optimization. Renew Energy 63:194–204
13. Askarzadeh A (2013) A discrete chaotic harmony search-based simulated annealing algorithm for optimum design of PV/wind hybrid system. Sol Energy 97:93–101
14. Merei G, Berger C, Sauer DU (2013) Optimization of an off-grid hybrid PV–wind–diesel system with different battery technologies using genetic algorithm. Sol Energy 97:460–473
15. Kumar R, Gupta RA, Bansal AK (2013) Economic analysis and power management of a stand-alone wind/photovoltaic hybrid energy system using biogeography-based optimization algorithm. Swarm Evol Comput 8:33–43
16. Das BK, Hoque N, Mandal S, Pal TK, Raihan Md (2017) A techno-economic feasibility of a stand-alone hybrid power generation for remote area application in Bangladesh. Energy 134
17. Amer M, Namaane A, M'Sirdi NK (2013) Optimization of hybrid renewable energy systems (HRES) using PSO for cost reduction. Energy Procedia 42:318–327
18. Bhattacharjee S, Acharya S (2015) PV–wind hybrid power option for a low wind topography. J Energy Convers Manag 89:942–954
19. Ghorbani N, Kasaeian A, Toopshekan A, Bahrami L, Maghami A (2018) Optimizing a hybrid Wind-PV-Battery system using GA-PSO and MOPSO for reducing cost and increasing reliability. J Energy 154:581–591
20. Ma T, Yang H, Lu L (2015) Study on stand-alone power supply options for an isolated community. Int J Electr Power Energy Syst 65:1–11
21. Kennedy J, Eberhart R (1996) Particle swarm optimization. In: Proceedings of IEEE international conference on neural networks, vol 4, pp 1942–1948
22. Chaturvedi K, Pandit M, Srivastava L (2009) Particle swarm optimization with time varying acceleration coefficients for non-convex economic power dispatch. Electr Power Energy Syst 31:249–257
23. Singh P, Pandit M, Srivastava L (2020) Comparison of traditional and swarm intelligence based techniques for optimization of hybrid renewable energy system. Renew Energy Focus 35:1–9
24. Borowy BS, Salameh ZM (1996) Methodology for optimally sizing the combination of a battery bank and PV array. IEEE Trans Energy Conv 11(2):367–373
25. Singh P, Pandit M, Srivastava L (2020) PSO-Based optimization of levelized cost of energy for hybrid renewable energy system. In: Pandit M, Dubey H, Bansal J (eds) Nature inspired optimization for electrical power system. Springer, Singapore, pp 31–42

# Chapter 30
# Applications of Graphene-Based Ink in Heating Purpose and a Prototype Using NodeMCU ESP8266

**Md Saif**

## 1 Introduction

This work presents one of the many possible applications of graphene and its unique properties. Graphene, also dubbed as a "super material", has researchers all over the world wanting to explore its uses even more [1]. The long list of graphene's miraculous traits sometimes makes it seem magical. It could surely lead to some very drastic implications for the future of physics and engineering. Graphene is a two-dimensional material that has high electrical conductivity, which is greater than silver and copper, high flexibility, good tensile strength. Along with this, graphene is also extremely lightweight, it is only one-sixth of copper's density [2]. It can act as an insulator as well as a superconductor with small changes that can be made in its lattice. Compared to the excessive use of silicon which will in future reach its saturation point, graphene possesses much better flexibility, stiffness and mechanical strength, which makes it an ideal choice to use in places such as flexible electronics gadgets or smart wearable technologies [3].

These properties of graphene have been used in this paper to explore the possibility of using a graphene-based ink with a paste-like consistency that can be coated on different surfaces. The conductive ink also contains a water base that further increases the electrical conductivity [4]. Currently, with the early-stage technology involved in the production of graphene, the job is quite tedious but when its large-scale production model is taken into consideration, it can reduce the cost of operations and storage drastically [5]. This will result in a positive trend of researches for inculcating the flexible and rugged material in our daily use gadgets and consumer electronics.

M. Saif (✉)
Tata Steel Limited, Jamshedpur, Jharkhand, India

© The Author(s), under exclusive license to Springer Nature Singapore Pte Ltd. 2022 
H. M. Dubey et al. (eds.), *Artificial Intelligence and Sustainable Computing*,
Algorithms for Intelligent Systems, https://doi.org/10.1007/978-981-16-1220-6_31

## 2 Layer Construction

For the prototype to be used for heating applications, we need a layering design that is electrically isolated and thermally conductive. This has been achieved by using a sandwich-type layering of polyimide, which has a melting point higher than 520 °C and is electrically insulated [6]. The upper and bottom-most layers are covered by polyimide and in-between these layers, the graphene-based ink has been applied, as shown in Fig. 1.

Below the sandwich layer of graphene ink and polyimide, there is an extra layer of insulating sealant that provides a second level of electrical isolation and the last layer consists of an aluminum foil on which all the layering is done.

In Fig. 2, we can see the actual representation of all the layering done over the aluminum foil using which heat will be produced. By using this technique, the main

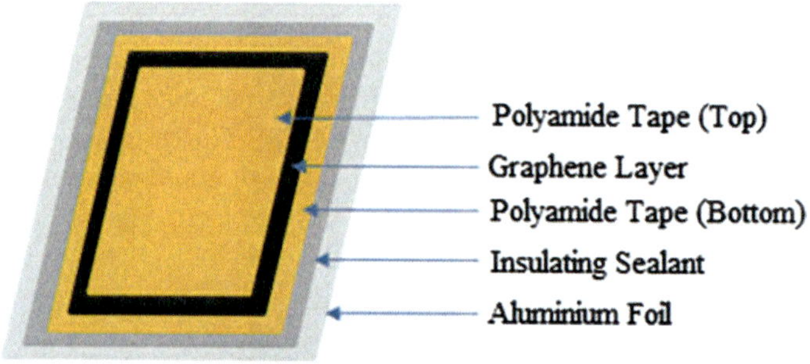


**Fig. 1** Layering representation of the prototype

**Fig. 2** Actual representation of the graphene layering

goal for a design that is electrically insulated and thermally conductive has been fulfilled [7]. Over the graphene ink, in between the sandwich layer, two straight lines of silver paste have been applied to increase the conductivity and area of the voltage application point. The voltage will be applied to the heating unit using copper strips, which are directly connected to the silver lines inside the heating unit.

## 3 Microcontroller Design

The control circuit of the prototype has been designed using a NodeMCU ESP8266 microcontroller and an LM35 temperature sensor, which can sense temperatures up to 150 °C. This microcontroller will allow the users to select between three types of modes, namely LOW, MEDIUM and HIGH corresponding to temperatures 90 °C, 120 °C and 140 °C, respectively, using a push button. Along with the two components mentioned above, the design of the controller also consists of an LM7805 voltage regulator and an NPN transistor BC547 [8]. The connection diagram of the controller with all its components and power supply is shown in Fig. 3.

The output of the BC547 alone is not sufficient to switch ON and OFF the relay so the 12 V supply has been connected to the second connection point of the relay to provide voltage to the operating coil. The two output connections of the relay are done using the COM and NO (normally open) ports. The white, green, red LEDs depict the LOW, MEDIUM and HIGH temperature modes respectively.

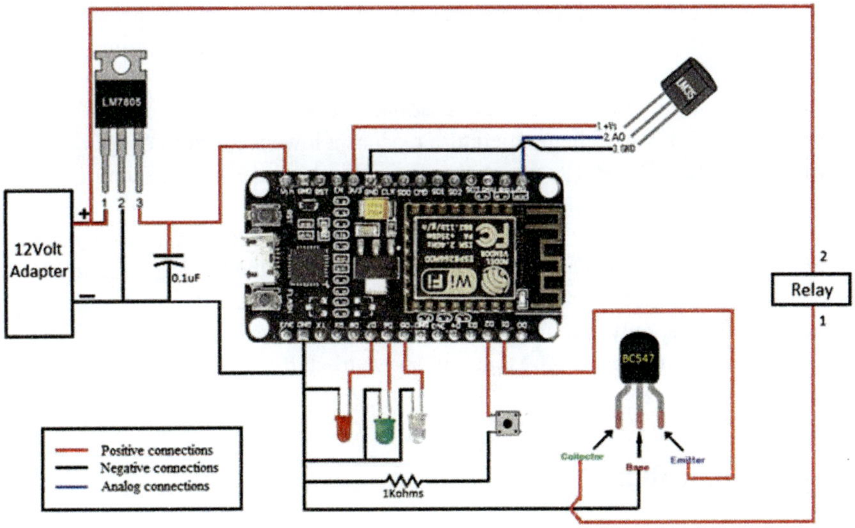


**Fig. 3** Circuit diagram for the temperature controller

**Fig. 4** Temperature controller with LM35

When the device is turned ON, the relay will also turn ON and a voltage will be applied across the graphene layer. This will produce heat due to the internal resistance of the graphene layer and the temperature will start increasing. For example, if the HIGH mode is selected then as soon as the temperature reaches 140 °C, the relay's output will be cut off and this will result in a decreasing nature of the temperature. After some moment when the temperature reaches a lower threshold of 130 °C, the relay will again turn ON and provides a voltage, thereby increasing the temperature. The device will further operate in this fashion.

Figures 4 and 5 show the actual controller, which has been designed using the LM35 temperature sensor. The head or the black portion of the LM35 can be inserted between the last layer of graphene ink and polyimide to measure its surface temperature and cut off the supply automatically when the ink surface has overheated. Table 1 shows the various modes with the auto-cut and auto-start temperatures.

## 4 Core Algorithm

The programming on the NodeMCU ESP8266 controller has been done such that as the temperature reaches 90 °C and the selection mode is LOW, the controller will cut off the voltage and the relay will switch off [9]. This will allow the surface temperature

**Fig. 5** Close up view of the controller

**Table 1** Modes of operation

| Mode of operation | Cutoff temperature (°C) | Auto start temperature (°C) |
|---|---|---|
| Low | 90 | 80 |
| Medium | 120 | 110 |
| High | 140 | 130 |

of the prototype to fall and as it reaches 80 °C, the controller will again switch on the voltage power supply making the surface temperature of the prototype rise again. This is the case in all other modes (LOW, MED, HIGH), with their respective maximum and minimum threshold depicted in Table 1. The algorithm below is a part of the whole program structure, which shows the logic of the temperature controller.

```
int outputFunction(int mode, int temperature){
switch(mode){
case 1:
if(temperature>90){
digitalWrite(OutMosfet,LOW); }
else{
if(temperature<80)
digitalWrite(OutMosfet,HIGH); }
break;
case 2:
if(temperature>120){
digitalWrite(OutMosfet,LOW); }
else{
if(temperature<110)
digitalWrite(OutMosfet,HIGH); }
break;
case 3:
if(temperature>140){
digitalWrite(OutMosfet,LOW); }
else{
if(temperature<130)
digitalWrite(OutMosfet,HIGH); }
break;
```

## 5 Results and Test Discussions

With the controller and graphene ink layer ready, it is now possible to take temperature and voltage observations. The reading for the time, temperature and voltages is shown in Table 2. The resistance of the graphene ink patch is around 4.8 ohms for a graphene patch of width 10 cm, height 5 cm and mass 10 grams.

The power dissipated from the graphene layer works on the principle same as the power dissipated through a resistor in the form of heat. Thus,

$$\text{Power Dissipated} = \text{Voltage} \times \text{Current} \tag{1}$$

$$\text{Power Dissipated} = (11.0 \times 2.29)\ \text{Watt}$$

$$\text{Power Dissipated} = 25.19\ \text{Watts}$$

The specific heat of graphene at room temperature is 710 J $(\text{kg K})^{-1}$ and mass of the graphene patch used is 10 grams or 0.01 kg [10]. The change in temperature '$\Delta$T' is 104 K, which is the difference between room temperature 36 °C or 309.15 K and desired temperature 140 °C or 413.15 K. Thus, the amount of heat energy 'Q' required to produce a certain change in temperature is given by:

**Table 2** Voltage versus temperature results

| Serial no. | Time (sec) | Voltage (volts) | Temperature (°C) | Current (amp) |
|---|---|---|---|---|
| 1 | 0 | 0 | 36 | – |
| 2 | 12 | 11.0 | 90 | 2.29 |
| 3 | 17 | 10.9 | 95 | 2.27 |
| 4 | 20 | 11.0 | 99 | 2.29 |
| 5 | 22 | 10.9 | 112 | 2.27 |
| 6 | 27 | 10.9 | 133 | 2.27 |
| 7 | 33 | 10.9 | 140 | 2.27 |
| 8 | 35 | 11.0 | 136 | 2.29 |
| 9 | 38 | 11.0 | 130 | 2.29 |
| 10 | 42 | 10.9 | 135 | 2.27 |
| 11 | 44 | 11.0 | 139 | 2.29 |
| 12 | 46 | 10.9 | 134 | 2.27 |
| 13 | 49 | 10.9 | 131 | 2.27 |

$$Q = mc\Delta T \tag{2}$$

$$Q = 0.01\,\text{kg} \times 710\,\text{J}\,(\text{kg K})^{-1} \times 104\,\text{K}$$

$$Q = 738.4\,\text{Joules}.$$

So, the time required for the graphene patch to reach a temperature of 140 °C from room temperature can be calculated by,

$$\text{Time} = \frac{Q}{\text{Power}} \tag{3}$$

$$\text{Time} = \frac{738.4\text{Joules}}{25.19\text{Watts}}$$

$$\text{Time} = 29.31\text{s}$$

Thus, it will take around 29.31 s for the temperature of the graphene patch to reach 140 °C from 36 °C theoretically. For the test of the prototype, the HIGH mode has been chosen, which will allow the temperature to remain in the band of 130 °C–140 °C as the voltage reaches around 11 volts, which is also applied from the adapter used to power the controller. It took around 33 s, which is very well near the theoretical value for the temperature of the graphene layer to increase up to 140 °C as shown in Table 2.

The relationship between time and temperature is shown in Fig. 6. The x-axis represents the time in seconds in which the graphene layer takes to reach a certain

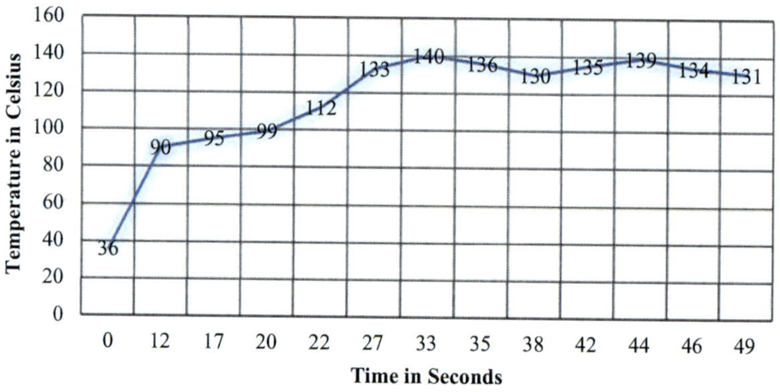


**Fig. 6** Time versus temperature curve

temperature and the y-axis represents the temperature in Celsius. It can be seen from the temperature curve that after the temperature of the graphene ink layer reaches 140 °C, it oscillates between the upper and lower bandwidth of the HIGH mode of the temperature controller. Figures 7 and 8 show the experimental results of the graphene ink tested for possible temperatures and the temperatures have been recorded using a temperature gun.

Several patch tests were conducted before finalizing the thickness and chemistry of the graphene patch used. The first few patches had the issue of burning, uneven and overheating, which had to be overcome before finalizing the patch. The most efficient way of applying the voltage ensuring a continuous supply of power was

**Fig. 7** Experimental result of the prototype

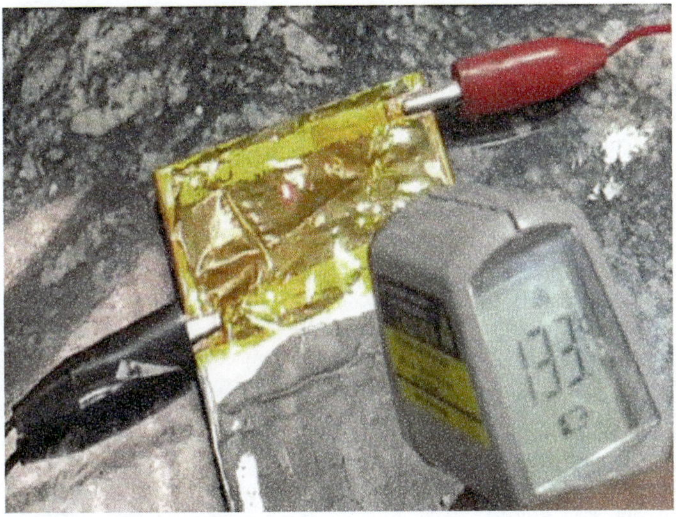

**Fig. 8** Close up view of the experimental result

also figured out. The graphene ink patch was tested vigorously for high stability and good heat dispersion.

## 6 Conclusion

A prototype with the desired qualities of an electric heater was implemented. The output voltages from the relay, LM7805 voltage regulator were checked and several tests were carried out using the prototype throughout a wide range of conditions. The electrically insulative and thermally conductive design aspect was successfully achieved through the sandwich layering concept. Throughout the experiment of the prototype, the graphene ink showed good durability and longevity, though after the completion of the tests upon inspection, there was some minor wear and tear in the surface of the ink after reheating about 24 times. The polyimide layer was also intact and withheld the temperature changes. The water base added to the graphene-based ink improved its conductive nature. The promising concept of using graphene in many other devices in our daily lives leads to a whole new set of convenience electronics products, which can be made much efficient and consumer-friendly.

## References

1. Dubey A, Dave S, Lakhani M, Sharma A (2016) Application of graphene for communication electronics and medical fields: a review. In: International conference on electrical, electronics,

& optimization techniques, pp 1–5
2. Simard-Normandin M, Ho Q-B, Rahman R, Ferguson S, Manga K (2018) Resistivity-strain analysis of graphene-based ink coated fabrics for wearable electronics. In: Pan Pacific microelectronics symposium, pp 1–8
3. Wang J, Liu Y, Fan Z, Wang W, Wang B, Guo Z (2018) Ink-based 3D printing technologies for graphene-based materials: a review. In: Advanced composites and hybrid materials, pp 1–33
4. Akbari M, Sydanheimo L, Juuti J, Vuorinen J, Ukkonen L (2014) Characterization of graphene-based inkjet-printed samples on flexible substrate for wireless sensing application. In: IEEE RFID technology and applications conference, pp 1–5
5. Li L, Wong H-SP (2018) Integrating graphene into future generations of interconnect wires. In: IEEE international electron devices meeting, pp 1–4
6. Klimchitskaya GL, Mostepanenko VM, Petrov VM (2014) Reflectivity properties of graphene and graphene coated substrates. In: International conference on next generation wired/wireless networking, pp 451–458
7. Zhang M, Wei X, Huang B, Long B (2017) Research of Water-Based Graphene Conductive Screen-Printing Ink and Its Property. In: Advanced Graphic Communications and Media Technologies, pp. 1–10. (2017)
8. Zhang G, Wang J, Wei A (2010) Study on the preparation and properties of novel water-based conductive ink. Pack Eng
9. Le T, Kaladosis V, Lin Z, Wong CP, Tentzeris MM (2012) Inkjet-printed graphene-based wireless gas sensor modules. In: Electronic components and technology conference, pp 1003–1008
10. Li QY, Xia K, Zhang J, Zhang Y, Li Q, Takahashi K, Zhang X (2017) Measurement of specific heat and thermal conductivity of supported and suspended graphene by a comprehensive Raman optothermal method. In: Royal society of chemistry—nanoscale, pp 1–10

# Chapter 31
# Improving Energy Efficiency of Ring Frame Motor of Textile Industry

**Saurabh Kumar Rajput, Dharmendra Kumar Dheer, and Sulochana Wadhwani**

## 1 Introduction

In India, textile mills are one of the highest employment generation sectors. After the agriculture, textile sector is providing the employment to the highest number of people. About 35 million people are employed under this sector. The economic growth of the country is depending on the textile sector. In terms of production, it contributes about 14% and contributes about 4% in GDP. The installed number of spindles is also the second largest (19.6%) in the world [1]. The energy requirement of the textile industries is very high because the workers do not want to work in the pollutic, dusty and noisy environment of the plant. In this way, heavy and automated machineries are used in the textile mills and these are high energy consumer of electrical energy. The improper use of machinery and negligence of energy conservation measures are also responsible for high energy requirements by textile mills [2]. The high energy is need is not only but also making atmosphere pollutic because the concentration of greenhouse gases is also increasing due to it. The spinning textile mill mainly produces thread from cotton; where ring frame machine uses the maximum part (about 37%) of total electricity. Rest of the electricity is used by open-end machines, blow room, carding, drawing, combing roving and winding [3]. The energy management is the requirement of textile industries for reducing energy wastage and to save the environment. There are many energy conservation practices, which are commonly used in Indian textile mills; spinning process includes the use of automatic power factor correction capacitor bank, energy-efficient motors and synthetic flat belts in ring frame machines. The weaving process includes the use of flat belt drive instead of V-belt drives, efficient H Plant fans and reduction in wastage

S. K. Rajput (✉) · D. K. Dheer
National Institute of Technology Patna, Patna, India

S. K. Rajput · S. Wadhwani
Madhav Institute of Technology and Science, Gwalior, India
e-mail: saurabh9march@gmail.com

© The Author(s), under exclusive license to Springer Nature Singapore Pte Ltd. 2022

H. M. Dubey et al. (eds.), *Artificial Intelligence and Sustainable Computing*, Algorithms for Intelligent Systems, https://doi.org/10.1007/978-981-16-1220-6_32

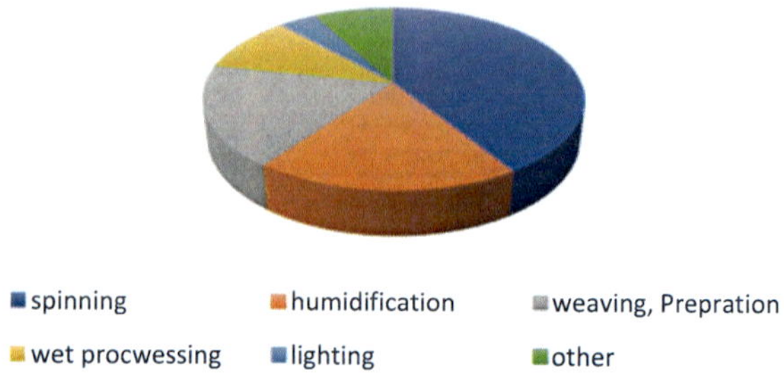


**Fig. 1** Breakup of electrical power in textile sector

of compressed air. Processing in Indian textile industries includes the reutilization of wastewater in dying process [4]. Energy consumption in textile industry is high but the reduction in wastage of energy is possible through different energy conservation methodologies [5–9]. Three-phase type induction principle-based motor is used in the ring frame process of textile mills. The efficiency of this motor can be improved by proper time to time required testing and by providing sophisticated rewinding to the motor [10, 11]. By load management and energy audits, energy should be conserved in textile mills. This is helpful in finding economic growth as well as energy efficiency [12, 13]. The various energy improving techniques in the industrial process by the use of standard loggers and meters are given in [14]. Figure 1 shows that the spinning process takes the highest amount of energy so there is a great opportunity for energy saving through improving the network of ring frame machine. The present study is based on the data collection from textile industry and analyzing it for improving energy efficiency of ring frame machine. In this study, some other possibilities of energy savings in Ring Frame machine of textile industries are investigated and that are analyzed for the selected spinning mill. These possibilities include—energy saving by optimizing ring frame suction tube size, replacement of the faulty suction tube of autoconer with small diameter caps for energy conservation. In addition to this, few studies are also performed for improvements in already applied energy conservation techniques and electrical energy savings in the industry are calculated. These studies are replacement of faulty motors by new and energy-efficient motor.

## 2 Methodology

In textile industries, the major part of the electrical energy is consumed by the ring frame section. The three-phase induction principle-based electric motors (also known as ring frame motors) are used for rotating the spindles at the required high speed for the production of yarn in the ring frame section. So, the present study is focused on finding the energy conservation measures in the ring frame motors. The steps involved in improving energy efficiency are shown by the flowchart (Fig. 2).

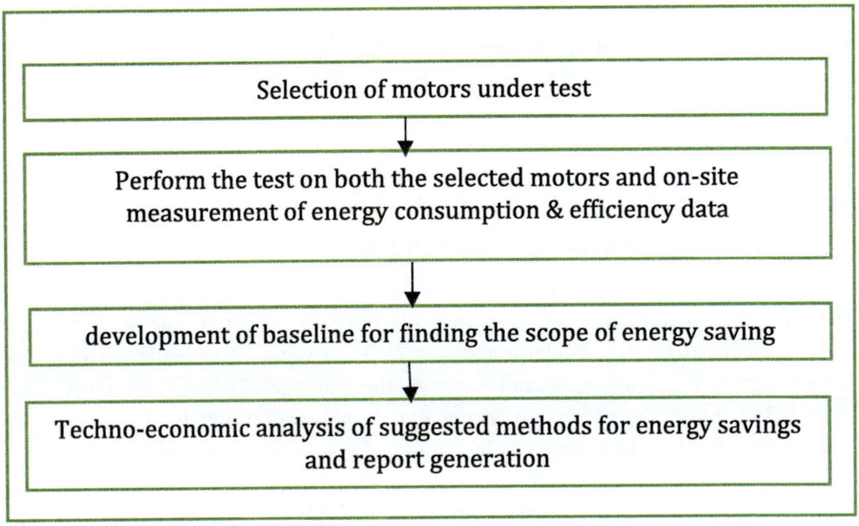


**Fig. 2** Steps for improving energy efficiency

## *2.1 Impact of Rewinding of Motor on Energy Saving*

The continuous operation of ring frame motor for yarn production is very common in textile mills, due to which the burning of motor winding takes place very frequently. In order to avoid the loss of production, the burnt motors are rewound and again continue their operation. This complete process misses the efficiency testing of the motor; however, after every rewinding, the efficiency of motor goes down. The use of an inefficient ring frame motor reduces the overall energy efficiency of ring frame machine section [15].

Considering a rewound, energy-inefficient motor; rated for 'X' kW and operating at average efficiency of $\eta 1$. If this motor is replaced by a new and energy-efficient motor having the same rating of 'X' kW and higher efficiency of $\eta 2$, then the energy (kWh) saving for the motor operation of 'h' hours per day is given by the formula given in Eq. (1), [16].

$$\text{Per day (kWh)energy saving} = Po\left[\frac{1}{\eta 1} - \frac{1}{\eta 2}\right] \times (h) \tag{1}$$

where Po is the electrical equivalent of mechanical power developed at shaft and is given by Eq. (2).

$$\text{Motor shaft power output } (Po) = (\text{loading of motor}) \times (\text{motor rating}). \tag{2}$$

If textile plant operates for 'd' number of working days per annum, the annual energy saving obtained by the energy-efficient motor is calculated by using the Eq. (3)

$$\text{Annual } (kWh)\text{energy saving } (E) = Po\left[\frac{1}{\eta 1} - \frac{1}{\eta 2}\right] \times (h) \times (d). \tag{3}$$

## 2.2 *Impact of Continuous Variable Load on the Performance and Energy Consumption of the Motor*

The percentage loading of ring frame motor plays a very important role in the conservation of electrical energy. Under loading as well as over-loading leads the motor to operate in poor efficiency zone and it causes the wastage of electrical energy. The loading of the motor depends on the actual electrical power drawn by the motor with respect to the motor rating, as given in Eq. (4), [16].

$$\text{loading of motor} = \frac{\text{Actual power drawn by the motor } (kW)}{\text{Rated input power } (kW)}. \tag{4}$$

In order to operate the motor in high efficiency region, it becomes important to select the proper size (rating, kW) of the motor, by keeping in mind that the motor should neither be under loaded and nor be over loaded. In the case of textile mill's ring frame motor operation, the connected load increases continuously in linear manner. In this situation, the motor is forced to operate in under-loading and over-loading conditions during the start and end of the production process respectively. In order to conserve the energy, the variable speed drive (VSD) is used along with the ring frame motor.

The textile mills come under the category of production-based industries, where production cannot be compromised on the cost of energy savings. So, for the energy-saving calculations, the electrical energy (kWh) consumption is measured with respect to the yarn production (kg) and is defined by the term 'UKG analysis' i.e. electrical units consumed for 1 kg of yarn production. In our study, the UKG analysis is performed and a comparison is done in between the ring frame motor operating with and without VSD.

If the ring frame machine is having 'N' number of spindles and the motor of efficiency 'η' is operating with a speed of 'n' rpm, then the production of yarn (in kg) for the yarn count of 'S' Ne is calculated by the formula given in Eq. (5).

$$\text{Ring frame production } (kg.) \text{ per shift} = \frac{0.0072 \times n \times N \times \eta}{TPI \times S} \tag{5}$$

where TPI is twist per inch and is given by Eq. (6)

$$TPI = (T.M.) \times \sqrt{S}. \tag{6}$$

## 3 Results and Discussion

The detailed study is performed with the ring frame motors in the textile plant and possibilities are identified for improving the energy efficiency. The study is divided into two parts.

- Energy saving by replacing burnt and rewound ring frame motors.
- Use of VSD for energy saving.

### *3.1 Energy Saving by Replacing Burnt and Rewound Ring Frame Motors*

During the study, it is found that the plant is using few rewound motors, which are more than three times rewound. One of these motors is selected for testing and is compared with the energy-efficient motor. The profile of the ring frame motor under test is given in Table 1.

The daily operation profile of the motor is given in Table 2. The motor is operating with 50% loading for 8 h and operating for 75% loading for 14 h, while the operation of motor is only for 1 h, when it is fully loaded. With the change in the loading pattern, the power (kW) is also changing (as given in Table 2).

**Table 1** Ring frame motor profile

| Term | Specification |
|---|---|
| Rating | 3 phase, 400 V, 50 Hz, 30 kW |
| Ignition protection | Without ignition protection |
| Application | Ring frame |

**Table 2** Daily operation profile of motor in plant

| Term | Specification | | |
|---|---|---|---|
| Loading (percentage) | 50 | 75 | 100 |
| Power (kW) | 15 | 22 | 30 |
| Operating hours per day | 8 | 14 | 1 |

**Table 3** Variation of efficiencies of motors with loading

| Term | Specification | | |
|---|---|---|---|
| Loading (percentage) | 50 | 75 | 100 |
| Old motor efficiency (percentage) | 85.5 | 86.7 | 84.8 |
| New motor efficiency (percentage) | 93.8 | 94.1 | 92.9 |

**Table 4** Energy-saving profile by replacing the rewound motor

| Term | Specification | | |
|---|---|---|---|
| Loading (percentage) | 50 | 75 | 100 |
| Energy saving (kWh) | 3725.73 | 8380.98 | 925.37 |
| Total energy saving (kWh) | | | 13032.08 |

The test results are given in Table 3. The two selected motors are compared according to the same operation profile as given in Table 2 and it is found that the efficiency of the motors depends on the two factors

- Loading pattern of the motor
- Rewinding of the motor

Both the motors are operating in the best efficiency zone, when these are loaded near or equal to 75% loading; while the efficiency of motors is poor in under-loading as well as over-loading region. It is also very clear from the study that after every rewinding, the efficiency of motors drops down. In our case study, the drop in the efficiency varies from 7.4% to 8.3%.

By considering the load profile, daily operation hours and annual working day (300 days), the annual energy saving is calculated and is shown in Table 4.

Considering the electricity price Rs. 8.00, the saving in the electricity bills is obtained as Rs. 1,04,256.64. The cost of new energy-efficient motor in the market may be taken as 98,619.00 (approximately). Now by considering the above-given data, the payback time of the replacing rewound motor is less than 1 year, so, it is suggested to replace all the old and inefficient motors with new energy-efficient motors for energy saving.

## 3.2 *Use of Variable Speed Drive (VSD) for Energy Saving*

The loading of ring frame motor varies from light load to full load condition. In the initial stage of doff, the motor is lightly loaded and it increases almost in linear manner during the production process. In this way the motor has to operate in poor efficiency zone and the overall energy efficiency the process gets reduced. The use of VSD may be a solution for this issue, so a test is performed on the 30-kW motor for which the specification details are given in Table 5.

The similar test is performed in the lab setup. A three-phase power analyzer is connected at the three-phase input supply terminals of the motor for the measurement of electrical units (kWh), as shown in Fig. 3 and the production of yarn (in kg) is calculated by using Eq. (5).

The test results show that the proper selection of speed pattern in the inverter reduces the overall energy consumption without sacrificing the yarn production.

During the study, it is also observed that the plant is using different size of ring frame suction tubes on the same machine (as shown in Fig. 4). So, a test is performed

**Table 5** Specifications and measurements of the test

| Term | Without VFD | With VFD |
|---|---|---|
| Motor rating | 3 phase, 400 V, 50 Hz, 30 kW | |
| Count | 2/45' S | |
| Delivery package | Cone | |
| Number of spindle | 168 | |
| R count | 22.5 | |
| TPI | 18.9 | |
| Actual speed | 9896 rpm | |
| Current output | 30.7 Amp | 36.93 Amp |
| Input kW on inverter | 18.55 kW | 17.67 kW |
| Production 100% | Without VSD:169 kg | 172 kg |
| Unit consumed | 445 kWh | 423.37 kWh |
| UKG | 2.64 | 2.46 |

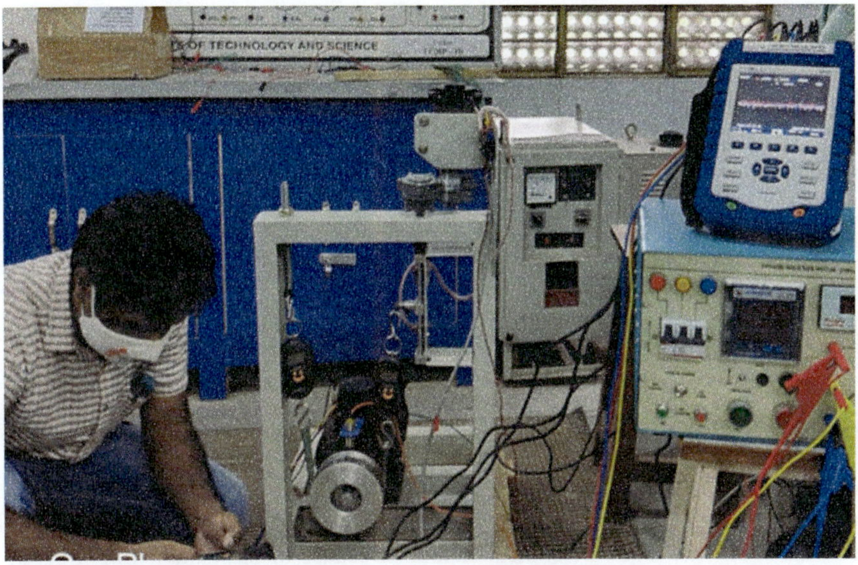


**Fig. 3** Real-time data measurement from the motor under trial

**Fig. 4** Different size suction tubes

**Table 6** Variation of pressure and power with suction tube size

| Parameter | Pressure of 8 mm tube (Pa) | Pressure of 6 and 8 mm tube (Pa) | Pressure of 6 mm tube (Pa) |
|---|---|---|---|
| 1st spindle | 352 | 420 | 590 |
| 250th spindle | 488 | 550 | 780 |
| 500th spindle | 1152 | 1195 | 1350 |
| Power (kW) | 5.09 | 4.96 | 4.19 |

to identify the optimum size suction tube for energy conservation. The details of the performed test are given in Table 6.

The pressure (Pa) power (kW) on ring frame machine is measured, which has 8 mm suction tube. The pressure on the 1st, 250th and 500th spindle of the machine was 352 Pa, 488 Pa and 1152 Pa, respectively, and the total power consumption was 5.09 kW. The study was repeated on the ring frame machine, which has 6 8 mm suction tube. In this case, the pressure on the 1st, 250th and 500th spindle of the machine was 420 Pa, 550 Pa and 1195 Pa, respectively, and the total power consumption was 4.96 kW. Finally, the test was performed on the ring frame machine, which has 6 mm suction tube and measured the pressure on the same 1st, 250th and 500th spindle, now the results were 590 Pa, 780 Pa and 1350 Pa, respectively, and the total power consumption was 4.19 kW. So, by optimizing the suction tube size, there is scope of energy saving and in the present study, energy up to 1, 32,480 kWh can be saved per year in one machine.

## 4 Conclusion

Increasing demand for energy in textile industries and price hiking are the two main factors that are responsible for high production costs. To reduce this cost, energy conservation is essentially required by textile industries. There are many energy conservation practices that are being implemented by industries. These techniques are either costly or require technical support. Because most of the textile mills in India are small and medium enterprises so they have limited resources to support these methodologies. During energy audit visits of several textile industries, some simple methodologies of energy savings are investigated and verified. By applying these techniques, an ample amount of energy can be saved with less investment. Following are the energy conservations obtained in the present case of study

- The efficiency of motor goes down after every rewinding. In our case of study, 13032.08 kWh energy is saved by replacing the three times rewound motor with energy-efficient motor.
- The simple payback time of replacing 30 kW rewound motor with energy-efficient motor is found less than 1 year during the study.

- The efficiency of motors is poor in under-loading as well as in over-loading region, so the VFD may be used with motor for energy saving.
- The results of the study are that the 4–6% energy is conserved by using variable speed drive (VSD) along with motor.
- The production (in kg) is also improved, when VSD is used with motor.
- The instantaneous power consumption is reduced from 5.09 kW to 4.19 kW by using 6 mm size suction tube instead of 8 mm.

The obtained energy savings are not fixed, it varies with the size of plant, operational practices and operating hours per day.

**Acknowledgements** The authors acknowledge NPIU and AICTE, New Delhi for providing the support under Collaborative Research Scheme (CRS id: 1-5729547560) projects of TEQIP III. Authors are also thankful to Er. Vikas Sharma (Accredited Energy Auditor) of NITRA Ghaziabad, India for his support and guidance.

## References

1. New Energy and Industrial Technology Development Organization (2009) A report on national survey on implementation of energy efficiency improvement technologies in textile sector
2. Dhayaneswaran Y, Ashok kumar L (2013) A study on energy conservation in textile industry. J Inst Eng India Ser B94:53–60
3. Hasanbeigi A (2010) Energy-efficiency improvement opportunities for the textile industry. Ernest orlando Lawrence Berkeley National Laboratory
4. Khude P (2017) A review on energy management in textile industry. Innov Ener Res 6
5. Goyal S (2011) Energy conservation in textile industry. Lambert Academic Publishing
6. Bhaskar M, Verma P, Kumar A (2013) Indian textile industries towards energy efficiency movement. Int J Environ Sci Dev Monit (IJESDM) 4:36–39
7. Hong G-B, Su T-L, Lee J-D, Hsu T-C, Chen H-W (2010) Energy conservation potential in Taiwanese textile industry. Energy Policy 38:7048–7053
8. Hasanbeigi A, Price L (2012) A review of energy use and energy efficiency technologies for the textile industry. Renew Sustain Energy Rev 16:3648–3665
9. Koc E, Kaplan E (2007) An investigation on energy consumption in Yarn production with special reference to ring spinning. Fibres Textiles in Eastern Europe 15:18–24
10. Rajput SK, Singh O (2017) Energy audit in textile industry: a study with ring frame motor. ICCCCM- 2nd IEEE International Conference on Control Computing Communication and Materials 2–5
11. Rajput SK, Godhar P (2015) AC motors burn out in textile industry—causes and impact on motor performance. In: 56th joint technological conference SITRA Coimbatore
12. Muhammad IY, Muhammad DZ, Tahir S, Rizv H (2018) Design and development of energy audit and load management system. Bahria Univ J Inf Commun Technol 11(II)
13. Balachander K, Amudha A (2019) Energy saving measures in Textile Mill. Int J Innov Technol Explor Eng (IJITEE) 8(8) ISSN:2278-3075

14. Kluczek A, Olszewski P (2017) Energy audits in industrial processes. J Cleaner Product 142
15. Rajput SK, Rani P, Sadhu PK, Sadhu M, Das N (2018) Energy conservation in textile industries by replacing rewound motors—an energy audit study. In: 2018 international conference on power energy, environment and intelligent control (PEEIC), Greater Noida, India, pp 820–824. https://doi.org/10.1109/peeic.2018.8665587
16. Energy efficiency in electrical utilities (2005) Guide book 3 for National Certification Examination for Energy Managers and Auditors, bureau of energy efficiency, India. www.bee-india.nic.in

# Chapter 32
# Contingency Analysis of Complex Power System Using Active Power and Voltage Performance Index

**Raghvendra Tiwari, Anamika Gupta, S. Chatterji, and Shivangi Shrivastava**

## 1 Introduction

Contingency analysis (CA) is a crucial step for the proper functioning of a power system. The security analysis of power system operation is more important for power system operation engineers to monitor an unwanted event's effect on the healthy part of the power system. The process of contingency analysis may be defined as identifying the voltage and active power limit violation during an outage, which can be avoided by taking some remedial actions to preserve the power system security. In a power system, the existence of a contingency means when an element or component fails to operate due to any reason, it may be a fault or external cause. The power system network element that might fail could be a transformer, generator, transmission line, substation, etc. Contingency analysis is categorized into four-part depending on the number of power system element outage are listed below

- Individual contingency (N-1)—In this kind of contingency, a single element of a network at a time is tripped. The N-1 contingent system or N-1 secure system signifies that the power system can continue to function within nominal limits if any one element fails to operate. The operational engineers are responsible for making sure that the system is at least N-1 contingent.
- Multiple contingency (N-K)—In this category, numerous elements of the power system are tripped. For N-k contingent system or N-k secure system, the system

R. Tiwari (✉) · S. Shrivastava
MNNIT Allahabad, Prayagraj, UP, India
e-mail: raghvendra@mnnit.ac.in

A. Gupta
RKGIT, Ghaziabad, UP, India

S. Chatterji
NITTTR, Chandigarh, India

© The Author(s), under exclusive license to Springer Nature Singapore Pte Ltd. 2022 
H. M. Dubey et al. (eds.), *Artificial Intelligence and Sustainable Computing*, Algorithms for Intelligent Systems, https://doi.org/10.1007/978-981-16-1220-6_33

can function properly without any constraints violation, although it is relatively more difficult and expensive to do.

- Overlapping of single contingency and generator outage (N-G-1)—a single contingency and one generator outage simultaneously.
- Overlapping of multiple contingencies with maintenance outage (N-m-k)—In this kind of contingency, 'm' elements are disassociated preceding a fault in the system, and 'k' elements are disjointed to clear the fault.

For contingency evaluation in a power system, there are some methods available in the literature. The recursion-based contingency analysis has been performed to improve the efficiency of the security analysis problem in [1]. The application of ANN in the contingency analysis problem is explained in [2]. To speed up the process of contingency analysis CA), the fast decoupled load flow (FDLF) method is used and only the reactive power performance index is taken into consideration [3]. To take the advantage of sensitivity factors [4] as well as a comparative analysis between newton-raphson and FDLF has been done in [5]. Since the contingencies are independent of each other so to improve the performance of the CA process, the concept of parallel contingency analysis has been introduced using open multi-processing std. in [6]. In [7] the author presents the improvement that is developed in the CA process using the node/breaker modal. In [8] the author discussed both N-1 and N-2 contingencies for a more practical approach. In [9] the author used a modified fast decoupled load flow method for CA. In [10] the author proposed the use of a power transfer distribution factor to solve this problem. In [11] the author investigated a real-time contingency analysis with transmission switching on real power system data. The application of fuzzy logic has been utilized in [12] to improve a CA procedure's performance.

In a power system, there are various kinds of outages that may occur. It may be possible that these outages have occurred sequentially or simultaneously. It becomes more challenging if, in a power system, more than two elements are outage at a time. In this paper, the severity level of an outage in the power system is determined using active ($PI_p$) and voltage ($PI_v$) performance indices [13, 14]. The rest of the paper is divided into five sections. Sections 2 and 3 give a brief description of the types of violations and operational conditions in the power system, respectively. Sections 4 and 5 explain the various stages and flow chart in the process of CA. Section 6 demonstrates the obtained simulation results. Remedial actions and conclusion is given in Sects. 7 and 8, respectively.

## 2 Types of Violations

In a power system, alternator outage and transmission line outages are the most promising type of contingencies. Because of such kinds of contingencies, various parameters of the power system are changed, but mainly two types of violations significantly impact the complete power system.

### 2.1 Under/Over Voltage Violation

A bus's operating voltage in a power system is generally in the range of 0.95pu-1.05 pu. Under/overvoltage violation mainly occurs when the bus operating voltages decreased or increased for a period of half-cycle to one minute. This shows the voltage at the bus is out of its nominal limits. Depending on the bus's voltage magnitude, the bus is specified as an under-voltage bus, when its voltage is below 0.95 pu, and an overvoltage bus, when its voltage is above 1.05 pu.

### 2.2 Line MVA Limit Violation

Line MVA limit violation occurs when the line power rating is exceeded beyond the prescribed limit/rating and it mainly happens in transmission line connecting between two buses. This kind of power violation occurs when the amplitude of the line current increases. If the line current crosses 80%-90% of the limit, it can be an alarming situation for a power system.

## 3 Power System Operational Condition

To perform contingency analysis, a power system is said to have three operating states.

### 3.1 Preventive State

This state is also known as a normal state. A normal state means that all equipment and components are operating within their specified limit. All the system parameters such as bus voltage, line MVA, etc., are within an acceptable range.

### 3.2 Emergency State

When some of the parameters operating limit are increased or decreased from the predefined limits, the state at which the power system exists is said to be in an emergency state.

### 3.3 Restorative State

When some of the power system components fail to operate, the system is said to be in a restorative state. This state's control action is to guide the system to a normal state once again by taking appropriate action. There are two types of failure events, namely transmission line outage and generation unit failure. The line outage simulation is more complicated than a generator outage because a line outage results in a system configuration change. In contrast, generator outages can be possible due to internal or external reasons. In line outages, the line resistance and reactance are set to a very large value. Whereas in the case of generator outage, the bus voltage magnitude is set to be unity and angle, reactive and active power generated by bus are set to be zero. The power system operating states are demonstrated in Fig. 1.

## 4 Contingency Analysis

Ejebe and Woolen Berg developed the idea of contingency. The main difference between contingency and outage is that contingency is related to an unpredictable event in the power system, whereas outage is related to the momentary interruption of operation. To perform contingency, we need to divide the whole process into three steps

- Contingency definition
- Contingency selection
- Contingency evaluation.

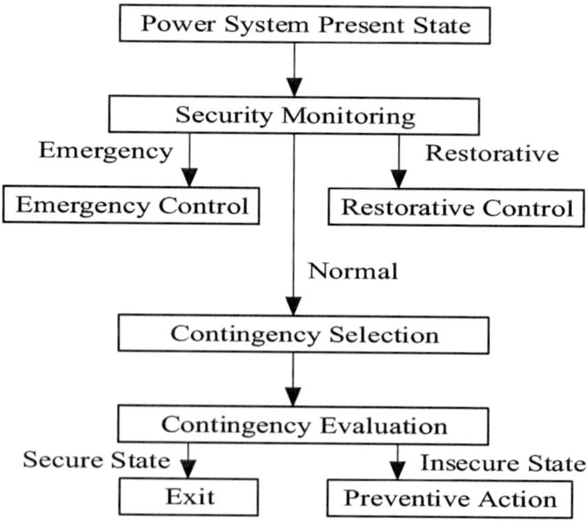


**Fig. 1** Power system operating states

Contingency definition is the first and crucial step of security analysis in which a set of possible contingency that may be present in a power system is considered. This step creates a list of contingency. In Contingency selection, there are mainly two methods are adopted.

- A direct method or screening method- This method uses an approximate network solution, and network monitor quantities are calculated. There are some techniques like the distribution factor method that are involved in this method.
- Indirect method or ranking method: In this method, different outages are ranked based on performance index (PI) or severity index (SI). PI's are defined in terms of network variables like voltages, the line's capacity, MW flow of lines, number of buses, number of lines connecting buses, etc.

Evaluation of performance index has been done as listed below

- Voltage performance index ($PI_v$): This index is used to indicate the magnitude of bus voltage violation [15].

$$PI_V = \sum_{i=1}^{N_{pq}} \left[ \frac{2(V_i - V_{Nom})}{V_{i,\max} - V_{i,\min}} \right]^2 \tag{1}$$

where $V_i$ = Voltage magnitude at $i$th bus

$V_{imax}$ and $V_{imin}$ = upper and lower limit of voltage at $i$th bus.

$V_{nom}$ = Nominal voltage of $i$th bus.

$N_{pq}$ = Total number of P-Q buses in the system

- Real power performance index (PIp): This index corresponding to a contravention of line real power flow [15].

$$PI_P = \sum_{i=1}^{L} \left[ \frac{P_i}{P_{i,\max}} \right]^{2n} \tag{2}$$

where $P_i$ = Real power of the $i$th line,

$P_{imax}$ = denoting maximum real power corresponding to the $i$th line,

n = Fixed component (= 1),

L = Number of transmission line in the power system

$$P_{i,\max} = \frac{V_i * V_j}{x} \tag{3}$$

where $V_i$ = $i$th Bus voltage.

$V_j$ = $j$th bus voltage.

x = reactance of line connecting $i$th and $j$th bus.

## 5 Procedure and Flow Chart of Contingency Analysis

Procedure to perform contingency analysis consists of the following steps-

(1) Fetch the provided system bus data and line data.
(2) Initiate the counter with zero value.
(3) Perform pre contingency analysis, that is, evaluate the system parameter without considering any contingency.
(4) Create a line outage, that is, a line is interrupted.
(5) For a particular line outage, run load flow analysis to find the MW power flow in remaining transmission lines and bus voltage after contingency state.
(6) To achieve the overall performance index, add two indexes, namely real power performance index and voltage performance index.
(7) Similarly, create all line outages one by one and repeat step 4 to step 6.

When the above process has been done all the contingencies are equipped with a suitable rank, which is based on performance index, that is, summation of $PI_p$ and $PI_v$. Load flow analysis is performed for the most critical line outage and the result is shown in the next section. The contingency analysis flowchart is shown in Fig. 2.

## 6 Simulation and Results

In this paper, two IEEE standard bus systems, that is, IEEE-5 and IEEE-14 bus system are considered to perform contingency analysis.

### *6.1 IEEE-5 Bus Test System*

Bus code 1and 0 is used for slack bus and load bus, respectively. The MW limits for bus B1, B2, B3, B4, and B5 are 155 MW, 110 MW, 95 MW, 95 MW, and 95 MW. The system consists of seven transmission lines (L1-L7) and the MW power flow in every line is calculated using the fast decoupled load flow method. For this system, base MVA = 100 and Base kV = 25 are taken. The Pre and post contingency state of the IEEE-5 bus system is drawn in Fig. 3 and Fig. 4, respectively. Table 1 shows the bus voltage magnitude before and after the line outage state of the IEEE-5 bus system. All the lines in the pre contingency state carry the active power flow in a specified limit as shown in Fig. 3. After that, for every line outage, fast decoupled load flow analysis is done. The $PI_p$ and $PI_v$ are also evaluated using (1) and (2) by considering only one-line outage sequentially. The ranking is assigned based on the severity index (SI) or performance index (PI), given by SI or PI = $PI_p + PI_v$. The largest PI value indicates the most severe. As shown in Table 2 the line L1 having the highest value of the severity index. Therefore, the transmission line connected between B1 and B2 is ranked first, that is, most severe whereas line L2 has the

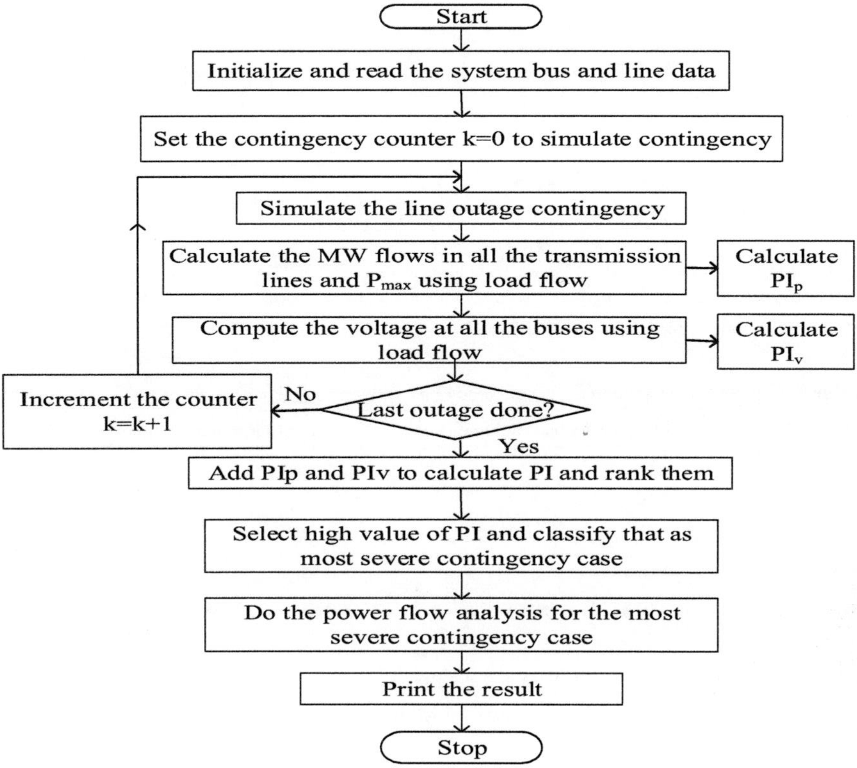


**Fig. 2** Contingency analysis flow chart

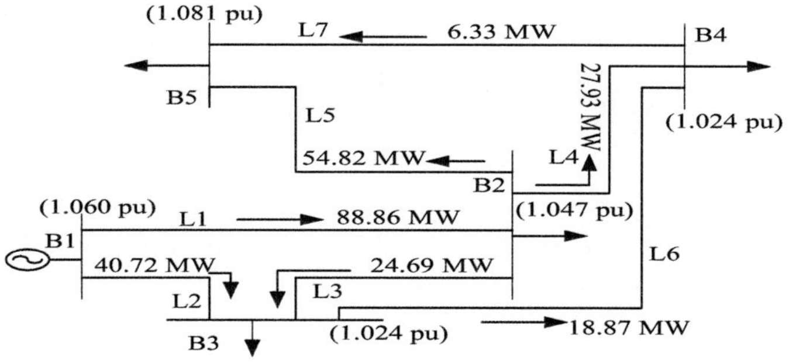


**Fig. 3** Pre contingency state of IEEE-5 bus system

lowest value of severity index then it is ranked seventh, that is, least severe. Figure 4 shows the real power flow of lines of the IEEE-5 bus system during line outage L1. Table 3 shows the active power flow during line outage L1 in the IEEE-5 bus system. Table 3 shows that during line outage L1 the line L2 is overloaded by 130% as their maximum capacity is 110 MW.

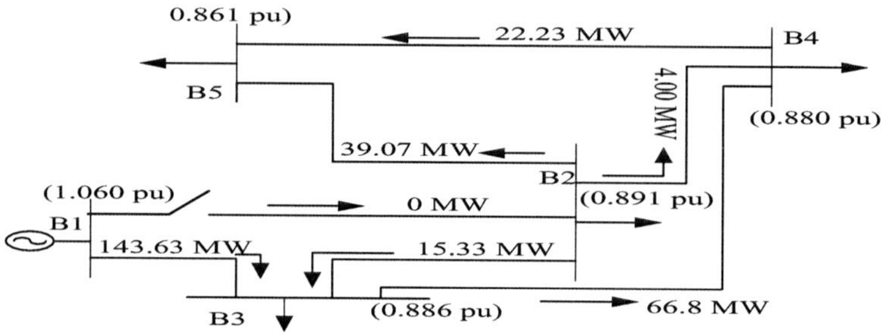


**Fig. 4** Post contingency analysis of IEEE-5 bus system

**Table 1** Pre contingency and post contingency bus voltage of IEEE-5 bus system

| Bus number | Pre contingency voltage (pu) | Post contingency voltage (pu) |
|---|---|---|
| B1 | 1.060 | 1.060 |
| B2 | 1.047 | 0.891 |
| B3 | 1.024 | 0.886 |
| B4 | 1.024 | 0.880 |
| B5 | 1.018 | 0.861 |

**Table 2** Performance index of IEEE-5 bus system

| Transmission Line | | Line number | APPI | VPI | PI or SI | Rank |
|---|---|---|---|---|---|---|
| From bus | To bus | | $PI_p$ | $PI_v$ | | |
| B1 | B2 | L1 | 0.2800 | 3.1916 | 3.4716 | 1 |
| B1 | B3 | L2 | 0.3619 | 0.2699 | 0.6318 | 7 |
| B2 | B3 | L3 | 0.3377 | 0.6557 | 0.9934 | 5 |
| B2 | B4 | L4 | 0.3790 | 0.6173 | 0.9963 | 4 |
| B2 | B5 | L5 | 0.4221 | 0.2653 | 0.6874 | 6 |
| B3 | B4 | L6 | 0.2995 | 0.8599 | 1.1594 | 3 |
| B4 | B5 | L7 | 0.3036 | 0.8799 | 1.1835 | 2 |

**Table 3** MW power flow for pre and post contingency state of IEEE-5 bus system during line 1 outage

| Line number | From bus | To bus | Before contingency MW flow | After contingency MW flow |
|---|---|---|---|---|
| L1 | B1 | B2 | 88.86 | 0 |
| L2 | B1 | B3 | 40.72 | 143.63 |
| L3 | B2 | B3 | 24.69 | 15.33 |
| L4 | B2 | B4 | 27.93 | 4.00 |
| L5 | B2 | B5 | 54.82 | 39.07 |
| L6 | B3 | B4 | 18.87 | 66.80 |
| L7 | B4 | B5 | 6.33 | 22.23 |

## *6.2 IEEE-14 Bus Test System*

In test case-II, the IEEE 14 Bus System is considered as drawn in Fig. 5. This system consists of a 1-slack bus, 9- PQ buses, and 4- PV buses. Bus coded as 0, 1, and 2 are defined as load bus, slack bus, and generator bus. The system has twenty transmission lines (l1-l20), and active power flow in each transmission line is obtained using fast decoupled load flow analysis. For this system, base MVA = 100, and Base KV-25 is taken. At bus B3, B6, and B8, three synchronous compensators are used for reactive power compensation. Table 4 shows the value of the performance index and ranking of the line outage of the IEEE-14 bus system.

The transmission line with the highest value of performance index, that is, line outage number L16 attains the first rank in terms of severity level, that is, most severe. Whereas the line having the lowest performance index, that is, line outage number L15 is the least severe. The most severe means when that particular line terminates the impact of this termination on the whole system is more. The Pre and post contingency state of the IEEE-14 bus system is shown in Figs. 5 and 6, respectively. Figure 6 shows the MW flow in corresponding transmission lines after the outage of line number16 which is connected between bus B9 and B10. Table 5 describes the real line power flow of the IEEE-14 bus system before and after line outage number L16. Table 6 describes the bus voltage value of the IEEE-14 bus system before and after the line outage number L16 (connected between bus B9 and B10), which is required to calculate $PI_p$ and $PI_v$.

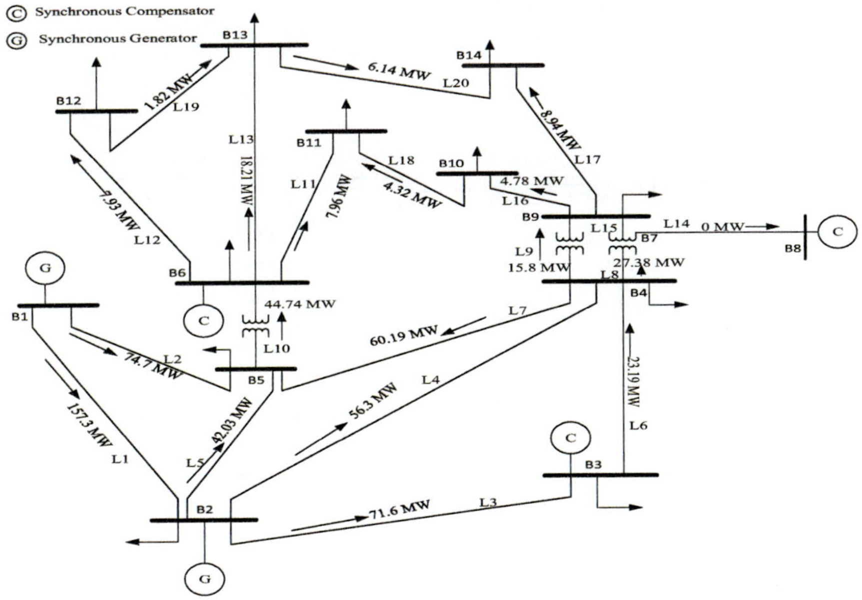

Fig. 5 Pre contingency state of IEEE-14 bus system

**Table 4** Performance index of IEEE-14 bus test system

| Line outage no. | From bus no. | To bus no. | $PI_p$ | $PI_v$ | Performance index (PI) | Rank |
|---|---|---|---|---|---|---|
| L1 | B1 | B2 | 1.1693 | 7.3032 | 8.4725 | 11 |
| L2 | B1 | B5 | 0.9807 | 7.6696 | 8.6503 | 10 |
| L3 | B2 | B3 | 1.1654 | 10.0014 | 11.1664 | 7 |
| L4 | B2 | B4 | 0.9999 | 7.3213 | 8.3212 | 12 |
| L5 | B2 | B5 | 0.9820 | 8.8759 | 9.8579 | 9 |
| L6 | B3 | B4 | 0.9640 | 13.2572 | 14.2212 | 2 |
| L7 | B4 | B5 | 0.9915 | 0.3566 | 1.3481 | 19 |
| L8 | B4 | B7 | 1.0747 | 1.1753 | 2.25 | 18 |
| L9 | B4 | B9 | 0.9807 | 10.5776 | 11.5583 | 4 |
| L10 | B5 | B6 | 1.2396 | 1.6047 | 2.8443 | 15 |
| L11 | B6 | B11 | 1.0142 | 9.5907 | 10.6049 | 8 |
| L12 | B6 | B12 | 1.0127 | 1.8089 | 2.8216 | 16 |
| L13 | B6 | B13 | 1.0569 | 1.3669 | 2.4238 | 17 |
| L14 | B7 | B8 | 1.0072 | 10.4518 | 11.459 | 6 |
| L15 | B7 | B9 | 1.0759 | 0.0844 | 1.1603 | 20 |
| L16 | B9 | B10 | 1.0114 | 13.3464 | 14.3578 | 1 |
| L17 | B9 | B14 | 1.0164 | 2.3482 | 3.3646 | 13 |
| L18 | B10 | B11 | 1.0030 | 10.5217 | 11.5247 | 5 |
| L19 | B12 | B13 | 1.0008 | 12.5538 | 13.5546 | 3 |
| L20 | B13 | B14 | 1.0076 | 2.2891 | 3.2967 | 14 |

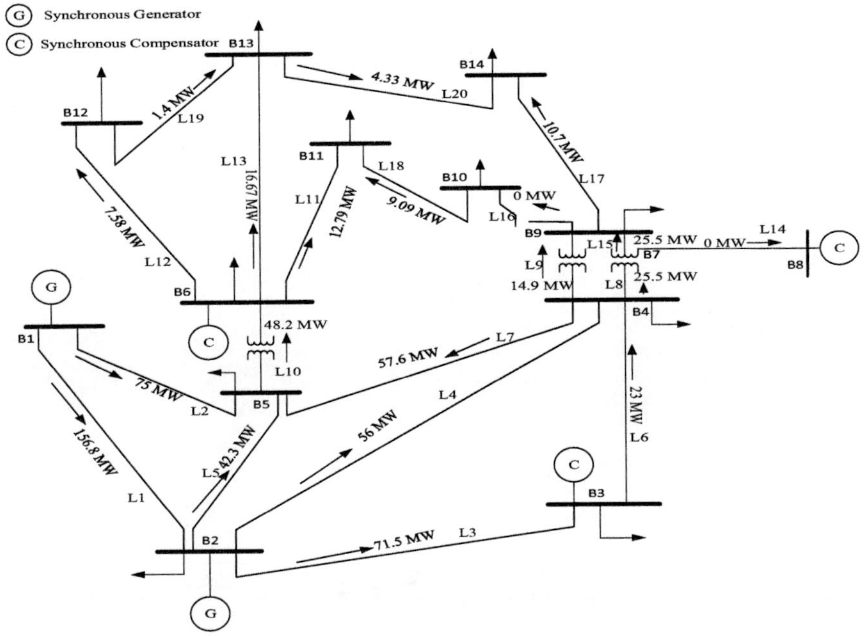


**Fig. 6** Post contingency state of IEEE-14 bus test system

# 7 Remedial Action

- Use of DVR
- Generation re-evaluation
- Use of tie-lines
- Islanding
- Distributed energy resources
- Load shedding

# 8 Conclusion

With MATLAB coding, the researchers can do load flow analysis of the IEEE-5 bus system and the IEEE-14 bus system by using the FDLF method. With the use of PI researchers can find the rank of the line outage. By determining the line outage number, operational engineers can take remedial action for a given power system. For the IEEE-5 bus system, the most critical line outage in the power system is line outage number 1 (connected between bus1 and bus2). Out of the seven-line outages, only one case lies under the category of most severe. For the IEEE-14 bus system, the most critical line outage present in the system is line outage number 16 (connected between bus9 and bus10).

**Table 5** Active power flow for pre and post contingency state of IEEE-14 bus test system during line-16 outage

| Transmission line no. | From bus no. | To bus no. | Before contingency | After contingency |
|---|---|---|---|---|
| | | | MW Flow | MW Flow |
| L1 | B1 | B2 | 157.3 | 156.8 |
| L2 | B1 | B5 | 74.7 | 75 |
| L3 | B2 | B3 | 71.6 | 71.5 |
| L4 | B2 | B4 | 56.3 | 56 |
| L5 | B2 | B5 | 42.03 | 42.3 |
| L6 | B3 | B4 | 23.19 | 23 |
| L7 | B4 | B5 | 60.19 | 57.6 |
| L8 | B4 | B7 | 27.38 | 25.5 |
| L9 | B4 | B9 | 15.8 | 14.9 |
| L10 | B5 | B6 | 44.74 | 48.2 |
| L11 | B6 | B11 | 7.96 | 12.79 |
| L12 | B6 | B12 | 7.93 | 7.58 |
| L13 | B6 | B13 | 18.21 | 16.67 |
| L14 | B7 | B8 | 0 | 0 |
| L15 | B7 | B9 | 27.39 | 25.5 |
| L16 | B9 | B10 | 4.72 | 0 |
| L17 | B9 | B14 | 8.94 | 10.7 |
| L18 | B10 | B11 | 4.32 | 9.09 |
| L19 | B12 | B13 | 1.82 | 1.4 |
| L20 | B13 | B14 | 6.14 | 4.33 |

**Table 6** Bus voltage for pre and post contingency of IEEE-14 bus system

| Bus no. | Before contingency voltage (pu) | After contingency voltage (pu) |
|---|---|---|
| B1 | 1.060 | 1.060 |
| B2 | 1.045 | 1.045 |
| B3 | 1.010 | 1.010 |
| B4 | 1.017 | 1.019 |
| B5 | 1.019 | 1.020 |
| B6 | 1.070 | 1.070 |
| B7 | 1.061 | 1.065 |
| B8 | 1.090 | 1.090 |
| B9 | 1.055 | 1.063 |
| B10 | 1.051 | 1.025 |
| B11 | 1.056 | 1.043 |
| B12 | 1.055 | 1.055 |
| B13 | 1.050 | 1.051 |
| B14 | 1.035 | 1.040 |

## References


1. Gusev S, Oboskalov V (2016) Recursion based contingency analysis of an electrical power system. In: 2016 international symposium on industrial electronics, INDEL 2016—proceedings, pp 1–4
2. Zhou DQ, Annakage UD (2010) Online monitoring of voltage stability margin using artificial neural network. IEEE Trans Power Syst 25(3):1566–1574
3. Dela Cruz EF, Mabalot AN, Marzo RC, Pacis MC, Tolentino JHS (2016) Algorithm development for power system contingency screening and ranking using voltage-reactive power performance index. In: 2016 IEEE region 10 conference (TENCON), Singapore, pp 2232–2235
4. Vykuka R, Nohacova L (2015) Sensitivity factors for contingency analysis. In: Proceedings of the 2015 16th international scientific conference on electric power engineering (EPE) 4, pp 551–554
5. Vykuka R, Noháčová L (2014) Fast-decoupled method for contingency analysis. In: Proceedings of the 2014 15th international scientific conference on electric power engineering (EPE), Brno, pp 35–38
6. Dong F, Xu X, Zhang X (2014) Parallel contingency analysis solution based on OpenMP. In: 2014 North American power symposium NAPS 2014, pp 1–4
7. Ramanathan R, Tuck B (2015) Contingency analysis using node/breaker model for operation studies. In: 2015 IEEE power & energy society general meeting, Denver, CO, pp 1–5
8. Ramesh S et al (2013) Special reprocessing logic for contingency analysis with special protection systems. In: 2015 IEEE power & energy society general meeting, pp 1–4
9. Vykuka R, Nohacova L (2014) Fast-decoupled method for contingency analysis. In: Proceedings 2014 15th international scientific conference on electric power engineering EPE 2014, pp 35–38
10. Mishra VJ, Khardenvis MD (2012) Contingency analysis of power system. In: 2012 IEEE students' conference on electrical, electronics and computer science: innovation for humanity, SCEECS, pp 1–4

11. Sahraei-Ardakani M, Li X, Balasubramanian P, Hedman KW, Abdi-Khorsand M (2016) Real-time contingency analysis with transmission switching on real power system data. IEEE Trans Power Syst 31(3):2501–2502
12. Elbarkouky MMG, Siraj NB, Fayek AR (2015) Fuzzy contingency determinator © a fuzzy arithmetic-based risk analysis tool for construction projects. In: Annual conference on North American fuzzy information processing society NAFIPS, September, pp 1–6
13. Schultis D-L, Ilo A, Schirmer C (2019) Overall performance evaluation of reactive power control strategies in low voltage grids with high prosumer share. Electr Power Syst Res 168:336–349
14. Shao Y, Huang S, Chen B, Xiao H, Jiang R, Qin FGF (2017) Performance evaluation indexes of distributed CCHP systems. Energy Procedia 142:2415–2422
15. Putri MDM, Afandi AN (2019) Performance index analysis (PIA) for N-1 contingency transmission in 150 kV electricity system. In: 2019 international conference on electrical, electronics and information engineering (ICEEIE), Denpasar, Bali, Indonesia, pp 1–5

# Chapter 33
# Optimal Placement of DG and Capacitor in Radial Distribution Networks Using Differential Evolution

Himmat Singh and Vijay Bhuria

## 1 Introduction

Balanced radial distribution generation normally works at high currents and low voltage, consequentially in low voltage profile and high power loss similar with transmission network [1].

Different evolution computing techniques have been implemented for optimal locations SC and DG to improve voltage profile and minimization real power loss [2–4]. modified differential evolution algorithm [5], genetic algorithm [6], particle swarm optimization [6], evolutionary programming [7] hybrid algorithm [8] artificial bee colony [9], teaching-learning optimization [10], krill herd algorithm [11] flower pollination algorithm [12], multi-objective particle swarm optimization [13], bacterial foraging optimization algorithm [14], Direct Search algorithm [15], analytical Hybrid [16], improved analytical [17], Improved Harmony Algorithm [18].

Outlined of this paper as shown: the problem formulation in Sect. 2, and proposed methodology in Sect. 3, proposed results and discussions in Sect. 4, and results obtained in Sect. 5.

Consider a single line diagram connected between receiving and sending end node of balanced distribution network as shown in Fig. 1.

H. Singh (✉) · V. Bhuria
Department of Electrical Engineering, Madhav Institute of Technology and Science, Gwalior, India
e-mail: ahirwar.himmat@gmail.com

© The Author(s), under exclusive license to Springer Nature Singapore Pte Ltd. 2022

H. M. Dubey et al. (eds.), *Artificial Intelligence and Sustainable Computing*, Algorithms for Intelligent Systems, https://doi.org/10.1007/978-981-16-1220-6_34

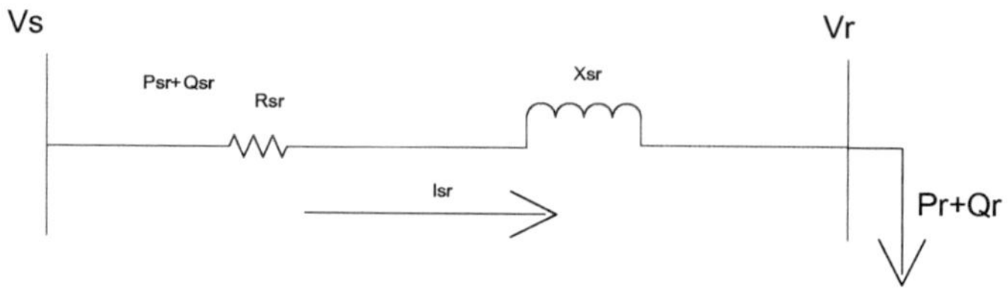


**Fig. 1** Equivalent diagram of radial distribution system

## 2 Problem Formation

The objectives considered for optimal allocation in radial distribution systems minimize real power loss and improve voltage profile at load buses power system network, subject to various equality and inequality constraints [8, 9]:

$$P_{sr} = P_r^f + P_r^l - P_r^{DG} + R_{sr}/V_s^2\left(P_{sr}^2 + Q_{sr}^2\right) \tag{1}$$

$$Q_{sr} = Q_r^f + Q_r^l - Q_r^{DG} - Q_r^c + X_{sr}/V_s^2\left(P_{sr}^2 + Q_{sr}^2\right) \tag{2}$$

$$V_q^2 = V_S^2 - 2(P_{sr}R_{sr} + Q_{sr}RX_{sr}) + \left\{\frac{R_{sr}^2 + X_{sr}^2}{V_s^2}\right\}\left(P_{sr}^2 + Q_{sr}^2\right) \tag{3}$$

Here
$P_{sr}$ = active power flow at sending end,
$Q_{sr}$ = reactive power flow at receiving end,
$R_{sr}$ = series resistance,
$X_{sr}$ = series Reactance,
$P_r^f$ = power flows at downstream branch,
$V_r$ = voltage magnitude at bus r,

The branch current flowing between two nodes s and r is given as:

$$I_{sr} = \sqrt{\frac{P_{sr}^2 + Q_{sr}^2}{V_s^2}} \tag{4}$$

The objective function is mathematically given as:

$$MinP_{Loss} = \sum_r I_{sr}^2 R_{sr} \tag{5}$$

**Equality Constraints**

The active and reactive power flow in all balanced radial distribution generation networks is the optimal location of SC and DG must satisfy Eqs. (1) and (2).

**Inequality Constraints**

Each bus the min and max limits of voltage given by

$$V_r^{\min} \leq V_r \leq V_r^{\max} \tag{6}$$

Thermal limit of current flow each branch given by

$$I_{rs} \leq I_{rs}^{\text{rated}} \tag{7}$$

## 3 Differential Evolution Algorithm

Differential evolution algorithm was introduced by Storn and Price in 1996. It is a stochastic population-based search optimization evolutionary algorithm for solving a nonlinear problem. It is an enhanced version of a genetic algorithm. This method can be applied for optimization of a discontinuous, nonsmooth, single as well as multi-model function. DE algorithm can find the nearest global solution with little tuning parameter. Its convergence is very fast as compared to other algorithms. It has been noticed that the performance of DE is considerably improved than that of other heuristic methods, such as GA, PSO, HMPSO, SDE, EPSDE, and other EC algorithms.

Recently Differential Evolution (DE) algorithm has been applied for various optimization problems. DE is an improved version of the genetic algorithm, which provides fast optimization. It is a simple population-based search algorithm, which is highly efficient in handling constrained optimization problems. This algorithm can take care of optimality on rough, discontinuous and multi-modal surfaces. Apart from this, the algorithm has been used to solve high-dimensional function optimization. The flow chart of the differential algorithm shown in Fig. 2

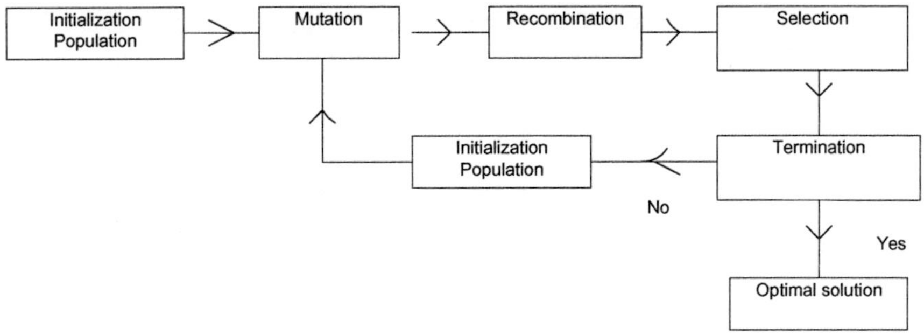


**Fig. 2** Optimization process of DE

### 3.1 *Implementations of DE for Optimal Location of DG and SC*

The proposed Differential Evolution algorithms have been formulated and implemented using Matlab software. The first step in various DE algorithms is creating an initial population of size *NP*. Each independent parameter of every individual in the population is assigned a value within its specified feasible region. Whenever, finding the independent variables, dependent variables, such as reactive power, voltages at load buses and line flows, etc., were calculated using the forward back Load Flow (FBLF) program. Similar to other evolutionary computing-based algorithms DE algorithms are able to solve only the unconstrained optimization problem. To apply these algorithms for constrained optimization problems, first, it is converted into an unconstrained problem. For solving of optimal location of DG and capacitor, the process can be concise in the following steps:

i. Run the initial load flow program
ii. Find optimal locations of DG and shunt capacitor
iii. Fixed capacitor capacity loading and DG capacity of the network
iv. Obtained result for minimum real power loss and voltage profile
v. Finally, run load flow program for the location of DG and capacity of shunt capacitor

## 4 Simulation Results

The proposed analytical differential evolution algorithm has been tested for solving the optimal allocation problem of the IEEE 12-bus balanced radial distribution system. This system consists of 12 kV voltage, number of buses 12 and 11 total branches with a load of the feeder of 455.71 kW and 413.04 kVAr. The proposed algorithms for five cases are considered in the following procedure

1. Placed shunt capacitor at a most favorable location.
2. Placed unity power factor DG at a most favorable location.
3. DG unity power factor and the shunt capacitor were placed at their most favorable location.
4. DG 0.9 power factor lag and the shunt capacitor were placed at their optimal location.
5. DG 0.75 power factor lag and the shunt capacitor were placed at their optimal location

The results of implemented Metaheuristic Differential Evolution algorithm provided for optimal location of DG and shunt capacitor is shown in Table 1. By the comparison of the results, it can be observed that the proposed DE algorithm gives a better voltage profile improvement (Tables 2 and 3).

**Table 1** Optimal location of DG and capacitor using DE

| Device | Optimal location | Differential evolution algorithm | | | |
|---|---|---|---|---|---|
| | | Size MW/MVAR | Real power | Voltage profile | Bus location of lowest voltage |
| SC | 9 | 0.210298 | 12.5842 | 0.313766 | 0.956298 at bus 12 |
| DG | 9 | 0.235504 | 10.7744 | 0.143276 | 0.983531 at bus 7 |
| SC&DG | 9<br>9 | 0.232224<br>0.212299 | 3.1520 | 0.060396 | 0.990805 at bus 7 |
| 0.9 *pf* DG | 9 | 0.272535<br>0.131998 | 4.49288 | 0.052760 | 0.991213 at bus 6 |
| 0.75 *pf* DG | 9 | 0.226722<br>0.199939 | 3.18572 | 0.071185 | 0.989928 at bus 7 |

**Table 2** Optimal location of DG and capacitor using PSO [19, 20]

| Device | Optimal location | Particle swarm optimization | | | |
|---|---|---|---|---|---|
| | | Size MW/MVAR | Real power | Voltage profile | Bus location of lowest voltage |
| SC | 9 | 0.210223 | 12.584 | 0.313795 | 0.956294 at bus 12 |
| DG | 9 | 0.235472 | 10.774 | 0.143310 | 0.983529 at bus 7 |
| SC &DG | 9<br>9 | 0.232434<br>0.212087 | 3.152 | 0.060257 | 0.990815 at bus 7 |
| 0.9 *pf* DG | 9 | 0.263423<br>0.131778 | 4.49288 | 0.052760 | 0.991213 at bus 6 |
| 0.75 *pf* DG | 9 | 0.215621<br>0.187834 | 3.12504 | 0.070132 | 0.96883 at bus 7 |

**Table 3** Comparison of CPU time using real power loss = 20.71 [19, 20]

| Device | Optimum location | Particle swarm optimization | | Proposed differential evolution algorithm | |
|---|---|---|---|---|---|
| | | No. of generations | Elapsed time (sec) | No. of generations | Elapsed time (sec) |
| SC | 9 | 20 | 197.294989 | 20 | 196.284986 |
| DG | 9 | 22 | 200.229419 | 22 | 199.219324 |
| SC&DG | 9<br>8 | 29 | 1223.780534 | 29 | 1222.680421 |
| 0.90 *pf* DG | 9 | 21 | 199.244364 | 21 | 198.232341 |
| 0.75 *pf* DG | 9 | 20 | 193.727058 | 20 | 192.636046 |

In this paper we consider voltage profiles different cases for balanced radial distribution network various buses are shown in Table 4. As can be noticed from Table 4, the voltage profile of optimal location of capacitor and DG obtained using DE algorithm is very close to as compared to those obtained using other GA algorithm (Fig. 3).

**Table 4** Voltage profile of 12-bus radial distribution

| Bus no. | Base case | SC | DG | SC&DG | 0.90 *pf* DG | 0.75 *pf* DG |
|---|---|---|---|---|---|---|
| 1 | 1.00000 | 1.000000 | 1.000000 | 1.00000 | 1.00000 | 1.00000 |
| 2 | 0.99433 | 0.995207 | 0.996566 | 0.997416 | 0.997461 | 0.997315 |
| 3 | 0.98903 | 0.990849 | 0.993687 | 0.995449 | 0.995545 | 0.99524 |
| 4 | 0.98058 | 0.984056 | 0.989528 | 0.992884 | 0.993077 | 0.992484 |
| 5 | 0.96982 | 0.975808 | 0.985311 | 0.99105 | 0.991399 | 0.99036 |
| 6 | 0.96654 | 0.973374 | 0.984265 | 0.990808 | 0.997626 | 0.994079 |
| 7 | 0.96375 | 0.971367 | 0.98369 | 0.995713 | 0.991264 | 0.989928 |
| 8 | 0.95531 | 0.965215 | 0.984052 | 0.993385 | 0.994495 | 0.992182 |
| 9 | 0.94728 | 0.960168 | 0.987452 | 0.999424 | 1.001335 | 0.997802 |
| 10 | 0.94446 | 0.95739 | 0.984752 | 0.996761 | 0.998673 | 0.995129 |
| 11 | 0.94356 | 0.956504 | 0.98389 | 0.995911 | 0.997823 | 0.994277 |
| 12 | **0.94335** | **0.956298** | **0.983531** | **0.990805** | **0.991213** | **0.990020** |

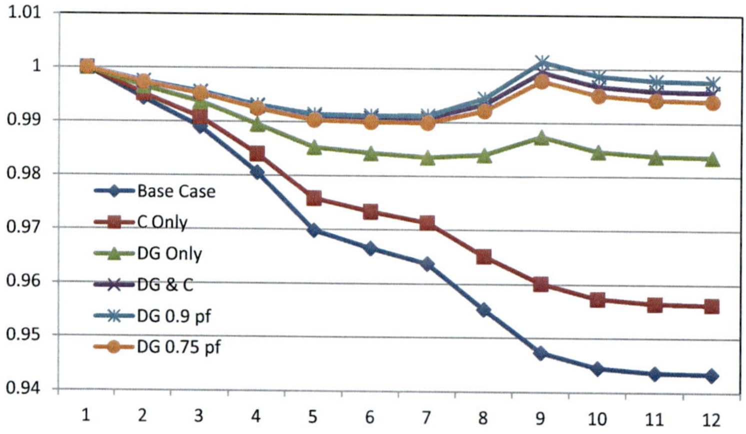


**Fig. 3** Voltage profile of 12-Bus BRDS [19, 20]

## 5 Conclusion

In the paper, DE and PSO is implemented to solve sizing and siting for capacitor and DG in a balanced distribution network. Both algorithms are tested on a standard IEEE 12-bus radial distribution system with integration of DG and capacitor. Simulation results demonstrate that DE has better performance in comparison to PSO in reducing line losses and voltage profile.

**Acknowledgements** The authors acknowledge the financial support provided project received under Innovative Research Scheme (IRS-2020) by TEQIP-III (File No. IRS/2020/R&D/CE/MKT/08/616) dated 09/03/2020 thanks to the Director, MITS Gwalior, India to carry out this research work.

## References

1. Ng HN, Salama MMA, Chikhani AY (2000) Classification of capacitor allocation techniques. IEEE Trans Power Deliv 15(1):387–392
2. Selim A, Kamel S, Jurado F (2019) Power losses and energy cost minimization using shunt capacitors installation in distribution systems. In: International renewable energy congress (IREC), Sousse, Tunisia, pp 1–6
3. Mohamed AA, Kamel S, Aly MM (2017) A simple analytical technique for optimal capacitor placement in radial distribution systems. In: Power systems conference (MEPCON), 2017 Nineteenth International Middle East: IEEE, pp 928–933
4. Kamel S, Mohamed M, Selim A, Nasrat LS, Jurado F (2019) Power system voltage stability based on optimal size and location of shunt capacitor using analytical technique. In: 2019 10th international renewable energy congress (IREC), Sousse, Tunisia, pp 1–5
5. Dixit M, Kundu P, Jariwala HR (2017) Incorporation of distributed generation and shunt capacitor in radial distribution system for techno economic benefits. Int J Eng Sci Technol 20(2):482–493
6. Zeinalzadeh A, Mohammadi Y, Moradi MH (2015) Optimal multi objective placement and sizing of multiple DGs and shunt capacitor banks simultaneously considering load uncertainty via MOPSO approach. Int J Electr Power Energy Syst 67:336–349
7. Sultana S, Roy PK (2014) Multi-objective quasi-oppositional teaching learning based optimization for optimal location of distributed generator in radial distribution systems. Int J Electr Power Energy Syst 63:534–545
8. El-Fergany A (2015) Optimal allocation of multi-type distributed generators using backtracking search optimization algorithm. Int J Electr Power Energy Syst 64:1197–1205
9. Kowsalya MIAM (2014) Optimal distributed generation and capacitor placement in power distribution networks for power loss minimization. In: 2014 international conference on advances in electrical engineering (ICAEE). IEEE, pp 1–6
10. Raju MR, Murthy KR, Ravindra K (2012) Direct search algorithm for capacitive compensation in radial distribution systems. Int J Electr Power Energy Syst 42(1):24–30
11. Kansal S, Kumar V, Tyagi B (2016) Hybrid approach for optimal placement of multiple DGs of multiple types in distribution networks. Int J Electr Power Energy Syst 75:226–235
12. Hung DQ, Mithulananthan N (2011) Multiple distributed generator placement in primary distribution networks for loss reduction. IEEE Trans Indus Electron 60(4):1700–1708
13. Sultana S, Roy PK (2016) Krill herd algorithm for optimal location of distributed generator in radial distribution system. Appl Soft Comput 40:391–404

14. Abdelaziz AY, Ali ES, Abd Elazim SM (2016) Flower pollination algorithm and loss sensitivity factors for optimal sizing and placement of capacitors in radial distribution systems. Electr Power Energy Syst 78:207–214
15. Ali ES, Abd Elazim SM, Abdelaziz AY (2016) Improved harmony algorithm and power loss index for optimal locations and sizing of capacitors in radial distribution systems. Int J Electric Power Energy Syst 80:252–263
16. Kumar A, Babu PV, Murty V (2017) Distributed generators allocation in radial distribution systems with load growth using loss sensitivity approach. J Instit Engineers (India) Ser B 98:275–287
17. Song YH, Wang GS, Johns AT, Wang PY (1997) Distribution network reconfiguration for loss reduction using fuzzy controlled evolutionary programming. Proc Gen Trans Distrib 144:345–50
18. Gallego RA, Monticelli AJ, Romero R (2001)Optimal capacitor placement in radial distribution networks. IEEE Trans Power Syst 16(4):630–7
19. Yadav PS, Srivastava L (2015) Optimal location of combined DG and capacitor for real power loss minimization in distribution networks. Int. J. Electr. Electron. Eng (IJEEE), ISSN- 2321-2055, 7(1)
20. Mohapatra A, Behera S, Nayak S, Panigrahi BK (2012) A study on DG and capacitor placement in radial distribution system. In: International Conference on Power Electronics, Drives and Energy Systems December16–19, 2012, Bengaluru, India, IEEE, https://doi.org/10.1109/PEDES.2012.6484268

# Chapter 34
# Optimal Sizing and Allocation of DG in Distribution System Using TLBO

**Vimal Tiwari, Hari Mohan Dubey, and Manjaree Pandit**

## Nomenclature

| | |
|---|---|
| TLBO | Teaching—Learning based optimization |
| RDS | Radial distribution system |
| MILP | Mixed—integer linear programming |
| MIP | Mixed integer program |
| SQP | Sequential quadratic programming |
| BSO | Backtracking search optimization |
| ACO | Ant colony optimization |
| ABC | Artificial bee colony |
| HSA | Harmony search algorithm |
| CSA | Cuckoo search algorithm |
| GWO | Grey Wolf Optimizer |
| SKHA | Stud Krill herd Algorithm |
| DFA | Dragonfly algorithm |
| ALO | Ant Lion optimization |
| WOA | Whale optimization algorithm |
| $DG$ | Distributed generation |
| $P_{loss}$ | Real power loss |
| $n_{br}$ | Total number of branches |
| $I_i$ | Current of branch $i$ |
| $R_i$ | Resistance of branch $i$ |
| $n$ | Total number of bus |
| $P_0$ | Slack's bus real power |
| $Q_0$ | Slack's bus reactive power |
| $Q_{loss}$ | Total reactive power loss |
| $P_{i(DG)}$ | Real power output |

V. Tiwari (✉) · H. M. Dubey · M. Pandit
Department of Electrical Engineering, MITS, Gwalior, India

© The Author(s), under exclusive license to Springer Nature Singapore Pte Ltd. 2022

H. M. Dubey et al. (eds.), *Artificial Intelligence and Sustainable Computing*, Algorithms for Intelligent Systems, https://doi.org/10.1007/978-981-16-1220-6_35

| | |
|---|---|
| $Q_{i(DG)}$ | Reactive power output |
| $P_{i(D)}$ | Real power demand at bus $i$ |
| $Q_{i(D)}$ | Reactive power demand at bus $i$ |
| $P_{i(DG)}^{min}$ | DGminimum capacity |
| $P_{i(DG)}^{max}$ | DG maximum capacity |

## 1 Introduction

The distribution system is a vital part of an electric power system, which established a link between the transmission systems at high voltage to the consumer's load at low voltage. RDS is one of the most preferred system in which the voltage is low at the far end, which reduces the performance of electrical equipment's. Out of the total power loss in the power system, the power loss in RDS is maximum.

Integration of DG at the optimal location is an efficient and attractive method to minimize power loss of the system and hence enhance the power quality and voltage profile of RDS [1, 2].

Various research works related to it have been reported in the literature using different methods [3–15]. In Ref. [3], the MILP method is used for DG installation of different types to curtail the capital cost and cost of operation per annum. In Ref. [4], the MIP approach was utilized to determine the optimal allocation of DG in a microgrid. Column and constraint generations (CCG) method were utilized for DG planning problem which includes the capital cost, fuel cost, operation and maintenance cost and emission released due to installation of DG. DG with storage system with objective is to minimize of cost of power generated and total energy loss per year was computed using SQP [5].

In Ref. [6], BSO were used for impact analysis of DG integration in the RDS for minimizing the real power loss and enhance voltage profile. A hybrid ACO with ABC is used for the solution of DG planning with an objective is to limit the power loss, total cost, total emission and enhance the voltage stability [7]. For solving the stochastic environment problem, a point estimate method is used and tested on the IEEE 33 and 69 RDS. In Ref. [8], a hybrid HSA method was utilized for analyzing optimal location and sizing of DG units along with shunt capacitors integrated to RDS to reduce power loss for the IEEE 33 and 118 RDS. Loss sensitive factor is used to determine the sensitive nodes. Hybrid PSO were used for DG planning problem and tested on the IEEE 33 and 69 RDS [9]. CSA were used to find the optimal sizing and location of DGs for RDS to reduce power loss and for improving the voltage stability index [10]. In Ref. [11], GWO is implemented for solving the problem of DG planning to minimize the total cost. SKHA is implemented for optimal sizing and location of DG units [12]. It is tested on IEEE 15, IEEE 33, IEEE 69 and 94 bus Portuguese RDS. In Ref. [13] DFA is implemented for optimal allocation of DG units and tested on IEEE 15, IEEE 33 and IEEE 69 bus RDS. ALO [14] and WOA

**Table 1** Different type of DG models [6]

| DG types | Power generation | Example |
|---|---|---|
| Type I | Inject real power | Photovoltaic arrays and fuel cells |
| Type II | Inject reactive power | Synchronous condensers and Capacitors |
| Type III | Inject both real and reactive power | Synchronous generator |
| Type IV | Inject real and absorb reactive power | Asynchronous generator |

[15] are implemented for solving the problem of DG planning in RDS and tested on IEEE 15, IEEE 33, IEEE 69 and IEEE 85 bus RDS.

The DG are classified into four types as shown in Table 1 [6].

In this paper, TLBO is implemented for finding out the optimal size of DG unit and its location in RDS, for the real power losses minimization and hence to improve the voltage profile of the system. Analysis has been done in the following scenario: (a) with no DG in the system (reference test case), (b) with single and (c) for multiple DGs installation at the same network. This paper is organized as follows: the problem formulation that includes objective function and various constraints is given in Sect. 2. The basic concept of TLBO and its implementation is presented in Sect. 3. The results and discussion are given in Sect. 4 and the concluding remarks are discussed in Sect. 5.

## 2 Problem Formation

Here the objective is to find out the optimal size of DG and its location in RDS to minimize the real power loss which is calculated as follows [7, 8, 12, 13]:

$$\text{Minimum}(F) = P_{\text{loss}} = \sum_{i=1}^{n_{br}} (|I_i|^2 * R_i) \tag{1}$$

Subjected to constraints as:

$$V_{\min} \leq |V_i| \leq V_{\max} \tag{2}$$

where, $\boldsymbol{V_i}$ is voltage at ith bus, with minimum voltage $\boldsymbol{V_{\min}} = 0.95$ pu and maximum voltage $\boldsymbol{V_{\max}} = 1.05$ pu.

Load generation balance should always be met i.e. incoming power should be equal to the outgoing power and can be given as follows [6, 7, 12]

$$\sum_{i=1}^{n} P_{i(DG)} = \sum_{i=1}^{n} P_{i(D)} + P_{loss} - P_0 \tag{3}$$

$$\sum_{i=1}^{n} Q_{i(DG)} = \sum_{i=1}^{n} Q_{i(D)} + Q_{loss} - Q_0 \tag{4}$$

$$P_{i(DG)}^{\min} \le P_{i(DG)} \le P_{i(DG)}^{\max} \tag{5}$$

## 3 Teaching-Learning Based Optimization (TLBO)

TLBO is a population-based optimization algorithm developed with the inspiration of the teaching and learning process in class [16]. The TLBO works in two-phase 'Teacher phase' and 'Learner phase'. In the teacher phase student learn from the teacher and in the learner phase student learn through interaction with fellow students. The teacher is selected from the class, which produces the best solution [17].

### 3.1 Teacher Phase

This phase refers to the improvement of the mean results of class for the subject taught by him. Among the entire population of NP, the best solution is to select as a teacher and he shares his knowledge to improve the mean results of the class. Therefore, the difference between the mean results of a $k$th learner and best result of teacher, $Mean\_Diff_{j,k,i}$ is given by,

$$Mean\_Diff_{j,k,i} = r_i\left(X_{j,kbest,i} - T_F M_{j,i}\right) \tag{5}$$

Here, $X_{j,kbest,i}$ and $M_{j,i}$ is the best and mean result of subject $j$th at iteration $i$th, $T_F$ is the teaching factor responsible to change the mean result of class and $r_i$ is the randomly selected value from 0 to 1. The value of $T_F$ is randomly decided either 1 or 2 by using the following expression:

$$T_F = \text{round}[1 + rand(0, 1)\{2 - 1\}] \tag{6}$$

$$X_{j,k,i}^{new} = X_{j,k,i}^{old} + Mean\_Diff_{j,k,i} \tag{7}$$

Here the updated value $X_{j,k,i}^{new}$ is accepted if they give better result as compared to the old value $X_{j,k,i}^{old}$. At the end of this phase, all accepted values become input for the learner phase.

### 3.2 Learner Phase

In this phase, learner enhances their knowledge from other learners by interacting with the fellow learners having more superior than him/her in terms of knowledge. In this phase, P and Q are two different learners selected randomly such that $X^{old}_{P,i} \neq X^{old}_{Q,i}$, where $X^{old}_{P,i}$ and $X^{old}_{Q,i}$ are the value of an updated function at the teacher phase end.

$$X^{new}_{j,P,i} = X^{old}_{j,P,i} + r_i\left(X^{old}_{j,P,i} - X^{old}_{j,Q,i}\right), if X^{old}_{P,i} < X^{old}_{Q,i} \tag{8}$$

$$X^{new}_{j,P,i} = X^{old}_{j,P,i} + r_i\left(X^{old}_{j,Q,i} - X^{old}_{j,P,i}\right), if X^{old}_{Q,i} < X^{old}_{P,i} \tag{9}$$

Here, $X^{new}_{j,P,i}$ is the updated value in this phase and get accepted only if the results are better compared to the $X^{old}_{j,P,i}$ obtained at the end of the teacher phase.

The flow chart of TLBO is shown in Fig. 1.

## 4 Results and Discussion

The algorithm of TLBO was applied to formulate the DG planning problem in MATLAB R2016a and executed on CPU with an i5 processor and 4 GB ram with speed 2.50 GHz. The problems are studies in four cases, case I without DG placement, case II with single Type I DG placement, case III with single Type II DG placement and case IV with single Type III DG placement.

### 4.1 Selection of Control Parameter of TLBO

Every metaheuristic needs proper tuning of their controls parameter to shows effective performance. TLBO is also governed by its control parameters, NP varies from 25 to 100 and tested on IEEE 85 RDS with single DG installation to obtain the best value of NP. The statistical results for different values of NP over 50 repeated trials, shown in Table 2. The mean power loss and standard deviation (SD) for different NP sizes are shown in Fig. 2. After analyzing the statistical results given in Table 2 the best value of the control parameter are NP = 50 and hence they are considered for further analysis.

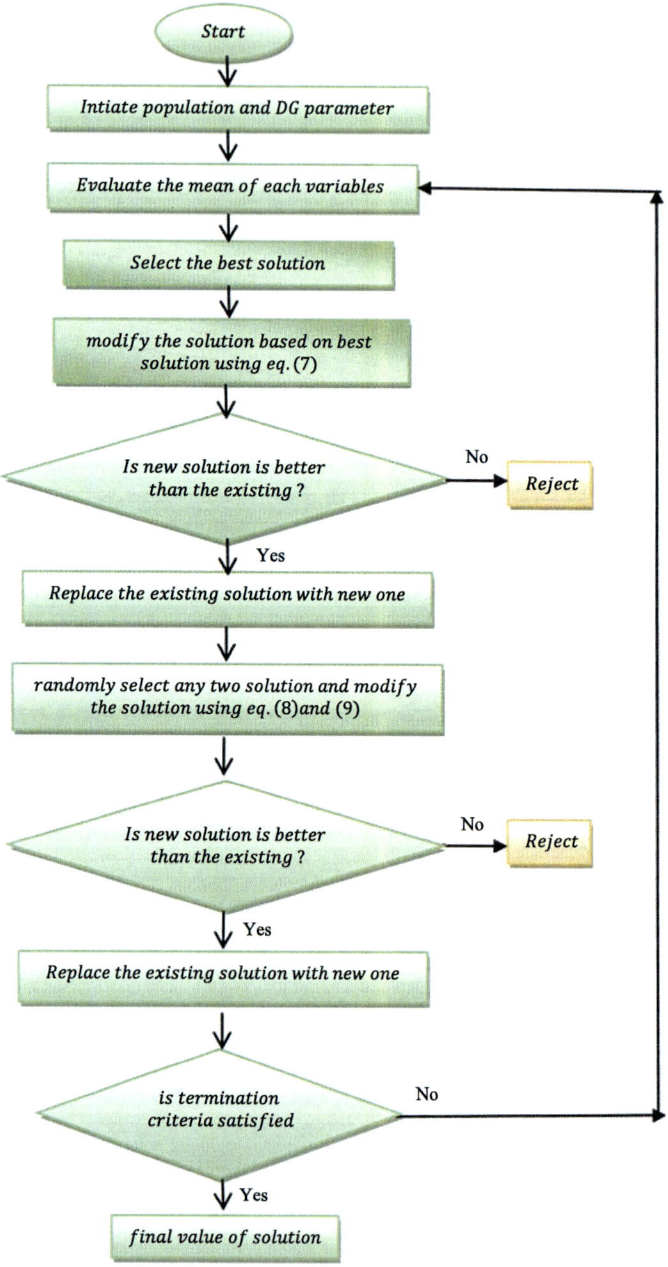


**Fig. 1** Flowchart of TLBO

**Table 2** Results for different values of control parameter NP

| NP | Power loss (kW) | | | SD | CPU times/iter (s) |
|---|---|---|---|---|---|
| | Min. | Mean | Max. | | |
| 25 | 188.8515 | 188.8515 | 188.8515 | 8.7737e-14 | 0.0474 |
| 50 | 188.8515 | 188.8515 | 188.8515 | 5.8593e-14 | 0.0727 |
| 100 | 188.8515 | 188.8515 | 188.8515 | 5.8708e-14 | 0.1324 |

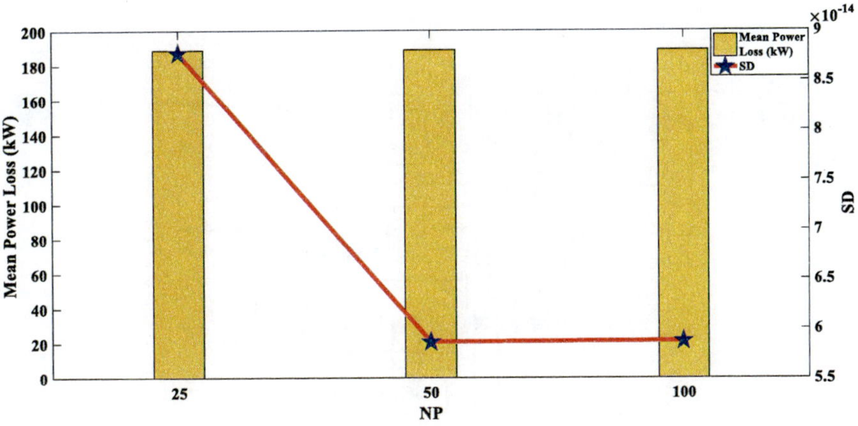


**Fig. 2** Mean power loss and SD for different NP sizes

## 4.2 *IEEE 85 Bus RDS*

The TLBO algorithm is used in the application of planning the DG installation in RDS and tested on IEEE-85 bus RDS with base voltage 11 kV [18] as shown in Fig. 3 having 84 branches and 85 buses. The total static load demand of this bus system is (2.570 + j2.622) MVA. The results obtained from simulation for both cases are given in Table 3. In case I, the real power loss is 316.1219 kW without DG installation with 0.8713 p.u. is the minimum voltage and maximum voltage is 0.9957 p.u.

### 4.2.1 Type I DG Installation

In this case the Type I DG is placed at bus 28 with the size of 1413.654 kW the real power loss is 188.8515 kW which is 40.26% less than case I. The 0.9221 p.u. is the minimum voltage and the maximum voltage is 0.9971 p.u. These results are compared with other results achieved by different methods and show better results as compared to ALO [14], WOA [15] given in Table 4. The convergence curve of the TLBO algorithm for DG installation is shown in Fig. 4.

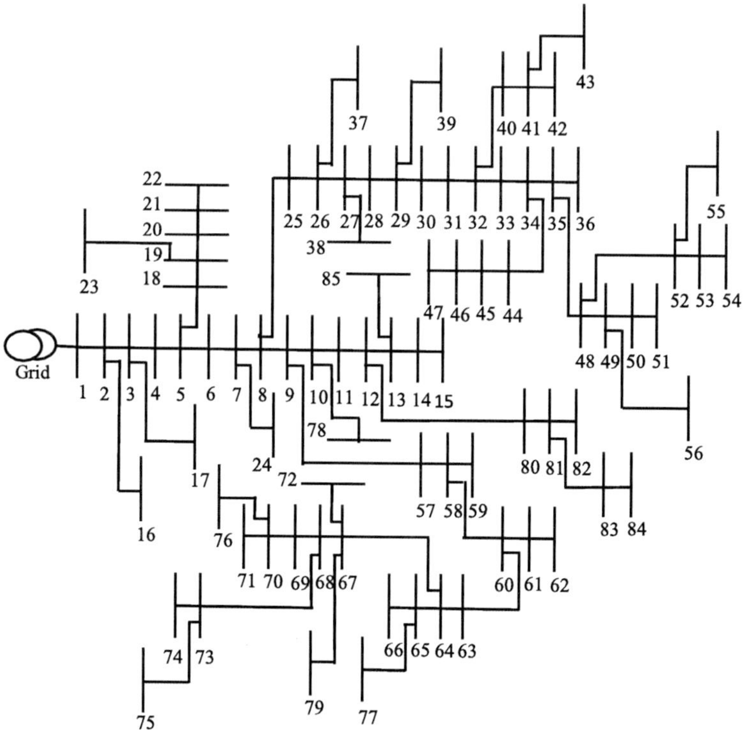


**Fig. 3** One-line diagram for IEEE 85 RDS

**Table 3** Simulation results for DG installation in IEEE 85 RDS

| Parameters | Without DG | With single DG | | |
|---|---|---|---|---|
| | | Type I | Type II | Type III |
| Power loss (kW) | 316.1219 | 188.8515 | 192.1046 | 113.376 |
| Loss reduction (%) | | 40.26 | 39.23 | 64.14 |
| DG size/location | | 1413.654 kW/28 | 1442.2155 kVAr/28 | 1413.6540 kVA/30 |
| Minimum voltage | 0.8713 | 0.9221 | 0.9056 | 0.9339 |
| Maximum voltage | 0.9957 | 0.9971 | 0.9967 | 0.9976 |

**Table 4** Comparison results for Type I DG installation in IEEE 85 RDS

| Algorithms | DG size(kW)/location | Power loss (kW) | Minimum voltage |
|---|---|---|---|
| Without DG | | 316.1219 | 0.8713 |
| ALO [14] | 946.347/55 | 224.049 | 0.9109 |
| WOA [15] | 910.075/54 | 227.105 | 0.9101 |
| **TLBO** | **1413.654/28** | **188.8515** | **0.9221** |

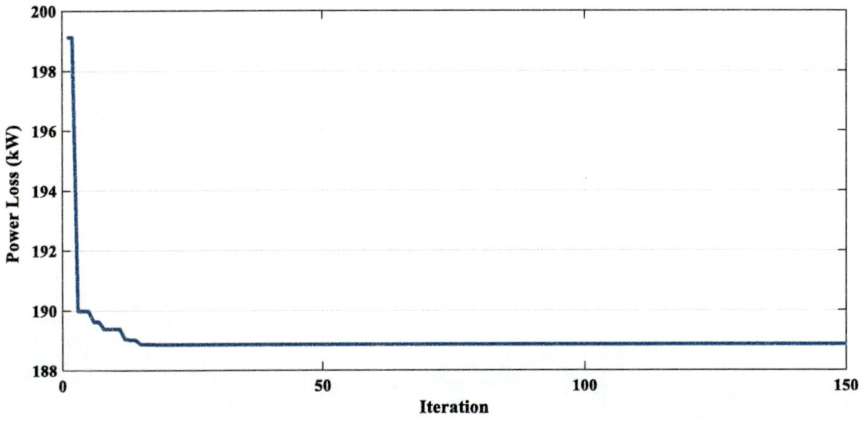


**Fig. 4** Convergence curve for Type I DG installation in IEEE 85 RDS using TLBO

**Table 5** Comparison results for Type II DG installation in IEEE 85 RDS

| Algorithms | DG size (kVAr)/location | Power loss (kW) | Minimum voltage |
|---|---|---|---|
| Without DG | | 316.1219 | 0.8713 |
| ALO [14] | 873.846/55 | 229.02 | 0.903 |
| WOA [15] | 837.285/54 | 232.136 | 0.9023 |
| **TLBO** | **1442.2155/28** | **192.1046** | **0.9056** |

### 4.2.2 Type II DG Installation

In this case, the Type II DG is placed at bus 28 with the size of 1442.2155 kVAr, with the real power loss of RDS is 192.1046 kW which is 39.23% less than case I. The 0.9056 p.u. is the minimum voltage and the maximum voltage is 0.9967 p.u. These results are compared with other results achieved by different methods and show better results as compared to ALO [14], WOA [15] given in Table 5.

### 4.2.3 Type III DG Installation

In this case, the Type III DG with power factor (pf) 0.9 is placed at bus 30 with the size of 1442.2155 kVAr, with the real power loss of RDS is 113.376 kW which is 64.14% less than case I. The 0.9339 p.u. is the minimum voltage and the maximum voltage is 0.9976 p.u. These results are compared with other results achieved by different methods and show better results as compared to ALO [14], WOA [15] given in Table 6. Figure 5 shows the impact on the voltage profile of IEEE-85 bus RDS without and with single type I, II, and III DG installation.

**Table 6** Comparison results for Type III DG installation in IEEE 85 RDS

| Algorithms | DG size (kVA)/location (pf = 0.9) | Power loss (kW) | Minimum voltage |
|---|---|---|---|
| Without DG | | 316.1219 | 0.8713 |
| ALO [14] | 1289/55 | 157 | 0.9255 |
| WOA [15] | 1243.18/54 | 162.329 | 0.9242 |
| **TLBO** | **1413.6540/30** | **113.376** | **0.9339** |

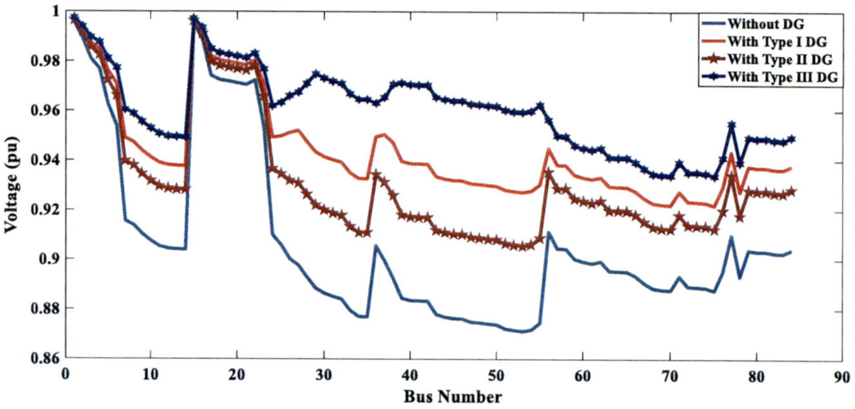


**Fig. 5** Voltage profile of IEEE 85 RDS without and with single DG installation

### *4.3 Multiple DG in IEEE 85 Bus RDS*

Simulation results are obtained and analyses to find out the optimal sizing and its best-suited location for multiple DG units in the same RDS. The simulation results obtained for different numbers of DG installed in RDS are given in Table 7. The size of DG and power loss for installation of different numbers of DG in RDS are shown in Fig. 6. After analyzing, as the number of DG unit installation in RDS is increased the maximum size of DG units is reduced with the total size of DGs for each case are nearly 1413.645 kW and power loss is reduced to 162.803 kW at the installation of 6 DG units. The minimum voltage of RDS is also enhanced up to 0.9350 p.u. for 6 DG unit installation. All DGs units are nearly equal to the same value for installation of 7 DG units.

## 5 Conclusion

In this paper, for integration of DG at optimal sized and its best-suited location in RDS, the TLBO optimization method has successfully been executed. The problem has been formulated to the minimization of RDS real power loss and hence gets a

**Table 7** Comparative results for different numbers of DG installation in IEEE 85 RDS

| No. of DG | DG Size (kW) | Location | Total DG size (kW) | Power loss (kW) | Minimum voltage | Maximum voltage |
|---|---|---|---|---|---|---|
| Without DG | | | | 316.1219 | 0.8713 | 0.9957 |
| 1 DG | 1413.654 | 28 | 1413.654 | 188.8515 | 0.9221 | 0.9971 |
| 2 DG | 728.4821 | 34 | 1413.654 | 168.3227 | 0.9347 | 0.9971 |
| | 685.1716 | 64 | | | | |
| 3 DG | 643.8929 | 34 | 1413.652 | 165.4367 | 0.9347 | 0.9971 |
| | 509.9317 | 67 | | | | |
| | 259.8271 | 80 | | | | |
| 4 DG | 367.0539 | 32 | 1413.654 | 163.8439 | 0.9348 | 0.9971 |
| | 341.4817 | 48 | | | | |
| | 258.7327 | 80 | | | | |
| | 446.3857 | 68 | | | | |
| 5 DG | 448.0956 | 32 | 1413.623 | 163.2433 | 0.9348 | 0.9971 |
| | 258.5427 | 52 | | | | |
| | 229.9754 | 64 | | | | |
| | 276.9064 | 69 | | | | |
| | 200.1024 | 81 | | | | |
| 6 DG | 247.0023 | 32 | 1413.245 | 162.8033 | 0.9350 | 0.9971 |
| | 205.5731 | 44 | | | | |
| | 207.6110 | 52 | | | | |
| | 291.7730 | 64 | | | | |
| | 244.0208 | 69 | | | | |
| | 217.2648 | 81 | | | | |
| 7 DG | 200.3474 | 12 | 1413.645 | 162.9475 | 0.9332 | 0.9971 |
| | 200.0000 | 32 | | | | |
| | 202.1941 | 44 | | | | |
| | 201.3699 | 52 | | | | |
| | 200.0986 | 61 | | | | |
| | 204.4179 | 67 | | | | |
| | 205.2171 | 69 | | | | |

better voltage profile. The TLBO algorithm used for the DG planning and tested on 85 bus RDS. The comparison of simulation results obtained with the TLBO algorithm with the results obtained from other optimization techniques reported in other literature. From the comparison, it observed that the TLBO algorithm provide superior results in term of reducing the power loss reduction and enhancement of voltage profile.

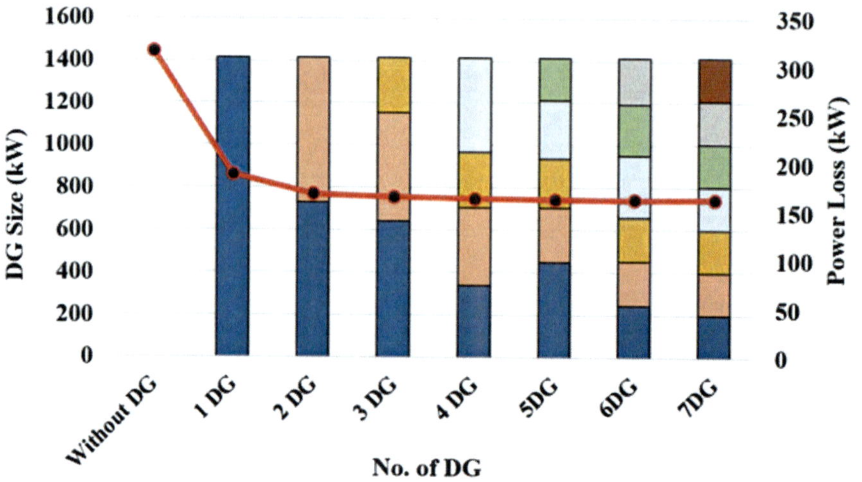


**Fig. 6** The DG units size and power loss for different numbers of DGs installation in IEEE 85 bus RDS

**Acknowledgements** This research work is financially supported by AICTE-RPS project File No. 8-228/RIFD/RPS/POLICY-1/2018-19 dated 20 March 2020. The authors are thankful to the Director MITS, Gwalior for providing essential facilities to carry out this work.

## References

1. Das B, Mukherjee V, Das D (2016) DG placement in radial distribution network by symbiotic organisms search algorithm for real power loss minimization. Appl Soft Comput 49:920–936
2. Oda ES, Abdelsalam AA, Abdel-Wahab MN, El-Saadawi MM (2017) Distributed generations planning using flower pollination algorithm for enhancing distribution system voltage stability. Ain Shams Eng J 8(4):593–603
3. Rueda-Medina AC, Franco JF, Rider MJ, Padilha-Feltrin A, Romero R (2013) A mixed-integer linear programming approach for optimal type, size and allocation of distributed generation in radial distribution systems. Electr Power Syst Res 97:133–143
4. Wang Z, Chen B, Wang J, Kim J, Begovic MM (2014) Robust optimization based optimal DG placement in microgrids. IEEE Trans Smart Grid 5:2173–2182
5. Sfikas EE, Katsigiannis YA, Georgilakis PS (2015) Simultaneous capacity optimization of distributed generation and storage in medium voltage microgrids. Int J Electr Power Energy Syst 67:101–113
6. El-Fergany A (2015) Optimal allocation of multi-type distributed generators using backtracking search optimization algorithm. Int J Electr Power Energy Syst 64:1197–1205
7. Kefayat M, Ara AL, Niaki SAN (2015) A hybrid of ant colony optimization and artificial bee colony algorithm for probabilistic optimal placement and sizing of distributed energy resources. Energy Convers Manage 92:149–161
8. Muthukumar K, Jayalalitha S (2016) Optimal placement and sizing of distributed generators and shunt capacitors for power loss minimization in radial distribution networks using hybrid heuristic search optimization technique. Int J Electr Power Energy Syst 78:299–319

9. Kansal S, Kumar V, Tyagi B (2016) Hybrid approach for optimal placement of multiple DGs of multiple types in distribution networks. Int J Electr Power Energy Syst 75:226–235
10. Nguyen TT, Truong AV, Phung TA (2016) A novel method based on adaptive Cuckoo search for optimal network reconfiguration and distributed generation allocation in distribution network. Int J Electr Power Energy Syst 78:801–815
11. Yahiaoui A, Fodhil F, Benmansour K, Tadjine M, Cheggaga N (2017) Grey wolf optimizer for optimal design of hybrid renewable energy system PV diesel generator-battery: application to the case of Djanet city of Algeria. Sol Energy 158:941–951
12. ChithraDevi S, Lakshminarasimman L, Balamurugan R (2017) Stud krill herd algorithm for multiple DG placement and sizing in a radial distribution system. Eng Sci Technol Int J 20(2):748–759
13. Suresh MCV, Belwin EJ (2018) Optimal DG placement for benefit maximization in distribution networks by using Dragonfly algorithm. Renew Wind Water Solar 5(1):4
14. Reddy PDP, Reddy VCV, Manohar TG (2018) Ant Lion optimization algorithm for optimal sizing of renewable energy resources for loss reduction in distribution systems. J Electr Syst Inf Technol 5(3):663–680
15. Reddy PDP, Reddy VCV, Manohar TG (2018) Optimal renewable resources placement in distribution networks by combined power loss index and whale optimization algorithms. J Electr Syst Inf Technol 5(2):175–191
16. Rao RV, Savsani VJ, Vakharia DP (2011) Teaching–learning-based optimization: a novel method for constrained mechanical design optimization problems. Comput Aided Des 43(3):303–315
17. Mohanty B (2015) TLBO optimized sliding mode controller for multi-area multi-source nonlinear interconnected AGC system. Int J Electr Power Energy Syst 73:872–881
18. Das D, Kothari DP, Kalam A (1995) Simple and efficient method for load flow solution of radial distribution networks. Int J Electr Power Energy 17(5):335–346

# Chapter 35
# Game-Theoretic Optimization for PV-Battery Fed Lighting Systems in DC Homes

**Ashima Kulshreshtha and Anmol Ratna Saxena**

## 1 Introduction

The utilization of renewable energy sources (RESs) has been an inclination for supplying load demand for over three decades [1]. In spite of its advantages of sustainability, it has various drawbacks such as intermittency and comparatively lower power capacity. With renewable energy sources, the nature of loads is also becoming critical by leaning towards automated operation and IOTs. DC distribution systems (DDSs) are introduced to interface such sources and loads and regulate them according to their requirement. These systems include grid-integrated, islanded, single-bus, or multi-bus topologies [2–5]. This paper considers DC Home as a system which are defined as single-bus standalone systems designed mainly for economical and residential purposes [6]. These systems deal with the power of few kilowatts with a fixed dc bus and are integrated with RES and energy storage systems (ESS).

DC homes have been raised as a reliable means for many under-developed and rural regions of the world with a lack of electricity. In DC Home, load sharing, enhancing stability, congestion management, improving power and voltage quality, modeling of different parameters, and simultaneous controlling of multiple components of DC Home are still some issues to be analyzed and improved [7, 8]. There are different architectures of DC Homes proposed as single-bus, multi-bus, or radial-bus architecture in literature. A decentralized control architecture is proposed for power management between multiple clusters of nanogrid [9], a non-isolated hybrid-modular dc-dc converter for DC Grid has been proposed for DC grids [10], a non-isolated fourth-order boost converter has been analyzed for regulating bus voltage of standalone dc systems [11]. Bus voltage regulation and power management of DCNG are obtained using the dc bus signaling approach [12].

The schematic view of the dc home lighting system proposed in this paper is shown in Fig. 1. It uses current-controlled light-emitting diodes (LEDs) for its several

A. Kulshreshtha (✉) · A. R. Saxena
National Institute of Technology Delhi, New Delhi, India

© The Author(s), under exclusive license to Springer Nature Singapore Pte Ltd. 2022 
H. M. Dubey et al. (eds.), *Artificial Intelligence and Sustainable Computing*,
Algorithms for Intelligent Systems, https://doi.org/10.1007/978-981-16-1220-6_36

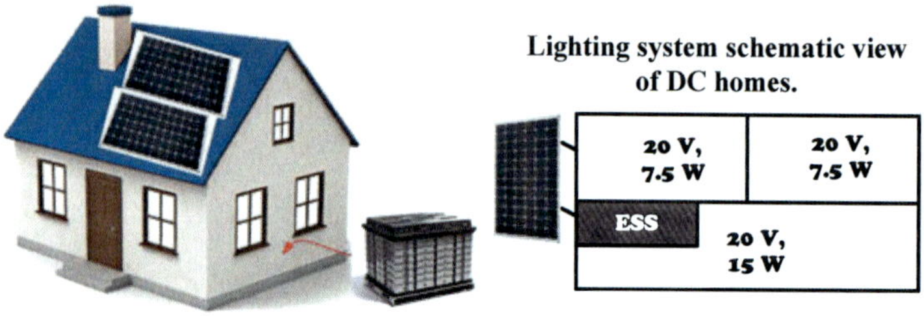


**Fig. 1** Schematic view of DC homes

advantages over conventional lighting sources. LED drivers are used to feed constant current to LEDs for effective illumination. DC-DC converters are the key component of the DC Homes which execute source, bus, and load regulation for reliability [7]. The performance of converters has to be robust, as deterioration of the performance of one converter will affect the performance of other interfacing converters. The performance needs to be considered at the converter and control stage of DC Homes. To operate multiple converters for obtaining multiple requirements, different control and optimization strategies have been evolved in literature [13–16]. Game theory is one of the optimization techniques which deal with multiple players with different costs. It is a branch of mathematics that is used to strategize in fields like economics, business markets, and biology where a number of variables called players deliberately work to minimize their cost function. Game theory has been implemented in load sharing [6], multi-team target assignment problem [17], electric vehicles [18, 19], small scale power systems [5, 20, 21].

A game-theoretic controller for DC Home is proposed in this paper for load sharing of a lighting system. Three cases have been shown based on the availability of solar photovoltaic sources and the requirement of the load. For each case, players work iteratively according to game-theory strategy so that the cost function of each player is minimized. During each case, equilibrium is derived for the simultaneous operation of all converters. The battery is either charged or discharged according to the power balance between solar power generated and load power consumed. Section 2 describes the architecture of DC Home and its different cases. Section 3 describes a generalized approach to game theory and a procedure for obtaining Nash equilibrium. Section 4 validates the game-theoretic centralized controller on a 30 W DC Home and Sect. 5 concludes the paper.

## 2 Topological Architecture of DC Homes

For a DC Home, Fig. 2 shows the basic architecture of lighting schematic of DC homes with Fig. 2a denoting the functions of source, load and bidirectional converters. Figure 2b shows Case 1 of DC Home, where solar power is more than the load power and additional power is consumed by the battery. Case 2 is illustrated

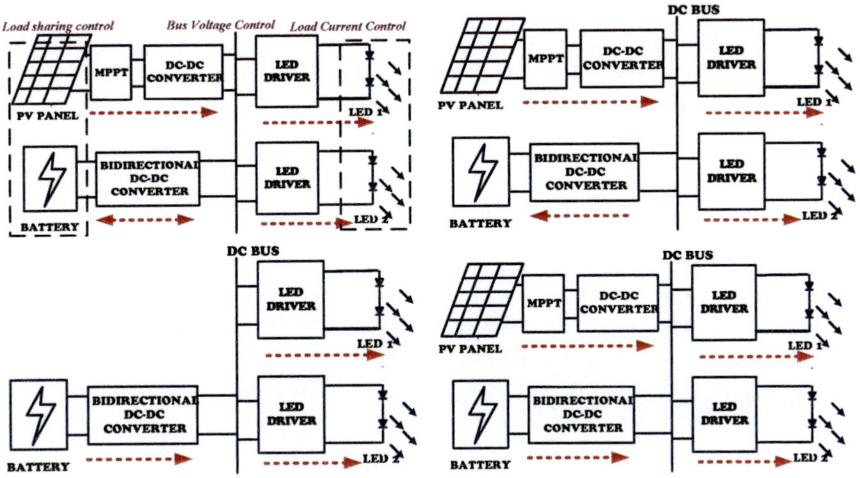


**Fig. 2** The architecture of lighting system used in DC homes

by Fig. 2c, where solar power is not available and only the battery is feeding the load and Case-3 is shown by Fig. 2c, where solar power is less than the load power, and battery feeds the additional required power. Battery is charged or discharged depending on the solar PV power.

### A. Source Converters for Solar PV System

For a solar PV panel, two converters are connected in a cascade. The first converter is to track and deliver maximum power point (MPPT) from the PV panel and the latter is to regulate bus voltage. A 30 W solar panel with PV terminal voltage at MPP 18 V is considered for DC Home lighting systems.

### B. LED Drivers

LEDs are current-fed semiconductor devices whose illumination varies with a slight variation in current. Converters are often used to regulate load voltage or source current however, in this paper, load converters regulate load current to feed LED. Digital current control of dc-dc converter was proposed using the state-space averaging technique to model the converters [22]. Two dc-dc buck converters are used as LED drivers to feed LEDs rated 15 W, 20 V and 7.5 W, 20 V.

### C. Bidirectional Converter for Battery

The bidirectional converter is used to charge and discharge the battery which aids power flow management of the DC Homes. Conventional two switches buck-boost bidirectional converter is interfaced to battery for efficient use. During charging case, converter plays as battery voltage-controlled buck converter and during discharging it act as a bus voltage-controlled dc-dc boost converter. During charging, $S_2$ is modulated and $S_3$ is OFF. During discharging, $S_3$ is modulated and $S_2$ is OFF. A 12 V Li-Ion battery is used as energy storage systems for interfacing to DC Home.

## 3 Game-Theoretic Approach to Obtain Nash Equilibrium

Game theory is a technique to find out the operating point of every variable in a multi-variable environment, where variables work to minimize their cost function irrespective of the cost function of other variables.

### A. Concept of Game Theory

A Multi-converter dc system comprises of multiple variables that interact in a multi-player environment. Variables are called players, $P_i$ ($i \in$number of converters). The game-theoretic algorithm derives equilibrium such that the cost function, $C_i$ of each player is minimized by (1). Here $r_i$ is the reaction of the player called action [21]. Each player works so that their cost function is minimized. The cost function is minimized using minima expression of differentiation given by (2).

$$C_i = f(r_i, r_{-i}) \tag{1}$$

$$\frac{\partial C_i}{\partial r_i} = 0 \tag{2}$$

The game theory derives a quiescent operating point at which the cost function of all players will be reduced. The game theory works iteratively by first deriving the minimum point of one variable using (2) keeping other variables constant. Further, the same procedure is implemented for other variables turn-by-turn without disturbing other players. This process recurs until the value of each player becomes saturated for increasing iterations. This saturation point is called as Nash equilibrium. The Nash equilibrium is mathematically given by Eq. (3).

$$C_{i(P_i^*, P_{-i}^*)} \leq C_{i(P_i^*, P_{-i})} \tag{3}$$

for $i \in$(Source and load converters).

### B. Game-Theoretic Central Controller for DC Home

Figure 3 shows the control architecture of DC Homes, where central controller based on the game theory provides references to the primary controllers of the PEICs. First level is the Power electronic interfacing converters level denoted by blue color, primary controllers of individual converter denoted by red and central controller providing input to each primary controller are denoted by green. Figure 4 shows the block diagram of the game-theoretic controller with input and outputs highlighted.

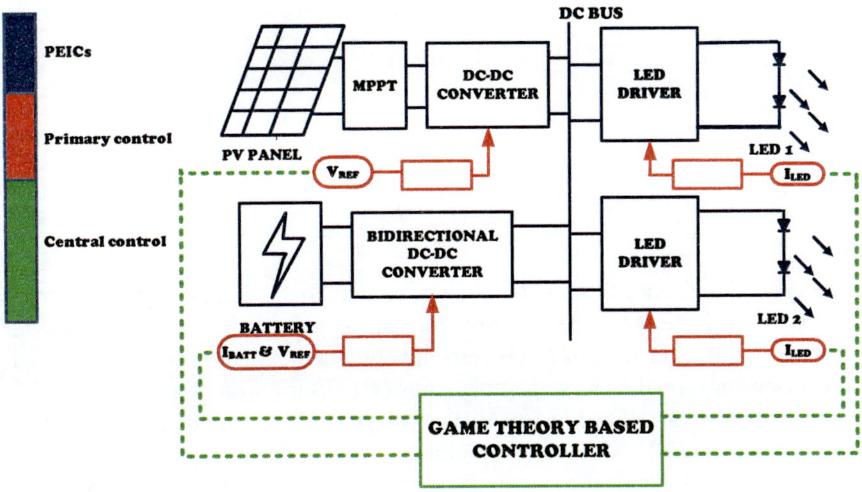


**Fig. 3** Control architecture of DC Home

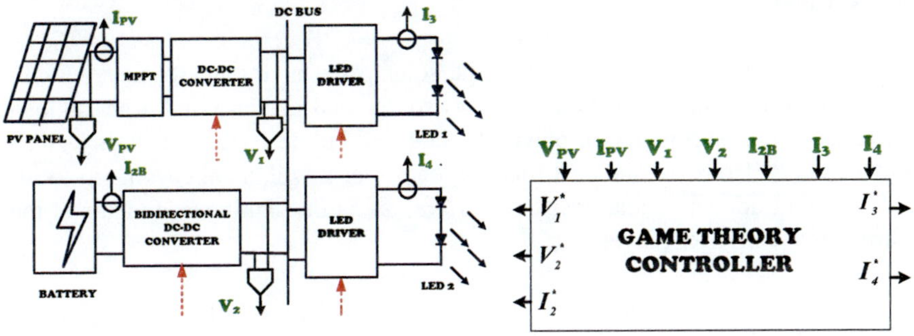


**Fig. 4** Input and Outputs of Game-theoretic controller

### (a) Case-I

During case-1, PV power is more than the load power, so surplus power will be consumed by the battery through the bidirectional converter. PV interfaced dc-dc converter is player-1, *LED* drivers are player-3,4 and bidirectional converter is 2. Player-1 will have a variable output voltage that is the bus voltage. Player-3 and 4 act as load resistance for the sake of simplicity as it consumes power. As the battery is consuming power, Player-2 will also act as load resistance. Player-1 will work to minimize the deviation of its output voltage i.e. bus voltage to 24 V and Player-2,3 and 4 will work towards consuming nominal current for appropriate illumination. Their cost functions are defined by (4)–(7).

$$C_1 = (1 - \omega_s)(V_1 - V_{BUS})^2 + \omega_s(V_{BUS} - 24)^2 \tag{4}$$

$$C_2 = (I_2 - I_{2N})^2 \tag{5}$$

$$C_3 = (I_3 - I_{3N})^2 \tag{6}$$

$$C_4 = (I_4 - I_{4N})^2 \tag{7}$$

where $\omega_S$ is significance factor determined according to the significance of player-1. $V_{BUS}$ is bus voltage, $I_2$ is the charging current through the battery and $I_3$ and $I_4$ are load current of $LED_1$ and $LED_2$ respectively. The nominal value $I_{2N}$, $I_{3N}$, and $I_{4N}$ is determined on the basis of voltage, power and series resistance of player-2,3 and 4. Cost function defined by eq. 4 where player 1 will have output voltage, that is, bus voltage = 24 V, with the least voltage deviation due to line losses. The cost function of the bidirectional converter is determined by its current derived by power difference between the source and load given by (5). Constant current LED requires nominal current to be fed, so its cost function is given by (6) and (7).

(b) **Case-II**

Case-2 deals with the condition when the solar panel is absent, so player-1 will not operate during case-2. Player-2 acts as source and player-3 and 4 act as load resistance. The cost function of player-3 and 4 will be the same as in case-1 given by (6) and (7). However, the cost function of player-2 will change from (5) to (8) as now it will not act as load resistance. Here, $\omega_B$ is the significance factor of the bidirectional converter-player-2.

$$C_2 = (1 - \omega_B)(V_2 - V_{BUS})^2 + \omega_B(V_{BUS} - 24)^2 \tag{8}$$

(c) **Case-III**

During case-3, both first and second player act as sources. But, here player-1 does not have sufficient power to feed LEDs so a battery interfaced bidirectional converter called player-2 will provide additional power to the LEDs. The cost function of player-1 is changed from (4) to (9).

$$C_1 = (1 - \omega_S)(V_2 - V_{BUS})^2 + \omega_S(V_{BUS} - 24)^2 + (P_1 - \frac{P_{1N}}{2})^2 \tag{9}$$

where $P_1$ is actually solar power and $P_{IN}$ is the reference nominal power.

The performance of power electronic converters is decided by the reference given to the primary controller. The basic mathematical foundations of dc-dc converters are defined by (10)–(13) [21].

$$P_{in} = P_O \tag{10}$$

$$I_{in} = MI_O \tag{11}$$

$$V_{in} = \frac{V_O}{M} \tag{12}$$

$$R_{in} = \frac{R_O}{M^2} \tag{13}$$

# 4 Results and Discussion

The lighting system of the DC home presented in the paper consists of one second-order boost converter as PV interface converter, 1 bidirectional second-order buck-boost converter interfaced with battery and two second-order buck converters as load converter as shown in Fig. 5. Table 1 provides the parameter rating of source, load and battery used in this system.

## A. Calculation of Operating Points of Players Using Game Theory

The cost functions obtained in Sect. 3(B) are minimized by varying the actions of corresponding variables. The minimization is mathematically obtained by differentiation however, in software it is developed using the min function of MATLAB toolbox. To lessen the iteration and computational efforts of the algorithm, all players are minimized simultaneously. Figure 6a shows the variation of all the involved

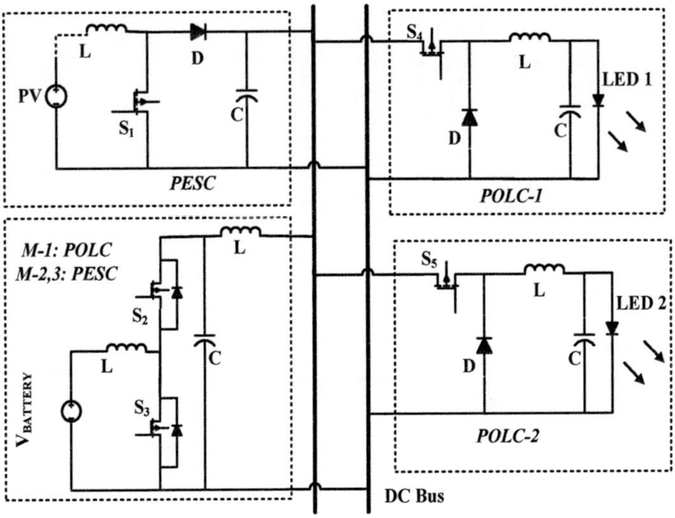


**Fig. 5** Architecture of DC Home Lighting system with Converter topologies

**Table 1** Specifications of source and loads of DC homes

| | Solar panel | Battery | LED 1 | LED 2 |
|---|---|---|---|---|
| Voltage (V) | 16 V | 12 | 20 | 20.1 |
| Current (A) | – | 0.5 | 0.75 | 0.377 |
| Resistance (Ω) | – | 24 | 26.47 | 53.38 |
| Power (W) | 30 | – | 15.2 | 7.55 |
| L | | 110 μH | | |
| C | | 220 μF | | |
| Element series resistance | | 0.1 Ω | | |
| Switching frequency | | 100 kHz | | |

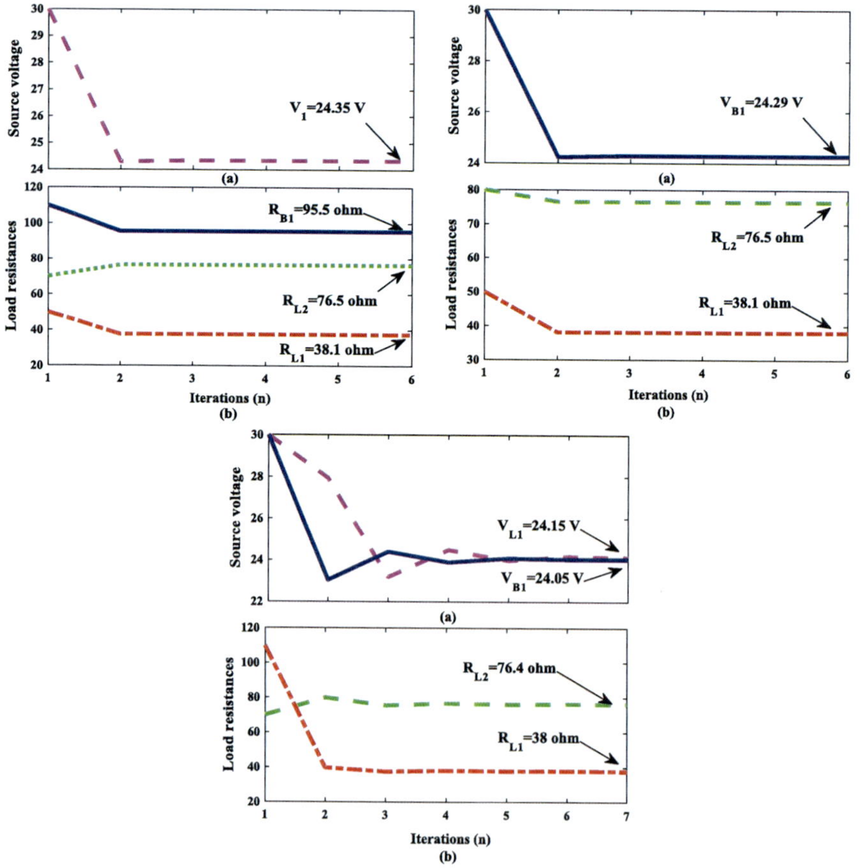


**Fig. 6** Variation of bus voltage of source converters and load resistance of load converter with iterations during **a** Case-1, **b** Case-2 and **c** Case-3

players with the iterations. The bus voltage and load resistance variation of players are shown separately for better observation. Nash equilibrium point is obtained in 5 iterations as $V_1 = 24.35$ V, $R_{L1} = 38.1\ \Omega$, $R_{L2} = 76.5\ \Omega$, $R_{B1} = 95.5\ \Omega$.

Figure 6b shows the variation of players 2, 3 and 4 with iterations in the absence of PV source converter. Load resistances are $R_{L1} = 38.1\ \Omega$, $R_{L2} = 76.5\ \Omega$, as in Case-I and battery voltage $V_{B1} = 24.29$ V. Figure 6c shows the variation of all the players i.e. player-1,2,3 and 4 with iterations during case-3. As in this case, PV interface and battery interface converter i.e. player-1 and 2 both act as source converter it took more iterations comparing to other cases. Nash equilibrium is obtained in seven iterations, in this case, using (6)–(9). The power generated by the solar panel is lessened in the case of half irradiation as 15 W. The parameters of each converter are given in Table 2 at the point of Nash equilibrium. These parameters show that all the obtained parameters are very close to their nominal value (Fig. 7).

## B. Simulated Results

Nash equilibrium obtained by game-theoretic controller provides the operating point of each converter. These operating points are validated using simulation on the PSIM

**Table 2** Specifications of each players at nash equilibrium

| | Solar panel | | Battery | | LED-1 | LED-2 |
|---|---|---|---|---|---|---|
| | Power(W) | $V_{BUS}$ (V) | $I_2$ (A) | $V_{BUS}$ | $I_3$ (A) | $I_4$ (A) |
| Case-1 | 29.87 | 24.01 | −0.256 | 24.05 | 0.636 | 0.3121 |
| Case-2 | – | – | 0.939 | 24 | 0.624 | 0.3124 |
| Case-3 | 15.65 | 23.96 | 0.304 | 23.98 | 0.625 | 0.3126 |

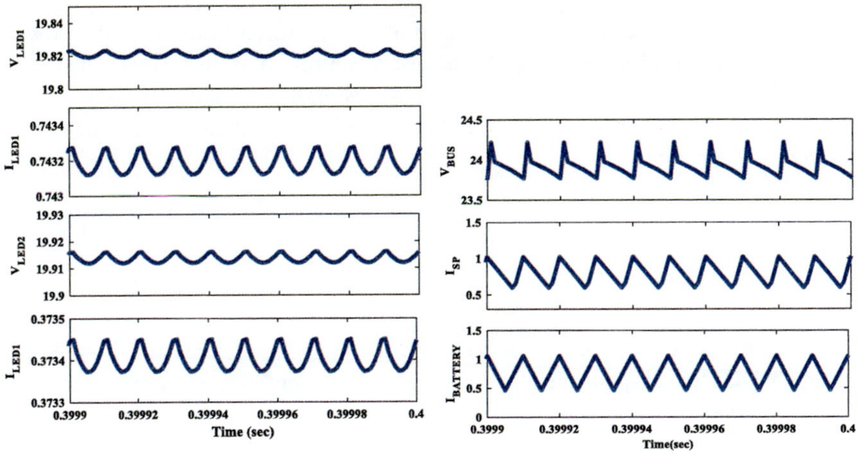


**Fig. 7** Steady-state ripples of LED voltage, $V_{LED1}$, $V_{LED2}$ ,current, $I_{LED1}$, $I_{LED2}$, bus voltage, $V_{BUS}$ PV source current, $I_{SP}$ and battery current, $I_{BATTERY}$

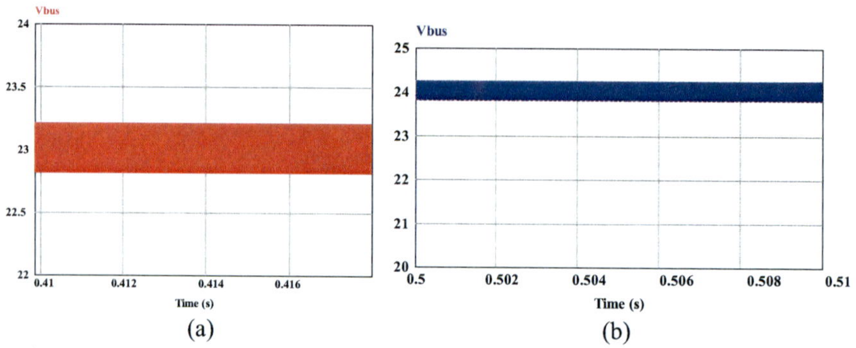


**Fig. 8** Performance of source converter for bus voltage regulation (**a**) without a controller, (**b**) with a controller

platform. Table 2 shows the voltage and current as parameters of the converter employed in DC Home. The steady-state ripples are shown in Fig. 7 to validate the stability of the system when multiple interfaced converters are coherently operating. The duty ratio is provided by the primary controller. Figure 8 shows the performance of the source converter to regulate bus voltage. In the absence of a central controller, the primary controller can have steady-state error shown in Fig. 8a, and with the game-theoretic controller primary controllers are tuned to exact 24 V. Figure 9 shows the performance of DC Home on the basis of power management, where without controller each source shares some power irrespective of the availability of the PV power or state of charge of battery as shown in Fig. 9a. With the game-theoretic controller, PV Power will provide more power as, during Case-1, PV power is given priority shown in Fig. 9b. The power-sharing is specified in detail in Table 3. The positive battery power denotes the discharged power and vice versa.

Figure 10 shows that during a change in the load current reference, PV power also changes from 30 to 20 W in absence of a central controller. However, when a game-theoretic controller is implemented as a central controller, PV power does not vary when load changes and the additional power is fed to the battery through a bidirectional converter.

## 5 Conclusions

This paper focussed on a central controller for DC Home based on game theory. The concept of DC Home is presented for standalone dc residential low power low voltage grid. Here, solar PV panel was the primary power source, and battery as a secondary power source. The game-theoretic central controller derives the equilibrium point of operation for each converter accurately and in very little time. Three different cases of DC Home were considered to validate the game theory. The Nash equilibrium is obtained using a game-theoretic algorithm and its time-domain analysis is done

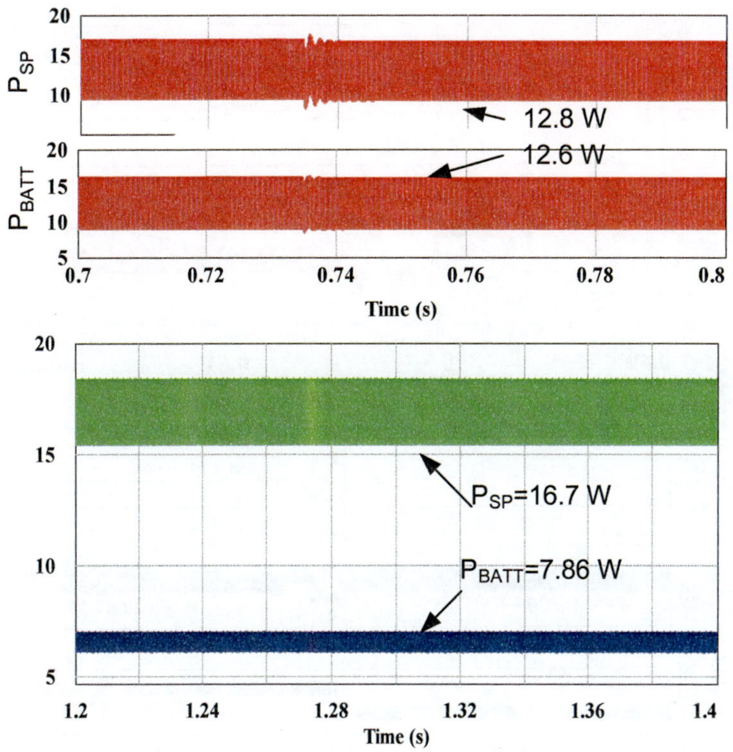


**Fig. 9** Performance of source converter for bus voltage regulation (**a**) without a controller, (**b**) with a controller

**Table 3** Power sharing (Case-wise)

| | PV power | Battery power (W) | LED-1 power (W) | LED-2 power (W) | Losses (W) |
|---|---|---|---|---|---|
| Case-1 | 28.87 W | −6.2 | 15.14 | 7.47 | 0.42 |
| Case-2 | – | 22.8 | 15 | 7.54 | 0.29 |
| Case-3 | 15.36 W | 7.29 | 14.99 | 7.49 | 0.16 |

using a simulation platform. The stable operation of each converter is shown by its steady-state ripple during the coherent operation of all converters. The dynamic operation is shown using transient changes in the power and load requirement. The game-theoretic controller stabilizes each converter in all the cases in very little time to avoid any fault or uncertainties in the DC Home.

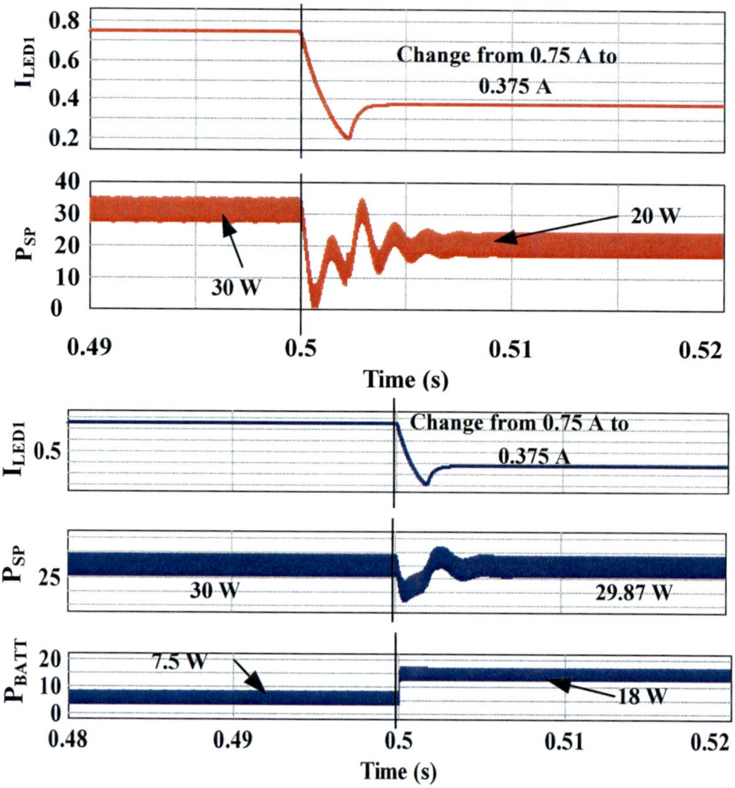


**Fig. 10** Performance of source converter for bus voltage regulation (**a**) without a controller, (**b**) with a controller

## References


1. Kwasinski A (2011) Quantitative evaluation of DC microgrids availability: effects of system architecture and converter topology design choices. IEEE Trans Power Electron 26(3):835–851
2. Korada N, Mishra MK (2017) Grid adaptive power management strategy for an integrated microgrid with hybrid energy storage. IEEE Trans Ind Electron 64(4):2884–2892
3. Khadse M, Jadhav V, Daware K, Sharma S, Rajwade M, Patil P (2016) Design and implementation of an indoor test bench for validation of solar MPPT algorithms. In: 12th ieee international conferences electron energy, environment communication computer control (E3-C3), INDICON 2015, pp 1–6
4. Bhosale R, Agarwal V (2019) Fuzzy logic control of the ultracapacitor interface for enhanced transient response and voltage stability of a dc microgrid. IEEE Trans Ind Appl 55(1):712–720
5. Ekneligoda NC, Weaver WW (2014) A game theoretic bus selection method for loads in multibus DC power systems. IEEE Trans Ind Electron 61(4):1669–1678
6. Kulshreshtha A, Saxena AR (2017) Optimized load sharing of LED drivers in DC homes using game theoretic approach. In: IEEE Region 10 annual international conference proceedings/TENCON, vol 2017–Decem, pp 1896–1901
7. Dragicevic T, Lu X, Vasquez JC, Guerrero JM (2016) DC microgrids—Part I: a review of control strategies and stabilization techniques. IEEE Trans Power Electron 31(7):4876–4891

8. Dragicevicˇ T, Lu X, Vasquez JC, Guerrero JM (2016) DC microgrids—Part II: a review of power architectures, applications, and standardization issues. IEEE Trans Power Electron 31(5):3528–3549
9. Nasir M, Jin Z, Khan HA, Zaffar NA, Vasquez JC, Guerrero JM (2019) A decentralized control architecture applied to dc nanogrid clusters for rural electrification in developing regions. IEEE Trans Power Electron 34(2):1773–1785
10. Elserougi A, Abdelsalam I, Massoud A, Ahmed S (2019) A non-isolated hybrid-modular DC-DC converter for dc grids: small-signal modeling and control. IEEE Access 7:132459–132471
11. Kulshreshtha A, Saxena AR, Veerachary M (2020) Non-isolated boost converter for power management in low voltage low power DC grids : design and interaction analysis. IEEE Access XX
12. Schonberger J, Duke R, Round SD (2006) DC-bus signaling: a distributed control strategy for a hybrid renewable nanogrid. IEEE Trans Ind Electron 53(5):1453–1460
13. Veerachary M, Saxena AR (2015) Optimized power stage design of low source current ripple fourth-order boost DC–DC converter: a pso approach;dc converter: a pso approach. IEEE Trans Ind Electron 62(3):1491–1502
14. Tahir M, Mazumder SK (2016) Event- and priority-driven coordination in next-generation grid. IEEE J Emerg Sel Top Power Electron 4(4):1186–1194
15. Wang F, Zhu T, Zhuo F, Yi H, Shi S, Zhang X (2017) Analysis and optimization of flexible MCPT strategy in submodule PV application. IEEE Trans Sustain Energy 8(1):249–257
16. Fu Y, Zhang Z, Li Z, Mi Y (2019) Energy management for hybrid AC/DC distribution system with microgrid clusters using non-cooperative game theory and robust optimization. IEEE Trans Smart Grid 11(2):1–11
17. Galati DG, Simaan MA (2007) Effectiveness of the Nash strategies in competitive multi-team target assignment problems. IEEE Trans Aerosp Electron Syst 43(1):126–134
18. Kaur K, Dua A, Jindal A, Kumar N, Singh M, Vinel A (2015) A novel resource reservation scheme for mobile phevs in V2G environment using game theoretical approach. IEEE Trans Veh Technol 64(12):5653–5666
19. Yin H, Zhao C, Li M (2016) A game theory approach to energy management of an engine–generator/battery/ultracapacitor hybrid energy system. IEEE Trans Ind Electron 63(7):4266–4277
20. Ekneligoda NC, Weaver WW (2012) Game-theoretic communication structures in microgrids. IEEE Trans Power Deliv 27(4):2334–2341
21. Weaver WW, Krein PT (2009) Game-theoretic control of small-scale power systems. IEEE Trans Power Deliv 24(3):1560–1567
22. Hankaniemi M, Suntio T (2006) Small-signal models for constant-current regulated converters. In: IECON Proceeding Industrial Electron Conference, pp 2037–2042

# Chapter 36
# Modeling the Effect of Quarantine and Isolation for COVID-19 Spreading

**Anurag Singh, Samriddhi Bhasin, Bhanu Gakhar, and Md. Arquam**

## 1 Introduction

Nowadays, the world is facing the fear of the spread of COVID-19, which belongs to an outsized family of viruses and can cause infection in humans. In humans, they're capable of inflicting metastasis infections starting from respiratory illness to a lot of severe diseases. COVID-19 is a contagious disease that is sparked off by a newly discovered virus, which appeared in December of 2019 in Wuhan, China. The disease spreads primarily from human to human; therefore, it is a contagious or communicable or infectious disease. A contagious disease is transmitted from one person to another through numerous ways that incorporate: interaction with blood and bodily fluids; inhaling an airborne virus; or by being bitten by an insect [1, 2].

SARS-CoV-2 is similar to SARS-CoV and MERS-CoV caused by a coronavirus. SARS-CoV is first appeared in Guangdong, China during 2002–2003 and proliferate in different parts of the world. MERS-CoV was seen in Saudi Arabia in 2012 [3]. Researchers and scientists have shown that the increasing rate of COVID-19 is almost double of the SARS and MERS [4]. We are aware of how to protect ourselves from the virus, and the disease it causes, by cleaning our hands completely, putting on masks and social distancing.

A. Singh (✉) · Md. Arquam
Department of Computer Science & Engineering, National Institute of Technology Delhi, New Delhi 110040, India
e-mail: anuragsg@nitdelhi.ac.in

Md. Arquam
e-mail: arquam@nitdelhi.ac.in

S. Bhasin
Department of Computer Science and Engineering, BPIT Delhi, New Delhi, India

B. Gakhar
Department of Information Technology, ADGITM Delhi, New Delhi, India

© The Author(s), under exclusive license to Springer Nature Singapore Pte Ltd. 2022 
H. M. Dubey et al. (eds.), *Artificial Intelligence and Sustainable Computing*, Algorithms for Intelligent Systems, https://doi.org/10.1007/978-981-16-1220-6_37

Therefore, the analysis of contagious disease is vital within the assurance of acceptable medical support and with the analysis of disease intervention and management programs, and therefore the finding of common-source outbreaks. An Outbreak is that the incidence of disease cases in more than traditional expectancy. If it's not quickly controlled, it will become an epidemic. At present, The two most studied model of epidemiological dynamics, are the deterministic model and the stochastic model with complex network transmission dynamics [5]. Scholars have already proposed the model to explain the spreading behaviour of SARS and MERS by considering various facets of deterministic models, stochastic models, and complex network transmission dynamics [4], but analyzing the spreading behaviour of COVID-19 is started and continue. Most of the proposed model on COVID-19 is either a mathematical model that does not consider the underlying network topology [6, 7] or a data-driven model without considering underlying network topology [8, 9]. To analyse the spreading patter in the human population, it is necessary to consider an underlying network topology that shows the spread of infection through contacts.

Hence, it is need time to analyse the spreading behaviour of COVID-19 by considering various underlying network topologies to depict the human contact network. To study the effect of COVID-19 by using the underlying network topology. Any contagious or infectious disease spread among human through their interaction. Hence for such diseases, human contact is important. We modify the classical SIR model considering various underlying network topologies to explain the spreading pattern of infectious disease.

The rest of this paper is structured as follows. Section 2 describes the background and related works. These related studies deal with epidemiological dynamics over various underlying network topologies and interaction patterns of the host population. Section 3 explains the proposed model and its algorithms are discussed in Sect. 4. Numerical simulations of the proposed model using quarantine and isolation compartment in the classical SIR model. Simulations are done on the proposed model using the real-world network and the analysis of the results is discussed in Sect. 4. Finally, the conclusions and future research directions are explained in Sect. 5.

## 2 Background and Related Work

In this section, the relevant current state of the art is explained that leads to our proposal. We explain the epidemiological dynamics on the network to elaborate on the spreading process in the human population in the combination of COVID-19 spreading behaviour in the human population, the effect of isolation and movement restriction.

Researchers studied an epidemic model focusing on the current outbreak of COVID-19. They derived a mathematical expression for basic reproduction rate $R_0$ for the COVID-19 by considering, demography, heterogeneous population, mobility pattern and permeability of households under confinement [10, 11]. In this work, [12] a greedy algorithm is proposed to choose the best link or best node to immunize and removed from the network. By removing the immunized node, the algorithm guarantees the prevention of outbreak. Salehi et al. studied and elaborated the spreading processes in multilayer networks exploring the dynamical processes over multilayer networks to study the epidemic spreading with control variables [13].

Researchers have developed targeted immunization strategies by considering contact network data due to its effectiveness at high vaccination coverage [14]. Authors studied the two crucial property that is used to describe the spreading of COVID-19, one is basic reproduction number $R_0$ and other is the individual variation in the number of secondary cases [15]. Radar et al. studied the spatial variables for cities to show the intensity of the spreading of COVID-19. They applied the metapopulation model for COVID-19 by considering spatial hierarchies [16].

Satorras et al. studied and elaborated the various epidemic spreading model considering different underlying network topologies [17]. Arquam et al. studied the impact of delay in the SIR model to become a node infectious. They explained that the delay in infection decreased the threshold of the spreading rate [18]. In this course of research and development, Qifang and his team represented the variations in demographics and severity between cases through symptom-based investigation and observation of close case contacts. They estimated the time of confirmation, isolation, and recovery as well as estimated the household secondary attack rate and reproductive number [19]. Bock et al. derived the $R_0$ to show the secondary infection and $R^*$ for outside contact infection by using the data of Germany and Poland of COVID-19 infection and proposed a mitigation strategy [20].

## 3 Proposed Model to Study the COVID-19 Over the Human Contact Network

Any communicable disease spreads when there is contact between the humans in the human contact networks. Here, we have taken the human contact network as an underlying network topology to study the epidemic spreading and its containment. A Network is a collection of nodes (vertices) along with identified pairs of nodes (edges or links), $G = (V, E)$, where $V$ denotes the set of nodes, and $E$ denotes the set of edges. In this section, we describe the spreading pattern and behaviour of COVID-19 over the network by modifying the classical SIR model.

A SIQHR model is considered for COVID 19 spreading over a human contact net-work in Fig. 1. A human population with fixed size $N$ is considered as the size of the network. It is divided into five compartments: *Susceptible* $S(k, t)$, *Infected* $I(k, t)$, *Recovered* $R(k, t)$, *Quarantine* $Q(k, t)$ *and Isolation* $H(k, t)$ population of degree $k$ at time $t$. A susceptible person is a healthy person those who had not been infected yet. An infected node may propagate the disease to a susceptible node with a rate $\alpha$.

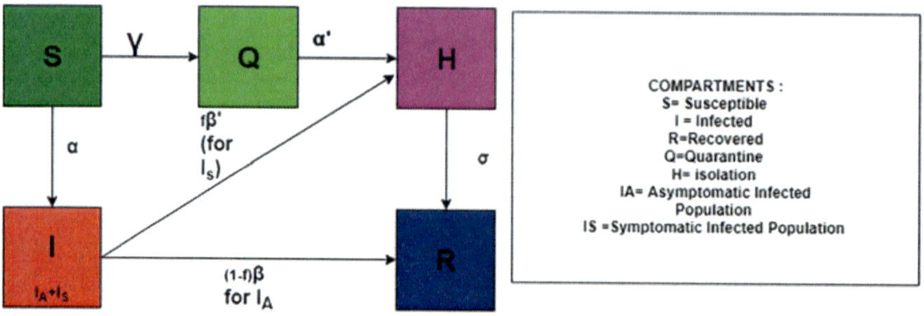


**Fig. 1** Block diagram of the SIQHR model. The human population is divided into 5 states: (1) Susceptible (S) (2) Infected (I) (3) Quarantine (Q) (4) Isolation (H) and (5) Recovered (R). Relevant parameters are the spreading rate $\alpha$ in susceptible population and $\beta$ is the recovery rate of infected population

The recovery rate of infected hosts is $\beta$. The recovered population cannot be infected again in this model. The susceptible population who may be infected due to some high number of contacts are sent into quarantine to avoid the contacts with the other persons into the human contact network with rate $\gamma$. The persons from quarantine may need to put in isolation if they found infected or symptoms of infection with a rate $\alpha'$. In this work, it is assumed that the whole population from the isolation will be recovered successfully and will not put any effect on the disease spreading further. *Quarantine* refers to the separation of the population who made contact with COVID-19 infected persons or contacted with a large fraction of the population (high rank of the centrality in the social contact network). Still, there is no development of clinical symptoms. On the other hand, *Isolation* is the fraction of those population which have the symptoms of the COVID-19. In the case of MERS and SARS mostly model consider Quarantine compartment without network model.

Some fraction of the infected population is not sent into Isolation can become infectious with the rate of $\alpha$. Some members from $S$ can be quarantined at the rate of $\gamma$. Quarantine is useful to slow down the contact rate by a factor. There may be the chance of spreading of COVID-19 by the isolated population. Symptomatic Infected people are required to send in the Isolation with the rate $f\beta'$ so that they can't make any contact with others in the human network. The quarantine population may also be infected after observing the symptoms of COVID-19 with the rate of $\alpha'$ and need to be isolated from this compartment. Some fraction of, $(1 - f)$ of the infected population may be recovered (will not be infected again or died). In the presented model, permanent recovery is considered. As per the current trend of COVID-19 infected population may be further categorized as the Asymptomatic and Symptomatic Infected population. Hence, the proposed model is shown in Fig. 1 as mentioned compartments and parameters for COVID-19.

Hence, here in the proposed work, a fraction of the population is identified for quarantine on the parameter of highly centralized into the social contact network. In reality, they may be such person who travels with a large number of passengers and may get into touch with the large fraction of people in the infected or red zone areas. Therefore, various centrality parameters are used to identify the highly contacted person and send them into quarantine which will result in the disconnection of those persons as a node onto the network. A study can be done with the removing of the various fraction of such population from the social contact network and study the infected population and find those fraction which will stop the outbreak of the disease. A human contact network is a heterogeneous network with a varying number of contacts for every node. Therefore, we need to create further groups of all the compartments based on its degree.

Where, $S(k, t)$, $E(k, t)$, $I(k, t)$, $R(k, t)$ represents the fractional Susceptible $S$, Exposed $E$ Infected $I$ and Recovered $R$ population belongs to the class of $k$ degree at time $t$ and fractional Quarantine $Q(k, t)$ and Isolated $H(k, t)$ population represents the persons those are coming from the class $E(k, t)$ and $I(k, t)$ respectively with degree $k$ at time $t$. As defined various compartment of the human population, rate equations of all the defined compartments are given by, The relative density of the population in each compartment at the mean-field level, satisfy the following set of coupled different equations:

$$S'(k,t) = -\gamma S(k,t) - \alpha S(k,t)\Theta(t) \tag{1}$$

$$Q'(k,t) = \gamma S(k,t) - \alpha' S(k,t)Q(k,t) \tag{2}$$

$$I'(k,t) = \alpha S(k,t)\Theta(t) - \beta(1 - f(t))I(k,t) - f(t)\beta' I(k,t) \tag{3}$$

$$H'(k,t) = \alpha' S(k,t)Q(k,t) + f(t)\beta' I(k,t) - \sigma H(k,t) \tag{4}$$

$$R'(k,t) = \beta(1 - f(t))I(k,t) + \sigma H(k,t) \tag{5}$$

where $P(k'|k) = \frac{kP(k)}{\langle k \rangle}$, $\langle k \rangle$ (for uncorrelation) is the average degree of the network and $P(k)$ is the probability of having degree $k$ of randomly chosen node and $\Theta(t)$ is the probability that any edge is coming from an infected node at time $t$, where.

$$\Theta(t) = \sum_{k'=1}^{n} P(k'|k)[(I_S(k', t) + I_A(k', t)]$$

$f(t)$ is used to represent a fraction of population sent for isolation at time t. In the previous work of COVID-19 modelling, all population is considered as fully mixed, and then interactions were studied. But, usually, the interactions will happen as per the underlying network structure. Therefore, in the proposed model, we need to use dynamical equations defined in Eqs. 1–5 over the underlying social contact network.

# 4 Results and Analysis

In simulations, we have extracted the parameter values from the data-set of COVID-19 spreading in India. We have used different synthetic network topology to explain the impact of contact network in epidemic spreading. The work is divided into three parts: (1) An epidemic spreading process over the contact network using classical SIR model [21] (2) Effect of quarantine population is studied on the final recovered population (3) Effect of isolation is studied on the Maximum spreading and final recovered population. In targeted immunization, we immunize those nodes whose degree is maximum to decrease the chance of an epidemic outbreak, while quarantined population is those nodes whose degree may be any value between [1, $k_{max}$].

## 4.1 *Epidemic Spreading Over Contact Network Using Classical Spreading Model*

It is the partial simulation of the proposed model (Fig. 1). Here, first, we simulate how COVID-19 spreads over the contact network. After that simulation setup is defined according to the classical SIR model Real data-set of Collaboration Network [22] is taken to visualise the real picture of epidemic spreading over networks with additional parameter values While statistics of Collaboration Network is mentioned in Table 1.

We use the various value of the spreading rate ($\alpha$) as 0.2, 0.5 and 0.8, and apply two values of recovery rate ($\beta$) as 0.1 and 0.9. Then spreading behaviour of disease is analysed for the various value of $\alpha$ and $\beta$. For Spreading rate, $\alpha = 0.2$ and recovery

**Table 1** Parameters of real and synthetic data set

| Name of network | No. of nodes | No. of edges | Degree | Clustering co-efficient | Degree distribution |
|---|---|---|---|---|---|
| Collaboration network (Scale Free Network) | 10000 | 53000 | $k_{max} = 208$, $k_{min} = 1$ and $k_{avg} = 5.7$ | 0.026 | Power law |

rate $\beta = 0.9$, a fraction of the infected population is increased and reaches to peak at $4th$ time steps with the value of 0.27. After that infection starts decreasing and vanishes at $14th$ time steps. In the meantime susceptible population starts decreasing and becomes constant after 14 time steps. No more infection spread in susceptible population after 10-time steps. The recovery process starts after $2nd$ time steps and is completed at $14th^{-}$time steps. The whole process is completed in 14-time steps as shown in Fig. 2.

We change the value of the Spreading rate as $\alpha = 0.8$ and recovery rate as $\beta = 0.1$ to see the spreading behaviour over a network. With this value, the Susceptible population starts decreasing when the infection spreads in the susceptible population and becomes stable after 16-time steps. In the meantime, the infected population increases and reaches a maximum at 4th time step with a value of 0.49. After that, the infection population decreases and vanishes after 16th time step. The recovery process starts after 3rd time step. The whole process is completed in 16-time steps as shown in Fig. 3. Increasing the spreading rate increases the infected population ,

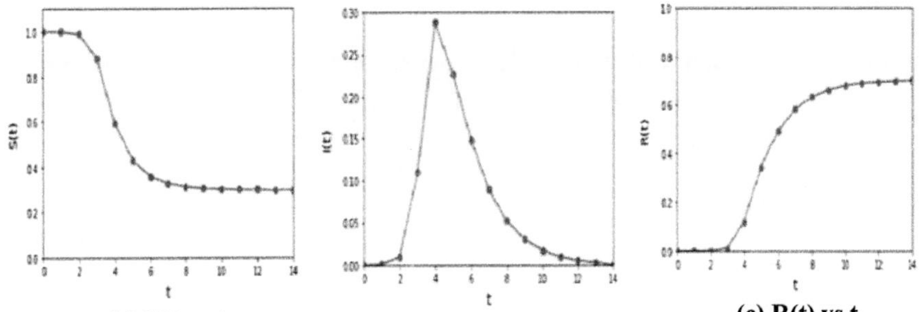

**(a) S(t) vs t** **(b) I(t) vs t** **(c) R(t) vs t**

**Fig. 2** Epidemic spreading versus time for $\alpha = 0.2$ and $\beta = 0.9$ for real world scale free network

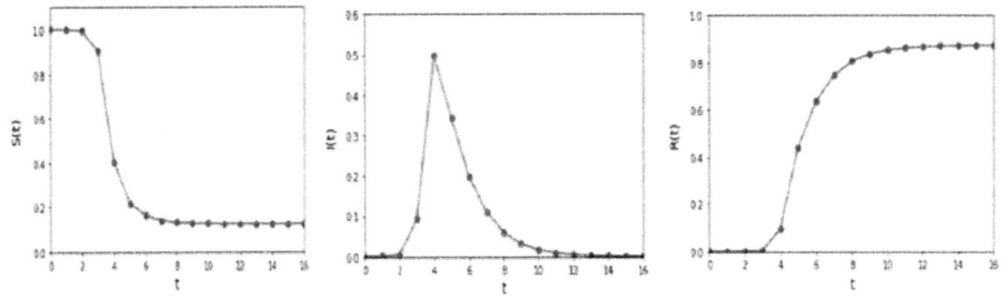

**(a) S(t) vs t** **(b) I(t) vs t** **(c) R(t) vs t**

**Fig. 3** Epidemic spreading versus time for $\alpha = 0.8$ and $\beta = 0.1$ for real world scale free network

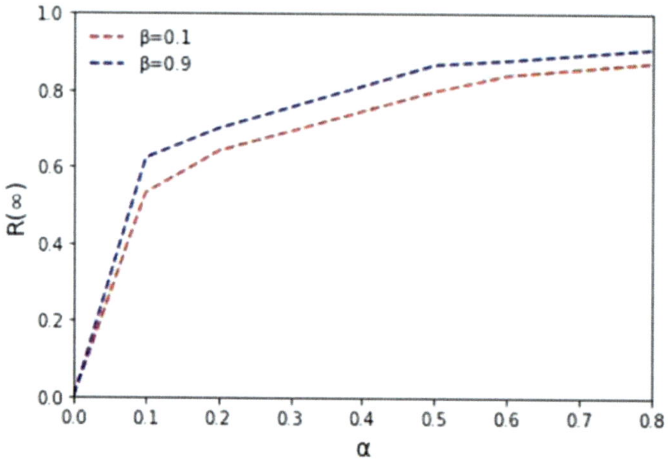


**Fig. 4** Final recovered population, R($\infty$) versus $\alpha$ at $\beta = 0.9$

resulting in fast decrements in a susceptible population. The final recovered population is recorded for the various value of spreading rate ($\alpha$) as 0.2, 0.5 and 0.8, and apply two values of recovery rate ($\beta$) as 0.1 and 0.9. Increasing the recovery rate, increases the fast recovery as shown in Fig. 4. For $\alpha = \{0.2, 0.5 \text{ and } 0.8\}$ and $\beta) = 0.9$, the fraction of recovered population is 0.7, 0.86 and 0.9 respectively.

## 4.2 *Effect of Quarantine*

In the proposed model some of the susceptible fraction of the population is quarantined with rate $\gamma$. There may be multiple reasons for choosing the population for quarantine. In the proposed work, the fraction of the population who is most important in the network needs to be quarantined. The idea is that if an important node exists in the network then it has a higher number of contacts and may be the most prone to get the COVID-19 effect. Therefore, various existing methods are used to find the most important nodes in the network, for example, Degree centrality (DC), Betweenness Centrality (BC), Closeness Centrality (CC) and Eigen Vector Centrality (EC) measures. The procedure is executed by using the steps mentioned in Algorithm 1.

**Algorithm 1** COVID-19 spreading with quarantine

**Input:** $(E, b)$ where $E$ is the edge list of the contact network and $b$ is the recovery rate.
The edge list of contact network is formed by:
i) Based on calculating the centrality.
ii) Based on the percent of the nodes removed.
**Initial Condition:** Initial conditions consist of $I(t), S(t), R(t)$ array indicators of initially infected nodes , susceptible nodes and recovered nodes. Adjacency Matrix of nodes.Time=0.
**Output:** array indicator of final recovered nodes $R(t)$, final infected nodes $I(t)$ and final susceptible nodes $S(t)$
**Method:**
for Edgelist :
  **for** *(for each node in edgelist)* **do**
    calculate degree, eigen vector, betweeness and closeness centrality
function (centrality)
  **for** *(various x percentage of nodes)* **do**
    **if** *( remove x percent nodes.)* **then**
    update edgelist
pass all the nodes of these centrality through function centrality
  call the SIR function for simulation one .

First, nodes are ranked by using the DC, BC and CC and EC. The fraction g of the total population is selected from the higher rank from the list for the quarantine. After sending the g fraction of the population into the quarantine, need to see the maximum infected population at any time ($I_{max}$) and total recovered population (R($\infty$)) into the network. I max is the maximum infected population during the course of epidemic spreading time. The total recovered population is calculated because it gives the estimate of the total infected population into the network which becomes recovered later. For the simulation same data-set as previous is used. The persons in the human contact networks are ranked using DC, BC, CC and EC. In Fig. 5, maximum infected people are plotted against the various fractions of the quarantine population. It is very interesting to find that if we will be able to quarantine 20% of the population in DC, BC and CC then the maximum infected people could be almost nil (Fig. 5a–c). Even, quarantine of the 15% population is enough to slow down the maximum infected population in BC and CC. In the case of EC, only 10% population is sufficient for quarantine to remove the infected population (Fig. 5d). Similarly, $R(\infty)$ is calculated against $g$. In Fig. 6, it is observed that quarantine of the 20% high ranked person are sufficient to stop the disease. In the case of BC and EC (Fig. 6b, d) the slowdown effect is higher than the DC and CC before g = 0.2 (Fig. 6b, d). The results validate that the person who makes higher contacts into the human contact network is most responsible for the COVID-19 spreading. If any way we will be able to quarantine a very less fraction of these persons then epidemic spreading can be stopped.

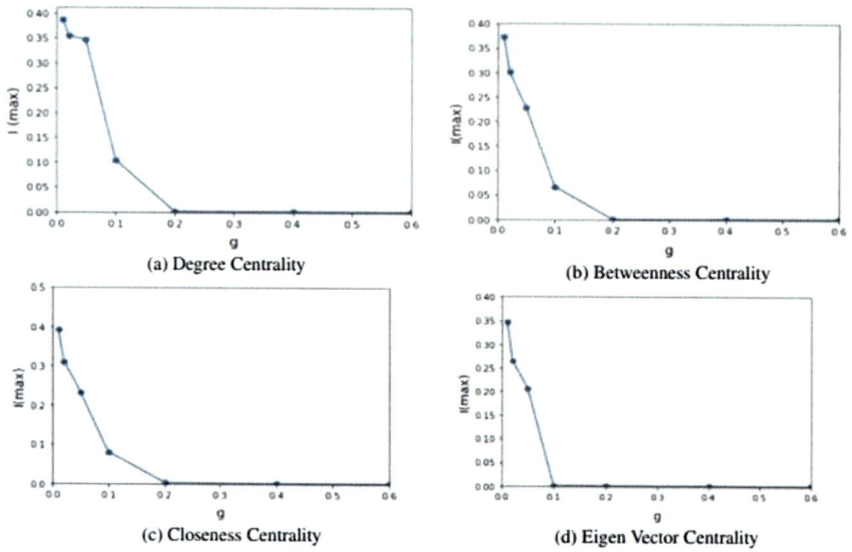


(a) Degree Centrality

(b) Betweenness Centrality

(c) Closeness Centrality

(d) Eigen Vector Centrality

**Fig. 5** Maximum infected population ($I_{max}$) at any time versus fraction of quarantine population (g) using different Centrality measures for, $\alpha = 0.6$ and $\beta = 0.4$

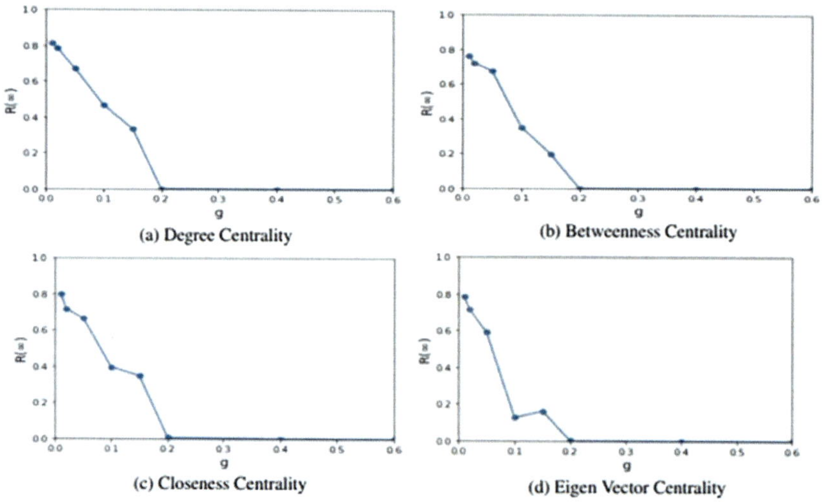


(a) Degree Centrality

(b) Betweenness Centrality

(c) Closeness Centrality

(d) Eigen Vector Centrality

**Fig. 6** Maximum Recovered population R($\infty$) versus fraction of quarantine population (g) using different Centrality measures for, $\alpha = 0.6$ and $\beta = 0.4$

## 4.3 Effect of Isolation

In the proposed model we considered the effect of the Isolated population from the infected population. Isolation means that the population who have the symptoms of COVID-19 are separated and able to spread further in the healthy population. Here, the infected population is identified and identified fraction of the population needs to be isolated. This isolation will help to slow the number of infected persons in the next time instant drastically. The same procedure is applied by using Algorithm 2.

**Algorithm 2** COVID-19 spreading using Isolation

**Input:** $(E, b)$ where $E$ is the edge list of the contact network and $b$ is the recovery rate.
**Initial Condition:** Initial conditions consist of $I(t), S(t), R(t)$ array indicators of initially infected nodes , susceptible nodes and recovered nodes. Adjacency Matrix of nodes. Time=0
**Output:** array indicator of the infected nodes $I(t)$
**Method:**

```
while ( spreader > 0 and previous-spreader ≠ spreader) do
    for Each node of spreader do
    if (Node ∈I(t)) then
        Check for the node been visited or not
        if ( visited == 0 ) then
            for (Each contact v of node u) do
                for (each neighboring node a pseudo random number(p) between 0 and 1 is
                  calculated) do
                    if (random number(p) < spreading rate ) then
                        let transmission of infection u ⟹ v occur with probability p
                        if (u ⟹ v) then
                            update S(t) and I(t)
                            if (visited == 1 or visited == 2) then
                                for (node w a pseudo random number(q) between 0 and 1 is
                                  calculated) do
                                  if ( random number(q) < recovering rate) then
                                      update state of u from infected to recovered with probabil-
                                      ity q
                                      if (recovery occur) then
                                          update R(t) and I(t) and for node w visited=2
    Time=Time+1; if (Time == 5 or Time ==11 or Time == 2 or Time == 4) then
        update I(t) isolate infected
```

For simulations same data-set is used as previous. The various parameter values are used for the simulations. Higher spreading rate and lower recovery rate are taken and vice versa. With $\alpha = 0.1$ and $\beta = 0.1$, maximum infection suddenly decreases from 12% to 2% after sending 75% infected population into isolation (Fig. 7a). For $\alpha = 0.1$ and $\beta = 0.9$ I max decreased from 10% to 3% after sending 70% of the population into isolation (Fig. 7b). When we consider $\alpha = 0.5$ then for $\beta = 0.1$ I

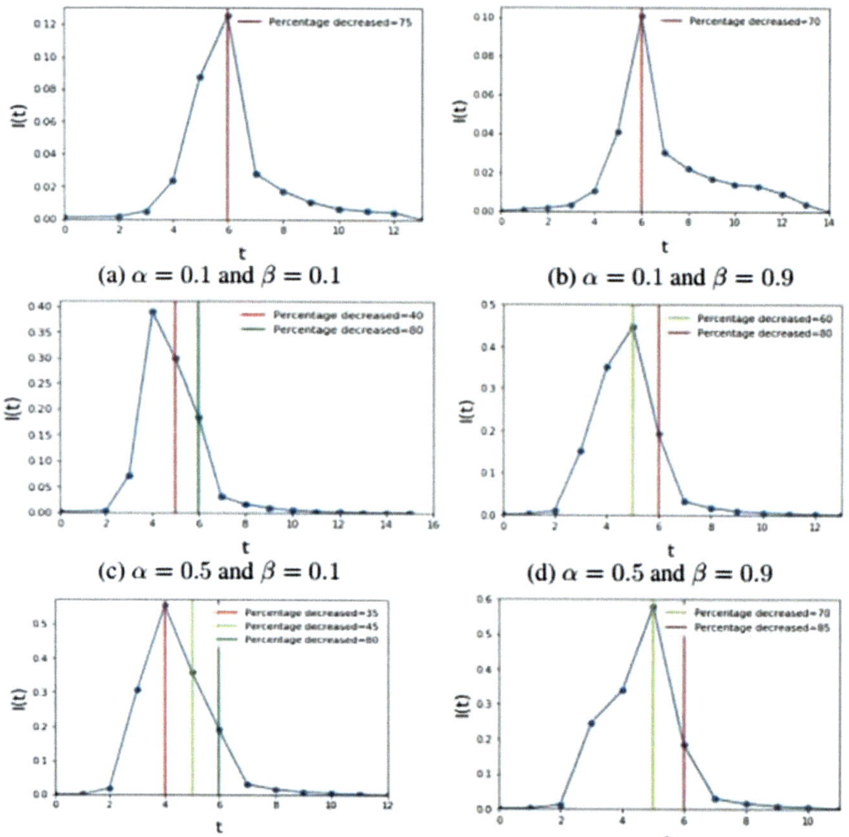


(a) $\alpha = 0.1$ and $\beta = 0.1$ (b) $\alpha = 0.1$ and $\beta = 0.9$

(c) $\alpha = 0.5$ and $\beta = 0.1$ (d) $\alpha = 0.5$ and $\beta = 0.9$

(e) $\alpha = 0.8$ and $\beta = 0.1$ (f) $\alpha = 0.9$ and $\beta = 0.9$

**Fig. 7** Infected Population versus time for various fraction for the isolated population

max decreased from 37% to 23% for 40% isolation and up to 4% with 80% isolation (Fig. 7c), and for $\beta = 0.9$, it decreased from 45% to 20% with 60% isolation and up to 5% with 80% isolation (Fig. 7d). Similarly, for $\alpha = 0.8$ with $\beta = 0.1$, $I_{max}$ decreased from 60% to 30% for 35% isolation and up to 2% for 80% isolation (Fig. 7e). For $\alpha = 0.9$ and $\beta = 0.9$, $I_{max}$ decreases from 60 to 18% with 70% isolation and 2% with 85% isolation of the infected population (Fig. 7f).

## 5 Conclusions and Future Work

The proposed model produces a preliminary representation of the dynamics of COVID-19. Here, underlying social contact network is considered. In the proposed work, we uncover the effect of quarantine and isolation on the COVID-19 spreading.

By using the various centrality measures important nosed are identified those make a higher number of contact and quarantined by various fractions and found that approx. 20% of those population is satisfactory to stop the infection. In other part, when COVID-19 is spreading in the meanwhile some fraction of isolated population is selected and isolated with different fractions. The drastic change in the spreading of the disease by using this very much helpful to slow down the spreading. In this work, proposed algorithms are not compared with the others, therefore in the proposed work algorithm evaluation is not presented. But may be considered as future work. In the full description, all the realistic parameters may also be considered as future work.

**Acknowledgements** This work is supported by Science and Engineering Research Board (SERB), DST, Government of India under MATRICS Short term special call on COVID-19 project (File. No. MSC/2020/000500).

## References

1. Arquam Md, Singh A, Cherifi H (2020) Impact of seasonal conditions on vector-borne epidemiological dynamics. IEEE Access 8:94510–94525
2. Arquam Md, Singh A, Cherifi H (2019) Integrating environmental temperature conditions into the sir model for vector-borne diseases. In: International conference on complex networks and their applications. Springer, pp 412–424
3. Zheng J (2020) Sars-cov-2: an emerging coronavirus that causes a global threat. Int J Biol Sci 16(10):1678
4. Liang K (2020) Mathematical model of infection kinetics and its analysis for covid-19, sars and mers. Inf Genet Evol 104306
5. Kwok KO, Tang A, Wei VWI, Park WH, Yeoh EK, Riley S (2019) Epidemic models of contact tracing: Systematic review of transmission studies of severe acute respiratory syndrome and middle east respiratory syndrome. Comput Struct Biotechnol J 17:186–194
6. Cao Z, Zhou S (2018) Dynamical behaviors of a stochastic siqr epidemic model with quarantine-adjusted incidence. Discr Dyn Nature Soc 2018
7. Safi MA, Gumel AB (2011) Mathematical analysis of a disease transmission model with quarantine, isolation and an imperfect vaccine. Comput Math Appl 61(10):3044–3070
8. Odagaki T (2020) Analysis of the outbreak of covid-19 in japan on the basis of an siqr model. medRxiv
9. Tang B, Xia F, Tang S, Bragazzi NL, Li Q, Sun X, Liang J, Xiao Y, Wu J (2020) The effectiveness of quarantine and isolation determine the trend of the covid-19 epidemics in the final phase of the current outbreak in china. Int J Inf Dis
10. Arenas A, Cota W, Gomez-Gardenes J, Gomez S, Granell C, Matamalas JT, Soriano-Panos D, Steinegger B (2020) Derivation of the effective reproduction number for covid-19 in relation to mobility restrictions and confinement. MedRxiv
11. Ball F, Sirl D, Trapman P (2010) Analysis of a stochastic sir epidemic on a random network incorporating household structure. Math Biosci 224(2):53–73
12. Bistritz I, Bambos N, Kahana D, Ben-Gal I, Yamin D (2019) Controlling contact network topology to prevent measles outbreaks. In: 2019 IEEE global communications conference (GLOBECOM). IEEE, pp 1–6
13. Salehi M, Sharma R, Marzolla M, Magnani M, Siyari P, Montesi D (2015) Spreading processes in multilayer networks. IEEE Trans Netw Sci Eng 2(2):65–83

14. Salathé M, Kazandjieva M, Lee JW, Levis P, Feldman MW, Jones JH (2010) A high-resolution human contact network for infectious disease transmission. Proc Natl Acad Sci 107(51):22020–22025
15. Riou J, Althaus CL (2020) Pattern of early human-to-human transmission ofwuhan 2019 novel coronavirus (2019-ncov), december 2019 to january 2020. Eurosurveillance 25(4):2000058
16. Rader B, Scarpino S, Nande A, Hill A, Reiner R, Pigott D, Gutierrez B, Shrestha M, Brownstein J, Castro M et al (2020) Crowdingand the epidemic intensity of covid-19 transmission. MedRxiv
17. Pastor-Satorras R, Castellano C, Van Mieghem P, Vespignani A (2015) Epidemic processes in complex networks. Rev Modern Phys 87(3):925
18. Arquam Md, Singh A, Sharma R (2018) Modelling and analysis of delayed sirmodel on complex network. In: International conference on complex networks and their applications. Springer, pp 418–430
19. Bi Q, Wu Y, Mei S, Ye C, Zou X, Zhang Z, Liu X, Wei L, Truelove SA, Zhang T et al (2020) Epidemiology and transmission of covid-19 in shenzhen china: aalysis of 391 cases and 1,286 of their close contacts. MedRxiv
20. Bock W, Adamik B, Bawiec M, Bezborodov V, Bodych M, Burgard JP, Goetz T, Krueger T, Migalska A, Pabjan B et al (2020) Mitigation and herd immunity strategy for covid-19 is likely to fail. MedRxiv
21. Hethcote HW (1976) Qualitative analyses of communicable disease models. Math Biosci 28(3–4):335–356
22. Rossi RA, Ahmed NK (2015) The network data repository with interactive graph analytics and visualization. In: AAAI

# Chapter 37
# Lung Cancer: Detection and Classification of Malignancies

**Shailesh Kumar Thakur, Dhirendra Pratap Singh, and Jaytrilok Choudhary**

## 1 Introduction

Lung cancer ranked the topmost disease for deaths in men and ranked third in women. About 18.4% of the total deaths due to Lung cancer [1]. Mostly, its cure depends upon the detection of the lung lesions from low-dose CT images [2] and effective diagnosing methods at the initial stage. The low-dose CT screening detects all of the tumors while screening and results in a 20.0% reduction in mortality and significantly increase the positive screening tests in comparison with traditional radiography techniques.

Mostly, Lung lesions at the initial stage are small and localized. They are very tiny in shape, approximately 3 mm in diameter. Based on CT screening, radiologists identify lesions and classify them as cancerous or non-cancerous. The number of 2D CT images is taken by the radiologists of 3D-Lung. This involves the meticulous examination of 3D-Lung. Due to the incorporation of lots of information in CTs, the examination suffers a lot due to human error. Hence, the diagnosis of Lung cancer has been automated using CAD systems as radiologists' assistance for the accurate detection and classification of the lesions.

In recent years, it has been shown that designing the CAD systems is a promising approach as a radiologists' assistance in interpretation and diagnosis of lesions from CT screening. Some of the CAD systems use handcrafted features to diagnose diseases. For example, [3–5] used Histogram of Oriented Gradients (HOG) techniques, [4, 6–9] used Gray-Level Co-occurrence Matrix (GLCM), [10] used Speeded-Up Robust Features (SURF), [3, 5, 6, 8] used Local Binary Patterns (LBP), [11] used the stationary wavelet transform and convergence index filter and [12] used the Genetic Algorithm(GA) for manual feature selection. However, the success of deep learning gave rise to data-driven CAD systems. Some of the recently proposed deep learning-based CAD systems for diagnosing Lung cancer are: CNN, [13] used

S. K. Thakur (✉) · D. P. Singh · J. Choudhary
Maulana Azad National Institute of Technology, Bhopal 462003, India

© The Author(s), under exclusive license to Springer Nature Singapore Pte Ltd. 2022 
H. M. Dubey et al. (eds.), *Artificial Intelligence and Sustainable Computing*, Algorithms for Intelligent Systems, https://doi.org/10.1007/978-981-16-1220-6_38

modified AlexNet, [14] used 2D-CNN with Taguchi parametric optimization, [15] used a combination of three CNN models (AlexNet, LeNet, and VGG-16), [16] used transfer learning that using an ensemble of weak models, etc. These models showed promising results but have certain limitations shown in [17] which are using multi-section CNN. Although this work shows good results, due to the lack of viewpoint invariance on CNN, it ignores lots of information that present in CT screening.

In contrast to the related work, we used a convolution-deconvolution Capsule architecture along with Efficient Network. Sabour et al. [18] proposed an architecture contain Capsules that communicate with higher-level capsules using a dynamic routing algorithm. Their model works on the principle of reverse rendering approach, that is, determine real-world objects through its subparts. It has various advantages like equivariance properties and requires very little data. We use a modified capsule network for segmentation and detection of lung nodules and further, apply a pre-trained efficient network (EfficientNet) for classification.

# 2 Architectures

## 2.1 CapsNet

CNN is an effective methodology in the current scenario and it already has been shown state-of-the-art performance in comparison with that of handcrafted features. Although various techniques like the stacking layers, changing activation unit, decreasing filter size, using pooling layers, regularization, and skipping connections made CNN get superhuman performance in many fields, but still, CNN is unable to correctly predict unseen images like a human brain. Hence, Sabour et al. [18] addressed the limitations of CNN and introduced a new concept of recognition known as CapsNet and the intuition behind this was to incorporate spatial information like spatial relationships, orientation, size, and perspective of features.

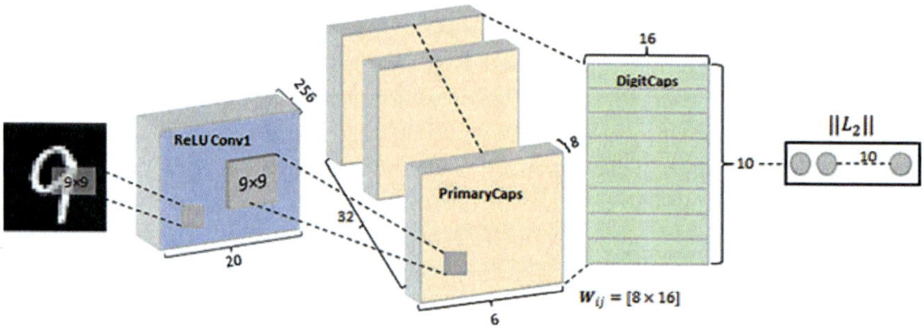


**Fig. 1** Encoder part of CapsNet

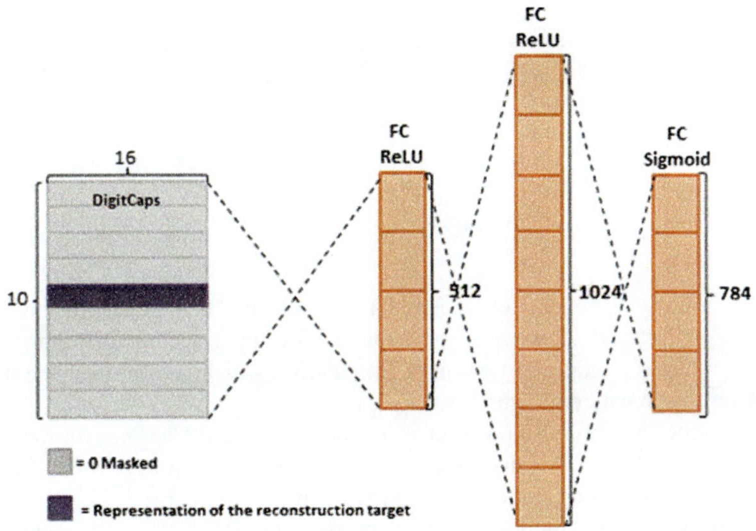


**Fig. 2** Decoder part of CapsNet

Figures 1 and 2 show the architecture of the encoder and decoder part of CapsNet. In the encoder part, it consists of several layers of capsules that encapsulate likelihood with properties of features. They introduced a shallow CapsNet that contains 2 convolution layers, 32 primary capsules with $6 \times 6 \times 8$ capsules which are fed into 10 higher layer capsules with 16 dimensions each, and at last, the loss, $L_C$ is calculated to give the class probability [18], $L_C = T_C \max(0, m^+ - ||v||)^2 + \lambda(1 - T_C)\max(0, ||v|| - m^-)^2$ where $T_C = 1$ for correct DigitCaps, hence the first part of loss function will be evaluated otherwise $T_C = 0$ for incorrect DigitCaps, hence the second part of the equation $(1–T_C)$ will be evaluated for incorrect DigitCaps and $m^+ = 0.9$ and $m^- = 0.1$. The constant $\lambda = 0.5$ used for numerical stability.

In the decoding phase, three fully connected layers are connected to higher layer capsules. This architecture showed a more promising result to classify the MNIST dataset. LaLonde et al. [19] showed that capsules achieved state-of-the-art performance in the segmentation of the LUNA-16 dataset.

## *2.2 FixEfficientNet*

In paper [20], the author systematically studied ConvNets model scaling and carefully identifies that balancing depth, width, and resolution that can lead to better performance and thereafter proposed a new scaling method that uniformly scales all these dimensions using compound coefficient and also demonstrates the effectiveness of this method on ResNet and MobilesNets. Figure 3 shows the EfficientNet also known as an efficient network that uses a compound scaling method to boost up the

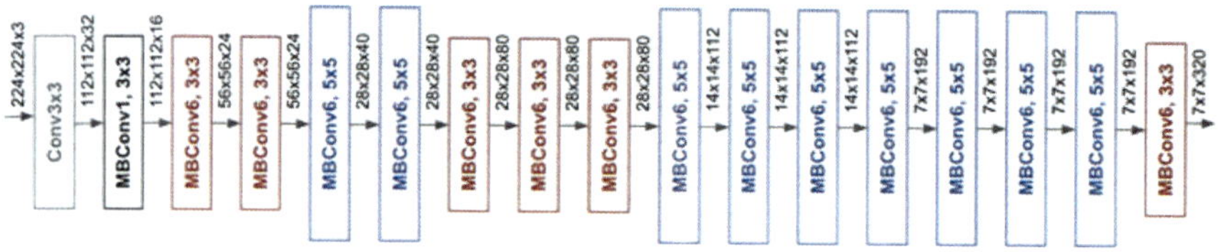


**Fig. 3** Baseline architecture of EfficientNet

CNN performance. EfficientNet-B7 was 8.4 × smaller and 6.1 × faster than the best existing ConvNet and shown state-of-the-art 84.3% top-1 accuracy on ImageNet. FixEfficientNet is a modified version of EfficientNet which incorporates the fixing of train-test discrepancy shown in paper [21].

In this paper, we proposed a framework using modified CapsNet for detection and FixEfficientNet for lung nodule classification. Our framework design is heavily inspired by two previous studies [17] and [19]. In the first paper [17], the authors introduced a lightweight model based on deep learning MVCNN which incorporates two view sampling approach to obtain cross-sections of lung nodules. The other related work [19] demonstrated the segmentation of Medical Images using UNet with higher accuracy.

## 3 Proposed Framework

The proposed framework performs tasks in four stages shown in Fig. 4. In the First stage, preprocessing is done and further generates masks from CT scans. The next phase involves segmentation using Modified CapsNet which is followed by labeling of detected nodules as TP or FP. And the final stage involves classification using FixEfficientNet that predicts cancer.

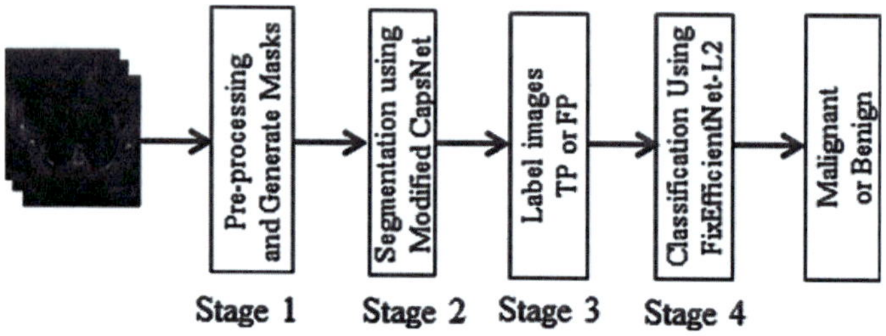


**Fig. 4** Proposed framework for detection and classification for nodules

## 3.1 Data Pre-processing

LIDC-IDRI [22] is a publicly available and more robust benchmark that has been extensively used in several studies in recent years as shown in our literature review [23]. For the early detection of cancer, we mainly focus on the nodules of 3 mm for the compilation and evaluation of the model. This benchmark is reviewed by four lung radiologists independently and marked with 5 levels, known as Highly Unlikely, Moderately Unlikely, Indeterminate, Moderately Suspicious, and Highly Suspicious shown in Fig. 5. As shown in [17], these levels are grouped into Malignant and Benign using an iterative approach. Radiologists annotated these datasets which are contained in the CSV(comma-separated file) file known as list3_2.csv that is present inside the LIDC-IDRI dataset. This file is composed of case, scan, the region of interest, volume, slice number, nodule IDs, nodules diameter, area, and coordinates(x and y locations).

Moreover, LIDC-IDRI contains 100–200 CTs of each case and each is of 512 × 512 resolution images. Hounsfield unit(HU) is used to measure the radio wave density to differentiate the nodules from the tissues of the lungs. Mostly the lung tissues are in the range of −500–0 HU but it could be range from −500 to 500 HU due to the existence of lung nodules as shown in Fig. 6.

The classification of malignancy is based on candidate features such as diameter, attenuation (MeanHU), eccentricity, calcification, and speculation. As shown in the following Fig. 7, the strongest predictor to detect cancer is nodule diameter. This visualization of LIDC datasets helps find the most informative features that can help in finding the likelihood of the features to predict cancer.

All the CT images are processed into nd-arrays of 512 × 512 and then normalized by subtracting mean and dividing with standard deviation. Also, the non-lung selected

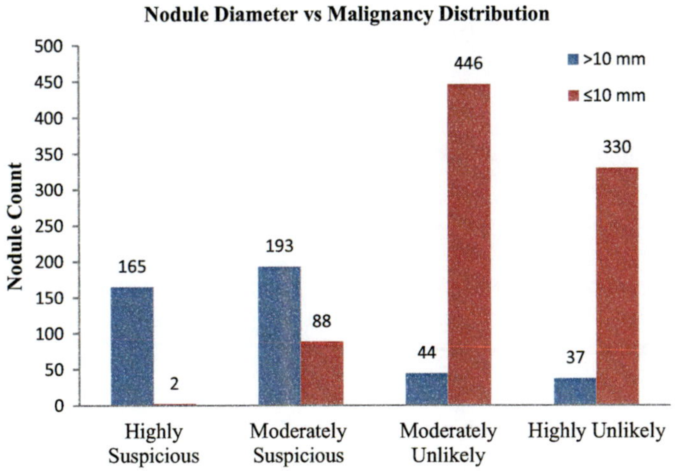


**Fig. 5** Nodules diameter versus malignancy distribution $\geq$ 3 mm in LIDC-IDRI

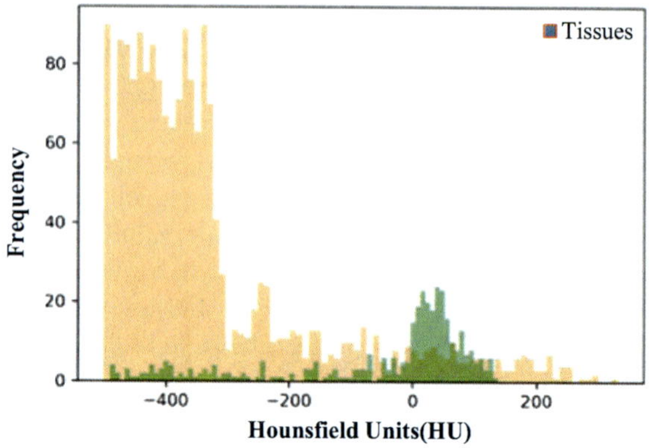


**Fig. 6** Radio density of lung nodule and tissues between −500 and 500 HU

outside of lung tissues that are under −500 HU are removed along with the nodules contained in 25 mm sq. area because these are less prone to malignancy. Figure 8 shows the cropped image of the lung nodule. Thereafter, we used the list3_2.csv file that is present inside the LIDC-IDRI dataset to get the coordinates(x and y locations) of lung nodules and generate the masks.

### 3.2 *Segmentation*

Pre-processed images and nodule masks generated during the first stage are split into 80:20 ratio and train the modified CapsNet model with the dice coefficient as a loss function.

$$DiceCoeff = \frac{2|X \cap Y|}{|X| + |Y|} \tag{1}$$

where |X| and |Y| are the cardinalities of two sets X and Y, respectively.

Figure 9 shows the network architecture for the segmentation tasks in which convolutional capsules are combined with UNet architecture that allow us to exhaust complete image information with a reduced number of parameters.

In this modified CapsNet, every child capsule is using locally constrained routing only to its parent capsules under a defined local window along with sharing transformation matrices for each grid member within a capsule. Segmentation involves the following steps:-

(a) A slice of 512 × 512 CT scan is passed through a 2D convolution layer and produces 16 feature maps using a 5 × 5 filter size. This will result in the formation of the single capsules type containing a set of 512 × 512 capsules and each having 16-dimensional vectors.

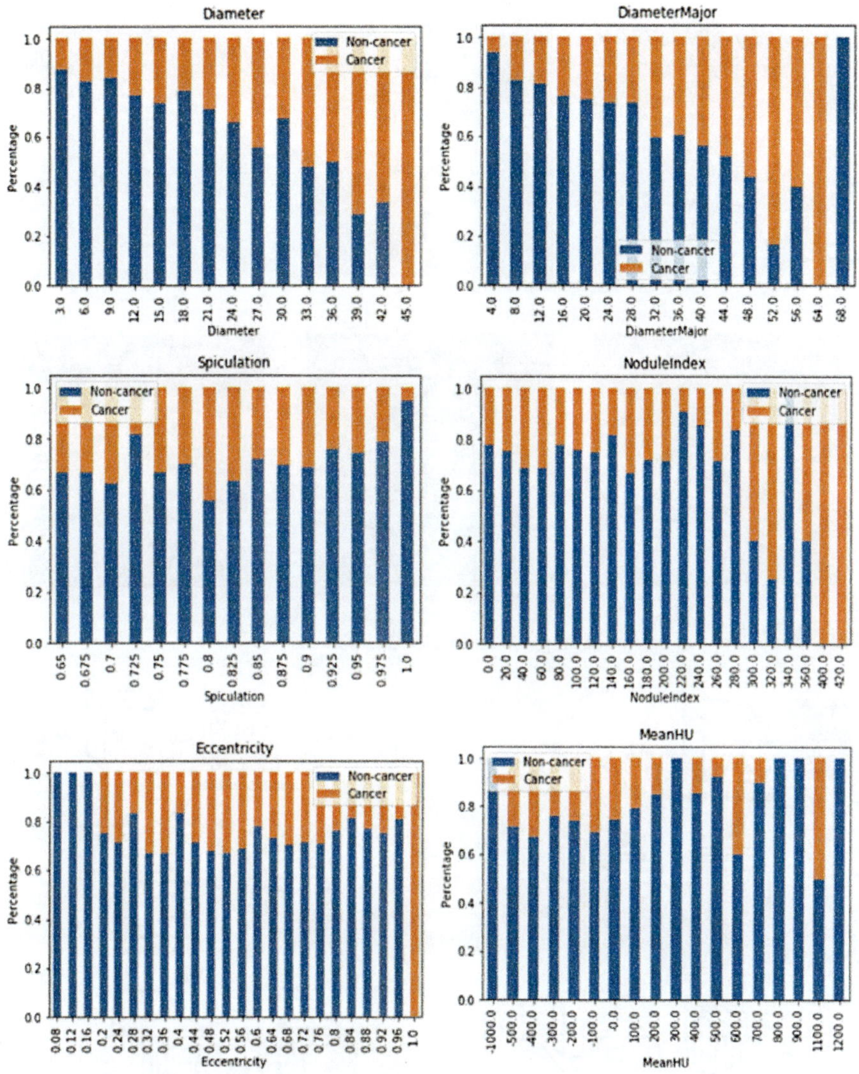


**Fig. 7** Feature visualization of LIDC-IDRI dataset

(b) At any layer L, every parent capsules receive "prediction vectors" $\{\hat{\mathrm{u}}_{xy|t_1^L}, \hat{\mathrm{u}}_{xy|t_2^L}, \ldots, \hat{\mathrm{u}}_{xy|t_n^L}\}$ from each type of capsules of child caps. And compute the weighted sum over these predicted vectors $p_{xy} = \sum_n \tau_{t_i^L} \hat{u}_{xy|t_i^L}$, where $\tau_{t_i^L|xy}$ is the routing coefficient obtained from dynamic routing. The routing softmax function computes these routing coefficients,

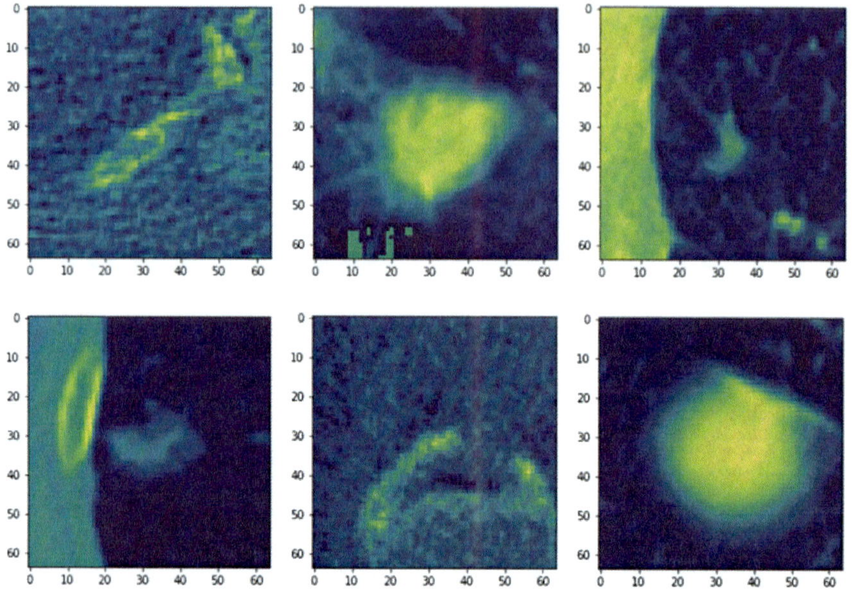

**Fig. 8** Some cropped images of lung nodules detected in patients

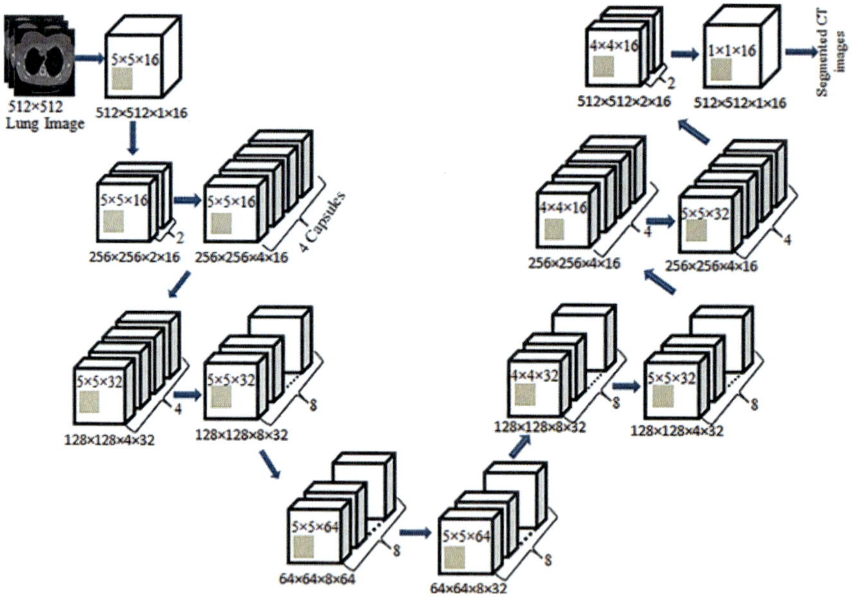


**Fig. 9** Segmentation using modified capsule network

$$\tau_{t_i^L|xy} = \frac{\exp\left(b_{t_i^L|xy}\right)}{\sum_k \exp\left(b_{t_i^L k}\right)} \tag{2}$$

where $b_{t_i^L|xy}$ are the initial logits that prediction vector routed to parent capsule.

(c) After performing dynamic routing between child capsules, we perform a non-linear squashing function to compute the output capsule.

$$v_{xy} = \frac{||p_{xy}||^2}{1 + ||p_{xy}||^2} \frac{p_{xy}}{||p_{xy}||} \tag{3}$$

where $v_{xy}$ is a non-linearity activation function for capsules at the spatial location $(x, y)$.

(d) Steps 2 and 3 repeated for each layer.

(e) Lastly, the final layer outputs a vector with 16 dimensions. The final segmentation mask is generated using the scalar product of a vector and $\hat{\mathrm{u}}_{xy|t_i^L}$.

## 3.3 Nodule Detection

In this stage, we load the Kaggle LUNA-16 dataset along with preprocessed images along with the weights in the model to generate candidate nodule masks to classify true positive and false positives and generates features such as area, eccentricity, the diameter of the major axis, mean HU value, diameter, and speculation. This stage involves the detection of the nodule in CT images with the intent of getting high TP and less FP. Figure 10 shows the example of the expected label which compared with the ground truth of CT images.

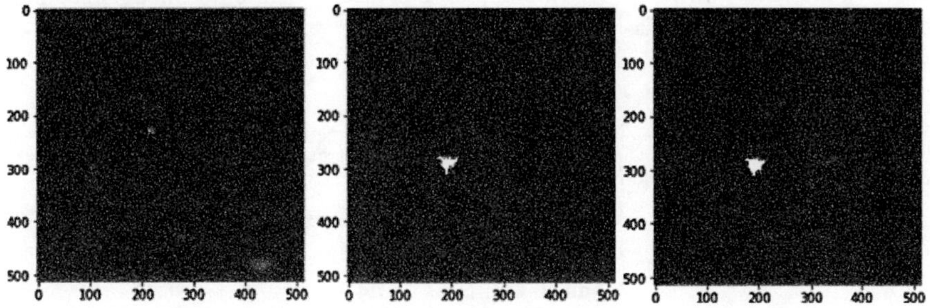


**Fig. 10** Processed CT images (left), ground truth label (center), predicted label (right)

### 3.4 Nodule Classification

The final Stage involves classification using pre-trained FixEfficientNet. It requires a large number of images to train and our dataset has a lesser number of nodules samples, hence we applied a transfer learning approach, that is, model is pre-trained on ImageNet used to avoid overfitting and accelerate the training process.

We replaced the last fully connected layer in FixEfficientNet with a fully connected layer which is randomly initialized for the task of binary classification. We need to tune around 66 M parameters. The main building block of FixEfficientNet has a mobile inverted bottleneck layer(MBConv) which has been first introduced in MobileNetV2. However, there are other versions of EfficientNet that are also available but FixEfficientNet has better top-1 and top-5 accuracy which shows state-of-the-art performance.

## 4 Result and Discussion

In the training and validation phase of our proposed model, the preprocessed data are fed with LIDC dataset in an 80:20 ratio and then we validate with the LUNA-16 dataset. We run our model on 10-fold with 0.0001 learning rate with stochastic gradient descent optimizer and further learning rate by a factor of 0.2 once the validation loss plateau is reached. It runs for 20 epochs and further compiles for 5 epochs for early stopping criteria using a learning rate of 0.001. We implemented the network with Keras library and trained it on Quadro K420 Nvidia GPU.

### 4.1 Detection Results

We evaluated the performance of the models using Accuracy. The results of the detection are shown in Table 1. We compare our results with other models such as MobileNet, InceptionV3, U-Net, and CapsNet.

**Table 1** Comparative analysis for lung nodule detection

| Model | Parameters (M) | Accuracy (%) |
|---|---|---|
| MobileNet [17] | 5.4 | 84.75 |
| Inception V3 [24] | 11.2 | 86.02 |
| U-Net | 31 | 98.44 |
| **Modified CapsNet** | **1.4** | **98.47** |

**Table 2** Comparative analysis for lung nodule classification

| Model | Accuracy(%) |
|---|---|
| ResNet [16] | 91.64 |
| EnsembleResNet [16] | 93.50 |
| MobileNet [17] | 93.18 |
| **FixEfficientNet** | **93.59** |

### *4.2 Classification Results*

We evaluated the performance of the models using Accuracy. The results of the classification are shown in Table 2. We compare our results with other models such as ResNet, Ensemble ResNet, MobileNet, and EfficientNet.

## 5 Conclusion

Our proposed framework is using the concept of Capsules to detect nodules and further, we apply pre-trained EfficientNet to classify CT images with fewer parameters to train compare to other models. We train these models on publicly available datasets: LUNA-16 and LIDC-IDRI and validate on these benchmarks. More importantly, the proposed framework is a lightweight model as there is a reduction in parameters around 1.4 M at the detection. During detection, we generate masks and candidate nodules with higher accuracy which is very much closer to ground truth nodules with convolutional and deconvolution capsule network We achieved an accuracy of 98.47% in nodule detection. Finally, the predicted nodules are classified into malignant and benign using a pre-trained efficient model known as FixEfficientNet, already trained on Imagenet and done binary classification through setting nodes of the final layer to two classes: 0(benign) and 1(malignant) and getting 93.59% in nodule classification. The proposed framework fundamentally increased the performance of CAD systems in the detection and classification of lung nodules and can be helpful in improves the current state-of-the-art performance.

## 6 Future Work

Future work may involve a study upon model sensitivity and model's FP reduction so that reduced FPs lead to an increase in the sensitivity of the model. And obtain the FROC curve which plots the sensitivity on the basis average number of FPs to evaluate the result of the models. As various researches are going on Lung cancer, it may help in designing a more robust model by changing the classification Model, that is, in place of EfficientNet, we can use MobileNetV3 and compare the final results in diagnosing the disease of Lung Cancer. We can test this framework and runtime on mobile test the result with existing models that are already built for mobile devices.

## References


1. Bray F, Ferlay J, Soerjomataram I, Siegel R, Torre L, Jemal A (2018) Global cancer statistics 2018: GLOBOCAN estimates of incidence and mortality worldwide for 36 cancers in 185 countries: global cancer statistics 2018 CA. Cancer J Clin 68
2. National Lung Screening Trial Research Team, Aberle DR, Adams AM, Berg CD, Black WC, Clapp JD, Fagerstrom RM, Gareen IF, Gatsonis C, Marcus PM, Sicks JD (2011) Reduced lung-cancer mortality with low-dose computed tomographic screening. N Engl J Med 365(5):395–409. https://doi.org/10.1056/NEJMoa1102873. Epub 2011 Jun 29. PMID: 21714641; PMCID: PMC4356534
3. Shen W, Zhou M, Yang F, Yu D, Dong D, Yang C, Zang Y, Tian J (2016) Multi-crop convolutional neural networks for lung nodule malignancy suspiciousness classification. Pattern Recognit 61
4. Dhara AK, Mukhopadhyay S, Dutta A, Garg M, Khandelwal N (2016) A combination of shape and texture features for classification of pulmonary nodules in lung CT images. J Digit Imaging 29
5. Shen W, Zhou M, Yang F, Yang C, Tian J (2015) Multi-scale convolutional neural networks for lung nodule classification bt – information processing in medical imaging, pp 588–599
6. Han F, Wang H, Zhang G, Han H, Song B, Li L, Moore W, Lu H, Zhao H, Liang Z (2014) Texture feature analysis for computer-aided diagnosis on pulmonary nodules. J Digit Imaging 28:99–115
7. Wei G, Cao H, Ma H, Qi S, Qian W, Ma Z (2017) Content-based image retrieval for lung nodule classification using texture features and learned distance metric. J Med Syst 42:13
8. Wei G, Ma H, Qian W, Han F, Jiang H, Qi S, Qiu M (2018) Lung nodule classification using local kernel regression models with out-of-sample extension. Biomed Signal Process Control 40:1–9
9. Sergeeva M, Ryabchikov I, Glaznev M, Gusarova N (2016) Classification of pulmonary nodules on computed tomography scans. Evaluation of the effectiveness of application of textural features extracted using wavelet transform of image. In: 2016 18th conference of open innovations association and seminar on information security and protection of information technology (FRUCT-ISPIT), pp 291–299
10. Mao K, Deng Z (2016) Lung nodule image classification based on local difference pattern and combined classifier. Comput Math Methods Med 2016:1091279
11. Li X, Shen L, Luo S (2018) A solitary feature-based lung nodule detection approach for chest X-Ray radiographs. IEEE J Biomed Heal Infor 22(2):516–524
12. Boroczky L, Zhao L, Lee KP (2006) Feature subset selection for improving the performance of false positive reduction in lung nodule CAD. IEEE Trans Inf Technol Biomed 10(3):504–511
13. Bhandary A, Prabhu A, Rajinikanth V, Krishnan P, Satapathy S, Robbins D, Shasky C, Zhang Y-D, Tavares J, Raja N (2020) Deep-learning framework to detect lung abnormality – a study with chest X-Ray and lung CT scan images. Pattern Recognit Lett 129:271–278
14. Lin C-J, Shiou-Yun J, Chen M-K (2020) Using 2D CNN with taguchi parametric optimization for lung cancer recognition from ct images. Appl Sci 10:2591
15. Togacar M, Ergen B, Cömert Z (2019) Detection of lung cancer on chest CT images using minimum redundancy maximum relevance feature selection method with convolutional neural networks. Biocybern Biomed Eng 40

16. Xie Y, Xia Y, Zhang J, Feng DD, Fulham M, Cai W (2017) Transferable multi-model ensemble for benign-malignant lung nodule classification on chest CT BT – medical image computing and computer-assisted intervention – MICCAI 2017, pp 656–664
17. Sahu P, Yu D, Dasari M, Hou F, Qin H (2019) A lightweight multi-section CNN for lung nodule classification and malignancy estimation. IEEE J Biomed Heal Infor 23(3):960–968
18. Sabour S, Frosst N, Hinton GE (2017) Dynamic routing between capsules
19. LaLonde R, Bagci U (2018) Capsules for object segmentation
20. Tan M, Le Q (2019) EfficientNet: rethinking model scaling for convolutional neural networks
21. Touvron H, Vedaldi A, Douze M, Jégou H (2020) Fixing the train-test resolution discrepancy: FixEfficientNet
22. Armato SG 3rd et al (2011) The lung image database consortium (LIDC) and image database resource initiative (IDRI): a completed reference database of lung nodules on CT scans. Med Phys 38(2):915–931. https://doi.org/10.1118/1.3528204
23. Thakur SK, Singh DP, Choudhary J (2020) Lung cancer identification: a review on detection and classification. Cancer Met Rev 39(3):989–998
24. Cao H, Liu H, Song E, Ma G, Xu X, Jin R, Liu T, Hung C (2019) Multi-Branch ensemble learning architecture based on 3D CNN for false positive reduction in lung nodule detection. IEEE Access 7:67380–67391

# Chapter 38
# Malicious Webpage Classification Using Deep Learning Technique

**Kushagra Krishna, Jaytrilok Choudhary, and Dhirendra Pratap Singh**

## 1 Introduction

In the modern era, the Internet is a major part of daily life. As per the data of 2017, approximately 57% of the world's population is connected over the Internet [1]. Internet is being used for internet banking, communication, shopping, entertainment, and several other non-commercial and commercial activities. Though it is acted as a boon to society by easing the life of people also it has proved to be a bane by exposing us to the extreme danger of getting attacked. Illegitimate users utilize malware programs to commit illegal things, that is, financial frauds or stealing private or sensitive information, from the legitimate users. As the number of attacks is growing every year, it becomes a hobby of tech-enthusiast, researchers and highly professional programmers to get easy profits [2] that are coming their way. Now, it has been full-fledged to be a multi-million dollar industry [3] where the technological tools are sold and bought for hacking in a similar way as the legal software industry.

The Web has now become vulnerable to several types of attacks such as XSS (Cross Site Scripting), hacking, drive-by-exploits, social engineering, phishing, DoS (denial of service), distributed DoS and several other types of attacks. Static and dynamic analysis are two common methods for the detection of malicious web pages [4, 5]. The static analysis examines the executable file without viewing the original file. At first, distinct static features are extracted from different web pages, then train a classifier and predict whether web pages are benign or malicious. Nowadays, due to the increase in popularity of URL shortening services, malicious URL can be easily hidden behind a short URL and the user vulnerable to different types of attacks when

K. Krishna (✉) · J. Choudhary · D. P. Singh
Maulana Azad National Institute of Technology, Bhopal, India

J. Choudhary
e-mail: jaytrilok@manit.ac.in

D. P. Singh
e-mail: dpsingh@manit.ac.in

© The Author(s), under exclusive license to Springer Nature Singapore Pte Ltd. 2022

H. M. Dubey et al. (eds.), *Artificial Intelligence and Sustainable Computing*,
Algorithms for Intelligent Systems, https://doi.org/10.1007/978-981-16-1220-6_39

accessing it. Using dynamic analysis, the webpage source code is executed in an emulated environment and if any one of the unknown instances includes malicious shell code, that instance is finally classified to be malicious.

Several researchers have applied different machine learning techniques for Malicious URL detection. Using Machine Learning approaches, a set of URLs are given as training data, and based on the statistical properties; the predictive model classifies a URL as malicious or benign. After getting the data set, features such as URL, lexical, contextual and visual features are extracted from data. The feature selection describes about different features used in the malicious web detection process. Nowadays while browsing sites, there are some malicious images also present on the site and for their feature are not needed the URL features, then the classification of images is needed using Neural network algorithms as neural network gives better performance than any machine learning algorithms. Several image processing techniques analyze malware images as gray-scale images. After malware, the binary file is visualized into a digital image. The malware binary detection converts the model into the multi-classifier model. In order to detect malware binaries, features can be extracted manually from malware images and then feed them into different classifiers such as SVM (support vector machine) or KNN (k-nearest neighbor's algorithm). To be more discriminative in extracting features, one can utilize CNN to automatically extract features and perform classification in an end-to-end fashion as shown in Fig. 1.

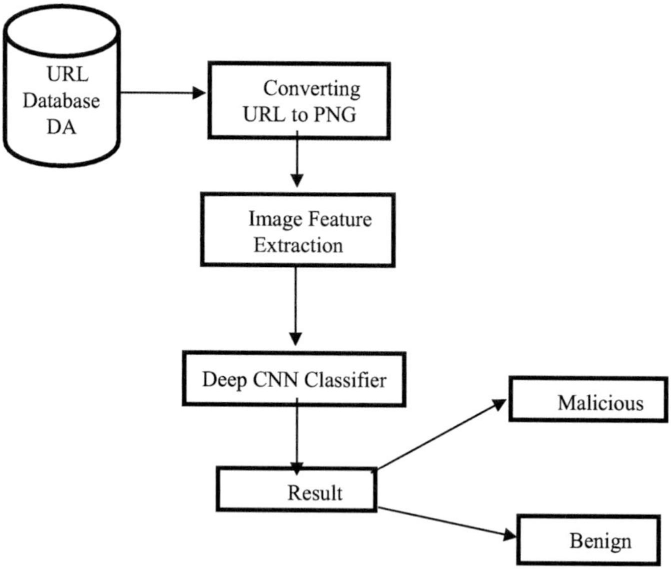


**Fig. 1** Malicious webpage classification

## 2 Related Work

Several machine learning techniques such as KNN, SVM, Naïve Bayes and Markov Detection Tree model has been used under supervised learning and under unsupervised learning K-means and Affinity Propagation is used for classifying malicious web page.

Kumar et al. [6] have proposed a multi-layer detection model that classifies webpages as malicious or benign using which is composed of a total of four layers. The first and second layers are used the Black and White list filter and Naïve Bayes filter to filter out the webpages that are already included in the list. The third layer consists of the CART decision tree filter [7, 8] and the final layer used an SVM filter. The accuracy obtained by this model is 79.55%. They used textual, lexical and semantic features for training the model.

Yue et al. [9] have proposed a machine learning models which performs better when the URL features are extracted. The extracted features are from three fields including HTML, JavaScript and URL feature. KNN classifier was used at two different stages with different value.

Apart from using textual features for classification using machine learning, Kazemian et al. [10] proposed an integrated technique that used the combination of URL, semantic and SURF features of the image to obtain better performance, but instead of using Machine Learning models such as KNN, SVM, Naïve Bayes and others for image classification, it is better to use Deep learning CNN.

CNN models on image data showed better results, so Kabanga et al. [11] proposed a technique based on CNN models that Converts malware images into a grayscale image using Deep Learning's CNN. It shows improvement in the process of classification of malware images as it is easier to use an image as input to a model that uses Deep Learning's CNN rather than using URLs and extracting textual features.

Nguyen et al. [12] have proposed a technique to visualize image malware as grayscale image. The accuracy obtained by this model is about 98%. The "Microsoft Malware Classification Challenge" held in 2015 was won by the technique proposed by Wang et al. [13], and the accuracy achieved was above 99%.

To classify Malicious URL [14–16], several techniques have been proposed such as Deep learning techniques, machine learning techniques, dynamic attack detection method, and cross-layer web detection approach. They used URL features for the classification. Word2Vec and TF-IDF features are two different techniques used for weighing the features that have been used for detecting malicious URLs. Word2Vec is a model that contains a two-layer neural network and is trained for the reconstruction of linguistic context of word. The SVM learning algorithm classification is based on TF-IDF and is compared with the logistic regression technique and CNN. Word2vec feature function is used for weighing features, it converts the texts features into a numerical value which can be easily given to CNN. 96% accuracy is achieved in detecting malicious URLs using this method.

A drive-by download is responsible for attacks that infect the hosts present on any network. Shibahara et al. [17] proposed a technique to detect Drive-by download attacks. Lists of all URLs that link to several other URLs are created and the malicious URL is detected using the EDCNN (Event de-noising CNN) technique. In EDCNN (Event De-noising CNN), the same combination of two malicious URLs features is extracted. The false alert generated by EDCNN is 47% less as compared to CNN (Convolution neural network) in detecting malware infection. The methods for malicious URL detection are mostly focused on fetching the IP addresses corresponding to the domain [5, 18]. The URL sequences were classified using three distinct approaches: individual-based approach, CNN-based approach and the last approach based on EDCNN. EDCNN lowers the false alert as compared to the individual approach and CNN. Also, the classification time of EDCNN = 0.07 is also less compared to the individual approach and CNN (0.21, 0.09). Thus, the EDCNN-based approach performs classification in a reduced time as compared to the individual-based approach and CNN.

# 3 Proposed Method

The proposed work classifies webpages into two categories malicious and benign. It first preprocesses images and then performs data augmentation.

## *3.1 Image Preprocessing*

The dataset has been taken from ISCX-URL-2016 containing URL links of malicious and benign URL. The dataset consists of around 25,000 collective URL links of malicious and benign. A CSV file is also provided with the correct y labels of the corresponding URL. This file has two columns URL link and the label '0' corresponding to benign URL and '1' corresponding to be malicious URL. As the data set contained URL links and the requirement for the method is malicious webpage images. The image data set is collected by giving the URL link data set as input to PhantomJS (a headless web-kit browser with JavaScript API) code which further downloaded the screenshot of the webpage and saved it in PNG (Potable Network Graphic) file format. The obtained image file are sorted as valid and invalid. As the images are of malicious web pages so some images obtained are blank and of no use, so they are removed from the dataset.

The VGG- 16 model requires a fixed input size of "224 * 244", so all the images are resized to the required dimensions, using the OpenCV2 library. Then the cropped images are randomly flipped horizontally and random RGB color shift is done to further augment the data set. The required image input size for this model is set at "224 * 224" but since the images obtained are considerably smaller than what the model is trained on and increasing their size introduces too many artifacts and it badly affects the accuracy. Thus images are resized to the size of "150 * 150" before passing them through the network. The data set is split into two distinct sets: validation dataset and testing dataset.

## 3.2 Data Augmentation

One common problem with neural networks is that they are very powerful and tend to overfit the training data. Overfit is the situation where the model performs extremely great on the training data but performs extremely poorly on unseen new data. Therefore, it has memorized the dataset rather than learning useful patterns that can be used for classification in general. Therefore, to overcome the problem of overfitting, we have done augmentation of images [19] by taking the relative size of each class, since it is very skewed as most of the images lie in benign class and very few lie in malicious class. More than 50% of the images belong to class 1: Neural networks are very sensitive to this kind of imbalanced distribution. The data augmentation operations that we performed are listed below.

- **Horizontal Flip**: Images are flipped horizontally.
- **Rotation**: Images are randomly selected and rotated between 0 and 90°.
- **Zoom**: Images are randomly stretched between (1/1.4, 1.5).
- **Translation**: Images are randomly shifted between −25 and +25 pixels.

# 4 Network Architecture

VGG-16 and VGG-19 network architecture are as follows.

## 4.1 VGG-16 and VGG-19

VGG-16 as the name suggests that it has 16 weighing layers but has a total of 20 layers. These layers are Max Pooling layers, Convolutional layers, fully connected layers and Activation layers. Thirteen convolutional layers, 5 Max Pooling layers and 3 Dense layers which sums up to 21 layers but only 16 weight layers. All the convolution layer of the network is of "3 * 3" with a stride of size of 1 and the same padding; and the pooling layers are all "2 * 2" pooling layers with a stride size of 2. The architecture of VGG-16 is shown in Fig. 2.

VGG-19 is an updated version of VGG-16 and it contains 19 layers. The architecture is almost similar to VGG-16 but the number of convolution layers is more as compared to VGG-16, but the rest of the size of the image taken as input and other functioning of each layer is similar to VGG-16. Thus, there is not much difference in the architecture of VGG-16 and VGG-19.

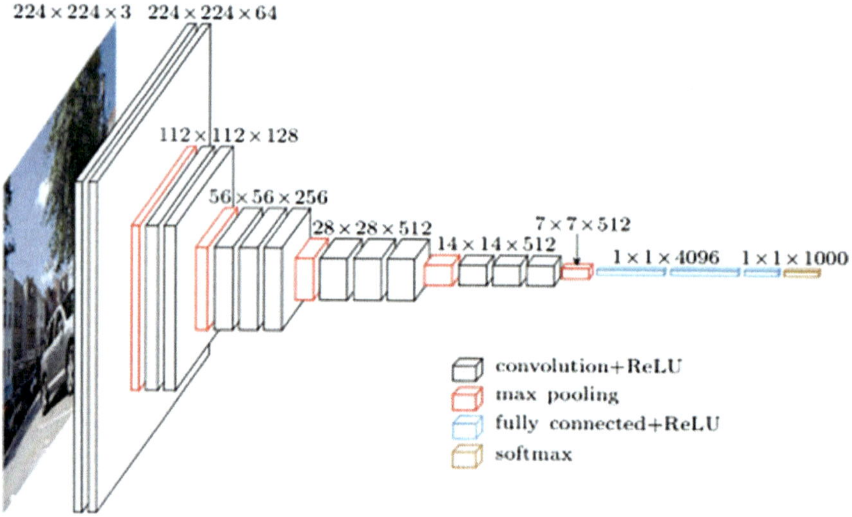


**Fig. 2** Architecture of VGG-16

As VGG-16 and VGG-19 is a pre-trained model, the output layer contains 1000 class, but the requirement is of only two classes i.e.: benign and malicious. So the dense layers are modified based on the requirement by changing the dropout values. The activation function used in the last output layer is "sigmoid", and the loss function is changed to "binary cross-entropy."

## 4.2 Training and Validation

The 16 layer VGG network has a total of around 1.9 million parameters. Both VGG-16 and VGG-19 have the same methods for training. Using the "Adagrad" optimizer, the training is done and the multinomial logistic regression is optimized. The batch size is set to 25. The training is regularized by weight decay and the dropout regularization for the first two fully connected layers (0.3 and 0.4). As the optimizer used is "Adagrad", thus the initial learning rate is already fixed as 0.01 and the value is reduced by a factor of 10 in case the validation set accuracy stops improving. Initially, the training and validation are performed for epoch = 50 and then further the epoch is increased to 150, and as the epoch value increased there is an increase in the model accuracy.

# 5 Experimental Results

## *5.1 Dataset*

Deep Neural network is data Hungary. So, it requires thousands and thousands of image data to learn all its parameters properly and thus avoid over-fitting. Over-fitting is the major concern during training the neural network. URL dataset (ISCX-URL-2016) of around 10,000 URLs has been taken into account. Firstly, the dataset is gathered and if it contains a URL link of the webpage, it is converted to an image (PNG format) using PhantomJS. After converting into an image, the valid images are retained and the invalid images are removed. The valid images are around 2500. The invalid images are those which have been removed from the server and keeping them in the dataset will lead to inaccurate results. After obtaining the final malicious webpage dataset, the preprocessing is done and the images are resized to the size of "224 * 224". Finally, these images are given as an input to VGG-16 and VGG-19.

For training the model, the dataset is split into two sets namely: training dataset and validation dataset. The training dataset is used to train the model while the validation set is used for testing the accuracy of the proposed model. Test accuracy of a model largely depends on the size (Number of images used for training) and the quality of images. Therefore, all noisy and corrupt data points must be removed from the dataset.

## *5.2 Performance Metrics and Evaluation*

The performance of the proposed model is measured by two performance metrics: mAP (macro Average precision) and F1 score. The Average Precision (AP) is one of the most common metric for evaluating the performance of image classification. It assumes that the labels contain bounding boxes and object class instances. It prevents the use of Receiver Operating Characteristic and Detection Error Trade-off. For a given category, ranked outputs (bounding boxes) are used to calculate the Precision/Recall.

Precision is defined as the number of positive samples above the given rank and Recall is defined as the proportion of positive samples ranked above the given rank. Detection outputs are assigned to ground truth in the order given by decreasing confidence scores. Detecting the same data twice is considered as false detections. The performance over the whole dataset is computed by averaging the Aps across all the categories. This is called the macro Average Precision or mAP. This is the metric used for the Performance evaluation in experiments. All the other networks use the same metric for performance evaluation. The model accuracy graph and model loss graph has been used to display the variation in model accuracy and loss in the model with respect to epochs. The other metric used for evaluating the proposed model is F1-score. The F1-score basically measures the test accuracy by calculating the harmonic mean of the precision and recall.

$$\text{F1 score} = 2 * \frac{Precision * Recall}{precision + recall} \tag{1}$$

$$\text{Accuracy} = \frac{\text{True Positive} + \text{False Negative}}{Total\ Count} \tag{2}$$

The proposed model accuracy and loss obtained for VGG-16 and VGG-19 are shown in Figs. 3 and 4. The VGG-16 model is run for epoch = 100 and the obtained proposed model accuracy and loss are shown in Fig. 3.

The VGG-19 model is run for epoch = 100 and the obtained proposed model accuracy and loss is shown in Fig. 4

The confusion matrix is further used for computing the performance metrics of both models. With the help of the confusion matrix mAP, precision and F1-score are computed. The confusion matrix obtained for VGG-16 is shown in Fig. 5.

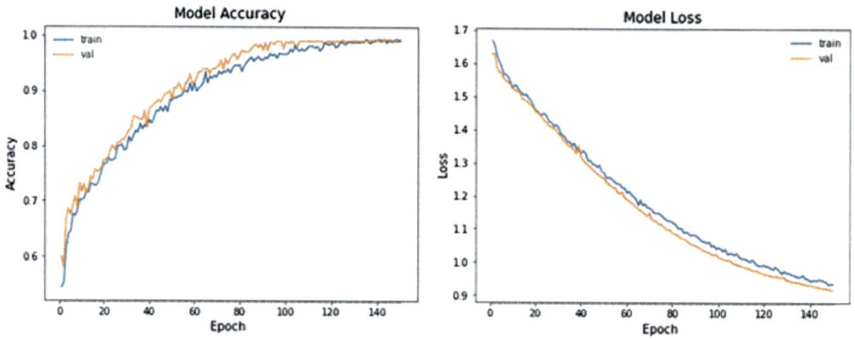


**Fig. 3** Training and validation accuracy and loss for VGG-16 network

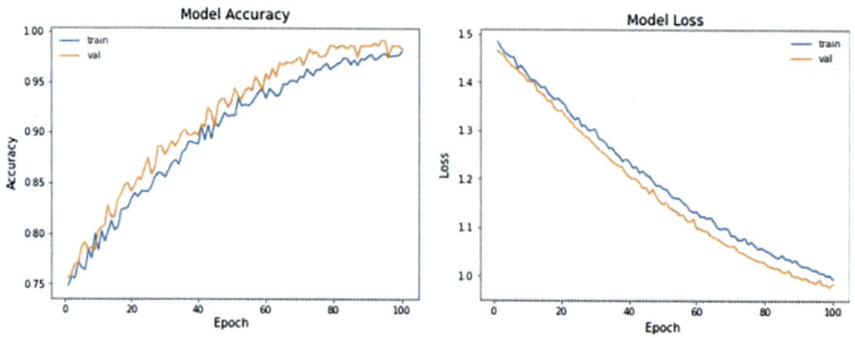


**Fig. 4** Training and validation accuracy and loss for VGG-19 network

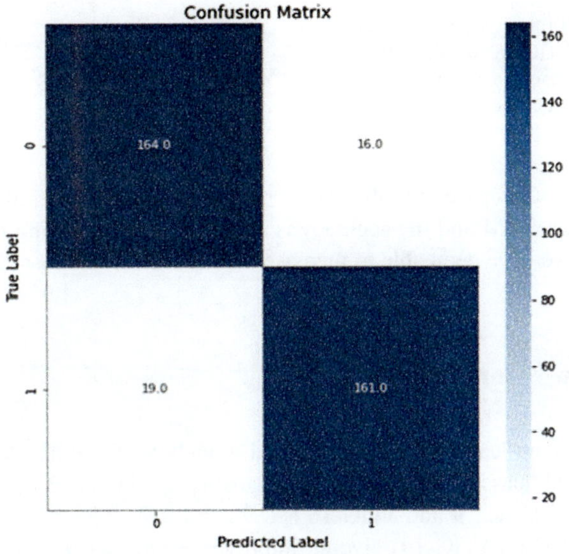


Fig. 5 Confusion matrix of VGG-16

The confusion matrix obtained for VGG-19 is shown in Fig. 6.

The F1-score of VGG-16 is close to 0.99, obtained mAP is equal to 0.99. The F1 score of VGG-19 is 0.98 and obtained mAP is almost equal to 0.98. Table 1 shows the comparison of VGG-16 and VGG-19.

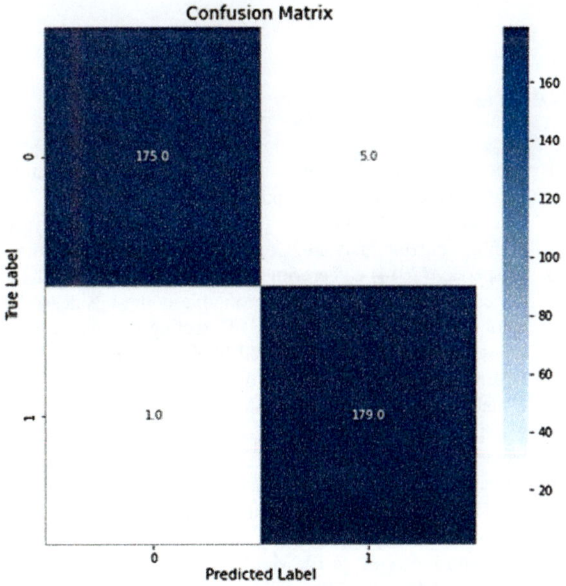


Fig. 6 Confusion matrix of VGG-19

Table 1 Comparison of VGG-16 and VGG-19 on the basis of F1-score and average precision

| Network model | F1-score | Average precision |
|---|---|---|
| VGG-16 | 0.99 | 0.99 |
| VGG-19 | 0.98 | 0.98 |

Compared to [10, 11] the result obtained is better as in [10] using image features on SVM and the accuracy is about 91%. While using the Deep learning technique, we have been able to increase the accuracy to almost 98%.

## 6 Conclusion

Deep-leaning CNN pre-trained models VGG-16 and VGG-19 have been presented for classifying web page images as malicious or benign. The obtained F1 score of VGG-19 and VGG-16 is 0.98 and 0.99. The new deep learning models (VGG-16 and VGG-19) give an improved performance as compared to previous models (ALEXNET, GOOGLENET and SVM) using features of image for classification. The performance can be further improved by changing in architectures of the base network model (VGG 16) and using a different base network. The different parameter tuning can be done for different combinations of learning rate and weight decay to reach convergence faster and get improved results. The performance with different optimization techniques and weight initialization can also be compared and studied further.

## References

1. Sanou B (2016) ICT: facts and Figures. http://www.itu.int/en/ITU-D/Statistics/Documents/facts/ICTFactsFigures2016.pdf
2. Ollmann G (2008) The Evolution of commercial malware development kits and colour-by-numbers custom malware. Comput Fraud Secur 2008(9):4–7
3. Kaspersky (2014) Cybercrime, Inc.: how profitable is the business? URL: https://blog.kaspersky.com/cybercrime-inc-how-profitable-is-the-business/15034/
4. Choi H, Zhu BB, Lee F (2012) Detecting malicious web links and identifying their attack types. In: 2nd proceedings of USENIX conference on Web application development. Berkeley: USENIX Association, CA, USA, pp 11–11
5. Canali D, Cova M, Prophiler GV (2011) A fast filter for the large-scale detection of malicious web pages. In: 20th proceedings of the international conference on World Wide Web. ACM, New York, USA, pp 197–206
6. Kumar R, Zhang X, Ahmad Tariq H, Khan RU (2017) Malicious URL detection using multi-layer filtering model. In: International computer conference on wavelet active media technology and information processing
7. Chipman HA, George EI, McCulloch RE (1998) Bayesian CART Model Search. J Am Stat Assoc 93(443):935–948
8. Steinberg D, Colla P (2009) CART: classification and regression trees. The top ten algorithms in data mining, pp 179–201

9. Yue T, Sun J, Chen H (2013) Fine-grained mining and classification of malicious web pages. In: 4th International conference on digital manufacturing & automation
10. Kazemian HB, Ahmed S (2011) Comparisons of machine learning techniques for detecting malicious webpages. Expert Syst Appl 42(3):1166–1177
11. Kabanga EK, Kim CH (2018) Malware images classification using convolution neural network. J Comput Commun 6:153–158
12. Nguyen A, Yosinki J, Clune J (2015) Deep neural networks are easily fooled: high confidence predictions for unrecognizable images. IEEE Comput Vision Pattern Recog (CVPR)
13. Wang X, Liu J, Chen X (2015) first place team: say no to over fitting. Winner of microsoft malware classification challenge
14. Abdi FD, Wenjuan L (2017) Malicious URL detection using convolution neural network. Int J Comput Sci Eng Inf Technol (IJCSEIT) 7(6)
15. Wang Y, Cai WD, Wei PC (2016) A deep learning approach for detecting malicious javascript code. Secur Commun Netw 9(11)
16. Xu L, Zhan Z, Xu S, Ye K (2013) Cross-layer detection of malicious websites. In: 3rd proceedings of ACM conference on Data and application security and privacy, pp 141–152
17. Shibahara T, Yamanishi K, Takata Y, Chiba D, Akiyama M, Yagi T, Ohsita Y, Murata M (2017) Malicious URL sequence detection using event de-noising convolutional neural network. In: ICC 2017 IEEE international conference on communications, pp 1–7
18. Antonakakis M, Perdisci R, Dagon D, Lee W, Feamster N (2010) Building a dynamic reputation system for DNS. In: 19th proceedings of USENIX security symposium, pp 273–290
19. Tanner MA, Wong WH (1987) The calculation of posterior distributions by data augmentation. J Am Stat Assoc 82(398):528–540

# Author Index



© The Editor(s) (if applicable) and The Author(s), under exclusive license to Springer Nature Singapore Pte Ltd. 2022
H. M. Dubey et al. (eds.), *Artificial Intelligence and Sustainable Computing*, Algorithms for Intelligent Systems, https://doi.org/10.1007/978-981-16-1220-6